REMEDIATION TECHNOLOGIES FOR SOILS AND GROUNDWATER

SPONSORED BY
Remediation Technologies for Soils and Groundwater Task Committee
of the Environmental Council

Environmental and Water Resources Institute (EWRI)
of the American Society of Civil Engineers

EDITED BY
Alok Bhandari
Rao Y. Surampalli
Pascale Champagne
Say Kee Ong
R. D. Tyagi
Irene M. C. Lo

Published by the American Society of Civil Engineers

Library of Congress Cataloging-in-Publication Data

Remediation technologies for soils and groundwater / sponsored by Remediation Technologies for Soils and Groundwater Task committee of the Environmental Council, Environmental and Water Resources Institute (EWRI) of the American Society of Civil Engineers ; edited by Alok Bhandari … [et al.].

p.cm.

Includes bibliographical references and index.

ISBN-13: 978-0-7844-0894-0

ISBN-10: 0-7844-0894-7

1. Soil remediation. 2. Groundwater--Purification. I. Bhandari, Alok. II. Environmental and Water Resources Institute (U.S.) Environmental Council III. Environmental and Water Resources Institute (U.S.) Remediation Technologies for Soils and Groundwater Task Committee.

TD878.R47 2007

628.5'5--dc22 2007001848

American Society of Civil Engineers
1801 Alexander Bell Drive
Reston, Virginia, 20191-4400

www.pubs.asce.org

ISBN 13: 978-0-7844-0894-0 ISBN 10: 0-7844-0894-7
Manufactured in the United States of America.

Preface

The value of our limited natural resources continues to appreciate as they are exploited to support and indulge the human species. The ever-growing demands for energy, security, food, healthcare and consumables have placed unprecedented pressure on our ecosystem and accentuated the need for sustainable management of the environment. Contamination of natural resources, including soils and groundwater, remains a major global ecological concern in the 21st century. It cannot be underemphasized that the health of our soils and groundwater is intimately tied to our well-being and to the wellness of other species that share our ecosystem.

This book presents a discussion of approaches and technologies that are most commonly deployed for the restoration of contaminated soils and groundwater. A need for an up-to-date text that summarized these technologies in an easy-to-read format was identified by the ASCE's Technical Committee on Hazardous, Toxic and Radioactive Waste in 2003. The committee envisioned a book prepared by a team of experts that would serve a reference for practicing professionals and could be equally effective as a text in an undergraduate or graduate classroom.

The organization of this book is based on the types of technologies used in the remediation of soils and groundwater. Chapter 1 introduces the topic of soil and groundwater remediation and summarizes the contents of the book. Chapter 2 presents a brief discussion of the fundamental physical, chemical and biological processes that are at play at a contaminated site and during site cleanup. Chapters 3 and 4 explore and analyze conventional physical and chemical site remediation technologies, respectively. Chapter 5 focuses on redox and precipitation processes, and associated engineering applications, and Chapter 6 takes a closer look at the concept and application of chemical reactive barrier technology.

Chapter 7 describes the processes and factors responsible for biotransformation of soil and groundwater contaminants. Chapters 8 and 9 discuss a variety of bioremediation technologies based on biotransformations mediated by bacteria, enzymes and fungi. Chapter 10 focuses on the application of phyto-processes to the cleanup of contaminated sites. Chapter 11 takes a closer look at processes and technologies used to remediate metal-contaminated soils. Finally, Chapter 12 discusses commonly followed approaches for long-term monitoring of contaminated and treated sites.

The editors acknowledge the hard work and patience of the all authors who have contributed to this book.

- AB, RS, PC, SKO, RDT and IL

Contributing Authors

Shankha Banerji, *University of Missouri - Columbia*

Alok Bhandari, *Kansas State University*

J. F. Blais, *Institut National de la Recherche Scientifique*

Satinder K. Brar, *Institut National de la Recherche Scientifique*

Pascale Champagne, *Queens University*

Ravi Damera, *ENSR Corporation*

Angela Kolz, *Iowa State University*

Keith Lai, *Hong Kong University of Science & Technology*

Irene Lo, *Hong Kong University of Science & Technology*

G. Mercier, *Institut National de la Recherche Scientifique*

N. Meunier, *Institut National de la Recherche Scientifique*

Kshipra Mishra, *Institut National de la Recherche Scientifique*

Say Kee Ong, *Iowa State University*

S. Bala Subramaniam, *Institut National de la Recherche Scientifique*

Rao Surampalli, *U.S. Environmental Protection Agency*

R. D. Tyagi, *Institut National de la Recherche Scientifique*

Mausam Verma, *Institut National de la Recherche Scientifique*

S. Yan, *Institut National de la Recherche Scientifique*

Contents

Chapter 1	**Introduction**	**1**
1.1	Background	1
1.2	Development and Use of Effective Remediation Technologies	2
1.3	Types of Remediation Technologies for Soils and Groundwater	3
Chapter 2	**Fundamental Processes**	**5**
2.1	Introduction	5
2.2	Solute Transport in Porous Media	5
2.3	Physicochemical Processes	12
2.4	Biological Processes	24
2.5	Conclusion	44
Chapter 3	**Physical Treatment Technologies**	**47**
3.1	Introduction	47
3.2	Free Product Recovery	47
3.3	Pump-and-Treat	51
3.4	Soil Vapor Extraction	55
3.5	Air Sparging	59
3.6	Groundwater Circulation Wells	62
3.7	Multi-Phase Extraction	66
3.8	Induced Fracturing	70
3.9	Soil Heating	73
3.10	Conclusion	76
Chapter 4	**Chemical Treatment Technologies**	**79**
4.1	Introduction	79
4.2	Precipitation	79
4.3	Chemical Oxidation and Reduction	83
4.4	Stabilization/Solidification	94
4.5	Adsorption and Ion Exchange	100
4.6	Electrochemical Processes	105
4.7	Chemical Leaching and Solvent Extraction	110
4.8	Soil Flushing	115
4.9	Soil Washing	123
4.10	Conclusion	126
Chapter 5	**Redox and Precipitation Technologies**	**133**
5.1	Introduction	133
5.2	Redox Reaction	134

5.3	Definition – Precipitation	141
5.4	Contaminants Amenable to Redox Processes and Precipitation	145
5.5	Redox and Precipitation in Natural Systems	163
5.6	Conclusion	166
Chapter 6	**Chemical Reactive Barriers**	**176**
6.1	Introduction	176
6.2	Mechanisms of Groundwater Remediation by PRBs	177
6.3	Geochemical and Hydraulic Behaviors of PRBs	184
6.4	Conclusion	190
Chapter 7	**Biodegradation**	**199**
7.1	Introduction	199
7.2	Definitions	201
7.3	Degradation Pathways and Mechanisms	202
7.4	Rates and Transformation	211
7.5	Case Studies	213
7.6	Conclusion	217
Chapter 8	**Bioremediation with Bacteria and Enzymes**	**223**
8.1	Introduction	223
8.2	Biosparging	224
8.3	Bioventing	232
8.4	Biostimulation	235
8.5	Bioaugmentation	237
8.6	Anaerobic Biotransformation	239
8.7	Aerobic Biotransformation	241
8.8	Biological Fixation	242
8.9	Enzyme-Catalyzed Treatment	244
8.10	Biological Reactors	246
8.11	Monitored Natural Attenuation	247
8.12	Conclusion	251
Chapter 9	**Bioremediation with Fungi**	**259**
9.1	Introduction	259
9.2	Saprotrophic Fungal Processes	260
9.3	Mycorrhizal Fungal Processes	274
9.4	Conclusion	280
Chapter 10	**Phytoremediation**	**290**
10.1	Introduction	290
10.2	Elements of Phytoremediation	291
10.3	Phytoremediation Applications	302
10.4	Phytoremediation of Inorganic Contaminants	317
10.5	Phytoremediation of Organic Contaminants	332
10.6	Design Considerations	342

10.7 Potential and Limits of Phytoremediation Technologies 344
10.8 Conclusion 348

Chapter 11 Metal Removal 353
11.1 Introduction 353
11.2 Metal Contaminants in Soil and Groundwater 354
11.3 Regulations Concerning Metal Contamination 358
11.4 Treatability of Soils Contaminated with Heavy Metals 359
11.5 Restoration of Soils Contaminated with Heavy Metals 361
11.6 Treatment of Groundwater Polluted with Heavy Metals 375
11.7 Conclusion 382

Chapter 12 Long-Term Monitoring 395
12.1 Introduction 395
12.2 Remediation Objectives 396
12.3 Performance Metrics and Monitoring Parameters 398
12.4 Groundwater Sampling for Performance Monitoring 404
12.5 Key Monitoring Parameters for Remediation Technologies 408
12.6 Conclusion 432

Editor Biographies 439
Index 441

CHAPTER 1

Introduction

Alok Bhandari

1.1 Background

The extensive contamination of soil and groundwater resources has been an undesirable consequence of the rapid industrialization experienced by the world's developed and developing economies since World War II. Post-war successes in enhancing the agricultural productivity of croplands led to an explosion of urban populations, which in turn, resulted in an economy devoted to mass production of consumables. This economic expansion was, to a great extent, fueled by energy resources including petroleum, coal and nuclear power. The post-war years also witnessed the manufacture and use of a variety of xenobiotic chemicals designed to maintain the growing standards of living by preservation of food and other consumables. The inappropriate use and disposal of natural and xenobiotic hazardous chemicals during the second half of the 20th century have led to massive contamination of soils and groundwater at sites across the United States and other countries. Today, these contaminated sites include those that are polluted with toxins and carcinogens including petroleum hydrocarbons, fuel additives, pesticides, heavy metals, radionuclides, explosives, and solvents.

In the United States, the Resource Conservation and Recovery Act (RCRA) of 1976 and the Comprehensive Environmental Response, Compensation and Liability Act (CERCLA or Superfund) of 1980 provided a major boost to cleanup efforts associated with contaminated soils and groundwater. While originally funded at $1.6 billion to clean 400 sites, CERCLA resulted in the discovery of additional sites expanding its budget to $27 billion by 1990 (NRC, 1999). The U.S. Environmental Protection Agency (EPA) predicted that the number of sites requiring cleanup could increase to 2,000 costing up to $500 billion. By the end of 2005, the Superfund program had completed construction at 966 or 62% of private and federal sites, and work was underway at 422 additional sites. More recent estimates expect the cost of environmental remediation of such sites to easily exceed $1 trillion (NRC, 1997, 1999).

1.2 Development and Use of Effective Remediation Technologies

Although sustained public and corporate funding of environmental cleanup efforts is an important factor in realizing the restoration of contaminated sites, an equally critical element is the availability of technologies capable of achieving the levels of treatment necessary for appropriate environmental restoration. The vast experience gained by engineers and remediation professionals with a variety of remediation technologies over the past two decades has resulted in greater acceptance of several technologies for soil and groundwater cleanup. However, most remediation engineers have continued to prefer time-tested technologies even when some of these technologies have repeatedly produced less-than-optimal restoration of contaminated sites. Preferred technologies, such as pump-and-treat, and excavation-and-burial, have often risked becoming 'cure-alls' for contaminated sites primarily due to the availability of extensive past-performance and cost data. It is clear that remediation professionals need to more aggressively explore the deployment of alternate effective technologies for site cleanup while the regulatory agencies need to create the appropriate framework that allow these technologies to be deployed and evaluated on the field.

The restoration of impacted soils and groundwater to ecologically sustainable levels requires development, selection and deployment of remediation technologies capable of responding to site-specific conditions. The development and adoption of new technologies is facilitated by a combination of targeted research funding and market demand for the technologies. While the market demand is influenced by project economics and cleanup requirements enforced by regulatory agencies, the research support for new technologies is linked to the general economy and competing public and corporate interests. Over the past few decades, the EPA's Superfund Innovative Technology Evaluation (SITE) program has been instrumental in facilitating field tests of a variety of new technologies for soil and groundwater cleanup. SITE projects have documented testing procedures, performance data and cost analyses from field-tests of several hundred innovative technologies. Information about these treatment and monitoring technologies is accessible at SITE's website: www.epa.gov/ord/SITE.

The National Research Council's Committee on Innovative Remediation Technologies recently offered several recommendations for stimulating the market demand for new remediation technologies. These recommendations are summarized in Table 1.1.

Table 1.1. The National Research Council's recommendations to stimulate market demand for new remediation technologies (NRC, 1997).

U.S. Securities and Exchange Commission (SEC) should clarify and strictly enforce requirements for disclosure of environmental remediation liabilities by all publicly traded U.S. corporations
SEC should enforce environmental liability reporting requirements through third-party environmental auditing
U.S. Congress should establish a program to allow companies to amortize reported remediation liabilities over a 20- to 50-year period
EPA should improve its enforcement of Superfund and RCRA requirements
Managers of federally-owned contaminated sites should contract remediation projects on a fixed-price basis and use independent peer review panels to check progress towards milestones
EPA should review procedures for approving remediation technologies and develop guidelines for increasing the consistency and predictability of these procedures among regions and across programs
Congress and EPA should assess the option of establishing national standards for soil and groundwater cleanup
The U.S. General Accounting Office should assess programs for licensing site professionals to select remediation technologies on behalf of environmental regulators
EPA should establish a national registry of contaminated sites and make it publicly available over the Internet

1.3 Types of Remediation Technologies for Soils and Groundwater

The successful cleanup of contaminated soils and groundwater depends on factors such as contaminant characteristics, local climatic conditions, site hydrogeology, and the type of technology used for site-restoration. Of these, the selection and deployment of appropriate site-specific and contaminant-specific cleanup technologies is the only factor under the influence of remediation professionals. It is, therefore, essential that remediation professionals have a strong understanding of various types of remediation technologies available in the market. Several such technologies are summarized in Table 1.2 and are the topics of discussion in the remainder of this text.

Table 1.2 Common soil and groundwater remediation technologies.

Physical treatment technologies	Free product recovery
	Pump-and-treat
	Soil vapor extraction
	Air sparging
	Groundwater circulation wells
	Multiphase extraction
	Induced fracturing
	Soil heating
Chemical treatment technologies	Precipitation
	Chemical oxidation and reduction
	Permeable reactive barriers
	Stabilization/solidification
	Adsorption and ion exchange
	Electrochemical processes
	Chemical leaching and solvent extraction
	Soil flushing
	Soil washing
Biological treatment technologies	Biosparging
	Bioventing
	Biostimulation
	Bioaugmentation
	Anaerobic biotransformation
	Aerobic biotransformation
	Biological fixation
	Enzyme-catalyzed treatment
	Saprotrophic fungal processes
	Mycorrhizal fungal processes
	Biological reactors
	Phytoremediation
	Monitored natural attenuation

References

NRC (1997) Innovations in Groundwater and Soil Cleanup, National Research Council, National Academy Press, Washington DC.

NRC (1999) Groundwater and Soil Cleanup, National Research Council, National Academy Press, Washington DC.

CHAPTER 2

Fundamental Processes

Pascale Champagne and Alok Bhandari

2.1 Introduction

Since no two contaminated sites are identical, remediation approaches for soil and groundwater cleanup have to be 'customized' for each scenario. The selection of an appropriate remediation technology is based on a careful analysis of the physical, chemical and biological factors affecting the speciation, transformation, and transport of contaminants in the geological media. This requires a thorough understanding of contaminant properties and a detailed characterization of the contaminated media. This chapter discusses the major physical, chemical and biological processes that affect the fate and transport of environmental contaminants in remediation systems. The chapter begins with a discussion of contaminant transport in soils and aquifers followed by sections describing various physicochemical, microbial and plant-based processes that can influence the remediation of contaminated sites.

2.2 Solute Transport in Porous Media

Chemicals that are soluble in water can be relatively mobile in the soil and subsurface environments. The ability of these contaminants to be transported with water can be exploited using appropriate remediation technologies. Soil and subsurface geological materials are comprised of porous media containing inter-connected voids that allow flow of water. The subsurface may comprise of a saturated zone where all voids are filled with water. This zone is typically overlain with a vadose zone where voids are not saturated and may contain air or water (Figure 2.1). The groundwater table forms a boundary between the saturated and unsaturated zones. A capillary fringe exists above the groundwater table where surface tension forces allow water to rise into small pores in the geological material above the water table. A saturated subsurface that contains sufficient groundwater to allow harvesting of the water for irrigation or potable use is called an aquifer. Aquifers are usually underlain by or sandwiched between relatively impermeable clays or bedrock.

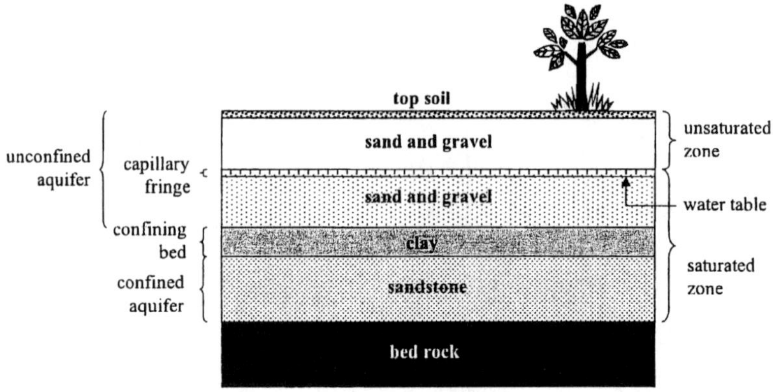

Figure 2.1 Major physical characteristics of the subsurface environment affecting groundwater flow and contaminant transport.

2.21 Groundwater Flow

Flow of groundwater occurs through interconnected pores in the aquifer media. The potential storage capacity of an aquifer is related to its porosity – the fraction of the total aquifer volume occupied by pores. The pore volume of an aquifer capable of transmitting fluid is represented by its effective porosity, n_e. The pore size distribution and interconnectivity of the pores are important characteristics controlling the movement of groundwater under a hydraulic gradient. The hydraulic gradient is the driving force for groundwater flow in an aquifer and is described by the change in hydraulic heads between two points that are a unit distance apart (*dh/dl*).

Groundwater flow velocities typically range from centimeters to tens of meters per day and follow a laminar flow regime. This allows the groundwater flow to be described by Darcy's law, which states that the velocity of groundwater, also known as the Darcy velocity or Darcy flux, is directly proportional to the hydraulic gradient:

$$v_D = K \frac{dh}{dl} \qquad \text{(Eq. 2.1)}$$

where, v_D, represents the Darcy velocity and the constant of proportionality, K, is the hydraulic conductivity or coefficient of permeability of the porous media. The actual velocity of groundwater or its average linear velocity v_i, is calculated by dividing the Darcy velocity by the aquifer porosity

$$v_i = \frac{K}{n_e}\frac{dh}{dl} \qquad \text{(Eq. 2.2)}$$

The hydraulic conductivity of a subsurface porous media represents it ability to transport water under a hydraulic gradient. Hydraulic conductivity is the flow rate of water through a unit cross-sectional area under the influence of a unit hydraulic gradient. The hydraulic conductivity of nonhomogeneous, anisotropic aquifers varies spatially and with the direction of measurement. Table 2.1 summarizes typical values of porosity and horizontal hydraulic conductivity for common subsurface materials.

Table 2.1 Porosity and hydraulic conductivity of common subsurface materials.

Subsurface Material	Porosity (%)	Hydraulic Conductivity (m/day)
Coarse sand	39	8×10^{-3} to 50
Medium sand	39	8×10^{-3} to 5
Fine sand	43	2×10^{-3} to 2
Sandstone	33 to 37	3×10^{-6} to 0.05
Dolomite and limestone	26 to 30	9×10^{-6} to 0.05
Shale	6	9×10^{-10} to 2×10^{-5}

The governing equation for three-dimensional transient groundwater flow through a control volume of a saturated anisotropic aquifer is given by (Mays, 2001):

$$\frac{\partial}{\partial x}\left(K_x\frac{\partial h}{\partial x}\right)+\frac{\partial}{\partial y}\left(K_y\frac{\partial h}{\partial y}\right)+\frac{\partial}{\partial z}\left(K_z\frac{\partial h}{\partial z}\right)=S_s\frac{\partial h}{\partial t}+W \qquad \text{(Eq. 2.3)}$$

where the left hand terms represent the Darcy flux in the x, y and z dimensions, K_x, K_y, and K_z are the hydraulic conductivities in the three dimensions of the anisotropic aquifer, S_S is the specific storage of the aquifer material, and W is the flow out of the control volume. S_S represents the volume of water that is released from storage from a unit volume of a saturated aquifer under a unit decline in hydraulic head. For a homogeneous, isotropic medium, Eq. (2.3) simplifies to

$$\frac{\partial^2 h}{\partial x^2}+\frac{\partial^2 h}{\partial y^2}+\frac{\partial^2 h}{\partial y^2}=\frac{S_s}{K}\frac{\partial h}{\partial t}+\frac{W}{K} \qquad \text{(Eq. 2.4)}$$

where, K is the hydraulic conductivity of the isotropic aquifer material. Under steady state flow conditions, Eq. (2.4) can be further simplified to

$$\frac{\partial^2 h}{\partial x^2}+\frac{\partial^2 h}{\partial y^2}+\frac{\partial^2 h}{\partial y^2}=\frac{W}{K} \qquad \text{(Eq. 2.5)}$$

Two- or three-dimensional flow nets based on Darcy's law are used to describe multi-dimensional transport of water in the subsurface. Flow nets such as those shown in Figure 2.2 are constructed from a specified set of streamlines and equipotential lines for a given set of boundary conditions. Equipotential lines are hypothetical curves that connect points of identical hydraulic heads in the aquifer. Streamlines are orthogonal to equipotential lines and describe the general direction of fluid flow. Flow nets are widely used to evaluate the impact of pumping on the level of groundwater and the direction of flow.

When soils are unsaturated, the flow of water occurs by capillary action under a driving force known as the capillary potential or capillary head. The potential of water in a capillary pore is inversely proportional to the pore diameter. Thus, when water levels are dropped, larger soil pores are drained first due to the lower capillary potential in these pores. All capillaries in unsaturated porous media do not participate in flow under a given set of conditions and the overall permeability of the unsaturated soil depends on the fraction and size of pores participating in flow (Reddi and Inyang, 2000). Therefore, the hydraulic conductivity of an unsaturated soil is directly dependent on its water content, which in turn, is related to the capillary potential.

2.22 *Solute Transport*

Contaminant transport in the subsurface can occur under normal hydraulic gradients or under gradients induced by injection or production wells during site remediation. The transport of dissolved contaminants in groundwater results from the combined action of advection, diffusion, and dispersion processes. Advection or convection is the process by which the contaminant is transported with the bulk fluid flow. Diffusion is a molecular-scale process that results in transport of solute molecules down a concentration gradient. Dispersive transport results from nonuniform flow velocities and mixing along flow paths in heterogeneous porous media, and contributes to solute transport at the advancing edge of a contaminant plume. This section presents a simple derivation of the equation of solute transport in aquifer systems based on the law of conservation of mass (Ogata, 1970; Bear, 1972; Freeze and Cherry, 1979).

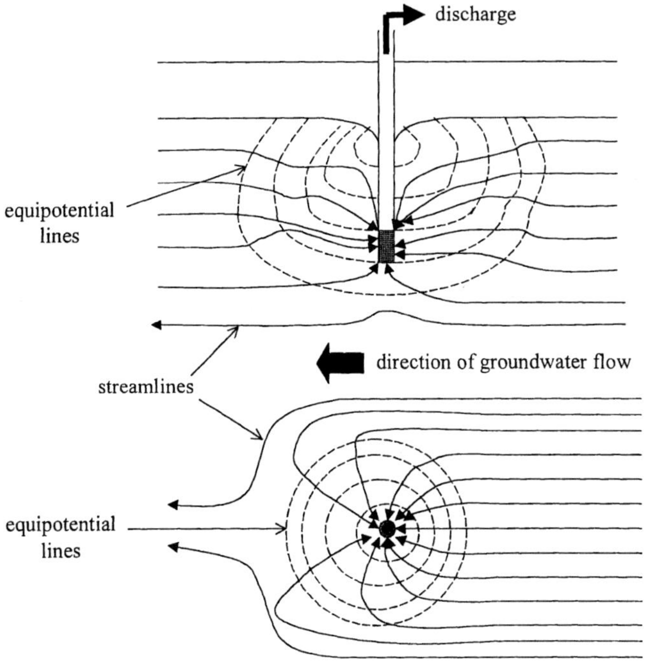

Figure 2.2 Sectional and plan views showing equipotential lines and streamlines associated with a single production well.

The advection driven, one-dimensional mass flux of solute in the i direction, F_i, is determined from the product of the solute concentration in water, C, and the Darcy velocity of the groundwater:

$$F_i = v_D C = v_i n_e C \quad \text{(Eq. 2.6)}$$

Solute masses entering and leaving a three-dimensional elemental volume ($n_e\ dx\ dy\ dz$) can be described by Eqs. (2.7) and (2.8), respectively:

$$F_x\ dz\ dy + F_y\ dx\ dz + F_z\ dx\ dy \quad \text{(Eq. 2.7)}$$

$$\left(F_x + \frac{\partial F_x}{\partial x}\right) dz\, dy + \left(F_y + \frac{\partial F_y}{\partial y}\right) dx\, dz + \left(F_z + \frac{\partial F_z}{\partial z}\right) dx\, dy \qquad \text{(Eq. 2.8)}$$

The rate of change of solute mass in the representative elementary volume is

$$\frac{\partial C}{\partial t}(n_e) dx\, dy\, dz = -\left(\frac{\partial F_x}{\partial x} + \frac{\partial F_y}{\partial y} + \frac{\partial F_z}{\partial z}\right) dx\, dy\, dz$$

or,

$$\frac{\partial C}{\partial t} = -\frac{1}{n_e}\left(\frac{\partial F_x}{\partial x} + \frac{\partial F_y}{\partial y} + \frac{\partial F_z}{\partial z}\right) \qquad \text{(Eq. 2.9)}$$

Substituting for F from Eq. (2.6) in Eq. (2.9) results in the three-dimensional equation for advective transport of a conservative solute in porous media:

$$\frac{\partial C}{\partial t} = -\left(\frac{\partial}{\partial x}(v\, C) + \frac{\partial}{\partial y}(v_y C) + \frac{\partial}{\partial z}(v_z C)\right) \qquad \text{(Eq. 2.10)}$$

In most aquifer systems groundwater flow rates are sufficiently slow to allow diffusive mass-transfer of solutes to become a significant contributor of overall contaminant transport. Diffusion is a near-molecular scale, concentration gradient driven mass transfer process. Solute molecules in an unmixed system can migrate from a volume of high concentration to one of low concentration to achieve maximum entropy and, consequently, optimum thermodynamic stability. Diffusive mass transfer continues to occur as long as a concentration gradient exists. Fick's first law describes the diffusive solute flux across a concentration gradient as:

$$F = -D_d\left(\frac{dC}{dx}\right) \qquad \text{(Eq. 2.11)}$$

where, F is the mass flux of solute per unit area per unit time, D_d is the solute diffusion coefficient in the fluid (m^2/s), C is the molar solute concentration (M), and dC/dx is the concentration gradient (M/m). The negative sign in Eq. (2.11) indicates that solute mass transfer occurs from a region of higher concentration to that of a lower concentration. When solute concentrations at a location also change with time due to macro-level mixing, Fick's second law states that

$$\frac{\partial C}{\partial t} = D_d \frac{\partial^2 C}{\partial x^2} \qquad \text{(Eq. 2.12)}$$

In porous media such as aquifer systems, the diffusion of contaminants is impacted by the tortuosity of diffusion paths. Tortuosity in porous media is the result of the natural packing of particles, which produces a random distribution of inter-connected pores providing flow paths for advective mass transport. An effective diffusion coefficient, D_e, is used to model diffusive mass transport of solute in porous media:

$$D_e = \omega D_d \tag{Eq. 2.13}$$

where ω is an empirically derived coefficient related to the tortuosity of the media.

Dispersive transport of solutes predominantly occurs under turbulent conditions that involve complex fluid motions and random fluctuations in fluid velocities. Dispersive transport is also observed in flow regimes that are laminar in the macroscale but associated with heterogeneous porous media comprising particles of irregular shapes and sizes. The irregular pore-connectivity in such media results in obstructions to fluid flow and produces a parabolic distribution of flow velocities. This phenomenon results in mechanical dispersion of the solute causing spreading of the contaminant near the advancing edge of the flow. Mechanical dispersion occurring in the direction of the flow is called longitudinal dispersion, while that normal to the flow is called transverse dispersion. The coefficients of longitudinal and transverse mechanical dispersions are defined as $\alpha_i v_i$ and $\alpha_j v_i$, respectively, where α_i is the solute's dynamic dispersivity in the longitudinal direction, α_j is its dynamic dispersivity in the transverse direction, and v_i is the average linear velocity in the i direction. Contaminant spreading resulting from the combined action of molecular diffusion and mechanical dispersion is known as hydrodynamic dispersion. The one-dimensional hydrodynamic dispersion coefficient in the i direction can, therefore, be represented as

$$D_i = \alpha_i v_i + D_e \tag{Eq. 2.14}$$

The overall transport of solute in porous media is a combination of advective and diffusive mass transport processes and the total mass of solute transported across a unit cross-sectional area in the i direction is represented by

$$F_i = v_i n_e C - n_e D_i \frac{\partial C}{\partial i} \tag{Eq. 2.15}$$

The negative sign before the dispersion term in Eq. (2.15) indicates that solute transport occurs in the direction of a lower concentration. Adding the dispersive transport term to Eq. (2.10) results in the overall three-dimensional advection-dispersion equation for solute transport in porous media:

$$\frac{\partial C}{\partial t} = -\left(\frac{\partial}{\partial x}(v\ C) + \frac{\partial}{\partial y}(v_y C) + \frac{\partial}{\partial z}(v_z C)\right) + \left[\frac{\partial}{\partial x}\left(D_x \frac{\partial C}{\partial x}\right) + \frac{\partial}{\partial y}\left(D_y \frac{\partial C}{\partial y}\right) + \frac{\partial}{\partial z}\left(D_z \frac{\partial C}{\partial z}\right)\right] \quad \text{(Eq 2.16)}$$

The one- and two-dimensional forms of Eq. (2.16) in a homogeneous medium where the average linear velocity is uniform in all directions are given by Eqs. (2.17) and (2.18):

$$\frac{\partial C}{\partial t} = D_x \frac{\partial^2 C}{\partial x^2} - v_x \frac{\partial C}{\partial x} \quad \text{(Eq. 2.17)}$$

$$\frac{\partial C}{\partial t} = D_x \frac{\partial^2 C}{\partial x^2} + D_y \frac{\partial^2 C}{\partial y^2} - v_x \frac{\partial C}{\partial x} \quad \text{(Eq. 2.18)}$$

Analytical solutions for different cases of the advection-dispersion equation are widely available in the literature (Fetter, 1999; Bedient et al., 1994; Reddi and Inyang, 2003).

2.3 Physicochemical Processes

Contaminants discharged into soil or groundwater environments can undergo a variety of transformations under the influence of physical and chemical processes. Physical and chemical interactions between a contaminant and the geological media are often tightly interwoven and jointly described as physicochemical processes. This section summarizes key physicochemical processes that affect the fate and transport of contaminants in soils and groundwater. A thorough understanding of such processes is essential to compare site-specific remediation approaches and develop treatment protocols that best exploit the local physics and chemistry.

2.3.1 Hydrolysis

A variety of pollutants in the subsurface environment can be transformed by hydrolysis reactions where the xenobiotics are cleaved into two parts in the presence of H_2O resulting in the introduction of a hydroxyl group (-OH) into the compound. This is distinct from a hydration reaction, in which water molecules are added to a substance, but no cleavage occurs. Hydrolysis reactions can occur spontaneously under certain environmental conditions (e.g., low pHs) or may be mediated by enzymes or inorganic catalysts. For example, the chemical hydrolysis of the pre-emergent, broadleaf herbicide atrazine, followed by biotransformation by soil microbes accounts for most its breakdown in soils.

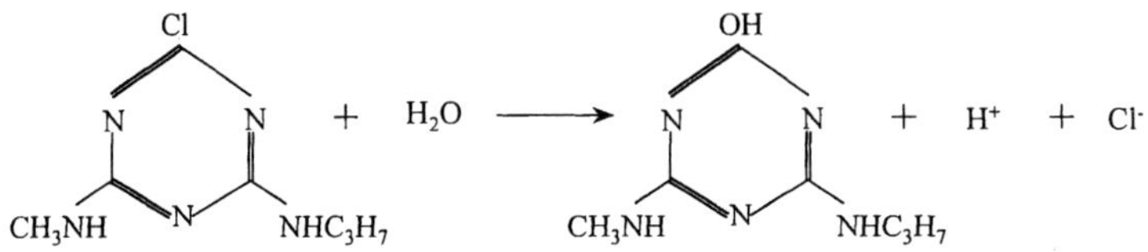

atrazine hydroxyl-atrazine

Atrazine readily hydrolyzes in acidic or basic aqueous environments but is stable at neutral pHs. Hydrolysis is an important reaction for many other organic contaminants in aqueous environments and can significantly impact the fate and transformation of these pollutants in soil and groundwater systems. The types of organic functional groups that are particularly susceptible to hydrolysis include alkyl halides, amines, amides, esters, carbamates, epoxides and nitriles (Lyman, 1990). The first-order kinetics of most hydrolysis reactions allows for the expression of hydrolytic half-lives of several contaminants.

2.3.2 Dissolution

Dissolution of contaminants into the aqueous phase associated with soil and groundwater systems, may be occur from gases, solids, or nonaqueous phase liquids (NAPL). Gases, such as carbon dioxide, hydrogen sulfide or methane, produced in the saturated subsurface environments may dissolve in groundwater. Similarly gaseous contaminants, such as gasoline constituents, encountered in the vadose zone may partition into groundwater in the capillary fringe or at the water table.

The air-water partitioning of solutes is described by Henry's law, which states that the concentration of a solute gas in a solution is directly proportional to its partial pressure above the solution:

$$C_e = K_H p \qquad \text{(Eq. 2.19)}$$

where, C_e is the aqueous phase solute concentration at equilibrium with a gas phase with a partial pressure p of the solute. K_H is the temperature-dependent Henry's law constant. K_H values and other chemical properties of commonly encountered soil and groundwater contaminants are summarized in Table 2.2.

Table 2.2 Selected chemical properties of commonly occurring soil and groundwater contaminants (Weber and DiGiano, 1996).

Compound	Henry's Law Constant (atm-m^3/mol)	Solubility (mg/L)	Octanol-Water Partitioning Constant (log)
Benzene	5.49×10^{-3}	1,770	2.13
Toluene	6.66×10^{-3}	530	2.73
Ethyl benzene	8.73×10^{-3}	169	3.15
o-xylene	5.27×10^{-3}	173	3.12
Chlorobenzene	3.71×10^{-3}	472	2.84
Methyl tert-butyl ether	9.15×10^{-4}	48,000	1.24
Phenol	4.0×10^{-7}	80,200	1.5
4-chlorophenol	5.6×10^{-7}	27,000	2.39
Naphthalene	1.08×10^{-3}	31	3.35
Phenanthrene	1.45×10^{-4}	1.1	4.52
Trichloroethylene	1.17×10^{-2}	137	3.4
Tetrachloroethylene	2.75×10^{-2}	150	3.4
Atrazine	6.2×10^{-6}	66.7	2.56

Dissolution of contaminants into groundwater may also occur from precipitated solids or geological deposits. For example, the dissolution of arsenic from geological strata has resulted in large-scale contamination of groundwater resources in parts of the Bengal basin in India and Bangladesh (Akai et al., 2004). The processes of dissolution and precipitation form a dynamic equilibrium where the extent of solubilization of a solid, AB, depends on its solubility product (K_{sp}) defined as:

$$K_{sp} = [A^+][B^-] \qquad \text{(Eq. 2.20)}$$

where, A^+ and B^- are the ions constituting the solid. Dissolution and precipitation reactions can result in the release or removal of large amounts of contaminants from the aqueous phase. Table 2.3 summarizes the solubility products of selected minerals of lead and copper.

Dissolution of organic contaminants into groundwater can also occur from bulk liquid or residual saturation in the unsaturated zone, or from light or dense non-aqueous phase liquids (NAPLs) in the saturated zone. Figure 2.3 illustrates the downward transport of a light NAPL (LNAPL) such as gasoline through the vadose zone, its accumulation near the groundwater table, and the dissolution of NAPL components into the groundwater.

Table 2.3 Solubility products of copper and lead minerals (Yen, 1999).

Mineral	**Dissolution Reaction**	K_{sp}
Cupric hydroxide	$Cu(OH)_2(s) \rightarrow Cu^{2+} + 2OH^-$	3.98×10^{-20}
Malachite	$Cu_2CO_3(OH)_2(s) \rightarrow 2Cu^{2+} + CO_3^{2-} + 2OH^-$	1.58×10^{-34}
Azurite	$Cu_3(CO_3)_2(OH)_2(s) \rightarrow 3Cu^{2+} + 2CO_3^{2-} + 2\ OH^-$	1.0×10^{-46}
Covellite	$CuS\ (s) \rightarrow Cu^{2+} + S^{2-}$	7.94×10^{-37}
Litharge	$PbO(s) + H_2O \rightarrow Pb^{2+} + 2OH^-$	5.01×10^{-16}
Cerussite	$PbCO_3(s) \rightarrow Pb^{2+} + CO_3^{2-}$	7.94×10^{-14}
Galena	$PbS(s) \rightarrow Pb^{2+} + S^{2-}$	3.16×10^{-28}

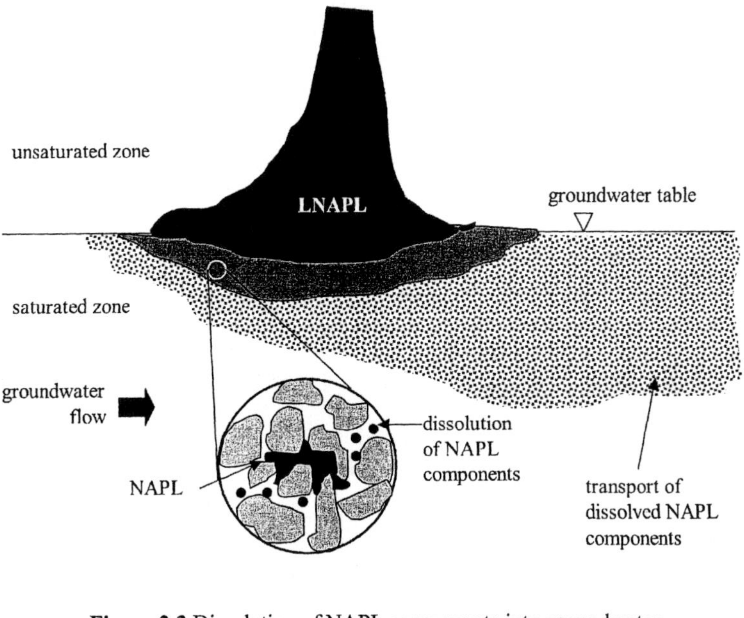

Figure 2.3 Dissolution of NAPL components into groundwater.

Due to the low solubility of NAPL components, contaminant dissolution from NAPL under laminar flow conditions typical of aquifers is mass-transfer limited, and occurs over long time scales producing large volumes of dilute, contaminated groundwater. The dissolution rate of NAPL components is controlled by the effective solubility of the dissolving organic compounds. Effective solubility is the theoretical aqueous solubility of a NAPL component in groundwater that is in chemical equilibrium with a multicomponent NAPL. The aqueous solubility of a NAPL component can be estimated from the product of its theoretical aqueous solubility and its mole fraction in the NAPL mixture based on principles of Raoult's law:

$$S_A = X_A S^A_{theoretical} \gamma_{NAPL} \qquad \text{(Eq. 2.21)}$$

where, S_A is the effective aqueous solubility of the NAPL component A, X_A is the mole fraction of A in the NAPL, $S^A_{theoretical}$ is the theoretical solubility of the pure subcooled component A in water, and γ_{NAPL} is the activity coefficient of A in the NAPL.

The dissolution rates of NAPL components generally decrease over time due to the reduction in their mole fractions in the NAPL mixtures. Typical time scales for the removal of BTEX and PAHs from blob zones and NAPL pools can be of the order of a few weeks to more than 10,000 years, respectively (Eberhardt and Grathwol, 2002).

2.3.3 *Volatilization*

Volatilization is the mass-transfer of contaminants from a solid or liquid phase into a gas phase. Contaminants that are dissolved in groundwater or sorbed to soil and aquifer materials, or are components of NAPL pools or residual blobs, can partition into the vapor phase. Solute mass-transfer is a function of the relative volatility of the contaminant and its solubility in water. The rate and extent of solute mass-transfer into the gaseous phase depends on the solutes water-air partition coefficient or its Henry's law constant (Table 2.1). Thus, a compound like benzene with a Henry's law constant of 5.49×10^{-3} (atm-m^3/mol) will partition preferentially into the vapor phase compared to phenol with $K_H = 4.0 \times 10^{-7}$ (atm-m^3/mol). Volatilization is a significant mass-transfer mechanism for contaminants with $K_H > 10^{-7}$. Mass-transfer of pure liquids and NAPL components into the vapor phase is governed by Raoult's law (Eq. 2.21).

Volatilization can account for significant losses of organic solutes from soil, groundwater, or NAPL. Volatilization of gasoline contaminants from saturated and unsaturated zones can result in vapor plumes in the vadose zone that can migrate offsite and accumulate under surface structures such as in basements of residential buildings. Such soil gases are health hazards and have the potential to cause explosions.

2.3.4 *Sorption*

Sorption can be described as a mass-transfer process that results in the migration of contaminants from a fluid phase to an adjacent solid surface, an interface or a second fluid phase (Adamson, 1990). The chemical potential of the solute remains unchanged in each phase and the solute is assumed to obey Henry's law. The term "sorption" includes absorption and adsorption processes. *Absorption* describes the inter-phase dissolution of solute molecules resulting from their complete mixing throughout the sorbent phase. Such dissolution or "partitioning" of a solute can occur between gas-gas, gas-liquid, liquid-liquid, gas-solid or liquid-solid phases. *Adsorption* is a surface phenomenon and refers to mass-transfer of a solute from a fluid to the surface of a solid or its accumulation at the interface between two phases. Adsorption processes are further described as physisorption if the solute-sorbent interactions are primarily through weak van der Waal's forces, or chemisorption if stronger chemical bond formation occurs between the solute and sorbent surfaces. Sorption processes can affect the mobility and biodegradability of pollutants in the soil and groundwater environment through complex combinations of absorption, physisorption and chemisorption reactions with soil components.

Figure 2.4 illustrates some common ways in which contaminant molecules interact with soils components. These include (i) surface adsorption, electrostatic interactions, hydrogen bonding, or ion-exchange of solute with the mineral surfaces, (ii) precipitation and crystallization of the contaminant on soil or sediment surfaces, (iii) hydrophobic interactions, sequestration, ion-exchange, proton-transfer, coordination reactions or covalent bond formation with soil or sediment organic matter, and (iv) diffusion into mineral or organic matter micropores.

The contaminant's distribution between the fluid and solid phases at equilibrium at a given temperature can be described using sorption isotherms such as those illustrated in Figure 2.5. Absorption of a solute into a three-dimensional phase is represented by Type I isotherms. Type I sorption behavior is described by a "partitioning-type" linear free energy relationship equating the concentrations of the solute in each of the two phases. For relatively small concentration ranges and dilute pollutant concentrations, the solute distribution at equilibrium can be represented by a simple linear phase distribution relationship expressed as:

$$\frac{q_e}{C_e} = K_d \qquad \text{(Eq. 2.22)}$$

where, q_e is the solid phase concentration, C_e is the fluid phase concentration at equilibrium, and K_d is a distribution coefficient analogous to a liquid-liquid partitioning coefficient such as K_{OW}. The distribution of solute between the two phases is, therefore, represented by the ratio of the solute concentration in each phase. Absorption of organic contaminants to natural soils containing diagenetically unaltered soil organic matter is rapid, reversible and noncompetitive and follows Type I sorption behavior (Huang and Weber, 1997).

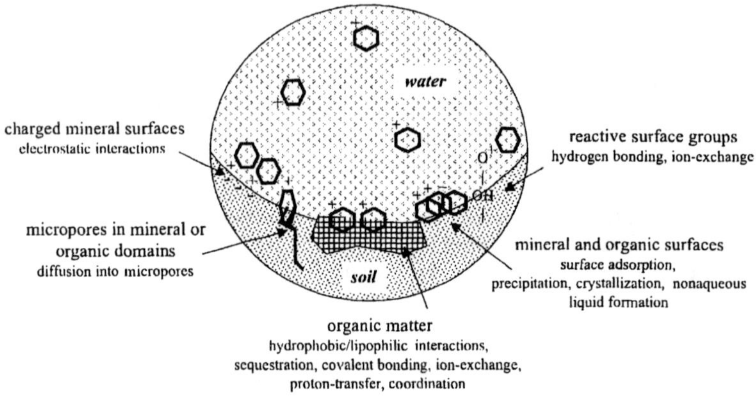

Figure 2.4 Physical-chemical interactions between dissolved organic solutes and soil components.

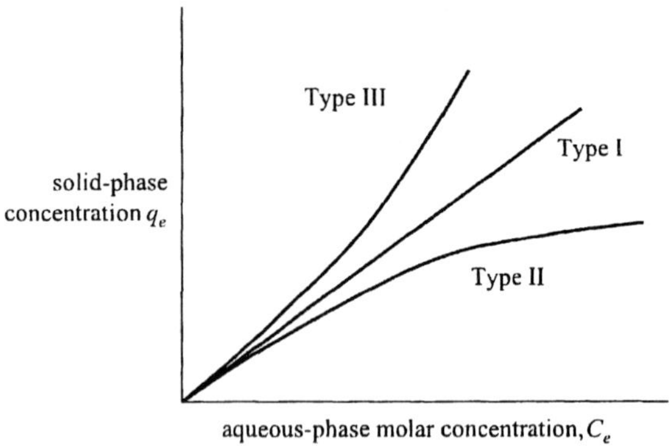

Figure 2.5 Sorption isotherms representing solute distribution between solid and fluid phases at equilibrium.

The Type II curve shown in Figure 2.5 describes an equilibrium relationship where solute transfer from a fluid phase to a solid surface decreases with increasing solute concentration. Type II sorption or *adsorption* occurs when solute molecules interact with a finite number of specific surface sites. In such cases, adsorbing solute molecules eventually saturate the surface producing a monolayer surface coverage as illustrated by the concave isotherm. Type II behavior is observed in the adsorption of organic contaminants on nonporous mineral surfaces. Contaminant adsorption is generally rapid if mass-transfer to the surface and into pores of the mineral or organic domains is faster compared to macroscopic solute transport in the bulk aqueous phase (Weber and DiGiano, 1996). Conversely, sorption is protracted if it is controlled by diffusion into intraparticle or intraorganic regions (Pignatello and Xing, 1996).

Type III sorption behavior results in a convex isotherm in which the tendency of the solute molecules to associate with a second phase, a surface or an interface increases with solute concentration. Type II behavior occurs when solute partitioning or adsorption results in the modification of the sorbent surface in a manner that allows more favorable solute transfer from the fluid phase. This phenomenon is commonly not observed in the interaction of organic solutes with soils and sediments. Irrespective of the type of sorption behavior, however, all isotherms appear linear at low solution concentrations and over narrow concentration ranges. A variety of models have been used to describe the phase distribution of organic solutes between gas/liquid and solid phases. Five widely used sorption models are summarized in Table 2.4.

Table 2.4 Commonly used sorption models for organic and inorganic solutes.

Sorption Model	Mathematical Representation	Coefficients
Partitioning Model	$q_e = K_d C_e$	K_d = linear phase distribution constant
Langmuir Model	$q_e = \dfrac{q_0 K_L C_e}{1 + K_L C_e}$	q_o = sorption capacity K_L = sorption energy factor
Brunauer-Emmett-Teller (BET) Model	$q_e = \dfrac{q_0 K_{BET} C_e}{(C_s - C_e)\left[1 + (K_{BET} - 1)(C_e/C_s)\right]}$	q_o = sorption capacity K_{BET} = sorption energy factor C_s = molar solubility
Freundlich Model	$q_e = K_F C_e^n$	K_F = specific capacity n = sorption energy and heterogeneity factor
Dual Mode Model	$q_e = K_d C_e + \sum_1^N \dfrac{q_{0,i} K_{L,i} C_{e,i}}{1 + K_{L,i} C_{e,i}}$	N = types of sorption-sites within heterogeneous SOM

2.3.5 *Ion Exchange*

Ion exchange is a reversible chemical reaction wherein an ion from solution is exchanged for a similarly charged ion attached on a solid surface. Ion exchange is also referred to as an adsorbent motivated sorption process. The presence of surface charges or counter-ions on soil particles, especially clays, gives rise to the soil's cation exchange capacity (CEC) or anion exchange capacity (AEC) allowing these particles to function as ion exchangers by exchanging surface H^+, Na^+, K^+, Ca^{2+}, Mg^{2+}, OH^-, Cl^-, NO_3^-, SO_4^{2-}, or PO_4^{3-} ions with ionic contaminants from the soil solution or groundwater. The ion-exchange process is competitive and depends highly on the pH and ionic strength of the solution. While expandable clays such as montmorillonite manifest high CECs (~ 1.0 mol/kg), non-expandable clays such as kaolinite exhibit exchange capacities of the order of 0.1 mol/kg.

Both ionizable organic contaminants and heavy metals can participate in ion exchange reactions with soil surfaces. The ion exchange of a cationic contaminant such as a heavy metal, M^+, to a soil surface saturated with Na^+ counterions can be expressed by Eq. (2.23):

$$M^+ + \text{Na–soil} \leftrightarrow Na^+ + \text{M–soil} \qquad \text{(Eq. 2.23)}$$

The equilibrium constant for this reaction is given by:

$$K_{ie} = \frac{[\text{M} - \text{soil}][\text{Na}^+]}{\left[\text{M}^+\right]\left[\text{Na} - \text{soil}\right]} \qquad \text{(Eq. 2.24)}$$

Due to their pH dependent surface charges, oxide minerals constitute the most important sorption domains for heavy metals in soils and subsurface materials. For example, arsenate and arsenite anions adhere electrostatically to mineral surfaces that are positively charged. These ion exchange reactions reach equilibrium within seconds to hours and have been observed to follow both Langmuir and Freundlich-type adsorption behaviors (Smith et al., 1999; Adriano, 2001).

2.3.6 *Complexation*

Short-range interactions between ions or molecules in soil solutions can also lead to complex formation:

$$aM^{m+} + bL^{n-} \rightarrow M_aL_b^q \qquad \text{(Eq. 2.25)}$$

where $q = am - bn$, M^{m+} is a metal ion, L^{n-} is a ligand ion or molecule and $M_aL_b{}^q$ is the soluble complex or ion pair. A ligand is an ion or molecule that is bonded to a central metal cation through a coordinate covalent bond and resides in its coordination sphere. There are three primary types of soluble complexes (Figure 2.6): a) a solvation complex comprised of the metal ion and its primary hydration sphere where H_2O is the ligand; b) a complex ion or inner-sphere complex is a soluble species in which a non-H_2O ligand has replaced a water of hydration; and c) an ion pair or an outer-sphere complex is formed when a ligand does not replace a water of hydration but is electrostatically attached to the outside of the hydration sphere (Essington, 2004). The formation of soluble complexes results in the loss of the original chemical properties of the metal ions.

Ion speciation and the stability of complexes have the potential to significantly impact the environmental fate of metallic contaminants. The equilibrium constant also known as the ion formation constant (K_f) for the complex formation reaction represented by Eq. (2.25) can be written as:

$$K_f = \frac{\left(M_aL_b^q\right)}{\left(M^{m+}\right)^a\left(L^{n-}\right)^b} \tag{2.26}$$

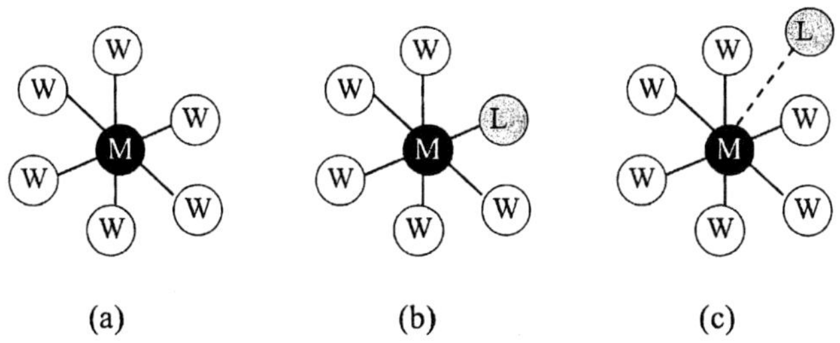

Figure 2.6 Three primary types of soluble metal ion complexes: (a) solvation complex, (b) inner-sphere complex, and (c) outer-sphere complex. M = metal ion, W = H_2O, L = ligand ion or molecule.

Complex stability is a function of the type of complex formed, the number of metals and ligands involved, and properties of component ions such as their radii, valences, polarizability and electronegativity (Essington, 2004). Complex stability is also affected by the number of bonds between the central metal ion and a complexing ligand. Ligands such as oxalate, carbonate, citrate and ethylenediamine tetraacetate (EDTA), which form multiple bonds with the central metal cation, yield polydentate complexes that are more stable than the monodentate complexes formed with ligands such as chloride, acetate and lactate (Figure 2.7). The mobility and bioavailability of metal ions in soil systems is strongly influenced by their tendency to form complexes with the carboxylic or phenolic moieties of soil organic matter.

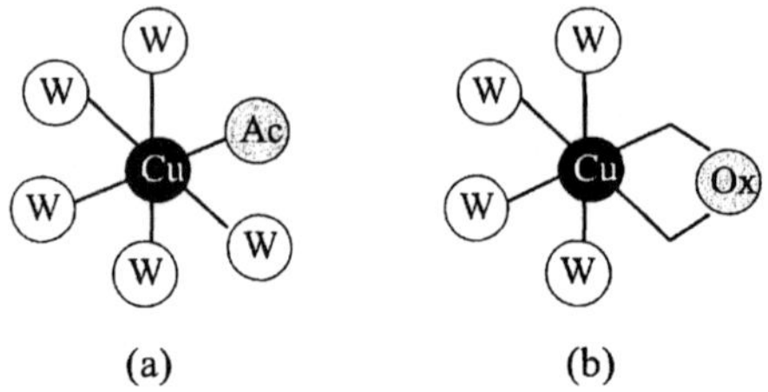

Figure 2.7 Formation of (a) monodentate and (b) bidentate complexes between copper ion (Cu) and acetate (Ac) and oxalate (Ox) molecules, respectively. W = H_2O.

2.3.7 Oxidation-Reduction

The biological and chemical fate of contaminants in the environment is often related to electron-transfer mediated oxidation-reduction (redox) reactions. Chemicals that are enriched in electrons are considered reduced and known as "electron donors" or reductants, while those deficient in electrons are oxidized and called "electron acceptors" or oxidants. Redox reactions result from electron transfers between reductants and oxidants, whereby the reductant is oxidized and the oxidant is reduced. Many inorganic reactions and all biological reactions in soils are redox reactions. Most redox reactions also result in the transfer of H^+ ions. Electron transfer involves the coupling of two half-reactions. The oxidation half reaction represents the release of electrons from an electron-rich chemical resulting in the generation of an oxidized species of the chemical. The reduction half reaction describes the acceptance of electrons by an electron-deficient chemical and the production of a reduced species. For example, the oxidation of phenol by chromate may be represented by two half reactions:

Oxidation half reaction: $1/28\ C_6H_6O + 11/28\ H_2O \rightarrow 3/14\ CO_2 + H^+ + e^-$

Reduction half reaction: $1/6\ Cr_2O_7^{2-}\ (aq) + 4/3\ H^+ + e^- \rightarrow 1/3\ Cr(OH)_3\ (s) + 1/6\ H_2O$

These half reactions represent single electron transfers. Phenol is mineralized to CO_2 and aqueous chromate is reduced to a hydroxide precipitate. The overall redox reaction can be written as:

$$1/28\ C_6H_6O + 19/84\ H_2O + 1/6\ Cr_2O_7^{2-}\ (aq) + 1/3\ H^+ \rightarrow 1/3\ Cr(OH)_3\ (s) + 3/14\ CO_2$$

or

$$3\ C_6H_6O + 19\ H_2O + 14\ Cr_2O_7^{2-}\ (aq) + 28\ H^+ \rightarrow 28\ Cr(OH)_3\ (s) + 18\ CO_2$$

The equilibrium constant K for single electron transfer half reactions is expressed as:

$$K = \frac{[\text{Red}]}{[\text{Ox}]\,[e^-]} \qquad \text{(Eq. 2.27)}$$

where [Red] and [Ox] refer to the molar concentrations of the reductant and oxidant species, respectively. Eq. (2.27) can be rewritten in terms of electron activity as:

$$[e^-] = \left\{\frac{1}{K}\frac{[\text{Red}]}{[\text{Ox}]}\right\} \qquad \text{(Eq. 2.28)}$$

or,

$$-\log[e^-] = \log K - \log\frac{[\text{Red}]}{[\text{Ox}]}$$

Replacing $-\log[e^-]$ and $\log K$ with pe and pe^o, the above reaction can be written as:

$$pe = pe^o - \log\frac{[\text{Red}]}{[\text{Ox}]} \qquad \text{(Eq. 2.29)}$$

The value of pe is an indicator of the local redox environment. Large values of pe favor the existence of electron rich (reduced) species while small values indicate the preference for electron-deficient (oxidized) species. At a pH of 7.0, $pe < 2$ is indicative of anoxic or reducing conditions, $2 < pe < 7$ represents suboxic conditions, and $pe > 7$ indicates oxic or oxidizing conditions (Knox et al., 1993). Redox conditions in soils and aquifers can vary widely over short distances and impact the fate of both organic and inorganic contaminants. The redox environment can affect the extents and rates of pollutant biotransformation and the mobility of metals in the subsurface.

Environmental redox conditions can also be characterized using the redox potential, E_H which has units of volts and is related to pe according to Eq. (2.30):

$$E_H = \frac{2.3RT}{F}\,pe \qquad \text{(Eq. 2.30)}$$

where R is the universal gas constant, T is temperature in degrees Kelvin, and F is the Faraday constant. The value $2.3RT/F$ is equal to 0.059 V at 25°C. Table 2.5 summarizes the pe and E_H values for selected redox reactions.

Table 2.5 Selected reduction half reactions and associated pe^0 and E_H values at 25°C.

Half Reaction	pe^0	E_H (volts)
$Co^{3+} + e^- \rightarrow Co^{2+}$	+33.1	+ 1.953
$Mn^{3+} + e^- \rightarrow Mn^{2+}$	+ 25.5	+ 1.508
$\frac{1}{2}\, NO_2^- + 3/2\, H^+ + e^- \rightarrow \frac{1}{4}\, N_2O\,(g) + \frac{3}{4}\, H_2O$	+ 23.6	+ 1.396
$1/5\, NO_3^- + 6/5\, H^+ + e^- \rightarrow 1/10\, N_2\,(g) + 3/5\, H_2O$	+ 21.1	+ 1.248
$\frac{1}{4}\, O_2\,(g) + H^+ + e^- \rightarrow \frac{1}{2}\, H_2O$	+ 20.8	+ 1.230
$\frac{1}{2}\, Fe_3O_4(s) + 3H^+ + e^- \rightarrow 3/2\, Fe^{2+} + 2H_2O$	+ 17.8	+ 1.053
$Fe(OH)_3(s) + 3H^+ + e^- \rightarrow Fe^{2+} + 3H_2O$	+ 15.8	+ 0.935
$1/8\, NO_3^- + 5/4\, H^+ + e^- \rightarrow 1/8\, NH_4^+ + 3/8\, H_2O$	+ 14.9	+ 0.881
$\frac{1}{2}\, NO_3^- + H^+ + e^- \rightarrow \frac{1}{2}\, NO_2^- + \frac{1}{2}\, H_2O$	+ 14.1	+ 0.834
$Fe^{3+} + e^- \rightarrow Fe^{2+}$	+ 13.0	+ 0.769
$\frac{1}{2}\, O_2(g) + H^+ + e^- \rightarrow \frac{1}{2}\, H_2O_2$	+ 11.6	+ 0.686
$1/3\, CrO_4^{2-} + 8/3\, H^+ + e^- \rightarrow 1/3\, Cr^{3+} + 4/3\, H_2O$	+ 8.6	+ 0.505
$\frac{1}{2}\, Pb^{4+} + e^- \rightarrow \frac{1}{2}\, Pb^{2+}$	+ 7.2	+ 0.423
$1/8\, SO_4^{2-} + 5/4\, H^+ + e^- \rightarrow 1/8\, H_2S + \frac{1}{2}\, H_2O$	+ 5.2	+ 0.308
$1/12\, C_6H_{12}O_6 + H^+ + e^- \rightarrow \frac{1}{4}\, C_2H_5OH$	+ 4.4	+ 0.260
$1/8\, SO_4^{2-} + 9/8\, H^+ + e^- \rightarrow 1/8\, HS^- + \frac{1}{2}\, H_2O$	+ 4.3	+ 0.254
$1/8\, CO_2(g) + H^+ + e^- \rightarrow 1/8\, CH_4(g) + H_2O$	+ 2.9	+ 0.172
$\frac{1}{2}\, AsO_4^{3-} + H^+ + e^- \rightarrow \frac{1}{2}\, AsO_3^{3-} + \frac{1}{2}\, H_2O$	+ 1.3	+ 0.078
$H^+ + e^- \rightarrow \frac{1}{2}\, H_2(g)$	0.0	0.0

2.4 Biological Processes

Biological processes also play a significant role in the mitigation of environmental pollutants. Terrestrial and aquatic ecosystems offer a diversified range of biological activities, where biotic processes in soil, sediment, surface water and groundwater can be mediated by microorganisms and higher plants that have adapted to the constituents present in these environments. Plant and microbial mitigation generally involves processes such as biological uptake, biological transformation and biologically mediated contaminant immobilization. While playing an instrumental role in the reduction and biogeochemical cycling of nutrients, biotic processes can also temporarily or permanently immobilize, accumulate or transform organic and inorganic constituents, resulting in modifications in the structural and toxicological properties of pollutants.

The interaction of vegetation, microorganisms and organic and inorganic constituents present in the soil, sediment, surface water or groundwater, can provide a desirable environment for enhanced growth and reproduction of microorganisms in the rhizosphere of higher plants, consequently enhancing their role in contaminant mitigation. An understanding of the biotic processes involved in the fate and transport of contaminants in natural systems is essential to compare and select biologically-based cleanup strategies and establish treatment protocols that are best suited for remediating contaminated sites. This section outlines microbiological and phytoremediation processes involved in the fate and transport of pollutants in natural systems for contaminated soil and groundwater remediation.

2.4.1 Microbial Processes

Plants and other organisms readily break down and immobilize many organic and inorganic compounds. However, microorganisms play the predominant role in the mineralization portion of biogeochemical reactions. Microorganisms are well suited to this task because of large numbers of microbial species that thrive in a variety of environments. Microbes can adapt to changes in environmental or substrate conditions, exhibit high metabolic and growth rates, and are collectively able to catalyze a wide range of metabolic reactions. The small size and mass of microorganisms make them ecologically significant, since they are easily carried by air and water currents to other locations, making them ubiquitous. They appear to thrive even under extreme conditions, where species distribution can sometimes be as diverse as that in temperate climates. Environmental and nutritional conditions determine which strains grow and reproduce successfully, and ultimately affect the generation of metabolic products and the level of xenobiotic compound degradation.

2.4.1.1 Microorganisms Involved in Remediation

Microorganisms involved in the mitigation of contaminants include: bacteria, algae and fungi. The majority of bacteria are single-celled organisms that are widely distributed in terrestrial and aquatic systems. Many bacteria absorb necessary nutrients from their environment, while others make their own nutrients via photosynthesis or other synthetic processes. Algae includes unicellular or multi-cellular organisms found in fresh and salt water aquatic environments that photosynthesize their own food by capturing energy from sunlight. Many fungi, such as yeasts and molds, are single-celled organisms, while others such as mushrooms are multi-cellular. They are widely distributed in soil and aquatic ecosystems as decomposers of dead organic matter that absorb nutrients from their environment. Protozoa are found in a variety of soil and aquatic environments. These single-celled organisms generally obtain food by engulfing or ingesting smaller microorganisms (Black, 1999).

As decomposers, bacteria, fungi and yeasts fill an important ecological role in the decay of organic matter in the environment. In waste treatment systems, bacteria are commonly found as aerobic and/or anaerobic consortia, where a variety of microorganisms are involved in pollutant mitigation. While protozoans are not significantly involved in the remediation of contaminants, they possess numerous highly efficient digestive enzymes, which may eventually be tapped for use in specific transformation reactions. Fungi have been found to degrade several recalcitrant organic contaminants. Algae are involved in metal recovery via biosorption, uptake and bioaccumulation, leading to the mitigation of waste streams with high metal content.

Microorganisms are important facilitators of the biogeochemical cycling of organic and inorganic nutrients. Most properties of microorganisms are directly attributable to their small size that ultimately affects their morphology, mobility, activity, ecological distribution, diversity and metabolic flexibility. There are also some limitations imposed by the small size of microorganisms when compared to plant or animal cells. Microorganisms tend to be less specialized and individual microbial cells cannot conduct as many functions as plant or animal cells.

The small size of microorganisms generally leads to a high surface area to volume ratio, which in turn is responsible for the relatively high metabolic rates observed in microorganisms, and allows for their extensive interactions with the environment. As the size and dimension of cells decrease, the metabolic rate (respiratory quotient) typically increases (Schlegel, 1988). Another measure of metabolic rate is the doubling or generation time. Small bacterial cells can double fairly rapidly, sometimes within a matter of 10 to 30 minutes under optimal growth conditions.

In microorganisms, extensive adaptability has evolved as a necessity imposed by their small size and constant exposure to and interactions with the environment. Most of their enzyme systems are able to change in response to variations in environmental conditions. This adaptability provides for efficient regulation of metabolism through induction and repression of genes that code for specific enzymes, broad substrate specificity of microbial enzymes, ease of gene transfer and mobilization, and collective metabolic versatility of microbial communities.

Regulatory mechanisms such as induction and repression play an important role in microbial cell physiology. An average microbial cell has several thousand structural genes which are able to encode approximately 100,000 protein molecules. However, not all structural genes are expressed at any given time. For instance, many important catabolic enzymes are produced only if the appropriate substrates are present. When the inductive substrates are no longer available, the enzymes are quickly broken down to make other proteins needed for the microorganism to adapt to new environmental conditions. Another reason for the metabolic flexibility observed in microorganisms is the ease with which gene transfer, including conjugation, transformation, transduction and transposition, occurs in microbial cells. Through the acquisition of new genetic elements, microorganisms can acquire new metabolic capabilities, when exposed to varying environmental conditions. Similarly, in the absence of selective forces, microorganisms can also readily lose certain genetic elements.

Microbial adaptability implies that, in many cases, microbial enzymes exhibit broader substrate specificity than their higher plant and mammalian counter parts. This broad substrate specificity is crucial in the biodegradation of many xenobiotic contaminants, where microbial enzymes normally adapted to using natural substrates, are also able to catalyze reactions using xenobiotic compounds. However, broad substrate specificity occasionally results in the generation of toxic or dead-end products. In general, various species of microorganisms can partially metabolize contaminants, generating metabolic products, which can subsequently be utilized for growth and energy production by other species of microorganisms. Hence, while individual cells may metabolize only a limited range of substrates, collectively, microorganisms can utilize most naturally occurring organic substrates, as well as many xenobiotic compounds.

2.4.1.2 Microbial Metabolism

Metabolism refers to cellular biochemical processes that involve energy transfer. Such processes include anabolism or reactions that require energy to synthesize complex molecules from simpler ones, and catabolism or reactions that release energy by breaking complex molecules into simpler ones. Microorganisms can be classified taxonomically (Table 2.6) or grouped into metabolic classes depending on their sources of energy and carbon (Figure 2.8) (Madigan et al., 1997). Heterotrophs use organic compounds as sources of carbon, while autotrophs employ inorganic CO_2 as a source of carbon for the synthesis of organic molecules.

While chemotrophic organisms obtain energy by oxidizing simple inorganic chemicals such as sulfides and nitrites, phototrophs use light as an energy source. Heterotrophic organisms can be further classified based on the compounds they utilize as the terminal electron acceptors during oxidation of organic substrates or other constituents. Strict aerobic heterotrophs use molecular oxygen as the terminal electron acceptor during oxidation of organic compounds:

$$\text{Organic Compound} + O_2 \rightarrow CO_2 + H_2O + \text{Energy} \qquad \text{(Eq. 2.31)}$$

Table 2.6 Taxonomic classification of microorganisms.

Characteristics		Bacteria	Algae	Protozoa	Fungi	Rotifers
Cellular Organization	Unicellular	X	X	X		
	Multicellular		X		X	*X*
Nuclear Organization	Prokaryotic	X	X			
	Eukaryotic		X	X	X	*X*
Autotrophic	Photo		X			
	Chemo	X				
Heterotrophic	Aerobic	X		X	X	*X*
	Anaerobic	X		X		
	Facultative	X				

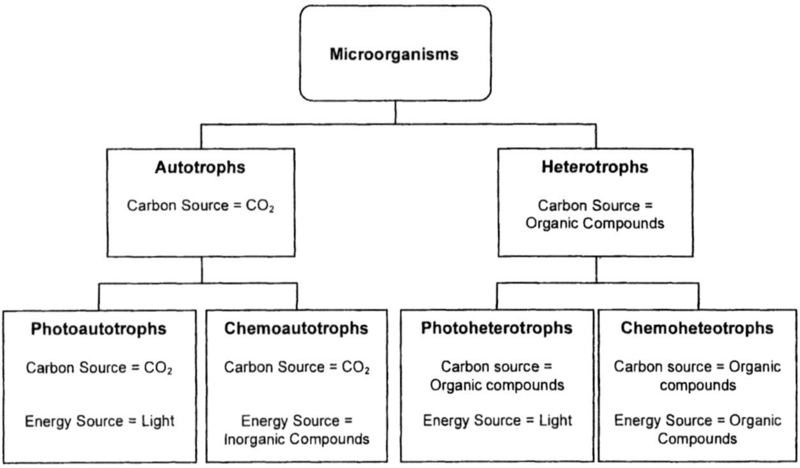

Figure 2.8 Microbial classification based on metabolism.

In the absence of oxygen, anaerobic heterotrophs oxidize organic compounds using other oxidized substances as terminal electron acceptors. This metabolic process is referred to as anaerobic respiration, and the general reactions involved are illustrated Eq. (2.32) for denitrification and Eq. (2.33) for sulfate reduction:

$$\text{Organic Compound} + NO_3^- \rightarrow CO_2 + N_2 + \text{Energy} \qquad \text{(Eq. 2.32)}$$

$$\text{Organic Compound} + SO_4^{2-} \rightarrow CO_2 + H_2S + \text{Energy} \qquad \text{(Eq. 2.33)}$$

Facultative heterotrophs comprise a group of bacteria that utilize oxygen, nitrate or sulfate for the oxidation of organic compounds depending on the availability of each electron acceptor. Another reaction that can be carried out by aerobic, anaerobic or facultative microorganisms is fermentation. In this case, organic compounds serve as electron acceptors during oxidation of organic substrates (Eq. 2.34).

$$\text{Organic Compound} \rightarrow \text{Organic Compounds} + CO_2 + H_2O + \text{Energy} \qquad \text{(Eq. 2.34)}$$

Methanogenesis is another anaerobic metabolic process that allows strictly anaerobic organisms to break down organic acids and produce methane gas (CH_4) using CO_2 as an electron acceptor (Eq. 2.35).

$$\text{Organic Acids} \rightarrow CH_4 + CO_2 + H_2O + \text{Energy} \qquad \text{(Eq. 2.35)}$$

Thus, a variety of microbial decomposers in soil and aquifer systems are able to utilize organic contaminants as sources of energy, converting them to gaseous CO_2 or CH_4. This provides an important biological mechanism for removal of a wide variety of organic compounds. The efficiency and rate of organic compound degradation by microorganisms is highly variable for different types of organic compounds. In general, when oxygen and an organic substrate are available, aerobic and facultative bacteria immediately start to biodegrade the organic substrate, until the oxygen available has been depleted. In the absence of oxygen, organic decomposition can be mediated by any number of other terminal electron acceptors, including NO_3^-, Mn^{2+}, Fe^{3+}, SO_4^{2-} and CO_2 (Figure 2.9). The redox potential decreases through the sequence of electron acceptors.

For E_H > 300 mV, the local environmental conditions are considered to be aerobic, at E_H < -100 mV they are anaerobic, while for intermediate redox values, the conditions are anoxic (Kadlec and Knight, 1996). O_2 is the acceptor for 400 mV < E_H < 600 mV. Between E_H values of 320 amd 340 mV, oxygen depletion is considered to be complete and nitrate reduction to N_2 begins to take place until an E_H of 220 mV is reached. Manganese reduction from managanic (Mn^{4+}) to manganous (Mn^{2+}) is prominent at E_H between 220 and 120 mV. Ferric (Fe^{3+}) to ferrous (Fe^{2+}) reduction takes place at E_H between 120 mV and -150 mV, followed by sulfate (SO_4^{2-}) reduction to hydrogen sulfide (H_2S) from -150 mV to -250 mV, and, finally, carbon dioxide (CO_2) is reduced to methane (CH_4) from -250 mV to -300 mV.

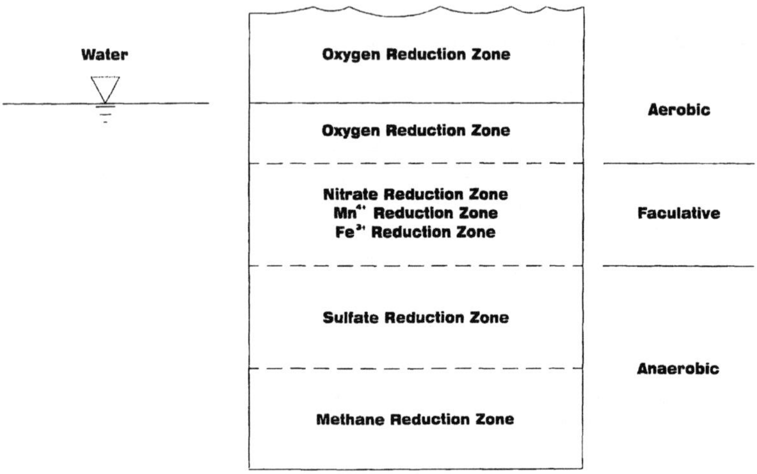

Figure 2.9 Redox zones in soil as a function of soil depth.

The primary reason for heterotrophs to decompose organic compounds is to gain energy for cellular purposes such as synthesis of macromolecules, respiration and motility, with some energy loss as heat. The amount of energy that can be biochemically extracted from a given quantity of organic matter depends on the type of terminal electron acceptor utilized. The greatest amount of energy is yielded when oxygen is used in the oxidation reaction. The least amount of energy is derived during strictly anaerobic metabolism. Thus, biological transformation of organic contaminants usually occurs most rapidly in the presence of oxygen, and more slowly when other electron acceptors are used.

Mixed substrate systems are characterized by the presence of a multiple sources of nutrients that serve similar physiological functions. In contaminated soil and groundwater, microbial growth and metabolism often occurs at low concentrations of diverse mixed substrates. Co-metabolism is an important degradation process where the fortuitous transformation of a non-growth-yielding substrate occurs in the presence of a growth substrate. The growth substrate provides energy for microbial growth and maintenance and reducing power, which allows for the degradation of the non-growth substrates. Metabolism of non-growth substrates does not provide energy or reducing power to the microorganism. However, the end products of the transformation can be mineralized or utilized by other microorganisms in a community. Co-metabolism is viewed as extremely important in the initial environmental transformation of many otherwise recalcitrant pollutants. It can be viewed as a consequence of the broad substrate specificity of some microbial enzymes, where in some cases, the growth substrates have a similar size and structure as the co-metabolites.

Autotrophic bacteria oxidize inorganic compounds for energy and use CO_2 as their source of cellular carbon. Nitrifying, sulfur and iron bacteria are among the most important autotrophs relevant to bioremediation technologies. Under low organic carbon loading, nitrifying bacteria such as *Nitrosomonas* and *Nitrobacter* spp., can oxidize NH_3 in two stages to produce energy and form NO_2^- and, subsequently, NO_3^- according to Eqs. (2.36) and (2.37):

$$NH_3 + O_2 \xrightarrow{Nitrosomonas} NO_2^- + \text{Energy} \qquad \text{(Eq. 2.36)}$$

$$NO_2^- + O_2 \xrightarrow{Nitrobacter} NO_3^- + \text{Energy} \qquad \text{(Eq. 2.37)}$$

Sulfur bacteria (such as *Thiobacillus* sp.) can tolerate pH levels as low as 1.0 and oxidize the weak acid hydrogen sulfide, H_2S, to a strong acid, H_2SO_4, as shown in the following equation:

$$H_2S + O_2 \xrightarrow{Thiobacillus} H_2SO_4 + \text{Energy} \qquad \text{(Eq. 2.38)}$$

Iron bacteria, such as the filamentous *Leptothrix* and *Crenothrix*, oxidize soluble inorganic Fe^{+2} to insoluble Fe^{+3}, resulting in precipitation of $Fe(OH)_3$. The filamentous bacteria carry out this reaction and deposit the oxidized iron, forming yellow or reddish brown colored slimes or effluents:

$$Fe^{2+} + O_2 \xrightarrow{Leptothrix\ or\ Crenotrix} Fe^{3+} + \text{Energy} \qquad \text{(Eq. 2.39)}$$

Table 2.7 summarizes general autotrophic and heterotrophic microbial metabolic reactions and the end products generated, as a function of electron acceptor availability (Suthersan, 1997).

2.4.1.3 Microbially-Mediated Processes

Microorganisms can play a significant role in the mitigation of organic and inorganic contaminants in the environment. Their metabolism is dependent upon the availability of electron donors and acceptors, as well as the essential nutrients and ions necessary for growth. Microbes have evolved a wide range of mechanisms for the immobilization, mobilization, uptake and transformation of organic and inorganic contaminants. A schematic of common microbially-mediated processes is illustrated in Figure 2.10 and a summary of these processes is presented in Table 2.8.

Table 2.7. Summary of autotrophic and heterotrophic microbial metabolism.

Metabolism	Electron Donor	Electron Acceptor	End Products
Aerobic Respiration	Organic substrates	O_2	CO_2, H_2O
	NH_4	O_2	NO_2^-, NO_3^-, H_2O
	Fe^{2+}	O_2	Fe^{3+}
	S^{2-}	O_2	SO_4^{2-}
Anaerobic Respiration	Organic substrates	NO_3^-	CO_2, H_2O, N_2, Cl^-
	Organic substrates	SO_4^{2-}	CO_2, H_2O, S^{2-}, Cl^-
	H_2	SO_4^{2-}	H_2O, S^{2-}
	H_2	CO_2	CH_4, H_2O
Fermentation	Organic substrates	Organic compounds	Organic compounds, CO_2, CH_4

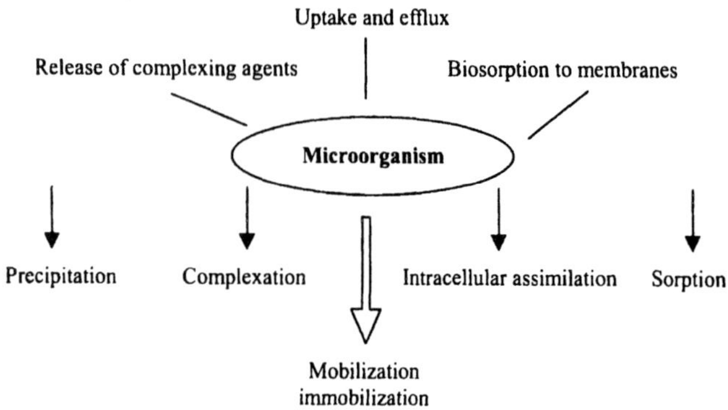

Figure 2.10 Microbial remediation mechanisms of organic and inorganic contaminants.

Microorganisms can temporarily or permanently immobilize pollutants, particularly inorganic contaminants, in the environment via biomineralization and biosorption. In the case of metallic pollutants, microorganisms provide a chemical environment that controls the nucleation and growth of crystalline metal species. The formation of these crystalline biominerals and non-crystalline precipitates can occur both within or outside the cell (Rabb and Feldmann, 2003).

A number of mechanisms grouped under the term biosorption, which include physical adsorption, ion exchange, complexation and precipitation, allow physical interactions between microbes and contaminants. Bacterial cell walls and envelopes, and the walls of fungi, yeasts, and algae are efficient metal biosorbents that bind charged groups. Cell walls are mainly composed of polysaccharides, proteins and lipids and have abundant binding groups such as carboxyl, sulfate, phosphate and amino groups. Physical adsorption takes place due to Van der Waals' forces and can also occur via electrostatic interactions between the inorganic ions in solution and the cell walls of the microorganisms (Kuyucak and Volesky, 1988).

Table 2.8 Selected microbiological processes.

Mechanisms	Process Goal	Media	Contaminants
Biomineralization	Permanent contaminant immobilization	Groundwater, surface water, wastewaters, sludges	Metals Inorganic anions such as HS^-, NO_3^-, NO_2^-, SO_4^{2-}
Biosorption • Precipitation • Complexation • Ion exchange • Adsortption	Temporary or permanent contaminant immobilization	Groundwater, surface water, wastewaters, sludges	Metals and radionuclides Charged inorganic and organic compounds
Bioaccumulation / Uptake	Contaminant extraction and accumulation from media to the microbial cell	Groundwater, soil, sediment, surface water, wastewater, sludges	Metals and radionuclides Some recalcitrant organic compounds
Transformation • Degradation • Mineralization • Detoxification • Activation	Change in contaminant structure to a more or less toxic end product	Soil, sludge, sediments, suface water, groundwater, sludges	Organic and inorganic compounds
Cometabolism	Change in contaminant structure to a more or less toxic end product	Soil, sludge, sediments, suface water, groundwater, sludges	Some recalcitrant organic compounds

The polysaccharides associated with microbial cell walls allow exchange of bivalent metal ions for the counter ions of the polysaccharides, resulting in the biosorptive uptake of heavy metals (Kuyucak and Volesky, 1988). Metal removal from solution can also occur through complex formation with amino and carboxyl groups found in cell wall polysaccharides. Microorganisms can mediate the precipitation of metals through the formation of less soluble non-crystalline metal species or complexes, depending on the availability of organic and inorganic anions (Rabb and Feldmann, 2003). Metal precipitation may take place both in the solution and on the cell surface. It may be dependent on microbial metabolism if, in the presence of toxic metals, the microorganism produces compounds that enhance the precipitation process. However, precipitation is not dependent on microbial metabolism, if it occurs as a result of a chemical interaction between the constituent and cell surface.

Cellular membranes of microorganisms are composed of hydrophobic lipid bilayers, which are practically impenetrable for charged compounds. The transport of inorganic and organic contaminants across microbial cell membranes can lead to intracellular accumulation, which is dependent on microbial cell metabolism. Potential mechanisms of flux across membranes can be ion pumps, ion channels, carrier-mediated transport, endocytosis, complex permeation, and lipid permeation. These processes can only take place with viable cells and in the case of metals, for instance, transport across microbial cell membranes may be mediated by the same energy dependent transport mechanism used to convey metabolically important ions (Raab and Feldmann, 2003). Permeabilization of cell membranes by these compounds can result in further exposure of intracellular binding sites and in an in increase passive accumulation.

The mobilization of organic compounds, metals and radionuclides occurs as a result of oxidation-reduction reactions, production of soluble mineral or organic acid metabolites or intermediates, and changes in abiotic conditions. For inorganic compounds that can exist in more than one oxidation state, enzymatic oxidation or reduction which impacts solubility may be a useful approach to remove the toxic inorganic species from solution. The use of complexation agents may be useful in mobilizing organic and inorganic compounds to facilitate their subsequent uptake. Microbial complexing agents can be low molecular weight organic acids and alcohols, high molecular weight ligands, siderophores, and metal-binding compounds. When microorganisms are grown in iron deficient media, they can produce siderophores, which are specific iron chelators, but are also capable of forming complexes with other metals and radionuclides, thereby increasing their aqueous solubilities. Siderophores are compounds that possess catecholate, phenolate or hydroxamate as their binding groups.

Microbial transformation of organic compounds and organometallic species, may be described by the terms degradation, mineralization, detoxification, activation or methylation. Degradation implies that the initial organic substrates no longer exists and have been metabolically broken down to organic and inorganic end products. Mineralization refers to the complete conversion of organic structures to inorganic forms such as CO_2, H_2O, anions and cations. Detoxification is the transformation of an organic compound to some intermediate form that is nontoxic or less toxic, while the process of forming more toxic or inhibitory end-products or intermediate products is known as activation (Suthersan, 1997). Reactions involved in such transformation processes typically involve intracellular or extracellular enzymes. Microorganisms may also produce organic acids (e.g., citric, oxalic, gluonic, fumaric, lactic and malic acids), which may chelate toxic metals resulting in the formation of metallo-organic molecules. Examples of microbial activation include the microbial formation of hydrides, which contain hydrogen or a mixture of methyl groups and hydrogen, from arsenic, selenium and other elements, as well as methylation, which is the biological synthesis of metal-carbon species where one or more methyl groups are transferred to a metal species. Subsequently, the toxic hydrides and/or methylated species are rapidly excreted as part of the microbial cell detoxification process (Raab and Feldmann, 2003).

2.4.2 Phytoremediation Processes

Phytoremediation processes make use of plants and interactions in the plant environment to uptake, transform and/or immobilize organic and inorganic pollutants in the contaminated soils and water. The vegetation-based treatment employs natural cycles within the plant and its environment. The selection of plant species is based on the knowledge of the contaminants of concern, plant adaptability to site-specific factors such as climate, depth of plant root structure, and soil type, and remediation objectives (US EPA, 2001).

2.4.2.1 Higher Plants and the Surrounding Rhizosphere

Vegetation characteristics which are sought in phytoremediation processes include the ability to extract or degrade contaminant to nontoxic or less toxic products, rapid growth rate, adaptability to local conditions, ease of planting and maintenance, and uptake of large quantities of water via evapotranspiration (US EPA, 2001). The types of plants involved in the remediation processes include species of algae in aquatic systems; floating, emergent and submerged macrophytes in wetland systems; hyperaccumulators for the treatment of metal contaminated environments; herbaceous and legume species for the remediation of pollutants in surface soils; and hybrid poplars, willow and cottonwood trees for the mitigation of groundwater contaminants or to provide hydraulic control.

The rhizosphere is a zone of increased microbial activity at the root-soil interface under the influence of the plant root. Rhizosphere soil microorganisms in plant ecosystems typically have well-established symbiotic relationships with the plants. Plants create microenvironments and attachments sites where communities of microorganisms can thrive along stems, in the root zone or in the rhizosphere. Certain plants contain porous and spongy tissue in their stems and roots that allows air to move quickly between the leaf surface and the roots. Oxygen released from plant roots oxidizes the rhizosphere environment and enabling microbial processes that require oxygen (Steinberg and Coonrod, 1994). Plants also release exudates into the surrounding soil which include sugars, alcohols, amino acids, and enzymes. The exudates and enzymes enhance microbial growth and the growth of mycorrhizal fungi. These fungi grow in symbiotic association with the plant and have unique enzymatic pathways that help degrade organics that could not be degraded by bacteria alone (Suthersan, 1997). The overall effect is a 100 to 1000-fold increase in the microbial biomass surrounding the roots compared with microbial populations in the bulk soil (Carman and Crossman, 2001; Crowley et al., 1997; Anderson and Coats, 1995).

2.4.2.2 Phytoremediation Mechanisms

Mechanisms involved in phytoremediation processes include rhizofiltration, rhizodegradation, phytodegradation, phytoaccumulation, phytostabilization, phytovolatilization and phytocontainment. These processes are described in more detail in this section. A summary of phytoremediation mechanisms, process goals and applications is presented in Table 2.9.

Phytoaccumulation

Phytoaccumulation, also known as phytoextraction, phytoconcentration, phytotransfer, phytomining or hyperaccumulation, is the use of plants to uptake, translocate and accumulate constituents from the plant root zone into above-ground plant tissue (Suthersan, 1997) (Figure 2.11). Pollutants that can be phytoaccumulated in plant tissues include metals, metalloids, radionuclides, perchlorate, BTEX, PCP and organic compounds that are not tightly bound to soils. These constituents are extracted by plants from the soil and pore water by cation pumps, absorption and other mechanisms (McCutcheon and Schnoor, 2003). Certain plants, called hyperaccumulators, can accumulate and tolerate very high levels (2 to 5%) of metals in their biomass (Carman and Crossman, 2001). Hyperaccumulation is a special case of phytoaccumulation that results in 100 times or greater than the normal plant accumulation of a particular element. For cadmium and other rare elements, this represents an accumulation greater than 0.01%, for most other heavy metals an accumulation greater than 0.1%, and for iron, manganese and other common elements an accumulation greater than 1% on a dry weight basis (McCrutcheon and Schnoor, 2003).

Phytodegradation

Phytodegradation, also known as phytoreduction, phytooxidation, phytolignification and phytotransformation is the biochemical transformation of extracted organic constituents via metabolic processes within the plant or external to the plant through the effect of plant-based enzymes (Figure 2.12). Contaminants are transformed into products that are subsequently used as nutrients and incorporated into plant tissue or by-products that can be broken down further by microorganisms (Suthersan, 2002, McCutcheon and Schnoor, 2003).

Table 2.9 Selected phytoremediation processes and their applications (US EPA, 2000).

Mechanisms	Process Goal	Media	Contaminants	Plants
Phytoaccumulation	Contaminant extraction and capture	Soil, sediment, sludges	Metals: Ag, Cd, Co, Cr, Cu, Hg, Mn, Mo, Ni, Pb, Zn; Radionuclides: ^{90}Sr, ^{137}Cs, ^{239}Pu, $^{238,234}U$	Indian mustard, pennycress, alyssum, sunflower, hybrid poplars
Phytodegradation	Contaminant destruction	Soil, sediment, sludges, groundwater, surface water	Organic compounds, chlorinated solvents, phenols, herbicides, munitions	Algae, stonewort, hybrid poplar, black willow, bald cypress
Phytostabilization	Contaminant containment	Soil, sediment, sludges	As, Cd, Cr, Cu, Hs, Pb, Zn	Indian mustard, hybrid poplars, grasses
Phytovolatilization	Contaminant extraction from media and release to air	Groundwater, soil, sediment, sludges	Chlorinated solvents, some inorganics (Se, Hg, and As)	Poplars, alfalfa, black locust, indian mustard
Rhizodegradation	Contaminant destruction	Soil, sediment, sludges, groundwater	Organic compounds: PH, PAHs, pesticides, chlorinated solvents, PCBs	Red mulberry, grasses, hybrid poplar, cattail, rice
Rhizofiltration	Contaminant extraction and capture	Groundwater, surface water	Metals, radionuclides	Sunflowers, indian mustard, water hyacinth
Phytocontainment	Contaminant degradation or containment	Groundwater, surface water	Water soluble organics and inorganics	Hybrid poplar, cottonwood, willow

Plant transformation pathways vary depending on the plant species, tissue type and nature of the contaminant. These pathways are generally categorized as reduction, oxidation, conjugation, and sequestration. Plants synthesize a large number of enzymes during primary and secondary metabolisms. Several of these enzymes are useful for phytodegradation including nitroreductase (for munitions and pesticides), dehalogenases (for chlorinated solvents and pesticides), phosphatases (for pesticides), peroxidases (for phenols), laccases (for aromatic amines), cytochrome P-450 (for pesticides and chlorinated solvents), and nitrilase (for herbicides). Contaminant degradation by these plant-based enzymes can occur in an environment free of microorganisms and even in environments that are not ideal for biodegradation (Suthersan, 2002).

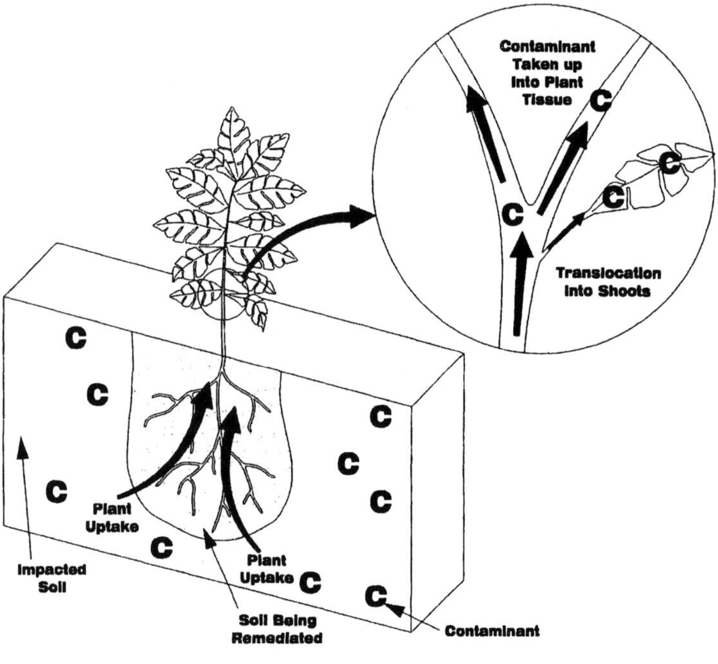

Figure 2.11 Phytoaccumulation of inorganic contaminants.

Phytostabilization

Phytostabilization is the use of plants to immobilize organic and inorganic constituents in the soil and groundwater through absorption and accumulation by roots, adsorption onto roots, or precipitation within the rhizosphere (Carman and Crossman,

2001) (Figure 2.13). This process is also known as biomineralization, phytosequestration and lignification and can be applied to metals, phenols, and chlorinated solvents. Plant exudates released to the root zone can increase the local soil pH and the soil oxygen content producing a significant effect on the redox conditions of the soil, promoting oxidation, and causing speciation and adsorption to form stable mineral deposits (Carman and Crossman, 2001). Humification, lignification and irreversible binding of some organic compounds can also occur as a result of plant functions (McCutcheon and Schnoor, 2003).

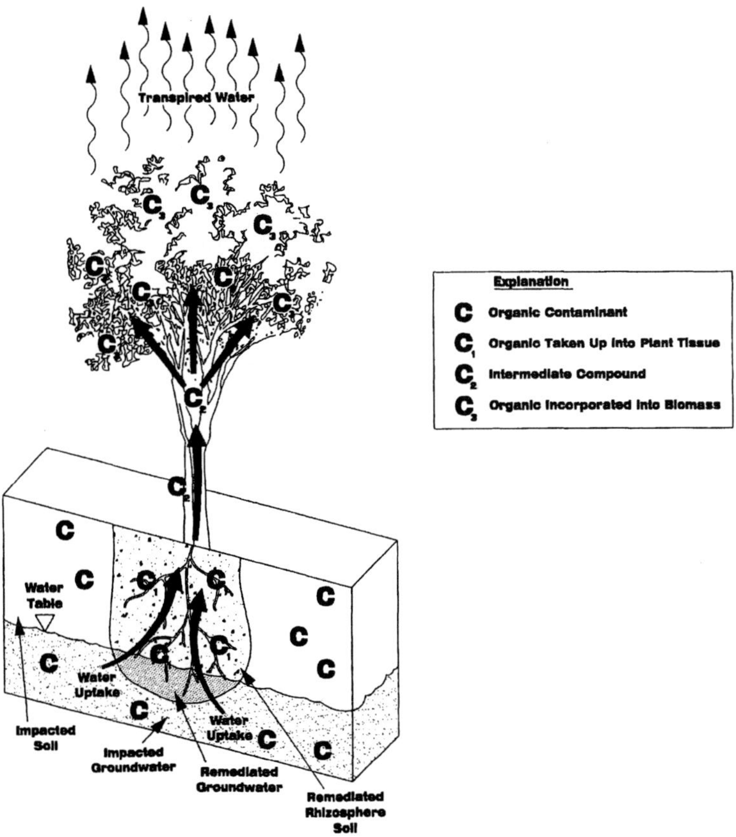

Figure 2.12 Phytodegradation of organic contaminants.

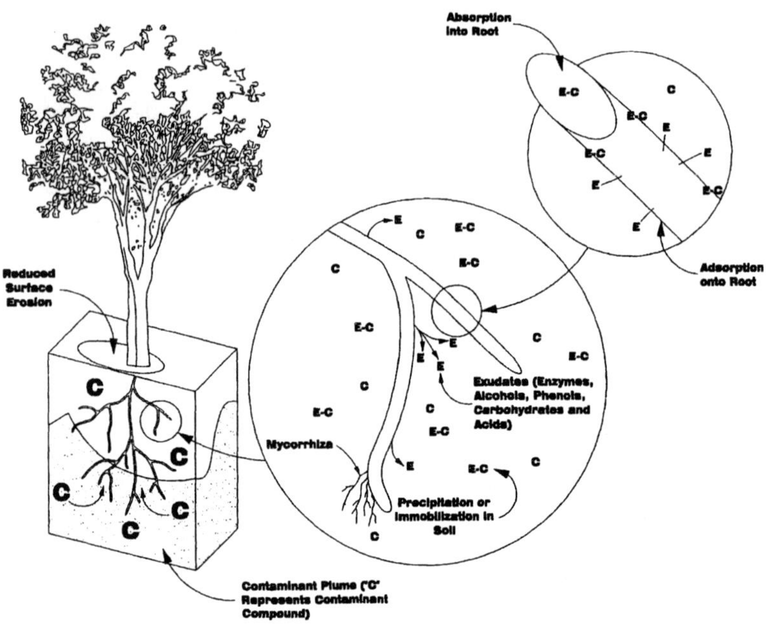

Figure 2.13 Phytostabilization of organic and inorganic contaminants.

Phytovolatilization

In this process, volatile inorganic and organic compounds are taken up, often re-speciated and transpired by the plant, as shown in Figure 2.14. This mechanism is also referred to as biovolatilization and phytoevaporation and typically applies to selenium, tritium, arsenic, mercury and chlorinated solvents (McCrutcheon and Schnoor, 2003). In some cases, organic compounds that are extracted, are subsequently either degraded or transformed within plant to form catabolic intermediates or end-products that can be volatilized from the plant tissue (Suthersan, 2002). Inorganic contaminants that are extracted and accumulated can be methylated and subsequently volatilized from the plant tissue (Carman and Crossman, 2001, Evans and Furlong, 2003).

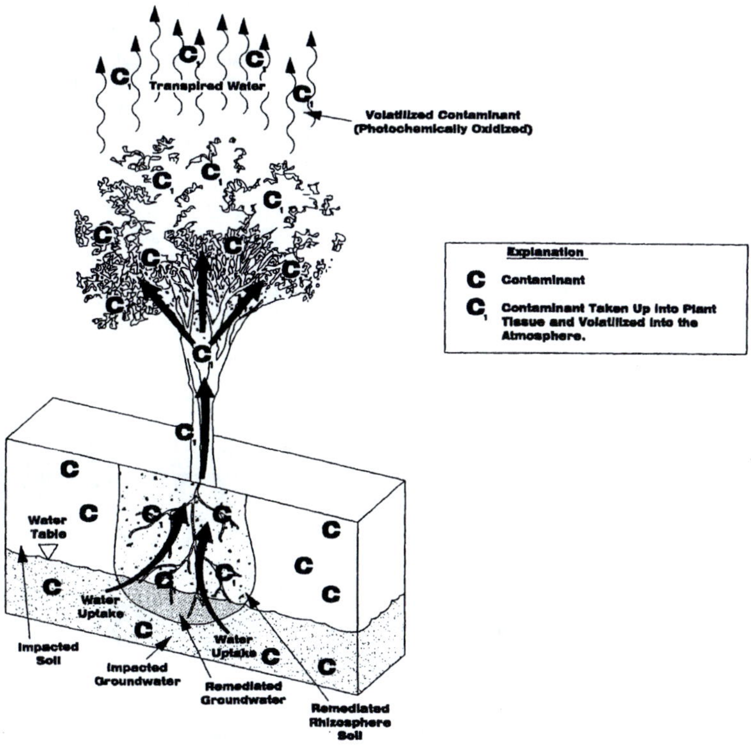

Figure 2.14 Phytovolatilization of organic and inorganic contaminants.

Rhizodegradation

Rhizodegradation, also called phytostimulation, rhizosphere biodegradation, enhanced rhizosphere biodegradation or plant-assisted bioremediation, can be applied to BTEX, petroleum hydrocarbons, PAHs, PCP, perchlorate, pesticides, PCBs and other organic compounds (McCrutcheon and Schnoor, 2003). The process involves the breakdown of organic constituents in the soil via microbial activity that is enhanced by plant-mediated processes within the rhizosphere (Figure 2.15). As much as 60% of the oxygen transported to the root zone of the plants can transferred to the rhizosphere generating aerobic conditions for the thriving microbial community associated with the root zone (Evans and Furlong, 2003). Plant exudates (amino acids, carbohydrates, nucleic acid derivatives, growth factors, carboxylic acids and enzymes), root necrosis and other processes provide the organic carbon and nutrients necessary to spur soil bacteria growth to as much as 20 times that is normally found in non-rhizosphere soil. The conditions also favor enzyme induction and cometabolic enzyme degradation by mycorrhizal fungi and other microorganisms in the rhizosphere (McCrutcheon and Schnoor, 2003). Plants can also enhance biodegradation by physical rooting mechanisms which loosen the soil in the root zone and facilitate oxygen and water transport (Suthersan, 2002).

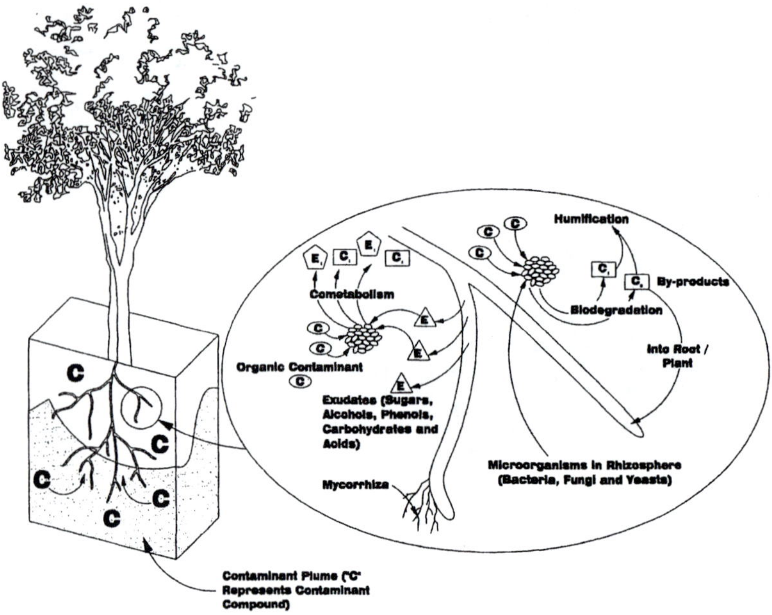

Figure 2.15 Rhizodegradation of organic contaminants.

Rhizofiltration

Rhizofiltration involves the uptake, absorption, adsorption, or precipitation of constituents in the soil solution around the root system onto or within the plant root, young shoots, fungi, algae and bacteria (Carman and Crossman, 2001) (Figure 2.16). The process is also referred to as phytofiltration, blastofiltration, phytosorption, biosorption, biocurtain, biofilter, and contaminant uptake for the removal of metals, radionuclides, organic compounds, nitrate, ammonium, phosphate and pathogens (McCrutcheon and Schnoor, 2003). Macrophytes and algae are known to accumulate metals and other toxic elements from solution. Plants that are naturally immobilized, such as attached algae and rooted plants, and those that can be easily separated from suspension, such as filamentous algae, macroalgae and floating plants possess high adsorption capacities (Suthersan, 1997). The process of rhizofiltration first involves the containment of the constituent via immobilization or accumulation on or within the plant or microorganism. Plant or microbial uptake, accumulation and translocation may follow depending on the contaminant. Plant root exudates may also lead to precipitation of some metals (Suthersan, 2002).

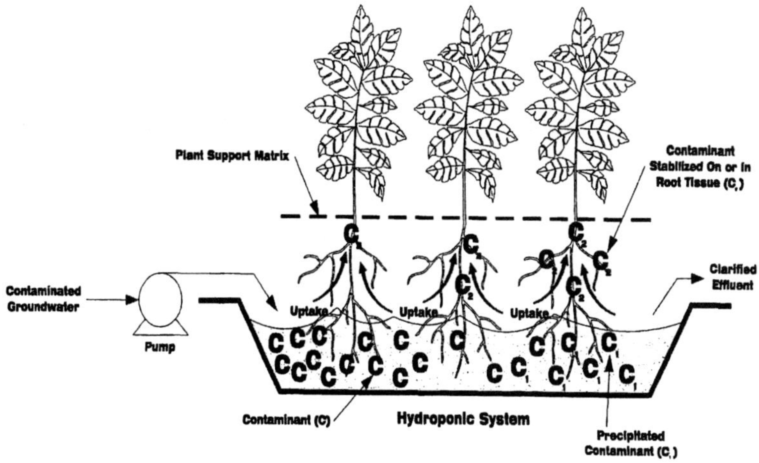

Figure 2.16 Rhizofiltration of inorganic contaminants.

Phytocontainment

Phytocontainment or hydraulic containment is the use of plants, particularly phreatophytes which can transpire large quantities of water, to control the migration and flow dissolved contaminants in pore water, shallow groundwater or wetlands (Carman and Crossman, 2001). An example of flow paths created by root uptake of water in phreatophytes is illustrated in Figure 2.17. This process is often applied in conjunction with rhizodegradation and phytodegradation to remediate shallow sites contaminated with water-soluble contaminants such as chlorinated solvents, MTBE, explosives, other organic contaminants, salts and some metals (McCrutcheon and Schnoor, 2003).

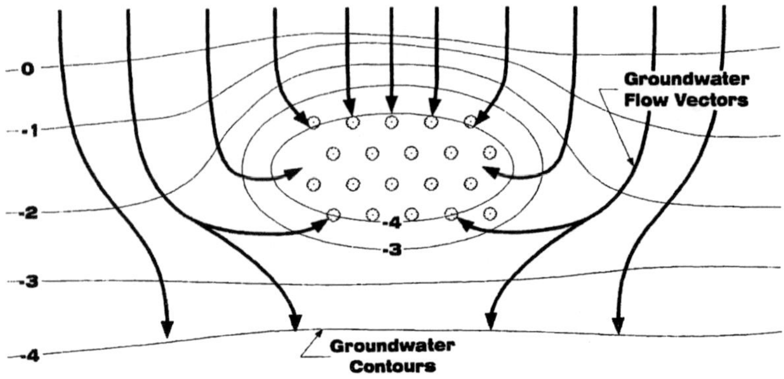

Figure 2.17 Phytocontainment of groundwater contaminants.

2.5 Conclusion

Many of the physico-chemical and biological processes described in this chapter work in conjunction with each other to alter the physical and chemical properties of contaminants in natural and engineered systems. These processes are crucial in the biogeochemical cycling of elements and the transformation or detoxification of contaminants in the environment. These processes depend upon soil and solution chemistry such as pH, redox conditions, temperature and salinity, which impact the rate at which the reactions or processes take place.Experiences from laboratory research and field implementation over the past few decades have enabled engineers and scientists to significantly advance their understanding of fundamental processes occurring at the particle scale in contaminated sites. Several best management practices for soil and groundwater remediation have emerged as a consequence of the improved knowledge of processes and growing public awareness of environmental issues. The subsequent chapters in this book illustrate how various treatment technologies can exploit these physical, chemical and biological processes to remediate contaminated soils and groundwater.

References

Adamson, A.W. (1990) *Physical Chemistry of Surfaces*, Wiley-Interscience, New York, NY.

Akai, J.; Izumi, K.; Fukuhara, H.; Masuda, H.; Nakano, S.; Yoshimura, T.; Ohfuji, H.; Anawar, H.M.; Akai, K. (2004) Mineralogical and geomicrobiological investigations on groundwater arsenic enrichment in Bangladesh. *Applied Geochemistry*, 19, 215-230.

Anderson, T. A. and Coats J. R. (1995) An Overview of Microbial Degradation in the Rhizosphere and its Implications for Bioremediation. Chapter 8 in *Bioremediation: Science and Applications*. H. D. Skipper and R. F. Turco Eds. SSSA Special Publication No. 43. Soil Science Society of America, Inc.; American Society of Agronomy, Inc.; Crop Science Society of America, Inc.; Madison, Wisconsin.

Bear, J. (1972) *Dynamics of Fluids in Porous Media*. American Elsevier, New York, NY.

Bedient, P.B.; Rifai, H.S.; Newell, C.J. (1994) *Ground Water Contamination: Transport and Remediation*. Prentice-Hall, Inc., Englewood Cliffs, NJ.

Black, J.G. (1999) *Microbiology: Principles and Explorations*. 4th Edition. Prentice-Hall, Inc., Upper Saddle River, NJ.

Carman, E. P. and Crossman, T. L. (2001) Phytoremediation. Chapter 2 in *In-Situ Treatment Technology*. 2nd Ed. E. K. Nyer, P. L. Palmer, E. P. Carman, G. Boettcher, J. M. Bedessem, F. Lenzo, T. L. Crossman, G. J. Rorech and D. F. Kidd (Eds.). Arcadis Environmental Science and Engineering Series. Lewis Publishers, Washington, D.C. pp. 392-435.

Crowley, D. E., Alvey, S. and Gilbert, E. S. (1997) Rhizosphere Ecology of Xenobitotic Degrading Microorganisms. Chapter 2 in *Phytoremediation of Soil and Water Contaminants*. E. L. Kruger, T. A. Aderson and J. R. Coats (Eds.). American Chemical Society. Washington. D.C.

Eberhardt C.; Grathwohl P. (2002) Time scales of organic contaminant dissolution from complex source zones: Coal tar pools vs. blobs." *J. Contam. Hydrol.* 59:45-66.

Ercole, C., Veglio, F., Toro, L., Ficara, G., Lepidi, A. (1994) Immobilisation of microbial cells for metal adsorption and desorption. In *Mineral Bioprocessing II*. Snowboard. Utah

Essington, M.E. (2004) *Soil and Water Chemistry*. CRC Press LLC, Boca Raton, FL.

Evans, G. M. and Furlong, J. C. (2003) *Environmental Biotechnology: Theory and Application*. John Wiley & Sons, Inc.

Freeze, R.A.; Cherry, J.A. (1979) *Groundwater*. Prentice Hall, Englewood Cliffs, NJ.

Huang, W.; Weber, W.J. (1997) A distributed reactivity model for sorption by soils and sediments. 10. Relationships between desorption, hysteresis, and the chemical characteristics of organic domains. *Environ. Sci. Technol.* 31, 2562-2569.

Internstate Technology and Regulatory cooperation Work Group (ITRC) (2001) *Thechnical and Regulatory Guidance Document – Phytotechnology*. 84 pp.

Jacob, D. L and Otte, M. L. (2003) Conflicting Processes in the Wetland Plant Rhizosphere: Metal Retention or Mobilization? *Water, Air, and Soil Pollution* 3:91-104.

Kadlec, R. H. and Knight R. L. (1996) *Treatment Wetlands*. Lewis Publishers, Washington, D.C.

Knox, R.C.; Sabatini, D.A.; Canter, L.W. (1993) *Subsurface Transport and Fate Processes*, Lewis Publishers, Boca Raton, FL.

Kuyucak, N., Volesky, B. (1988) Biosorbents for Recovery of Metals From Industrial Solutions. *Biotechnol Left.* 10(2):137-142.

Lyman, W.J.; Reehl, W.F.; Rosenblatt, D.H. (1990) *Handbook of Chemical Property Estimation Methods*, American Chemical Society, Washington DC.

McCutcheon, S. C. and Schnoor, J. L. (2003) *Phytoremediation: Transformation and Control of Contaminants.* Wiley and Sons, Inc.

Madigan, M.T.; Martinko, J.M.; Parker, J. (1997) *Brock Biology of Microorganisms.* 8th Edition. Prentice-Hall, Inc., Upper Saddle River, NJ.

Mays, L.W. (2001) In *Water Resources Engineering*, John Wiley & Sons, Inc., New York, NY.

Ogata, A. (1970) Theory of dispersion in a granular medium. U.S. Geoligical Survey Professional Paper 411-I.

Pignatello, J.J.; Xing, B. (1996) Mechanisms of slow sorption of organic chemicals to natural particles. *Environ. Sci. Technol.* 30: 1-11.

Raab, A., Feldmann, J. (2003) Microbial transformation of metals and metalloids. *Science Progress.* 86(3):179-202.

Reddi, L.N.; Inyang, H.I. (2003) In *Geoenvironmental Engineering*, Marcell Dekker, New York, NY.

Schlegel, H.G. (1988) *General Microbiology.* Cambridge University Press, Cambridge.

Smith, E.; Naidu, R.; Alston, A. M. (1999) Chemistry of arsenic in soils: I. Sorption of arsenate and arsenite by four Australian soils. *J. Environ. Qual.* 28:1719-1726.

Steinberg, S. L. and Coonrod, H. S. (1994) Oxidation of the root zone by aquatic plants growing in gravel-nutrient solution culture. *Journal of Environmental Quality* 23: 907-913.

Suthersan, S. S. (2002) *Natural and Enhanced Remediation Systems.* Lewis Publishers, Inc. Washington, D.C.

Suthersan, S. S. (1997) *Remediation Engineering: Design Concepts.* Lewis Publishers, Inc. Washington, D.C.

US EPA. (2001) *Brownfields Technology Primer: Selecting and Using Phytoremediation for Site Cleanup.* Solid Waste and Emergency Response Technology Innovation Office. Washington, D.C. EPA 542/R/01/006.

US EPA (2000) *Introduction to Phytoremediation.* National Risk Management Research Laboratory. Cincinnati, Ohio. EPA 600/R/99/107

Weber, W.J.; DiGiano, F.A. (1996) In *Process Dynamics in Environmental Systems.* John Wiley and Sons. New York, NY.

Yen, T.F. (1999) In *Chemical Principles for Environmental Processes.* Prentice Hall, Upper Saddle River, NJ.

CHAPTER 3

Physical Treatment Technologies

Ravi Damera and Alok Bhandari

3.1 Introduction

Physical treatment technologies take advantage of physicochemical properties of the contaminant and the affected media to remediate soil and groundwater. The most significant physicochemical properties of a contaminant in this context include its density, solubility, viscosity, and volatility. Physical properties of the media that can be exploited during site remediation include the physical state (solid, liquid or gas) of the media, its bulk density, moisture content, permeability, porosity, particle size distribution and ability to conduct heat or electric current. Most conventional physical treatment technologies take advantage of the physicochemical properties of both the contaminant and the media. Such remediation technologies include free product recovery, pump and treat, soil vapor extraction, air sparging, groundwater circulation wells, multi-phase extraction, induced fracturing and soil heating. This chapter discusses the application of the above-mentioned physical treatment technologies for soil and groundwater remediation.

3.2 Free Product Recovery

Free product or free phase represents light nonaqueous phase liquids (LNAPLs) or dense nonaqueous phase liquids (DNAPLs) present in sufficient volumes in the subsurface to saturate the geologic media and allow flow of the NAPL into recovery wells or excavation trenches. Subpart F of 40 CFR specifies that the corrective action for sites contaminated by petroleum hydrocarbons or hazardous substances should include removal and recovery of free product, wherever applicable, to the maximum extent practicable as determined by the implementing agency. Free product removal should be accomplished using protocols that minimize the spread of contamination into previously uncontaminated zones, and adequately manage the treatment, discharge or disposal of recovery byproducts in compliance with applicable regulations. Typically, only about 50% of the free phase in the subsurface is recoverable while the rest remains as a residual phase that continues to serve as a long-term source of groundwater contamination.

A free product recovery plan for sites contaminated by petroleum products or other NAPLs should seek answers to the following questions:

- Does sufficient site data exist to evaluate the need for free product recovery?
- Is free product recovery necessary, and is the recovery approach consistent with the comprehensive corrective action plan?

Free product recovery may be necessary when estimates indicate the presence of high volumes (> 200 gallons) of free product. Recovery may also be considered for permeable aquifers with hydraulic conductivities > 10^{-3} cm/sec, when thick accumulations (> 1 foot) are detected in wells, or when the free product is located in proximity of surface water or groundwater (EPA 1996a).

The design of an effective free product recovery system requires a thorough geological characterization of affected media including its porosity, permeability, heterogeneity and anisotropy. The effective porosity of the subsurface media can be used to estimate the total mass of spilled product within the affected volume. Rates of groundwater flow and migration of free products can be estimated from the permeability of the aquifer. Free products tend to migrate faster in heterogeneous porous media by preferential flow through the most permeable pathways. Nonaqueous phases in the subsurface generally move in the direction of groundwater flow. However, their migration rates are retarded due to their higher viscosity and the lower relative permeability of the porous medium. Table 3.1 summarizes key properties of the fluid and geologic media that affect subsurface migration and free product recovery.

The design phase for a free product recovery system is often initiated by seeking answers to the following questions (EPA 1996a):

- What is the spatial distribution of the free product in the subsurface?
- What is the expected mass of free product in the contaminated zone?
- What mass/volume of the free product is recoverable?
- How quickly/easily can the free product be recovered?

First, the areas where free product accumulations are the greatest are located and the spatial distribution of the free product pool is mapped out. This can be accomplished by measuring free product in excavation or test pits, estimating contaminant levels in geological samples collected from soil borings, or determining free phase thickness in monitoring wells or well points. Other factors that can affect the delineation of free phase in the subsurface include the duration and volume of release, the depth to groundwater, and the direction of groundwater flow.

An estimate of the volume or mass of the free phase in the subsurface can help in the selection and performance of a free product recovery system. Volume estimates may be based on information about release events from inventory records, analysis of NAPL in samples collected from soil borings, and thickness of free product in monitoring wells or well points. Volume of the free product pool may also be estimated by extrapolating from the free product volume recovered. Recoverable free product may account for only 20 to 50% of the total volume originally released. The anticipated free product recovery rates can be estimated from field tests such as bail down or pumping tests, or using multiphase flow analysis.

Table 3.1 Fluid and porous media properties affecting free product migration and recovery.

Property	Significance
Specific Gravity	Specific gravity helps characterize the free phase as a LNAPL or DNAPL. It is also needed to estimate the mass of free product in the subsurface.
Viscosity	Lower viscosity free products are more easily recovered.
Interfacial Tension	Interfacial tension is inversely related to pore size. It determines how easily a geological media will be wetted with a fluid. It also controls the height of capillary rise in porous media.
Capillary Pressure	Free product movement in the subsurface tends to occur where capillary pressures are low. Capillary pressure is inversely related to saturation.
Relative Permeability	Relative permeability of the geological media controls the mobility of the free phase in the subsurface. It is directly proportional to saturation. In binary systems, the permeability of each fluid is reduced due to the presence of the other.
Wettability	Most geological material are preferentially wet in the following order: water > free product > air
Saturation	The degree of saturation controls the mobility of the free phase. Saturation levels are also used to estimate the mass of free and residual product
Residual Saturation	Residual saturation is the minimum saturation level below which continuous flow of the free product ceases.

Widely used approaches to recover free product from subsurface contaminant pools include skimming systems, recovery with water table depression, soil vapor/groundwater extraction, and dual phase (liquid and vapor) recovery and separation. The selection of a particular approach is often guided by site conditions, specific remedial objectives or design constraints. Skimming systems (Figure 3.1) are used for short-term remedial actions where the free phase is removed from a well or trench without recovering groundwater. These systems work effectively to remove free product exits from permeable subsurface zones such as utility bedding or buried underground open structures. Skimming systems may be passive such as filter canisters or absorbent bailers or active such as belt skimmers, pneumatic pumps or floaters.

Systems for free product recover by water table depression can remove free phase LNAPL by directing the product toward pumping wells inside the plume area. Water table depression is controlled to a necessary minimum that is sufficient to direct the plume to the point of collection without producing a large smear zone. These recovery systems work best in regions of high permeability. Separation of water from free product may be necessary after recovery. Key design elements to consider while selecting or installing these recovery systems include estimation of number, location and depth of wells, pump selection, recovery rates, and post-extraction processing. While single pump recovery systems extract both water and NAPL simultaneously, dual pump systems can optimize the cone of depression and achieve maximum product recovery by using separate wells.

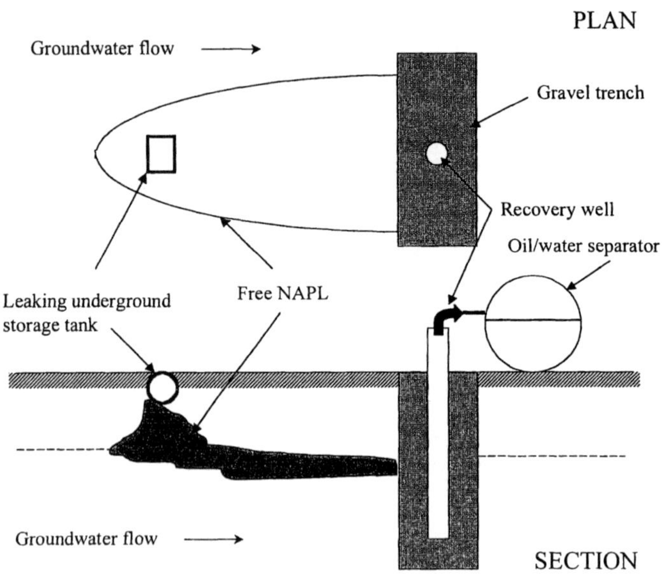

Figure 3.1 Skimming system for LNAPL free product recovery includes an interceptor trench and a recovery well.

Soil vapor/groundwater extraction systems utilize a combination of soil vapor extraction with conventional water table depression techniques. Groundwater pumping exposes the smear zone in the capillary fringe and vapor extraction allows mass-transfer of LNAPL into the gas phase. Since volatilization of hydrocarbons is the primary removal mechanism, this approach works best for volatile contaminants. Dual-phase recovery systems extract free product, vapor and groundwater simultaneously by vacuum enhanced pumping techniques. This technique minimizes drawdown and smearing of product.

Recovery of free phase DNAPLs is complicated by the fact that these products sink to the bottom of the aquifer. Chemical flooding methods such as enhanced dissolution and physical mobilization are often used to flush the free product out of soil pores before recovery using extraction wells. Enhanced dissolution methods utilize cosolvents or solubilizing surfactants to increase the solubility of the DNAPL. Physical mobilization of DNAPL by reducing interfacial tension is accomplished through mobilizing surfactants or miscible alcohol floods. Jawitz et al (2000) utilized *in-situ* alcohol flushing to enhance solubilization and extraction of tetrachloroethylene (PCE) in a pilot-scale study at a former dry cleaning facility in Jacksonville, Florida. The test zone was flushed with 34,000 liters of an ethanol-water mixture (95:5 v/v) over a period of three days. Approximately 43,000 liters of PCE (62%) were extracted from the test zone. Post flushing groundwater concentrations of PCE were reduced to 8% of their pre flushing values. Kuhlman (2002) described creosote recovery from an aquifer 102 feet underground using steam injection. Nearly 140,000 kg of creosote was recovered in the first 6 weeks. Vaporization and mineralization of the DNAPL upon contact with steam were responsible for free product recovery removal.

3.3 Pump-and-Treat

Pump-and-treat is the most widely used remediation technology. Pump-and-treat has been used both as a stand-alone treatment system and in conjunction with complementary technologies. Conventional pump-and-treat methods focus on the extraction of contaminated groundwater to the surface for subsequent treatment. Such systems have been used in about 75% of Superfund clean up actions where groundwater was contaminated (NRC, 1994). The treated groundwater may be re-injected into the subsurface or discharged into a receiving water body or a municipal wastewater collection system. The years of experience from the operation of these groundwater cleanup systems has resulted in a significantly improved understanding of the advantages and limitations of conventional pump-and-treat systems. It is now understood that the pump and treat has a limited advantage in most scenarios as the cost of remediation and the length of time required for cleanup escalates exponentially with the desired extent of removal (Figure 3.2).

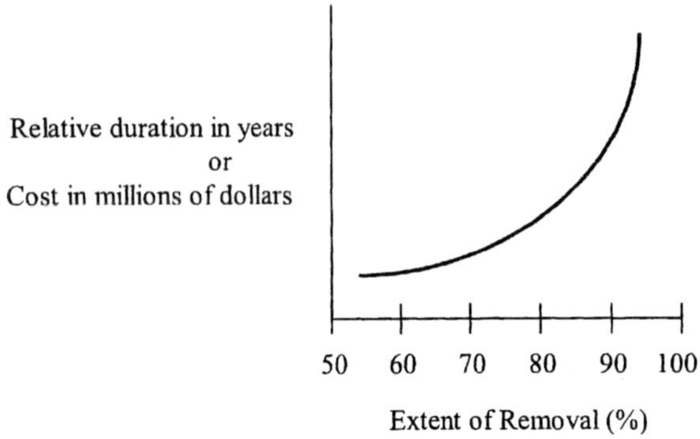

Figure 3.2 Impact of desired extent of cleanup on the relative cost and duration of a conventional pump-and-treat remedial action (NRC, 1994).

Nevertheless, pump-and-treat systems continue to constitute essential components of most modern groundwater remediation efforts. An important design objective of a groundwater extraction system may be the hydraulic control of groundwater to prevent offsite migration of the contaminant plume during restoration efforts. Properly located extraction wells can remove water from the aquifer by creating a capture zone for migrating contaminants. As water is extracted, a capture zone curves develops upstream of the well. Groundwater inside the capture zone is extracted by the well while that outside is not. In its simplest form, the envelope surrounding a capture zone curve may be defined by the following equation (Javandel and Tsang, 1986):

$$y = \pm \frac{Q}{2Bv} - \frac{Q}{2\pi Bv}\tan^{-1}\frac{y}{x} \qquad \text{(Eq. 3.1)}$$

where, x and y are the coordinates describing the envelope (Figure 3.3), B is the aquifer thickness (m), v is the Darcy velocity and Q is the well pumping rate (m^3/day).

The aquifer is assumed to be homogeneous, isotropic, uniform in cross-section, infinite in width, and with an insignificant drawdown compared to the aquifer thickness. For this simple analysis, the extraction well is assumed to extend through the entire thickness of the aquifer and is screened to extract uniformly at every level. As seen from Figure 3.3 and Eq. 3.1, at large values of x, the ordinate, y is equal to $Q/2Bv$, and the width of the capture zone envelope is, Q/Bv. Similarly, the width of the capture zone along the y-axis, where $x = 0$ is $Q/2Bv$. A general equation for the positive half of the capture zone curve for n equally placed wells along the y-axis and arranged symmetrically across the x-axis is given by (Masters, 1996):

$$y = \pm \frac{Q}{2Bv}\left(n - \frac{1}{\pi}\sum_{i=1}^{n}\phi_i\right) \qquad \text{(Eq. 3.2)}$$

where ϕ_i is the angle in radians between a horizontal line and through the ith well and a point on the capture zone envelope.

Once groundwater is extracted, a variety of physical, chemical and biological processes can be used to treat the contaminated water. These processes include adsorption, volatilization, precipitation, oxidation-reduction, and biotransformation and are discussed elsewhere in this text. However, the effective extraction of contaminated groundwater at the site requires that the contaminant sources be identified, located, and controlled. Wherever possible, source removal should be implemented to prevent further contamination. A thorough study of the site should include assessing the type and extent of contamination and a three-dimensional characterization of the subsurface including hydraulic conductivity, particle size distribution and the sorption potential of aquifer material. Such a characterization may help estimate realistic cleanup goals by classifying the site according to the National Research Council's criteria described in Table 3.2. For pump-and-treat systems to work effectively, the system should be managed dynamically onsite utilizing continuously updated site-specific information such as extraction rates, water levels in monitoring wells and contaminant concentrations in observation wells (EPA 1996b).

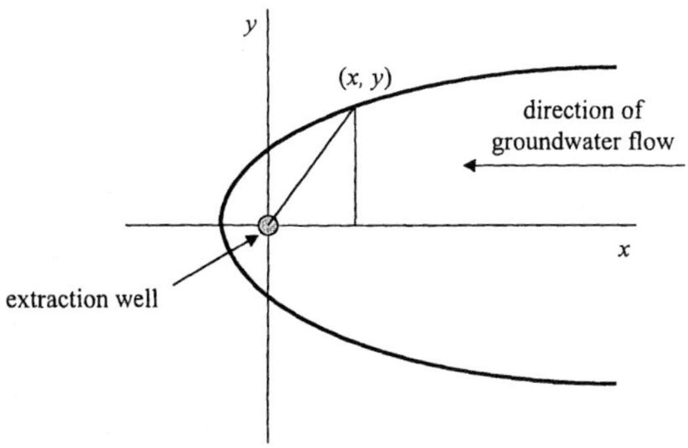

Figure 3.3 Two-dimensional capture zone envelope for a well extending the entire depth of an aquifer and pumping at a constant rate to extract groundwater equally at all levels.

Table 3.2 Realistic cleanup goals: Classification of contaminated sites based on hydrogeology and contaminant chemistry (NRC, 1994).

Site Classification	Site Description
Class A	Cleanup to health-based standards is likely to be feasible using current technology. These sites can include homogeneous aquifers containing dissolved contaminants
Class B	Technical feasibility of complete cleanup is uncertain. These sites include a wide range of hydrogeological settings and contaminants type that are not characterized as classes A or C.
Class C	Full cleanup to health-based standards is not likely to be technically feasible. These sites include fractured-rock aquifers contaminated by LNAPL or DNAPL.

Pump-and-treat systems are severely limited by the extensive contaminant tailing and rebound problems often associated with the technology. Tailing represents the progressively slow decline in aqueous contaminant concentration in the extracted water with pumping duration (Figure 3.4). Tailing of contaminants results in long treatment times. At longer treatment periods, larger and larger volumes of water have to be extracted to remove the smaller and smaller concentrations of pollutant. At several sites, pump-and-treat systems would have to run for hundreds of years before contaminant concentrations have receded to acceptable cleanup levels.

Several laboratory and field studies also report that when tailing occurs, aqueous contaminant concentrations eventually stabilize at levels that are generally above cleanup goals. Contaminant rebound refers to the momentary increase in aqueous contaminant concentration observed when a well is re-started after a period of no pumping. Rebound is also known to occur when pumps are restarted after a treatment system appears to have achieved the cleanup standard. The reasons for tailing and rebound during pump-and-treat operations often relate to mass-transfer limitations from aquifer surfaces and NAPL pools. Other factors may include precipitate dissolution in case of inorganic solutes, contaminant diffusion into low permeability media, and variations in groundwater velocity (EPA 1996b).

The capital cost of free-phase petroleum recovery using a pump-and-treat system installed between 1992 and 1994 at a fuel dispensing area in Fort Drum, Watertown, NY was estimated to be $958,780 (DENIX, 2006). Approximately 36% of this cost was associated with labor. Total annual operating costs at this site were $ 129,440. The estimated average operation and maintenance costs of 79 pump-and-treat systems installed at Superfund sites were approximately $ 570,000 with a median cost of $ 350,000 (EPA 2001a). Another EPA study reposted that the average capital costs for 32 pump-and-treat sites was \$ 5,900,000 with a median cost of \$ 2,000,000 (EPA 2001b). The average annual operating cost per volume of groundwater treated was \$32 per thousand gallons.

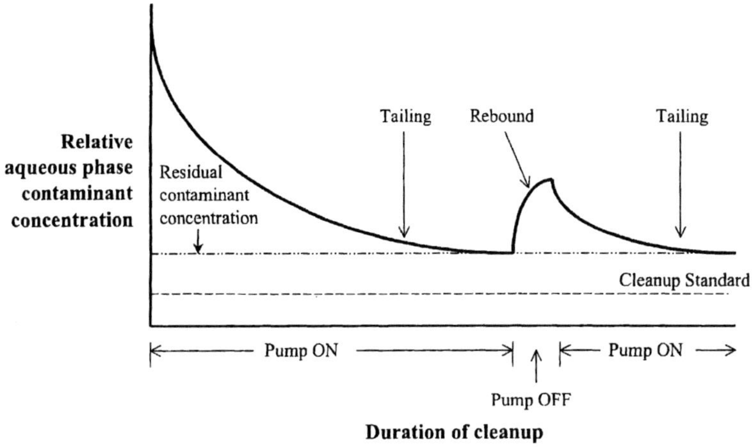

Figure 3.4 Tailing and rebound behavior typically observed in pump-and-treat systems (Cohen et al., 1994).

3.4 Soil Vapor Extraction

Soil vapor extraction (SVE), also known as *in situ* soil venting or *in situ* volatilization, is a vacuum extraction technology that is analogous to pump-and-treat processes because it removes organic contaminants into a moving fluid phase. In the case of SVE, a pressure gradient is induced in the unsaturated zone and air serves as the medium for the removal of volatile contaminants from the affected area. Contaminants with low vapor pressures partition into readily into the moving air and are brought to the surface by means of extraction wells and piping. SVE can remove a variety of contaminants existing in the sorbed phase, as residuals or as LNAPLs. Figure 4.5 illustrates a simple SVE system for vadose zone remediation with provision for contaminant recovery or treatment.

A significant advantage of SVE is that its effectiveness in removing volatile and semi-volatile contaminants has been proven in various pilot and field scale applications.

Table 3.3 presents a summary of advantages and limitations of SVE. It is recommended that the feasibility of SVE at a particular site be assessed by a detailed assessment of the site and contaminant characteristics, the SVE system design, and the operating and monitoring requirements.

Table 3.3 Advantages and limitations of soil vapor extraction systems (EPA, 2002).

Advantages	Limitations
Proven technology, demonstrated experience	Air emission permits may be required
Minimal site disturbance	On site containment or treatment of extracted vapors necessary
Effective at hard to access sites, e.g., under buildings	Effective for unsaturated zones only
Relatively short treatment times	Not very effective in clayey or low permeability soils
Efficient and cost-effective	High (> 90%) contaminant removal efficiencies unachievable
Can be combined with other technologies	Lack of guidelines for optimal design, installation and operation

The vapor pressure is the most relevant contaminant property controlling its removal effectiveness by SVE. Generally, solutes with vapor pressures higher than 65 Pa (0.5 mm of Hg) or Henry's Law constants greater than 10,000 kPa (100 atm) can be successfully removed by SVE. These contaminants include chemicals such as BTEX, and naphthalene. In the case of petroleum hydrocarbons, gasoline, kerosene and diesel components with boiling points less than 300°C can be removed by vapor extraction.

Soil and vadose zone properties of significance for SVE systems include moisture content, groundwater depth, permeability, soil structure, and stratification. The presence of soil water in the unsaturated zone can reduce the effectiveness of SVE by preventing the flow of air through the contaminated area. Fine-grained soils produce thicker capillary fringes that can also reduce the effectiveness of SVE. Contaminated zones that are affected by groundwater fluctuations are more difficult to treat by vapor extraction. SVE is most effective at sites with groundwater tables greater than 3 meters deep but may also be implemented at sites with shallower water tables. Vacuum extraction is not applicable at sites with groundwater surfaces less than 1 meter deep. SVE holds great promise in soils with low moisture contents and intrinsic permeability values greater than 10^{-10} cm^2. A fractured soil may manifest high permeability due to preferential flow paths for the air but result in ineffective or slow treatment of the unfractured media. Similarly, stratification of soils with different permeabilities can also result in non-uniform remediation of the contaminated site.

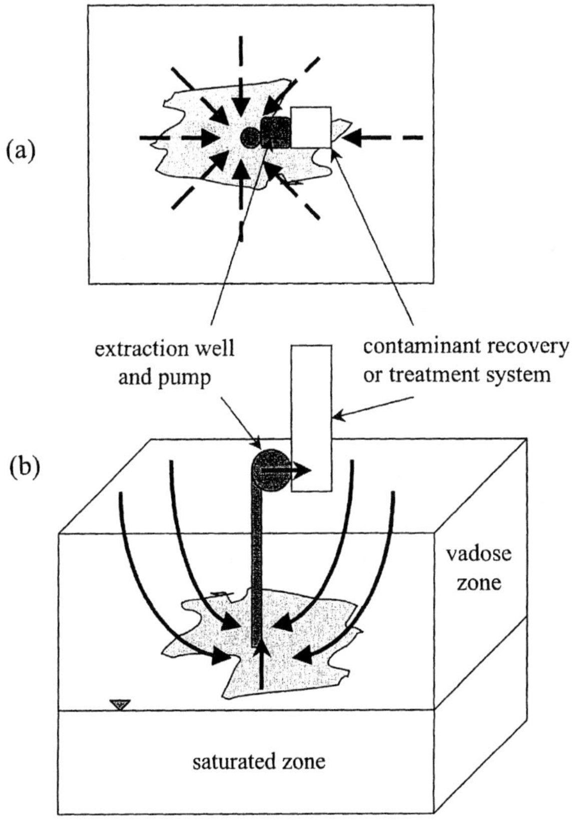

Figure 3.5 A simple soil vapor extraction system with provision for contaminant recovery.

SVE systems may include system components such as extraction wells, injection wells or air inlets, piping or air headers, air blowers or vacuum pumps, flow meters, vacuum gauges, and controllers, sampling ports, air-water separators, vapor recovery or treatment systems, and a cap (Bedient et al., 1994). The extraction wells usually completely penetrate the vadose zone and consist of slotted pipe placed in a permeable packing. The packing is grouted near the ground surface to prevent air flow through the trench or well casing. Air inlets or injection wells are constructed to allow air to be drawn into the vadose zone at the desired locations. These components strategically direct the flow of air to obtain optimal contaminant removals.

The radius of influence (ROI), wellhead vacuum, and vapor extraction flow rate are important design parameters in SVE systems (EPA 2002). ROI is defined as the maximum distance from an extraction well at which contaminant removal is feasible. The ROI helps determine the number and spacing of extraction wells. ROI depends on site properties but can be estimated from the distance from the extraction well at which a vacuum of at least 0.1 inches of water (0.19 mm of Hg or 25 Pa) is observed. Wellhead vacuum is the vacuum pressure at the top of an extraction well that produces the desired vapor extraction flow rate from the well. Wellhead vacuum pressures depend on sol permeability and can range from 3 to 100 inches of water vacuum (0.75 to 25 kPa). Vapor extraction flow rate is defined as the volumetric rate of soil vapor that can be extracted from an extraction well. Typical vapor extraction rates range from 20 to 200 m^3/hr per well.

For radial flow, the governing equations for SVE systems may be written as (Johnson et al., 1988; 1990)

$$\frac{1}{r}\frac{\partial}{\partial r}\left(r\frac{\partial P'}{\partial r}\right)=\left(\frac{n_e \mu}{kP_{\text{atm}}}\right)\frac{\partial P'}{\partial t} \qquad \text{(Eq. 3.3)}$$

where P' denotes the pressure difference from the reference pressure (P_{atm}), k is the intrinsic permeability of the soil, μ is the vapor viscosity, n_e is the soil porosity, r is the radial distance from the well, and t is time. Under appropriate boundary conditions, a solution to equation 4.3 can be represented by Eq. 3.4:

$$P'=\frac{Q}{4\pi m(k/m)}W(u) \qquad \text{(Eq. 3.4)}$$

where, m is the thickness of the vadose zone, $u = (r^2 n_e \mu)/(4kP_{\text{atm}} t)$ and $W(u)$ is the well function of u.

Contaminant removal during vapor extraction depends on the rate and extent of solute partitioning between water and air as predicted by Raoult's Law:

$$P_i = x_i t_i P_i^0 \qquad \text{(Eq. 3.5)}$$

where, P_i is the vapor pressure of the ith component (atm) in the soil gas, x_i is the mole fraction of component i in air, $P^0{}_i$ is the vapor pressure of air at the ambient temperature, and t_i is the activity coefficient for component i.

The costs associated with installing and operating a SVE system are specific to a contaminated site. These costs depend on the types of contaminants, extent and duration of contamination, and characteristics of the affected media. The most important design components that impact the cost of an SVE system include the number and depths of installed wells, level of vacuum required, capacity of blowers, type of off-gas and water treatment, and the anticipated duration of operation.

A full-scale SVE system was used to clean up 13,500 m^3 of sandy soil at Camp LeJeune, North Carolina (EPA, 1999). The soil had been contaminated with volatile organic chemicals (VOCs) including TCE, PCE and benzene. The air-permeability at this site ranged between 1.2 and 2.8 x 10^{-7} cm^2. The SVE system included eight vertical extraction wells that penetrated 15 to 16 feet below surface and a horizontal well for injection of air. The vacuum extraction unit consisted of a positive displacement blower (42.5 m^3/min at 0.5 atm vacuum). The unit also consisted of a vapor-liquid separator, a liquid transfer pump, a silencer, particulate filters, a discharge stack and a control panel. The SVE system also consisted of 32 soil probe clusters to measure soil vapor concentrations of the contaminants. Each cluster was comprised of a shallow and deep probe installed 6 feet and 12 feet below surface, respectively. While flow rates at individual well-heads ranged between 0.62 and 3.74 m^3 at 0.1 to 0.23 atm of vacuum, the total flow rate in the SVE system averaged 11.6 m^3 at 0.19 atm of vacuum. The capital cost associated with this SVE system was $222,455, and the operation and maintenance costs were $247,485. This averaged treatment cost was $35/$m^3$ of soil treated.

3.5 Air Sparging

Air sparging, also known as *in-situ* air stripping or *in situ* volatilization, is used in conjunction with SVE to remove volatile contaminants from the saturated zone. In this case, the air is injected into the saturated zone at a point below the affected area in the aquifer (Figure 3.6). Air bubbles rise within an inverted-cone zone through the contaminated water and strip the aqueous contaminants into the gas phase. The vapor extraction system then removes the air and contaminants from the soil. A contaminant recovery or treatment system can be used to manage the extracted contaminants.

Air sparging systems have found the widest use in the treatment of petroleum contaminated capillary fringe zones and dissolved plumes resulting from leaking underground storage tanks (Leeson et al., 2002). Table 3.4 summarizes the key advantages and limitations of this technology. Factors that can impact the effectiveness of air sparging at a particular site include soil heterogeneity, air flow regime, depth of contamination, and contaminant characteristics. Soils with low intrinsic permeability are difficult to treat by air sparging and soil heterogeneity can prevent the uniform distribution of air. Non-uniform flow of air may also lead to uncontrolled migration of contaminants. Contaminants with low Henry's Law constants may be difficult to strip from the aqueous phase. The low number of field implementations of air sparging means that design, installation and operation have been largely empirical.

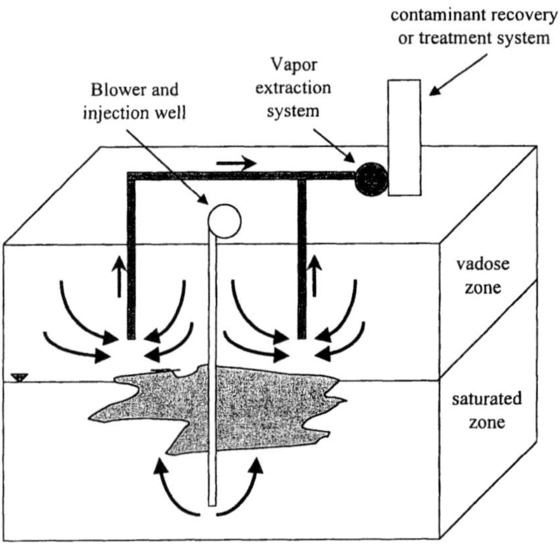

Figure 3.6 Air sparging system with vapor extraction.

The feasibility of air sparging as a treatment technology for a contaminated site can be assessed by a four step process (EPA 2002): (1) initial screening of air sparging effectiveness; (2) detailed investigation of air sparging effectiveness; (3) evaluation of air sparging system design; and (4) evaluation of operation and monitoring plans. The initial screening comprises of determination of the existence of free product at the site, evaluation of soil permeability, an analysis of the types of contaminants present, and an investigation of nearby underground structures. A follow-up detailed investigation is conducted by determining the exact composition and concentration of the contaminants and evaluating the Henry's Law constants, boiling points, vapor pressures, solubility, and phase partitioning behavior of individual components. This second step also involves a more thorough characterization of the subsurface including determining the intrinsic permeability, soil structure and stratification. The concentrations of dissolved iron (Fe^{2+}) and manganese (Mn^{2+}) in groundwater are measured. These metals can be oxidized to insoluble species during air sparging resulting in occlusion of soil pores and reduction in permeability.

Table 3.4 Advantages and limitations of air sparging systems (EPA, 2002).

Advantages	Limitations
Minimal site disturbance	Not useful for NAPLs
Cost-effective	Not useful for confined aquifers
Relatively short treatment times	Soil heterogeneity and low permeability may reduce effectiveness
Management of extracted water not required	Potential for uncontrolled movement of contaminants
Can be combined with other technologies such as SVE and biodegradation	Thorough pilot testing and monitoring required
	Lack of guidelines for optimal design, installation and operation

Pilot studies play an essential role in evaluating the design of air sparging/SVE systems. A pilot test is conducted by installing a single sparging well and several extraction wells (EPA 2002). Monitoring points are established to delineate the vapor plume, evaluate vapor generation rates, and observe the changes in groundwater table (EPA 2002). Design of the air sparging/SVE system consists of determining the ROI for the air sparging wells. The ROI is defined the maximum distance from the sparging well where sufficient sparge pressure and air flow can be induced to strip contaminants from groundwater into soil gas. ROI helps determine the number and spacing of sparge wells.

The sparging air flow rate is an important design parameter and is defined as the air flow rate necessary to provide sufficient aqueous-to-gas phase mass transfer of the contaminant. Typical air flow rates range from 5 to 50 m^3/h per injection well (EPA 2002). Sparging air pressures of 70 to 100 kPa are commonly used at injection wells (EPA 2002). The accompanying vapor extraction system is designed for a greater air flow and ROI than the air sparging system.

The duration of operation of an air sparging system may depend on the cleanup goals, the volume of groundwater to be treated, characteristics of the aquifer, the concentration, distribution and Henry's law constant of the contaminant, the radius of influence of air sparging well, and the contaminant diffusion and desorption rates. An operating air sparging/SVE system requires constant monitoring for optimal performance. Monitoring the contaminant levels in the groundwater and soil gas can provide critical information for assessment of remedial progress.

Cometabolic air sparging (CAS) was investigated at the McClellan Air Force Base, CA to remediate a site contaminated with chlorinated aromatic hydrocarbons including TCE, DCE, vinyl chloride and trichloroethane (ESTCP, 2001). The CAS process involved the subsurface injection of a gaseous organic growth substrate (propane) with air to promote cometabolic transformation of the target contaminants. The remediation system included sparge wells, SVE wells, multi-level soil and groundwater monitoring points, an air injection system, a vacuum extraction blower unit, a propane gas storage and injection system and off-gas treatment units. The costs for the CAS system ranged from \$ 184/m^3 for a 1-year operation to \$264/m^3 for a 5-year treatment. Approximately 69% of the 1-year costs were capital costs.

3.6 Groundwater Circulation Wells

Ground-water circulation wells (GCW) technologies, also known as in-well vapor stripping, in-well air sparging or re-circulation wells, are designed to remove contaminants from groundwater and saturated soils. A GCW typically consists of two hydraulically separated well screens: a lower screen through which groundwater is drawn into the well; and an upper screen through which treated water is released to the formation. Groundwater treatment occurs within the well (below ground surface) during its travel between the screens. The continuous extraction and recharge in a single well at different levels creates a three-dimensional circulation cell around the GCW (EPA, 1998a). Figure 3.7 illustrates the GCW circulation patterns at different groundwater velocities under ideal conditions (NRL, 1999).

Figure 3.8 shows a typical double-cased GCW system with the lower screen toward the bottom of the aquifer and the upper screen across the water table. Pressurized air is injected into the inner casing to air lift the groundwater entering the well. As the air-water emulsion rises in the well casing, dissolved VOCs volatilize out of the water into the surrounding air bubbles. The air-water emulsion is forced out through the upper screen by a separator or deflector placed in the inner casing. Vacuum is applied to the outer casing to collect the VOC-laden air for treatment aboveground. Majority of the water released to the formation through the upper screen travels vertically downward to the lower screen and reenters the well to replace the air-lifted water. A part of the treated water follows the natural gradient and migrates to downgradient of the circulation cell, while an equal amount of untreated groundwater captured from upgradient makes up for it. Based on literature review (NRL, 1999), typical volume of upgradient water captured is estimated to be approximately 15% of the volume of the water circulated in the cell at any time.

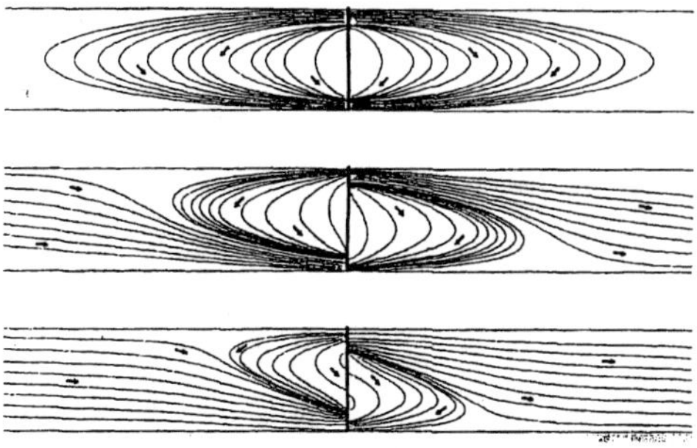

Figure 3.7 GCW circulation patterns under ideal conditions at different horizontal velocities (a) 0.0 m/day, (b) 0.30 m/day, and (c) 1.0 m/day (NRL, 1999).

The original design included stripping of VOCs from contaminated groundwater while pumping water through the well, with the intent being treatment of groundwater contaminated with VOCs. Several subsequent applications extended this technology to deliver and mix oxygen and other chemical amendments to promote bioremediation of various organic compounds and stabilization of metals. GCW is a patented, vendor-based technology. The four commercially available GCW systems are NoVOCs™, Unterdruck-Verdampfer-Brunnen (UVB), Density Driven Convection (DDC), and C-Sparger® system (USDOE, 2002). These systems employ different variations in pumping and/or water treatment components of the generic system discussed above. Some of these systems are also available in downward flow configuration and in designs to treat unconfined and confined aquifers alike.

The applicability of GCW technology to a site depends on many factors such as contaminant type, contaminant distribution, and hydrogeology. The generalized applicability of GCW technology is summarized in Table 3.4. The economics of the GCW system often depend on creating the largest circulation cell possible to minimize the number of wells required to treat a given plume area. The radius of a circulation cell depends on many parameters including the thickness of aquifer, geometry of well screens, pumping rates, natural groundwater velocity, and aquifer anisotropy. Maximum reported radius for circulation cells is approximately 2 to 3 times the distance between the upper and lower screens. Thus thinner aquifers limit the circulation cell radius. The circulation cell radius increases with pumping rates and decreases with higher natural groundwater velocity (see Figure 3.7). Lower anisotropy (< 3) results in short-circuiting between upper and lower screens and reduces the size of the circulation cell. Higher anisotropy (> 10) results in little or poor circulation (NRL, 1999).

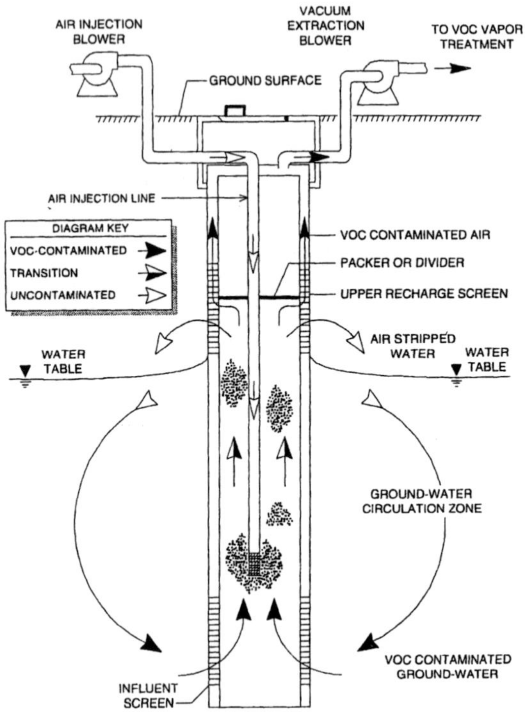

Figure 3.8 A typical groundwater circulation well configuration (GWRTAC, 1997).

A comprehensive technology assessment conducted by the Naval Research Laboratory noted that while GCW is based on sound principles, the anisotropies found at most contaminated sites will not be conducive to the successful implementation of the GCW technology (NRL, 1999). However, if significant vertical flow and circulation can be established this technology offers many advantages as noted in Table 3.5.

Table 3.4 General applicability of GCW technology (NRL, 1999).

Parameter	Applicability
Contaminant Type	
VOCs	XXX
SVOCs	XXX
Metals	XX
Radionuclides	X
Cleanup Strategy	
Source Treatment	XXX
Plume Reduction	XX
Plume Interception	XX
Unsaturated Thickness	
0 – 5 ft	X
5 – 1,000 ft	XX
Saturated Thickness	
0 – 5 ft	X
5 – 115 ft	XX
> 115 ft	X
Aquifer Characteristics	
Porous Media	XX
Fractured Media	X
Karst	X
Background Flow Velocity	
Low (< 0.001 ft/d)	XXX
Moderate (0.001 – 1.0 ft/d)	XX
High (>1.0 ft/d)	X
Horizontal Hydraulic Conductivity	
Moderate (0.03 – 1.0 ft/d)	XX
High (>1 ft/d)	XXX
Anisotropy Ratio	
Anisotropic (3 – 10)	XX
Highly Anisotropic (>10)	X
Aquifer Chemistry	
High Fe in Water	X
High Ca/Mg in water	X

XXX – Good potential for success; XX – Moderate potential; X – Limited/no potential

The U.S. Army Corps of Engineers conducted a pilot investigation of a GCW at the Nebraska Ordnance Plant, Mead NE (EPA 2001c). The dual GCW system included an in-well air stripper to treat groundwater containing TCE and UV technology to treat groundwater containing cyclotrimethylene trinitramine (RDX). The pilot scale system demonstrated contaminant mass removal rates of > 96%. The air stripping component used a 12 inch well to remove groundwater from 60 to 70 feet below surface. TCE concentrations were observed dropped from 7,000 ppb to 110 ppb after a single pass through the air stripper. The UV system used a 6 inch well to extract groundwater which passed a UV contactor. RDX concentrations reduced from 70 ppb to below detection in the recharge zone of the well. Full-scale treatment costs for the GCW system were similar to pump-and-treat at $1,900,000, but annual operation and maintenance costs were nearly half at $44,000.

Table 3.5 Advantages and Limitations of GCW (USDOE, 2002).

Advantages	Limitations
Minimal aboveground space needs, water handling, or water discharge	High anisotropies reduce the effectiveness of circulation
Vertical groundwater flow facilitates flushing of contaminants from source zones	Unaccounted vadose zone flushing may mobilize and spread contaminants beyond treatment zone
Reduced permitting requirements	Co-current air stripping process generally used in GCW is not an efficient treatment process
Low impact on groundwater levels facilitating its use near sensitive environments such as wetlands	Potential for well fouling due to possible geochemical changes
Less aboveground treatment and resulting process wastes	Ineffective sealing between influent and effluent screens may cause short circuiting
Depth-dependent operational costs are limited compared to pump-and-treat	Application of GCW to thin treatment zones may not be cost-effective
Increase in dissolved oxygen concentration promotes aerobic biodegradation	Different treatment techniques needed for non-volatile contaminants
Facilitate reagent delivery and compatible with other technologies such as SVE	Difficulty in monitoring the geometry and dynamics of circulation cells

3.7 Multi-Phase Extraction

Multi-phase extraction (MPE) is the most widely used extension of the SVE technology. MPE has been increasingly used as a source remediation technology at many sites due to its ability to achieve simultaneous extraction of vapor-phase, dissolved-phase, and non-aqueous phase contamination all in one borehole. In MPE systems all three contaminant phases are carried in one common conduit (drop tube or slurp tube) under applied vacuum (USACE, 1999). When a high vacuum (up to 25 in-Hg) is applied to the drop tube, soil vapor entering from the vadose zone entrains groundwater and NAPL at the tip of the drop tube. Alternatively, vacuum can be directly applied to 2 to 4-inch diameter wells screened across vadose and saturated zones to entrain groundwater at the well screen. A typical MPE system with a drop tube is shown in Figure 3.9.

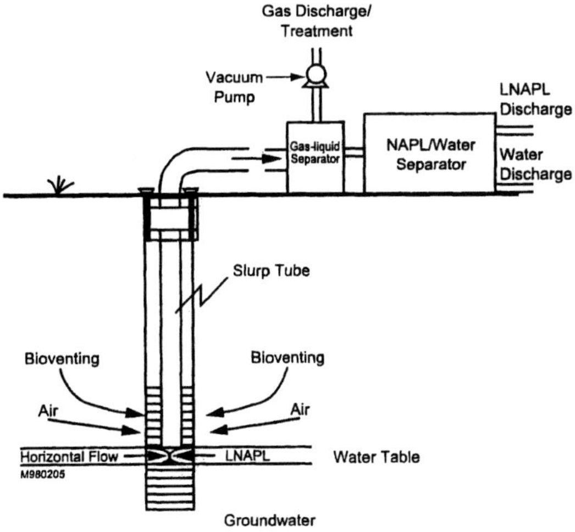

Figure 3.9 Schematic of multi-phase extraction system (AFCEE, 1994).

Vacuum application during MPE increases the hydraulic gradient toward the extraction well without creating a substantial drawdown. Effective drawdown is equal to the sum of the induced vacuum and the physical drawdown. This concept is illustrated in Figure 3.10. Water and/or NAPL flow to the well can be effectively increased as a result of the higher gradients. Since drawdown is controlled, MPE greatly reduces the risk of LNAPL smearing. Due to the fact that the induced air flow supplies oxygen to subsurface and could promote aerobic biodegradation, MPE is also referred to as bioslurping. Water, NAPL, and air extracted from the MPE wells are separated in above ground phase separators, treated as required by the applicable regulations, and then discharged into the environment.

Like SVE, MPE is also suitable for remediation of VOCs and some aerobically biodegradable SVOCs. In addition, MPE can be effectively used to recover NAPL, LNAPL in particular. MPE is most applicable to formations with moderate permeability (10^{-3} to 10^{-5} cm/s). In tighter formations, capillary pressures will be too high and result in a smaller radius of influence. MPE costs escalate rapidly in more permeable formations due to increased water production. Extraction well yields of 5 gpm or less are considered suitable for entrainment extraction (EPA, 1997). Theoretical maximum suction lift is 33.9 ft of water. However, the lighter-than-water mixture of entrained liquids can be lifted to higher levels. Liquid extraction from a depth of 210 ft at one location was noted in the literature (NFESC, 1998). Several advantages and limitations of using MPE are listed in Table 3.6.

The first step in MPE system design involves determining the radius of influence, gas and liquid flow rates. This information is then used to design wells, well spacing, and aboveground equipment such as vacuum pumps, phase separators, and other treatment units. Two primary approaches to MPE design are computer modeling using multi-phase flow models and field pilot testing. Of these, field pilot testing is more commonly practiced, although a combination of limited field testing and computer modeling lends a better design approach. The specific remedial objective of the MPE system (enhancement of SVE, NAPL recovery, or enhancement of groundwater yield) must be understood clearly prior to conducting the pilot test and full-scale design, as optimal recovery of one phase could greatly affect recovery of the other phases. The applicability of MPE process for LNAPL recovery is well understood and documented compared to DNAPL recovery.

Dual phase extraction was conducted at the U.S. Defense Supply Center Richmond's Acid Neutralization Pit site for remediation of soil and groundwater contaminated with PCE and TCE. The impacted soil consisting of silty clay, fine and coarse grained sand and interlayered gravel extended from the surface to 25 feet below. The depth of water table ranged from 10 to 15 feet. The transmissivity of the upper aquifer ranged between 35 to 47 m^2/day with a hydraulic gradient of 0.001 to 0.002 m/m. The treatment system consisted of 12 dual phase extraction wells and six air injection wells located in a rectangular grid pattern. Each extraction well consisted of a sealed casing to maintain vacuum for vapor extraction, and a submersible pump for groundwater extraction. Extraction wells penetrated to a depth of 22 to 28 feet below surface. A low-pressure blower was utilized for air injection. Extracted groundwater was subjected to air stripping. Off-gas from the stripper and vapors from the SVE system were vented into the atmosphere. The cost of pilot and aquifer testing was $134,092. Engineering design of the DPE system cost $73,198. Equipment cost was $205,743. Startup costs were $24,309 and the cost for one year of operation and maintenance was $101,148. The total per unit cost for the treatment was $0.03/gallon for the 17 million gallons of groundwater recovered.

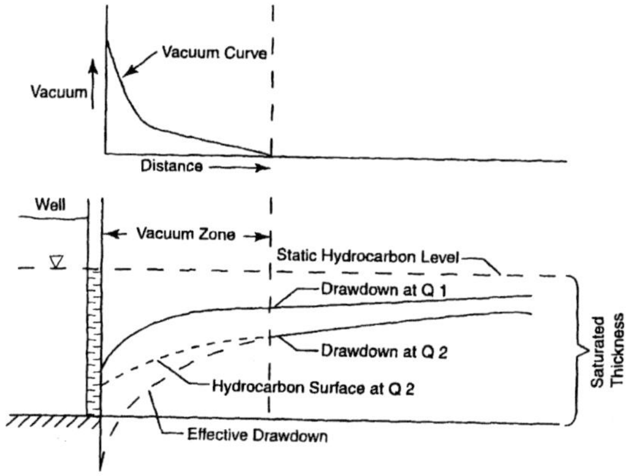

Figure 3.10 Drawdown observed during multi-phase extraction (USACE, 1996).

Table 3.6 Advantages and limitations of multi-phase extraction.

Advantages	Limitations
Effective at increasing water and vapor extraction rates in moderate permeability soils	Increased water production in higher permeability formations results in higher treatment costs
Increased radius of influence compared to SVE or pumping	Depending on equipment configuration, emulsification of NAPL may complicate phase separation and/or water treatment
Enhances NAPL recovery	MPE in shallow formations could lead to short circuiting
Effective for capillary zone remediation	Higher capital and operational costs compared to SVE or pumping
Promotes aerobic biodegradation of certain contaminants	
Minimizes drawdown and resulting smear zone	

3.8 Induced Fracturing

Most in-situ remediation technologies involve circulation of carrier fluids (water, air or other gases) either to deliver amendments or recover contaminants. Low permeability formations severely limit the circulation of these carrier fluids, and hence effectiveness of the intended remedy. Fractures can be induced in low-permeability silts, clays, silty sands and bedrock materials such as shale, limestone and sandstone to enhance their permeability. The fracturing technology is not contaminant specific, as fracturing alone does not treat or remove any contaminants. A network of induced fractures helps improve the advective transport and shorten the diffusive transport pathways. As a result, delivery of carrier fluids and recovery from these low-permeability formations becomes more efficient. Fracturing can enhance the effectiveness of a variety of remediation approaches including SVE, MPE, and air sparging. Fracturing technologies can be readily extended to deliver amendments for bioremediation, emplace oxidants for in-situ chemical oxidation, zero-valent iron for abiotic degradation of chlorinated VOCs and graphite to improve electrokinetics. Table 3.7 presents a summary of key advantages and limitations of the induced fracturing process.

Table 3.7 Advantages and limitations of induced fracturing.

Advantages	Limitations
Increases permeability of tight formations	High injection pressures and rates for certain techniques may require special safety precautions
Increases radius of influence	Monitoring deep fractures is difficult
Enhances vapor and fluid recovery	Displacement or differential settlement could damage surface structure/utilities
Reduces formation heterogeneities	Creating or intersecting vertical fractures could enhance DNAPL migration
Helps effective delivery of liquid and granular amendments	Not applicable to normally consolidated formations
Useful as retrofit technique	

Three types of fracturing techniques (hydraulic, pneumatic, and blast) are commercially available for environmental applications. As the names imply, hydraulic fracturing uses pressurized liquids, pneumatic fracturing uses pressurized air, and blast fracturing detonates high explosives to propagate fractures (GWRTAC, 2002). Typically, a fluid is injected into a borehole until the injection pressure exceeds a critical value. Figure 3.11 shows a typical plot of injection pressure versus time obtained during laboratory tests (EPA, 1993). This plot indicates that the fracture propagation is initiated at the break in the slope. The injection pressure at this point (P_{df}) is the critical pressure required to initiate fracturing.

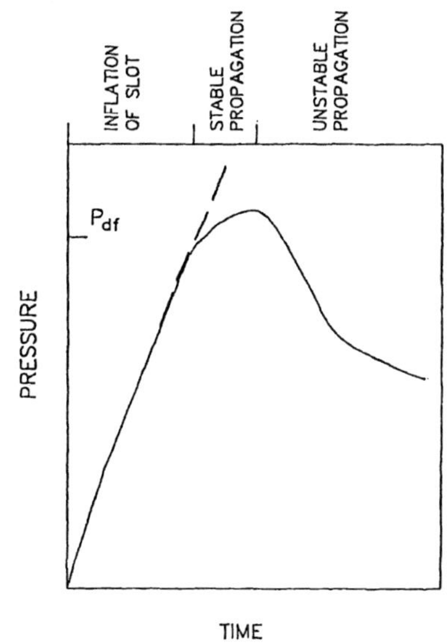

Figure 3.11 Typical pressure response with time during fracturing (EPA, 1993).

Fracture initiation pressure depends on the several factors including confining stresses, toughness of the enveloping formation, injection rate, size of incipient fractures and pores and defects in the borehole wall. The following generalized expression for predicting initiation pressure (P_{df}) is based on Schuring and Chan (1992) and consists of four distinct components (formation overburden, formation cohesion, fluid compressibility, and hydrostatic pressure).

$$P_{df} = C_1 \gamma Z + C_2 c' Cot\phi + \beta + P_0 \qquad \text{(Eq. 3.6)}$$

where, C_1 and C_2 are site-specific coefficients, Z is the depth of overburden, γ is the bulk density of the overburden, c' is the cohesion, ϕ is the angle of internal friction, β is correction factor for fluid compressibility, and P_0 is the hydrostatic pressure (zero for vadose zone). For hydraulic fracturing, assuming water to be incompressible sets β to zero.

In hydraulic fracturing, a high-pressure water jet (18 L/min and 20 MPa) is used to cut approximately a 6-inch, disk-shaped notch at the required depth in a borehole. The purpose of the notch is to reduce the fracture initiation pressure and ensure that the fracture starts in a horizontal plane. Water is then injected into the borehole at a rate of 20-40 L/min, to initiate a hydraulic fracture at the notch (Murdoch and Slack, 2002). A slurry of cross-linked guar gum gel and sand is pumped into the borehole to propagate the fracture. The fracture propagates a moderate rate of 0.1± m/sec and a single injection normally takes 15 to 30 minutes (GWRTAC, 2002). The viscous guar gum is capable of suspending 10 to 15 lb of sand per gallon of gel (EPA, 1994). An enzyme is added to break the guar gum several hours after the injection, leaving sand within the fracture to prop the fracture open.

Pneumatic fracturing involves injecting high volumes (25 to 50 m^3/min) of air or a gas at high pressure (0.5 to 2.0 MPa) into an isolated section of a borehole over a 15 to 20 second interval (EPA, 1994). Fractures propagate at a rapid rate of ~ 2.0 m/sec. A series of gas cylinders are used to provide the high volume and high pressure needed for pneumatic fracturing. Proppants are typically not used with pneumatic fracturing and the process relies on "self-propping" fractures. Self-propping is thought to be a combination of asperities along the fracture surface and shifting of geological media. The low viscosity of air results in rapid leak-off (fluid loss into the formation), thus requiring a high volume of flow to propagate the fractures. The fracture initiation pressures in pneumatic fracturing are approximately twice those required for hydraulic fracturing due to the higher compressibility of air (Schuring and Chan, 1992).

All reported cases of blast-enhanced fracturing for remediation purposes have been applied to bedrock formations primarily to enhance pump-and-treat systems. Subsurface fractures are created by placing explosives in open bedrock boreholes and detonating them. Such detonations create an intensely fractured area of bedrock that is essentially rubble (referred as fracture trench). The fracture trench acts as an effective groundwater sink from which increased volumes of groundwater can be pumped (GWRTAC, 1996). Fractures propagate at the fastest rate (300± m/sec) in blast fracturing. Blast fractures are self-propping similar to pneumatic fractures. Emplacement of amendments by hydraulic and pneumatic fracturing processes is well developed; however, it is under development in case of blast fracturing.

The cost of hydraulic and pneumatic fracturing depends on the number of fractures produced per borehole (ITSR, 1998). EPA estimates the cost of a single hydraulic fracture to be between $950 and $1,425 (EPA, 1993). Approximately 35% of the costs are associated with labor. The daily cost of 4 to 6 fractures was estimated to be

$5,700 (EPA, 1993). Cost of pneumatic fracturing at a technology demonstration project in New Jersey was estimated at $308/kg of TCE removed. Pneumatic fracturing costs can also be estimated as $10 to $22/m^3 of soil treated (ISTR, 1998). Fracturing reduces overall costs of remediation by increased mass removal of contaminants and more rapid cleanup.

3.9 Soil Heating

Soil heating is a remediation approach that can be combined with SVE to improve contaminant removal from saturated and unsaturated zones at polluted sites. This technology employs six-phase electrical heating of soils to remove contaminants that are strongly sorbed to soil matrices and may be difficult to treat using SVE alone. In this approach, three-phase electricity from a conventional utility power transformer is converted into six phases, and each phase is delivered to a single electrode placed along a hexagonal pattern (Figure 3.12). Six-phase soil heating (SPSH) uses electrical resistive heating to raise soil temperature and facilitate desorption of the contaminant into the gas phase. Strategically placed extraction wells remove the gas phase for on-site treatment. The amount of heating achieved is a function of the soil moisture content. Low soil permeability and high moisture contents result in better heating. Steam resulting from vaporization of soil moisture helps in removing strongly bound contaminants.

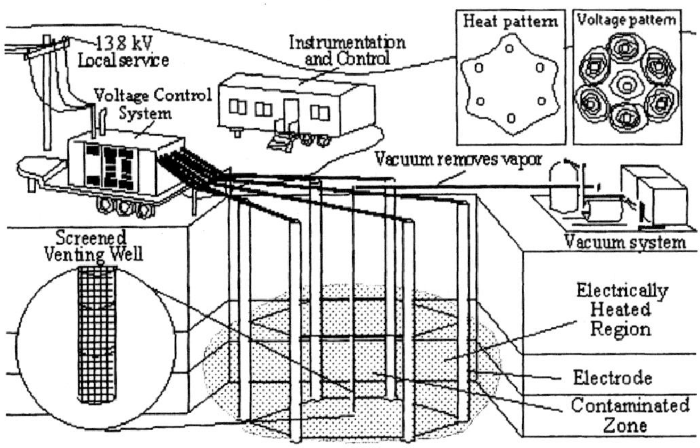

Figure 3.12 Schematic illustrating the application of six-phase soil heating (SPSH) to a contaminated soil (FRTR, 2006).

SPSH was evaluated for DNAPL removal during a field test conducted at the Groundwater Remediation Field Laboratory at Dover Air Force Base, Delaware (Peurrung and Bergsman, 1997). The ground water table at the site was located 25 feet below surface and formed an aquifer of thickness between 5 to 7 feet underlain by a dense clay layer. The subsurface soil consisted of sand and gravel with thin layers of clay and silt. Electrodes were inserted to a depth of 35 feet along a circle of 30 feet diameter. This formed a treatment zone of approximately 42 feet diameter by 15 feet thickness. The system design allowed heating of the entire saturated depth and 5 feet of the unsaturated depth. On-site equipment included a conventional utility power transformer, a collection header, a vacuum blower, a condenser and knockout box, and granular activated carbon drums to manage off gas and condensate. Most of the DNAPL introduced to the site was removed from the soil in 21 days. During this time, 136,000 kW-hrs of energy was used and 29,000 gallons of condensate was removed from the site. At $0.07/kW-hr, this represented an energy cost of approximately $16/m^3 soil treated.

Soil heating can also be achieved by using electromagnetic energy in the radio frequency band (EPA 1995). Radio frequency heating (RFH) is often employed in conjunction with SVE to improve contaminant removal. The soil is heated using an array of exciter electrodes inserted in the contaminated soil. Energy is applied to this array of electrodes. Two rows of ground electrodes are positioned on either side of the exciter electrodes forming a triplate capacitor. The application of energy to the central array of electrodes, results in dielectric heating of the surrounding soil. Soil temperatures can rise to over 300°C. The heating results in vaporization of contaminants and moisture, which are subsequently removed by vapor extraction wells.

The Illinois Institute of Technology Research Institute evaluated RFH at Kelly Air Force Base in San Antonio, TX (EPA 1995). A 40-kW radio frequency generator was used as the energy source. Exciter electrodes consisting of 2.5 and 4-inch diameter copper pieps were installed in 10-inch bore holes to a depth of 19.5 feet. Four exciter electrodes were installed 2.5 feet apart and two rows of eight ground electrodes (2-inch diameter aluminum pipes) were positioned on either side of the exciter array. The RFH system raised the temperature of the target zone to 150°C. The system was operated for a 9-week duration followed by a 2-month cool down period. This study found that concentrations of total recoverable petroleum hydrocarbons were reduced by an average of 95%.

Soil heating can also be achieved by the injection of steam into the subsurface (EPA 1998b). This technology was first developed by the petroleum industry for the enhanced recovery of product from geological formations. In the case of remediation, steam injected above or below the groundwater table heats the soil and water around the injection well. Some heated water moves away from the injection well. Eventually the surrounding water is heated to the temperature of the steam and is vaporized *in situ*. Steam injection results in the development of three distinct zones in the subsurface: the steam zone, the variable temperature zone, and the ambient temperature zone. The steam and hot water travel through the contaminated media and displace contaminants that are either sorbed to soil particles or present as NAPLs. The hot water reduces the viscosity of the NAPL making it easier to be displaced.

Both contaminant and soil types are important factors that can impact the success of steam injection. The Henry's Law constant of the contaminant will indicate its likelihood to volatilize. Steam injection is most appropriate for the removal of trapped lenses of volatile or semivolatile contaminants that may otherwise not be removed by hot water displacement of vacuum extraction. Steam injection is much more likely to be successful for sandy or gravelly soils with low organic matter contents than for finer soils with higher organic matter. Subsurface heterogeneity can significantly impact remediation efficiency using steam injection.

The primary design parameter for a steam injection process is the steam injection rate. Other design considerations include injection pressure, temperature and steam quality. Injection pressures should be controlled to prevent fracturing of soil, which can lead to loss of steam to the surface. Higher injection rates can result in faster heating of the subsurface. However, once steam breakthrough occurs, improved thermal efficiency can be achieved by reducing the injection rate. Greater heating rates result in greater recoveries and better energy efficiency. Higher steam quality, defined as the proportion of the total water in the vapor phase, produce better contaminant extraction efficiencies.

Steam stripping has been used as a remediation technology in the Netherlands since the early 1980's. In the U.S., a small-scale pilot demonstration project was conducted in San Jose, CA, in 1988 where surface spills and leaking underground storage tanks had caused the release of several VOCs (EPA 1998b). SVE and steam injection were performed cyclically. This approach was effective in rapidly removing a considerable portion of the contaminants. However, low permeability zones appeared to retain contaminants at high levels.

Steam injection was evaluated to recover JP-5 fuel at the Naval Air Station Lemoore in Lemoore, CA in 1994 (EPA 1998b). The remediation system consisted of eight extraction wells that enclosed two injection wells. The concentration of total petroleum hydrocarbons in the unsaturated zone within the treatment area was reduced from approximately 100,000 ppm to 20 to 50 ppm. However, contaminant levels remained at 20,000 ppm at and below the groundwater table. Contaminant concentrations increased at the interface of the surface clay layer and underlying silt due to the upward migration of contaminated vapors.

3.10 Conclusion

This chapter described eight commonly used physical treatment processes for soil and groundwater remediation. Physical treatment of contaminated subsurface media remains popular because of predictable end results and continued public acceptance. These processes are also being used in combination with chemical or biological treatment techniques. Pump-and-treat remains a common plume control method. It is expected that in coming years research and improved understanding of physicochemical processes will bring newer technologies to the market while improving the performance of existing remediation methods. Greater insights will also lead to the effective deployment of remediation approaches that combine the best characteristics of "active" physical processes and "passive" chemical or biological techniques.

References

AFCEE (1997). *Test Plan and Technical Protocol for Bioslurping.* Air Force Center for Environmental Excellence, Battelle Press, Columbus, OH.

Bedient, P.B.; Rifai, H.S.; Newell, C.J. (1994) In *Ground Water Contamination: Transport and Remediation.* Prentice Hall, Inc. Englewood Cliffs, NJ.

Cohen, R.M.; Vincent, A.H.; Mercer, J.W.; Faust, C.R.; Spalding, C.P. (1994) Methods for Monitoring Pump-and-Treat Performance. U.S. Environmental Protection Agency Pub. No. 600/R-94/123. R.S. Kerr Environmental Research Laboratory, Ada, OK.

DENIX (2006). Defense Environmental Network & Information Exchange, URL: www.denix.osd.mil/

EPA (1993) "Hydraulic Fracturing Technology – Technology Evaluation Report." U.S. Environmental Protection Agency Pub. No.540/R-93/505, Risk Reduction Engineering Laboratory, Cincinnati, OH.

EPA (1994) "Alternative Methods for Fluid Delivery and Recovery." U.S. Environmental Protection Agency Pub. No.625/R-94/003, Office of Research and Development, Washington, DC.

EPA (1995) "IITRI Radio Frequency Heating Technology" U.S. Environmental Protection Agency Pub No. 540/R-94/527a, Office of Research and Developent, Washington, DC.

EPA (1996a) "How to Effectively Recover Free Product At Leaking Underground Storage Tank Sites: A Guide for State Regulators." U.S. Environmental Protection Agency Pub. No. 510-R-96-001, Office of Research and Development, Washington DC.

EPA (1996b) "Pump-and-Treat Ground-Water Remediation. A Guide for Decision Makers and Practitioners." U.S. Environmental Protection Agency Pub. No. 625/R-95/005, Office of Research and Development, Washington DC.

EPA (1997) "Analysis of Selected Enhancements for Soil Vapor Extraction." U.S. Environmental Protection Agency Pub. No. 542/R-97/007, Office of Research and Development, Washington DC.

EPA (1998a) "Field Applications of In Situ Remediation Technologies: Groundwater Circulation Wells." U.S. Environmental Protection Agency Pub. No. 542/R-98/009, Office of Solid Waste and Emergency Response, Washington DC.

EPA (1998b) "Ground Water Issue. Steam Injection for Soil and Aquifer Remediation." U.S. Environmental Protection Agency Pub No. 540/S-97/505. Office of Research and Development, Washington, DC.

EPA (1999) "Cost and Performance Summary Report. Soil Vapor Extraction at Camp LeJeune Military Reservation, Site 82, Area A, Onslow County, North Carolina." U.S. Environmental Protection Agency CERCLIS ID Number NC6170022580, Office of Solid Waste and Emergency Response, Technology Innovation Office, Washington, DC.

EPA (2001a) "Groundwater Pump and Treat Systems: Summary of Selected Cost and Performance Information at Superfund-financed Sites." U.S. Environmental Protection Agency Pub No. 542-R-01-021b. Office of Solid Waste and Emergency Response, Washington, DC.

EPA (2001b) "Cost Analyses for Selected Groundwater Cleanup Project" U.S. Environmental Protection Agency Pub No. 542-R-00-013. Office of Solid Waste and Emergency Response, Washington, DC.

EPA (2001c) "GCWs used with air stripping and UV technologies to remove TCE and RDX." In Ground Water Currents. U.S. Environmental Protection Agency Pub No. 542-N-01-007. Office of Solid Waste and Emergency Response, Washington, DC.

EPA (2002) "How to Evaluate Alternative Cleanup Technologies for Underground Storage Tank Sites: A Guide for Corrective Action Plan Reviewers." U.S. Environmental Protection Agency Pub. No.510-R-04-002, Office of Solid Waste and Emergency Response, Washington DC.

ESTCP (2001) "Use of Cometabolic Air Sparging to Remediate Chloroethene-Contaminated Groundwater Aquifers." Environmental Security Technology Certification Program Report No. CU-9810, U.S. Department of Defense.

FRTR (2006) Remediation Technologies Screening Matrix and Reference Guide, Version 4.0. Federal Remediation Technologies Roundtable. URL: www.frtr.gov

GWRTAC (1996) "Technology Overview Report: Blast-Enhanced Fracturing." Groundwater Remediation Technologies Analysis Center Pub. No.TO-96-01, Pittsburgh, PA.

GWRTAC (1997) "Technology Overview Report: In-Well Vapor Stripping." Groundwater Remediation Technologies Analysis Center Pub. No.TO-97-01, Pittsburgh, PA.

GWRTAC (2001) "Technology Evaluation Report: Fracturing Technologies to Enhance Remediation." Groundwater Remediation Technologies Analysis Center Pub. No.TE-02-02, Pittsburgh, PA.

ITSR (1998) "Hydraulic and Pneumatic Fracturing." Innovative Technology Summary Report No. 1916 and 1917. U.S. Department of Energy, Office of Environmental Management.

Jawitz, J.W.; Sillan, R.K.; Annable, M.D.; Rao, P.S.C.; Warner, K. (2000) "In-situ alcohol flushing of a DNAPL source zone at a dry cleaner site." *Environ. Sci. Technol.* 34, 3722-3729.

Kuhlman, M.I. (2002) "Analysis of the steam injection at the Visalia Superfund Project with fully compositional nonisothermal finite difference simulations." *J. Haz. Mat.* 92, 1-19.

Leeson, A.; Johnson, P.C.; Vogel, C.M.; Hinchee, R.E.; Marlee, M.; Peargin, T.; Bruce, C.L.; Amerson. I.L.; Coonfare, C.T.; Gillespie, R.D.; McWhorter, D.B. (2002) "Air sparging design paradigm." Battelle Corp., Columbus OH. Website of the Environmental Security Technology Certification Program (ESTCP): http://www.estcp.org/documents/techdocs/Air_Sparging.pdf

Masters, G.M. (1996) *Introduction to Environmental Engineering and Science*. Prentice Hall, Upper Saddle River, NJ.

Murdoch L.C., and Slack W.W. (2002) "Forms of Hydraulic Fractures in Shallow Fine-Grained Formations." *Journal of Geotech. And Geoenv. Engg.* 128(6), 479-487.

NFESC (1998). "Application Guide for Bioslurping – Volume II: principles and Practice of Bioslurping." Naval Facilities Engineering Service Center, Pub. Technical Memorandum TM-2301-ENV, Port Hueneme, California.

NRC (1994). *Alternatives for Ground Water Cleanup*. National Research Council. National Academy Press.

NRL (1999). "Groundwater Circulating Well Technology Assessment." Naval Research Laboratory, Pub. No.NRL/PU/6115-99-384, Washington DC.

Peurrung, L.M>, and Bergsman, T.M. (1997) Applications Analysis Report: Six-Phase Soil Heating of the Saturated Zone, Dover Air Force Base, Delaware." Project Report for Pacific Northwest National Laboratory: http://www.cesiweb.com/dover.htm

Schuring, J.R.; Chan, P.C. (1992) "Removal of Contaminants from Vadose Zone by Pneumatic Fracturing ." New Jersey Institute of Technology, Prepared for the United States Geological Survey, Department of Interior.

USDOE (2002). "Innovative Technology Summary Report: In-Well Vapor Stripping." U.S. Department of Energy, Pub. No.DOE/EM-0626, Office of Environmental Management, Washington DC.

USACE (1999). "Multi-Phase Extraction – Engineer Manual" U.S. Army Corps of Engineers, EM 1110-1-4010, Department of the Army, Washington DC.

CHAPTER 4

Chemical Treatment Technologies

Say Kee Ong and Angela Kolz

4.1 Introduction

There are more than 700,000 different chemicals known to be in use across the world (Raven and Berg, 2006). In the U.S., more than 500 tons of approximately 3,000 different chemicals (excluding polymers and inorganic chemicals) are produced or imported each year (US EPA, 1998). While most of these chemicals are not hazardous and, when released into the environment, degrade readily several are highly toxic, recalcitrant and persistent in the environment. A variety of remediation technologies that are based on physical, chemical and biological processes can be applied to clean sites contaminated by hazardous chemicals. Among these, chemical technologies are used to chemically destroy environmental pollutants, especially organic chemicals, to harmless and innocuous compounds. Inorganic chemicals such as heavy metals that cannot be degraded chemically can be immobilized in the soil and aquifer environment to minimize exposure to ecoreceptors. This chapter discusses the application of chemical remediation technologies such as precipitation, chemical oxidation and reduction, stabilization and solidification, adsorption and ion-exchange, electrochemical processes, chemical leaching and solvent extraction, soil flushing, soil washing.

4.2 Precipitation

Precipitation is a common treatment method for waste streams containing heavy metals. Aqueous streams that can be treated using precipitation include industrial wastewater containing arsenic (As), barium (Ba), chromium (Cr), lead (Pb), copper (Cu), nickel (Ni) and zinc(Zn) and groundwater contaminated with these pollutants. Precipitation involves the addition of chemicals or precipitants to alter the physical state of a dissolved metal such that its solubility is exceeded and a precipitate is formed. The process can be reversed when the activity of the dissolved species is less than its solubility or environmental conditions such as pH and redox are changed. Formation of precipitates proceeds at a finite rate with the initial spontaneous formation of a fine nuclei (nucleation) followed by deposition of the precipitate constituent ions onto the nuclei (crystallization). This leads to agglomeration and ripening of the precipitate into a stable form. Sufficient reaction time must be allowed for the reaction to occur accordingly.

4.2.1 Process Fundamentals

The solubility product (K_{sp}) of a compound is a measure of its solubility in a given solution. The dissolution of a solid $A_ZB_Y(s)$ into its constituent ions A^{Y+} and B^{Z-} can be expressed by the following equation:

$$A_ZB_Y(s) \leftrightarrow zA^{Y+} + yB^{Z-} \qquad \text{(Eq. 4.1)}$$

The solubility product is the product of the activities of the species involved in the reaction:

$$K_{sp} = \{A^{y+}\}^z\{B^{z-}\}^y \qquad \text{(Eq. 4.2)}$$

where A is the metal or cation, B is the anion, and z and y are number of molecular units in the compound.

Table 4.1 summarizes typical values of K_{sp} for different lead compounds along with the estimated lead concentration in solution assuming 1 x 10^{-3} moles/liter of anion is present. Eq. 4.2 assumes equilibrium conditions for the dissolved metal in the absence of other anions and cations. The presence of similar ions in solution may impact the solubility of a metal through the "common ion effect". This occurs when a solution containing an ion that is similar to one of the ions from the dissolution of a precipitate causes the solubility of the precipitate to be less than when the precipitate dissolves in pure water.

Table 4.1 Solubility products of lead compounds.

	K_{sp}	Pb^{2+} (moles/liter)	Concentration* (mg/L)
$Pb(OH)_2$	$10^{-14.3}$	5 x 10^{-9}	0.001
$PbSO_4$	$10^{-7.8}$	1.58 x 10^{-5}	3.27
PbS	$10^{-27.0}$	1 x 10^{-24}	2.07 x 10^{-19}
$PbCO_3$	$10^{-13.1}$	1 x 10^{-11}	1.6 x 10^{-5}

*Assuming 1 x 10^{-3} moles/liter each of [Cl^-], [OH^-], [SO_4^{2-}], and [S^{2-}], [CO_3^{2-}]

In most situations, cations and anions do not exist as free ions in solution but instead interact with other ions or molecules to form complex ions or coordination compounds to remain in solution. Complex ions can be cationic, anionic or neutral. Examples of complexes include hydro-complexes, and cyano- and ammonium complexes formed in the presence of cyanide and ammonium ions.

Formation of hydro-complexes is dependent on the solution pH. Hydro-complexes of metals are amphoteric and result in minimum solubility for the metal hydroxides. The pH range for the minimum solubility varies with the constituents present in the water. Typically, metals are fairly soluble under acidic conditions, have solubility minima between pH 9 and 11, and are fairly soluble at elevated pHs (> 11). Therefore, although the solubility product of a compound provides a good approximation of the solubility of a compound, to obtain a precise estimate of the metal solubility, all complex species and the pH of the solution must be taken into consideration.

Figure 4.1 shows the solubilities of several metal hydroxides in equilibrium with the precipitate over the pH range of 0 to 14. Chemicals that are commonly added to form metal hydro-complexes for subsequent precipitation as hydroxides include lime ($Ca(OH)_2$), magnesium hydroxide ($Mg(OH)_2$), and sodium hydroxide (NaOH). Lime is the least expensive of the three and is widely used. It is usually purchased in the form of quicklime (CaO) and slaked and slurried to form calcium hydroxide before application.

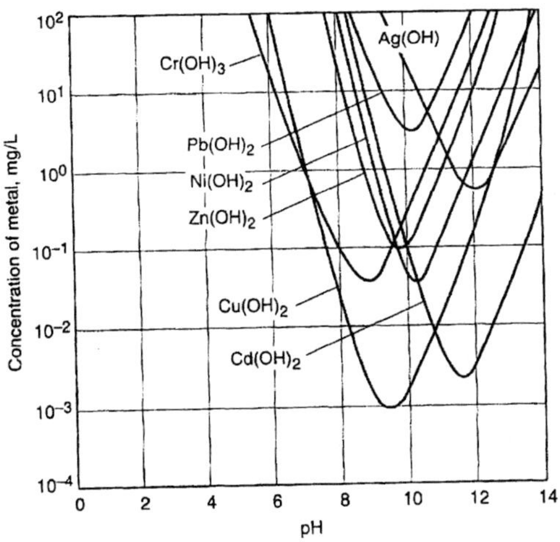

Figure 4.1 Solubilities of metal hydroxides as a function of solution pH (US EPA, 1983).

Metals can also be precipitated as sulfides as shown in Figure 4.2. Chemicals used for sulfide precipitation are sodium sulfide or bisulfide (Na_2S or NaHS). Extremely low aqueous-phase concentrations of metal can be obtained using sulfide precipitation. This is illustrated by the very low K_{sp} values of lead sulfide as compared to lead hydroxide (Table 4.1). Because of the possibility of generating hydrogen sulfide (H_2S) gas, which is poisonous, sulfide precipitation is generally controlled to occur under alkaline conditions.

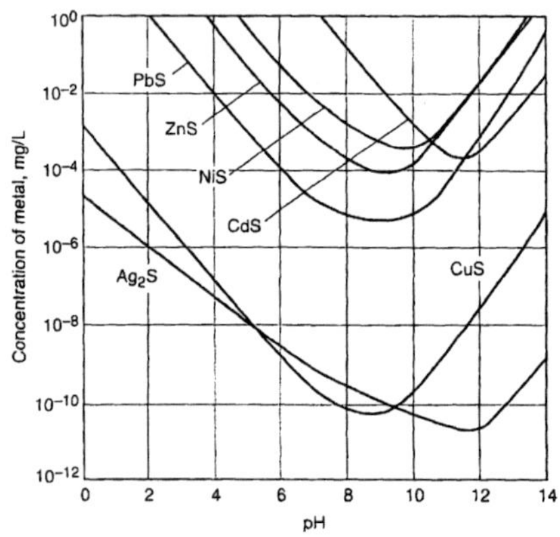

Figure 4.2 Solubilities of metal sulfides as a function of solution pH (US EPA, 1983).

4.2.2 Process Application

A typical flow diagram of a precipitation system is illustrated in Figure 4.3. The following must be considered while designing an above ground metal precipitation system to treat contaminated groundwater:

- A rapid mixing tank of detention times between 3 to 5 minutes is needed where coagulants such as hydroxides or ferric salts are added to initiate the coagulation and precipitation. When using lime as a precipitant, a reaction tank of detention time between 10 to 20 minutes may be needed since lime reacts slower than sodium hydroxide, alum or ferric chloride.

- A pre-reaction tank may be needed for pH adjustment.

- A slow mix tank or flocculation tank with a detention time of 20 to 30 minutes is needed for agglomeration of flocs.

- To aid in the coagulation, organic polymers or inorganic polymers such as activated silica may be added before the slow mix to aid in the flocculation.

- The precipitates are settled out in a settling tank or clarification tank. For most heavy metal precipitates, the clarifier surface overflow rate should be maintained between 500 and 1500 gpd/ft^2. The sludge can be reprocessed to recover metals or dewatered and stabilized before disposal.

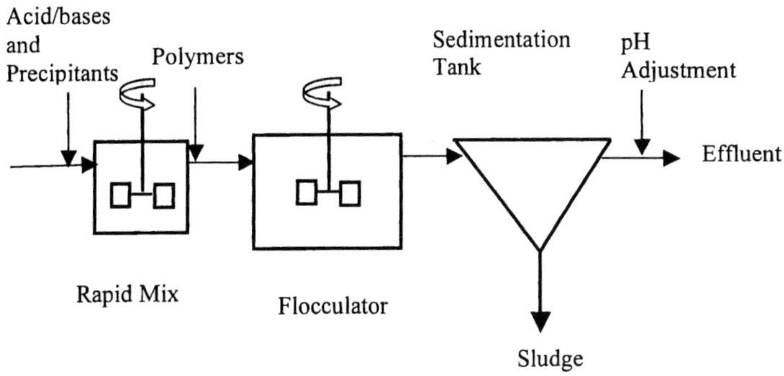

Figure 4.3 Schematic diagram of a chemical precipitation system.

Typical capital costs (1993 costs) for 75 to 250 liters/minute (20 to 65 gpm) packaged metals precipitation systems can range from approximately \$85,000 to \$115,000. Operating costs excluding sludge disposal may range from \$0.08 to \$0.18 per 1,000 liters (\$0.30 to \$0.70 per 1,000 gallons) of groundwater containing up to 100 mg/L of metals. Sludge disposal will add about \$0.13 per 1,000 liters (\$0.50 per 1,000 gallons) of groundwater treated (NEESA, 1993). The costs of a precipitation system treating chromium-contaminated groundwater with a flow rate of 5,625 liters/minute (1500 gpm) was estimated to be \$1.13 million (US EPA, 2000). This cost included groundwater extraction, chromium reduction, metals precipitation using sodium hydroxide and discharge into a river about 4,000 feet from the treatment system. Operating and maintenance costs were estimated to be \$240,000 per year.

4.3 Chemical Oxidation and Reduction

The objective of chemical oxidation/reduction system is to detoxify toxic pollutants to harmless or less toxic products by utilizing oxidizing or reducing agents.

Chemical oxidation is a well-established technology that is capable of destroying a wide range of organic compounds such as chlorinated hydrocarbons, aromatic contaminants and inorganics such as cyanide. Chemical reduction, however, has limited applications but can be effective in the destruction of certain hazardous compounds.

4.3.1 Process Fundamentals

Oxidation is a reaction where there is a loss of electrons while reduction is a reaction where there is a gain in electrons. An approach in remembering these principles is by using the acronym - OIL RIG (where O stands for Oxidation and L for Loss of electrons and R for Reduction and G for Gain in electrons). The reactant that loses the electrons is the reducing agent which reduces the reactant that receives the electrons. The reactant gaining electrons is the oxidizing agent and it oxidizes the reactant that gives up the electrons.

Common oxidizing agents used in the oxidation of hazardous wastes include ozone (O_3), hydrogen peroxide (H_2O_2) and chlorine. Table 4.2 lists several chemical oxidants and their oxidation potentials E° (V). A chemical with a high E° is a strong oxidant. Included in Table 4.2 is a relative comparison of the oxidation potential of each oxidizing agent with chlorine. In recent years, the application of chemical oxidation has advanced by using various enhancements such as ultraviolet light (UV), ultrasound (US), enzymes and inorganic catalysts, or a combination of oxidants to improve oxidation of hazardous compounds. The in most advance oxidation processes (AOPs) objective is to generate hydroxyl radicals (E^o = 2.80 V) which are strong oxidizing agents but with extremely short half-lives. Table 4.3 summarizes the different combinations of oxidants and enhancements that are commonly used to treat hazardous organics. The more common AOP applications utilize Fenton's reagent, O_3/UV, H_2O_2/O_3/UV, and metal catalysts (TiO_2 or ZnO)/ O_3 or H_2O_2/UV.

4.3.2 Process Applications

4.3.2.1 Chlorine

Chlorine is the most commonly used oxidizing agent. It is available as a gas or sold as a liquid in the form of sodium or calcium hypochlorite. When dissolved in water, chlorine forms hypochlorous acid (HOCl) and hydrochloric acid (HCl). Hypochlorous acid dissociates to form hypochlorite ion. Chlorine is relatively inexpensive in comparison with other oxidizing agents and chlorination equipment is widely available. However, chlorine is heavier than air, extremely toxic and can form chlorinated compounds during reaction with organics that may be more toxic than the original compound. For example, destruction of phenol using chlorine can result in the formation of chlorinated phenolic compounds which are more toxic than phenol. Another common application of chlorine in hazardous waste treatment is the destruction of cyanide in water. This process is called alkaline destruction. The reaction is sensitive to pH, and a pH greater than 10 is maintained. In above ground systems, the destruction of cyanide can be facilitated in a two-step process using separate tanks in a continuous mode. The two steps can be chemically illustrated using equations 4.3 and 4.4.

Cyanide is oxidized to less toxic cyanate

$$NaCN + Cl_2 + 2NaOH \rightarrow NaCNO + 2NaCl + H_2O \quad \text{(Eq. 4.3)}$$

Cyanate is further oxidized to carbon dioxide and nitrogen

$$2NaCNO + 3\,Cl_2 + 4\,NaOH \rightarrow 2CO_2 + N_2 + 6NaCl + 2H_2O \quad \text{(Eq. 4.4)}$$

Table 4.2 Typical oxidizing agents and their relative oxidizing power as compared to chlorine (Haas and Vamos, 1995).

Redox Reaction	**Oxidizing Agent**	**E° (V) at 25° C**	**Relative Oxidative Power**
$F_2 + 2e^- = 2F^-$	F_2	3.06	2.55
$OH\bullet + H^+ + e^- = H_2O$	$OH\bullet$	2.80	2.05
Atomic oxygen	O	2.42	1.78
$O_3 + 2H^+ + 2e^- = H_2O$	O_3	2.07	1.52
$H_2O_2 + 2H^+ + 2e^- = H_3O_2^+$	H_2O_2	1.76	1.30
Perhydroxyl radical	$HO2\bullet$	1.70	1.25
$MnO_4^- + 4H^+ + 3e^- = MnO_2 + 2H_2O$	MnO_4^-	1.68	1.24
$HClO_2 + 3H^+ + 4e^- = Cl- + 2H_2O$	$HClO_2$	1.57	1.15
$HOCl + H^+ + 2e^- = Cl^- + H_2O$	HOCl	1.49	1.10
$Cl_2 + 2e- = 2Cl^-$	Cl_2	1.36	1.00
$HBrO + H^+ + 2e^- =$	HBrO	1.33	0.97
$Br_2 + 2e^- = 2Br^-$	Br_2	1.07	0.79
$HIO + H^+ + 2e^- = I^- + H_2O$	HIO	0.99	0.73
$I_2 + 2e^- = 2I^-$	I_2	0.54	0.40

Stable complexes of cyanide such as ferrocyanides are not resistant to chlorine oxidation. Cyanide complexed with copper, nickel and precious metals can be oxidized by chlorine, but reacts more slowly than free cyanide. Excess chlorine is needed to oxidize these cyanide complexes. Typical reaction times for the first step of the process are between 15 to 60 minutes while reaction times for the second stage is typically between 30 to 60 minutes. About 7 lbs of chlorine or 7.5 lbs of hypochlorite is needed for every lb of cyanide (NCMS, 1994).

Table 4.3 Common advanced oxidation systems for hazardous organic contaminants.

Chemical Oxidation Systems	**Chemical Oxidation Systems with Irradiation**
H_2O_2/Fe^{2+} (Fenton's reagent)	O_3/ultraviolet (UV)
O_3/H_2O_2	H_2O_2/UV
	O_3/H_2O_2/UV
	H_2O_2/ultrasound (US)
	Metal catalyst/UV
	Metal catalyst/O_3/UV
	Metal catalyst/H_2O_2/UV

4.3.2.2 Ozone

Ozone is an allotropic form of oxygen. Ozone is generated by passing air or pure oxygen through a narrowly-spaced electrode subjected to a high voltage where high energy corona discharges dissociate oxygen molecule into atomic oxygen. Ozone is formed when atomic oxygen combines with oxygen molecules.

$$O_2 + O \leftrightarrow O_3 \text{ and } \quad 3O_2 \leftrightarrow 2O_3$$

The gas stream generated from air by this process typically contains about 0.2 to 3 percent ozone by weight and up to 1 to 6 percent, if pure oxygen is used. Treatment of aqueous contaminants requires the transfer of ozone from gas to liquid phase. Ozone is sparingly soluble in water with a Henry's law constant, K_H, of 0.082 atm•m^3/mole at 25° C (Glaze, 1987). Ozone decomposes rapidly in aqueous solutions after reacting with impurities such as organic compounds or particulates. The half-life for ozone is approximately 18 minutes in groundwater and about 10 minutes in lake water (Rice and Netzer, 1982). Ozone does not produce a persistent toxic residual and decomposes to oxygen resulting in an increase in dissolved oxygen concentrations in the treated water. Ozone is generally more effective in an alkaline solution.

In aqueous solution, organic pollutants are oxidized by direct reaction with molecular ozone and/or by indirect oxidation by free radicals, primarily hydroxyl (OH•) or hydroperoxide (HO_2•) radicals (Hoigne and Bader, 1976). Reactions of ozone in aqueous solutions are illustrated in Figure 4.4. Aqueous solutions may contain initiators such H_2O_2, Fe^{2+}, UV, and humic acid which promote decomposition of ozone to form free radicals, and promoters such as organic compounds, which react with OH• to form various radical species, resulting in complicated sequences and propagation of chain reactions. A chain reaction may be quenched when hydroxyl radicals react with scavengers to form secondary compounds which do not promote further chain reactions. Scavengers include bicarbonate and carbonate ions.

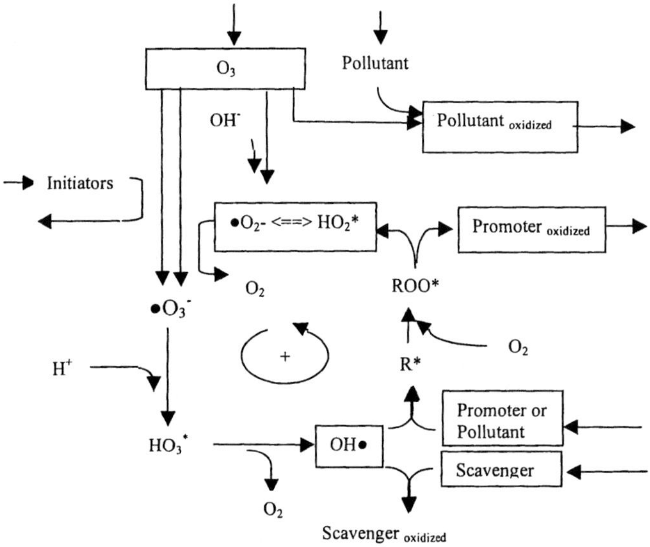

Figure 4.4 Proposed mechanism of ozone oxidation. R* represents oxidized metabolites of the pollutant, O_3^- is the ozonide ion, and O_2^- is the superoxide ion (Staehelin and Hoigne, 1985).

The reaction rate constant for the destruction of organics by OH• is typically several orders of magnitude greater than for O_3 alone. Molecular ozone oxidation can be termed as "selective" with second order rate constants (with respect to ozone and reduced species constant) between 1 and 10^3 $M^{-1}s^{-1}$ (Hoigne amd Bader, 1983). Kinetics of OH• reactions are several orders of magnitude faster than molecular ozone oxidation (10^8 – 10^{12} $M^{-1}s^{-1}$) (Hoigne and Bader, 1976; Farhataziz and Ross, 1977). Examples of estimated second order rate constants for oxidation of selected hazardous compounds by molecular ozone and hydroxyl radicals are summarized in Table 4.4.

Table 4.4 Second order reaction rates for common organic pollutants (US EPA, 1998, Cater et al., 1990, Dussert, 1997).

Compound	**Molecular Ozone** ($M^{-1}s^{-1}$)	**Hydroxyl Radical** ($M^{-1}s^{-1}$)
Acetylenes	50	10^8 - 10^9
Alcohols	10^{-2} to 1	10^5 to 10^9
Alkanes	10	10^9
Aromatics	10^{-2}	10^6 to 10^9
Carboxylic Acids	1^{-3} to 10^{-2}	10^7 to 10^9
Chlorinated alkenes	10^{-1} to 10^3	10^9 to 10^{11}
Ketones	1	10^9 to 10^{10}
Nitrogen-containing organics	10 to 10^2	10^8 to 10^{10}
Olefins	1 to 450 x 10^3	10^9 to 10^{11}
Phenols	10^3	10^9 to 10^{10}
Sulfur-containing organics	10 to 1.6 x 10^3	10^9 to 10^{10}

4.3.2.3 Hydrogen Peroxide

Hydrogen peroxide is slightly more powerful than chlorine but less powerful than ozone (see Table 4.3). To enhance the oxidizing power of H_2O_2, hydroxyl (OH•) radicals are generated through irradiation by ultraviolet light or ultrasound and metal catalysts. High intensity, medium-pressure, broad band UV lamps are typically used although high intensity, xenon flash lamps with adjustable spectral output to match the absorption characteristics of H_2O_2 have also been used. Low-pressure mercury vapor UV lamps are not the best choice as the lamp has peak emission at 259 nm while the maximum absorbance of UV radiation by H_2O_2, occurs at about 220 nm.

$$H_2O_2 \xrightarrow{hv} 2OH\bullet \qquad \text{(Eq. 4.5)}$$

Hydroxyl radicals can also be generated by using metallic catalysts such as ferrous ions (Fenton's reagent) or titanium oxide:

$$Fe^{2+} + H_2O_2 \rightarrow Fe^{3+} + OH\bullet + OH^- \qquad \text{(Eq. 4.6)}$$

The OH• formed can react with and initiate oxidation of organic pollutants present in a waste stream or react with Fe^{2+} to produce Fe^{3+} as illustrated in Eq. 4.7. Eq. (4.6) implies that Fenton's reagent is strongly dependent on solution pH. In fact, in acidic conditions, OH• radicals are the predominant reactive species. This process is effective at pH levels less than or equal to 3.0.

$$Fe^{2+} + OH\bullet \rightarrow Fe^{3+} + OH^- \qquad \text{(Eq. 4.7)}$$

If Fe^{3+} ions are used with H_2O_2, reactions are improved considerably by irradiation with near-UV radiation and visible light (Ruppert et al., 1994). A point to consider is that since many contaminated waters show high absorbance at wavelengths below 300 nm and because H_2O_2 absorbs UV light at 254 nm poorly, treatment by UV/H_2O_2 may not be effective for some situations. An approach used to overcome this limitation is to use ferrioxalate and H_2O_2 as ferrioxalate has high absorption coefficient at wavelengths above 200 nm, absorbs light strongly at longer wavelengths (up to 450 nm) and generates OH• with a high quantum yield (Zepp et al., 1992).

4.3.2.4 Ozone/Hydrogen Peroxide/UV Systems

Combinations of O_3 and/or H_2O_2 along with UV irradiation have been successfully applied to enhance hydroxyl free radical formation during oxidation of organic compounds. Examples of commercially available systems include Calgon perox-pure UV/H_2O_2, US Filter UV/O_2/H_2O_2, Calgon Rayox® UV/H_2O_2, Magnum CAV-OX® UV/H_2O_2, WEDECO UV/H_2O_2, WEDECO UV/O_3, and Matrix UV/TiO_2. Flow diagrams for two systems, US Filter UV/O_2/H_2O_2 and Calgon perox-pure UV/H_2O_2, are presented in Figures 4.5 and 4.6, respectively. Hydraulic retention times of the two systems are about 5 minutes. Results of pilot studies using the two systems for the treatment of groundwater contaminated with chlorinated compounds are summarized in Table 4.5. For some of the chlorinated compounds, more than 99% destruction was observed. The reactions for a UV/O_2/H_2O_2 system are speculated to proceed by H_2O_2 initiating the decomposition of O_3 through a single electron transfer to form hydroperoxide ion (HO_2^-). The hydroperoxide ion then reacts with ozone to produce ozonide ions (O_3^-) and hydroxyl radicals (HO•). The reaction steps are speculated to consist of the following (Aieta et al., 1988):

$$H_2O_2 \rightarrow HO_2^- + H^+ \qquad \text{(Eq. 4.8)}$$

$$HO_2^- + O_3 \rightarrow O_3^- + HO_2\bullet \qquad \text{(Eq. 4.9)}$$

$$HO_2\bullet \rightarrow H^+ + O_2^- \qquad \text{(Eq. 4.10)}$$

$$O_2^- + O_3 \rightarrow O_3^- + O_2 \qquad \text{(Eq, 4.11)}$$

$$O_3^- + H^+ \rightarrow HO_3 \qquad \text{(Eq. 4.12)}$$

$$HO_3 \rightarrow OH\bullet + O_2 \quad \text{(Eq. 4.13)}$$

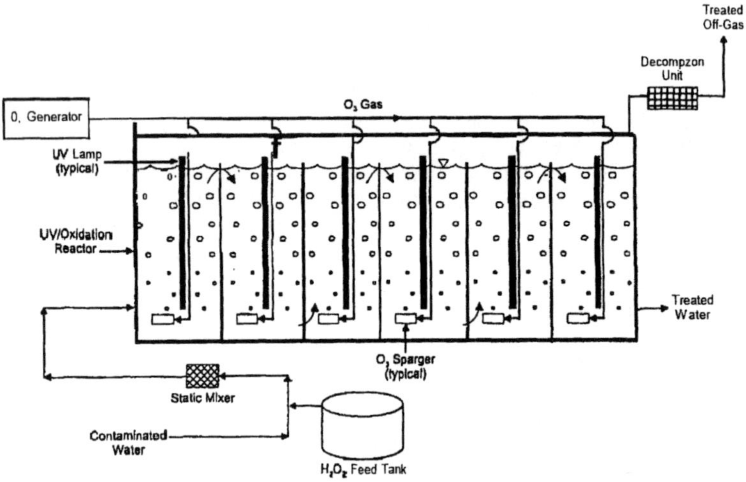

Figure 4.5 Flow diagram of US Filter Ozone/H_2O_2/UV system (US EPA, 1998).

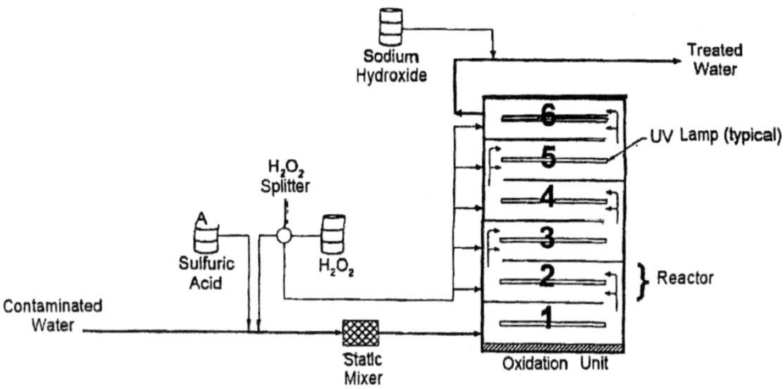

Figure 4.6 Flow diagram of Calgon UV/H_2O_2 system (US EPA, 1998).

Table 4.5 Pilot study results for Calgon UV/H_2O_2 and US Filter O_3/H_2O_2/UV systems (US EPA, 1998).

System	Compounds	Initial Concentration (μg/L)	Percent Removal (%)
Calgon UV/H_2O_2	Benzene	52	> 96
	Chlorobenzene	3,100	> 99.9
	Chloroform	41 – 240	93.6 to > 97
	1,1-Dichloroethane	120 – 400	> 95.8 to > 99.5
	1,2-DCA	22	> 92
	Tetarchloroethylene	63 – 2,500	> 98.7 to > 99.9
US Filter O_3/H_2O_2/UV	1, 1-Dichloroethane	9.5 to 13	65
	1,1,1- trichloroethane	2 to 4.5	87
	Trichloroethylene	50 to 520	>99

4.3.2.5 Ultrasound

Although ultrasound has been used for decades in the cleaning of surfaces, it is only in recent years that ultrasound has been explored as a means for the destruction of the hazardous organic compounds (Bremner 1990). Aqueous solutions subjected to ultrasound with frequencies in the range of 15 kHz to 10 MHz, experience an effect known as "acoustic cavitation." The mechanical vibration of sound waves creates a successive compression and expansion cycle which causes formation of small vapor-filled bubbles. The number of bubbles formed is estimated to be in the range of 4 x 10^8 bubbles/sec/m^3 with sizes ranging from 0 to 200 microns and a lifetime of approximately 10 microseconds (Suslick and Hammerton 1986). The cyclic compression and expansion modes cause these bubbles to grow and contract. During the expansion phase, the surface area of the bubble is greater than during the compression phase, resulting in an increase in the size of the bubble size over many cycles. Eventually, a critical size is reached depending on the ultrasonic frequency whereby the internal pressure of the bubble is unable to withstand the external pressure of the surrounding solution. An implosion occurs with the bubble collapsing resulting in high local pressures and temperatures (Suslick, 1990). Temperatures as high as 5000° C within the bubble have been estimated (Suslick et al., 1986) and with pressures near 1000 atm (Mason and Lorimer, 1988). The water vapor in the bubble during the collapse decomposes into hydrogen, and hydroxyl- and hydroperoxyl-radicals, which in turn oxidize organic compounds present in the aqueous solution. To further enhance the destruction rate, oxidizing agents such as hydrogen peroxide and ozone may be added.

Studies by Lin et al. (1996) showed that 99% of chlorophenol (100 mg/L) was removed in 360 minutes by using ultrasound at 20 kHz with 200 mg/L of hydrogen peroxide. First order degradation rates in argon-saturated aqueous solutions of 1.95 x 10^{-4} and 1.95 x 10^{-5} mol/L of carbon tetrachloride were found to be 3.3 x 10^{-3} and 3.9 x 10^{-3} s^{-1}, respectively while presence of ozone did not affect the degradation rate significantly (Hua and Hoffmann, 1996). Koszalka et al. (1992) obtained 98% reduction of CCl_4 in the presence of air or argon with chloroform, dichloromethane and chloromethane as initial reaction products followed by methane and chlorine/chloride. Kruger et al. (1999) observed complete destruction of 1,2-dichloro ethane (350 μg/L) in groundwater after 60 minutes with pseudo-first-order rate constants of 0.062, 0.063, and 0.044 min^{-1}, for sonication frequencies of 361, 620, and 1086 kHz, respectively, at 105 W.

4.3.2.6 Wet Air Oxidation

At elevated pressures and temperatures but below the critical point of water (374° C and 218 atm), organic wastes can be oxidized in the liquid phase in the presence of air. The temperatures and pressures of wet air oxidation systems are typically in the range of 150 - 325° C and 100 - 200 atm, respectively. Wet air oxidation is attractive when waste streams are too dilute to incinerate and too concentrated to be treated by biological treatment. It is effective for waste streams with COD in the range of 20,000 - 200,000 mg/L and suitable for waste with high refractory content. COD removal between 75 to 90% can be achieved. The end products consist of simpler forms of biodegradable compounds such as acetic acid and inorganic salts along with the formation of carbon dioxide and water. Depending on the waste, further treatment of the waste stream may be needed. The wet air oxidation process consists of a pre heater where heat is transferred from hot oxidized effluent to the incoming feed. Figure 4.7 shows a schematic of a typical wet air oxidation process. The incoming heated feed is mixed with air or oxygen at the bottom of the reactor and oxidation takes place within the reactor. The hydraulic residence time of the fluid in the reactor is 15 to 120 minutes. Liquids and noncondensables are separated in the separator drum and discharged separately. In a typical wet air oxidation system, heat may be recovered to produce steam or hot water. An example of wet air oxidation applications is the Zimpro Process which operates at a temperature of 150 - 350° C and at a pressure of 20 to 200 atm. Table 4.6 provides the residence time and destruction efficiencies for several organics using wet air oxidation.

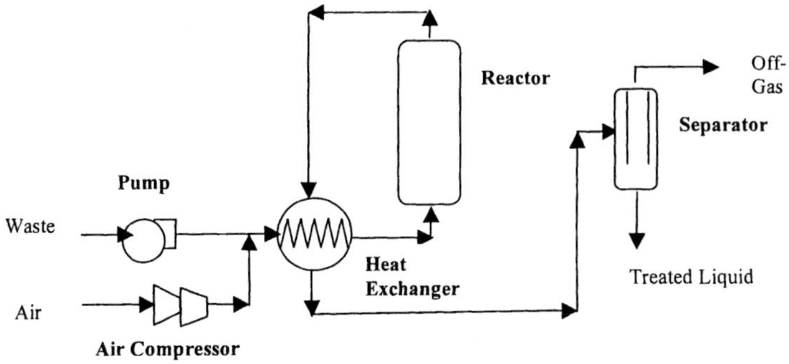

Figure 4.7 Schematic diagram of a wet air oxidation system.

Table 4.6 Destruction efficiencies of organic compounds using wet air oxidation at a temperature of 550° F, pressure of 1500 psig, and residence time of 30 minutes (Copa and Gitchel, 1988; Modell, 1989).

Compound	TOC Destruction (%)
o-Cresol	84
Carbon tetrachloride	99
Pentachlorophenol	94
DDT	60
Aniline	79
Diethylene Glycol	66
Ethylene Glycol	24

The reaction mechanisms occurring during wet air oxidation are still not well understood. It appears that formation of free radicals such as OH• may be the cause of oxidation of organic compounds. Oxidation is assumed to occur as follows:

hydrocarbon + oxygen → alcohol
alcohol + oxygen → aldehyde
aldehyde + oxygen → acid
acid + oxygen → carbon dioxide + water

The system is adaptable to a wide variety of oxidizable materials and water acts as a heat sink to control the temperature within the reactor. Special alloy materials are needed for the reactor due to the high corrosivity of the reactions resulting in high maintenance costs of the system.

4.3.2.7 Supercritical Water Oxidation

An extension of wet air oxidation is the supercritical water oxidation where the system is operated at temperatures and pressures above the critical point. Beyond the critical point, a single phase fluid exists with properties between those of the liquid and gas phases. Under critical fluid conditions, the solubility of organics is enhanced while the solubility of inorganics in the fluid is decreased by three to four orders of magnitude. The almost complete miscibility of oxygen and organics maximizes the interfacial diffusion of organics and oxygen. In addition, the low viscosity and gas-like properties of the fluid enhance contact between the target organics and the oxidizing radicals. A typical supercritical water oxidation system is described schematically in Figure 5.8. The process has been effectively demonstrated at pilot and developmental scales with demonstration units ranging from 400 kg/hr to 2,700 kg/hr (64.8 t/day) for destruction of chemical warfare materials (Killilea, 1998). Residence times needed for reactions may be as short as several minutes at temperatures of 600 - 650° C. More than 99.9% conversion of EPA priority pollutants has been achieved in pilot-scale plants with retention times less than 5 minutes (Gloyna and Li, 1996). The process is cost effective for aqueous wastes with organic concentrations in the range of 1 to 20 percent by weight. The system is limited to treatment of liquid wastes or solids less than 200 μm in diameter. Formation of char during reaction may impact the oxidation time of the organics while separation of inorganic salts during the process may be a problem. Typical materials for the reactor are Hastelloy C-276 and Iconel 625 (high nickel alloys), which can withstand high temperatures and pressures and the corrosive conditions. The residence time and destruction efficiencies for several organics using supercritical oxidation is summarized in Table 4.7.

4.4 Stabilization/Solidification

Stabilization/solidification (S/S) is one of the best demonstrated available technologies (BDATs) recommended for the treatment of hazardous wastes before disposal at a secured landfill. Development of S/S for hazardous wastes has its roots in the application of solidification to radioactive wastes (Conner, 1990). S/S is widely used for the treatment of residues from other hazardous waste treatment processes such as ash from thermal treatment, sludge from industrial treatment plants, especially for waste with inorganic constituents, and treatment of contaminated soils and hazardous waste sites including in situ applications. S/S can be used as a temporary or permanent treatment solution to minimize the risk and transport of the contaminants. Use of S/S can minimize risk by:

- reducing the solubilities of hazardous constituents in the waste by pH adjustment, sorption or precipitation; decreasing the surface area of the waste material across which transfer or loss of contaminants can occur; and
- altering the physical characteristics of the waste, for example, through absorption of free liquids.

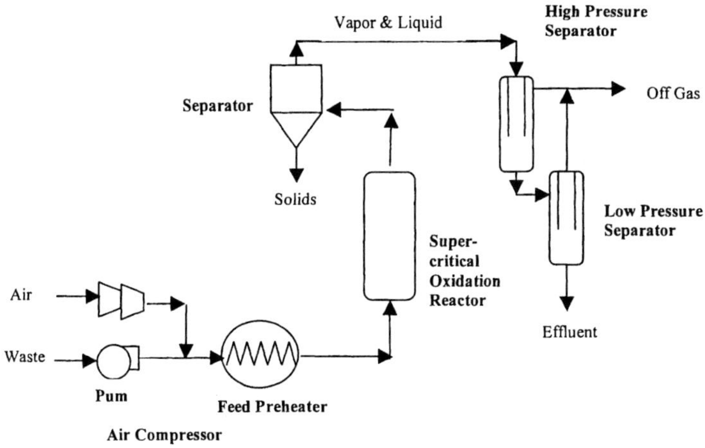

Figure 4.8 Schematic diagram of a supercritical water oxidation system.

Table 4.7 Destruction of organics using supercritical oxidation (Modell, 1989).

Compound	Temp. (° C)	Residence Time (mins)	Destruction Efficiency (%)
Cyclohexane	445	7	99.97
o-xylene	450	3.6	99.97
1,1,1-trichloroethane	495	3.6	99.99
1,2,4-Tetrachlorobenzene	495	3.6	99.99
DDT	505	3.7	99.997
PCB 1234	510	3.7	99.99
2,4-dinitrotoulene	513	0.5	99.992

The widespread application of S/S technology is due to the widely available, common and inexpensive additives and reagents used. The resulting solidified material may require little or no further treatment if proper conditions are maintained. However, the volume of treated material may increase due to the addition of reagents.

Cumulatively, S/S, both in situ and ex situ applications, is the second most common type of source control treatment technology implemented at Superfund sites, representing about 24 % of all source control treatment technology projects (US EPA, 2000). The cost to treat a cubic yard of waste by S/S at 29 Superfund sites ranged from \$8 to approximately \$1,200 with an average cost of \$264 per cubic yard (US EPA, 2000).

4.4.1 Process Fundamentals

Stabilization is a process by which binders (additives or reagents) are added to hazardous waste to convert the waste and its constituents into their least mobile or toxic form (US EPA, 1997d). The process is sometimes known as fixation. An example of stabilization includes the addition of lime or sulfide to metal-containing aqueous streams to precipitate the metals.

Solidification is a process by which binders (additives or reagents) are added in sufficient quantities to form a solidified mass or "monolith" with high structural integrity (generally compressive strength) and to decrease the permeability of the contaminated soil and minimize exposure of the hazardous constituents to outside conditions (US EPA, 1997d). Contaminants present in the contaminated soils do not necessarily interact chemically with the reagents. Instead they are mechanically locked or encapsulated within the solidified mass. Contaminant loss is limited largely by decreasing the surface area exposed to the environment and isolating the contaminants from environmental influences by encapsulating the waste particles.

Processes involved in stabilization and solidification include the following (US EPA, 1997d):

- **Precipitation.** Soluble forms of hazardous constituents are made less soluble by forming precipitates of hydroxides, sulfides, carbonates and phosphates within the stabilized mass depending on the binder used. Precipitation is generally utilized for metallic constituents. The solubility of the metal precipitates depends on the pH and the redox potential of the environment.

- **Microencapsulation**. The crystalline structure of the solidified matrix at a microscopic level is used to trap the hazardous waste constituents. In this particular situation, release of hazardous constituents occurs only when the matrix degrades to very fine particles. Proper mixing of the binders and the waste constituents plays an important role in the microencapsulation process. Improper mixing may result in some proportion of the waste to become macroencapsulated rather than microencapsulate.

- **Macroencapsulation**. Hazardous waste constituents are trapped within a larger structural matrix such as discontinuous pores within the stabilizing matrix. Exposure of the waste constituents to the surrounding environment is generally the result of degradation of the solidified matrix from compressive stresses, expansion and contraction due to temperature of presence/absence of moisture.

- **Absorption**. Waste constituents are transferred into the matrix of the binders in the same way as a sponge takes up water. For example, free liquids in the waste are absorbed by the binders to minimize the movement of the dissolved waste constituents. Binders such as fly ash, cement kiln dust, lime kiln dust, clay materials, sawdust and hay and straw have certain level of absorbing capabilities.

- **Adsorption**. Waste constituents are chemically bound or "fixed" or adsorbed to the surface of the binders within the matrix. Depending on the nature of the bonding, the waste constituents may be tightly bound to the matrix and are less likely to be released into the environment.

4.4.2 Binders

Binders for inorganic wastes are typically cement, pozzolans, lime, silicates, phosphates, proprietary chemicals such as organoclays and organically modified lime. Binders used for organic wastes include thermosetting organics and thermoplastics. For a total of 59 Superfund projects, 94% of the projects used inorganic binders (cement, flyash, lime, silicates, sulfur) while only 3% of the projects used organic binders (asphalt, organoclays and activated carbon) (US EPA, 2000). Three percent of the projects used a combination of inorganic and organic binders.

4.4.2.1 Cement-Based S/S

The most common S/S approach is to mix the waste materials with Portland cement and water. Cement hydrates to form a crystalline structure consisting of alumino-silicates which microencapsulates the waste constituents while at the same time, precipitates of metal hydroxides are formed which are much less soluble than ionic species of the metals. In some cases, small amounts of fly ash, sodium silicate, bentonite, or proprietary additives are added to enhance processing of the waste. Cement based stabilization has been applied to plating wastes containing various metals such as cadmium, chromium, copper, lead, nickel and zinc. The presence of organic contaminants may interfere with the hydration process, reducing final strength and reducing the crystalline structure resulting in an amorphous material. Additives such as organically modified clays may be used to adsorb the organic contaminants. Some of the advantages of cement-based S/S are that cement is easily handled, can be pumped and can be used for a large variety of hazardous wastes. The operating cost of this technology is relatively low as cement is widely available and used in the construction industry. The alkalinity of cement is capable of neutralizing acidic wastes. However, cement is sensitive to the presence of certain contaminants which may affect setting and hardening of the material.

4.4.2.2 Pozzolan-Based S/S

Pozzolan is a material that produces a cementitious material when mixed with water and chemicals such as lime (US EPA, 1997d). The cementitious materials are usually siliceous and aluminosilicate minerals. Pozzolanic materials include fly ash, ground blast furnace slag and cement kiln dust. For example, a typical composition of fly ash is 45% SiO_2, 25% Al_2O_3, 15% Fe_2O_3, 10% CaO, 1% MgO, 1% K_2O, and 1% Na_2O. The mixed pozzolan material can vary from a soft fine-grained material to a hard cohesive material similar in appearance to cement. Compared to cement reactions, pozzolanic reactions tend to be slower and, therefore, require a longer time to stabilize. Wastes stabilized/solidified with pozzolan materials can include plating sludges containing various metals such as lead, zinc, waste acids, and oil sludges (US EPA, 1997d). Unburned carbon in fly ash may adsorb organics from the waste.

4.4.3 Process Design

There is no established design protocol for S/S process. Bench-scale feasibility studies are typically conducted to assess the initial suitability of a binder or a combination of binders to achieve the desired physical strength and leaching requirements. In many cases, binders selected from bench-scale tests are used in pilot-scale tests to ensure that the binders and the waste can be thoroughly mixed before full-scale implementation.

A S/S system is fairly simple and Figure 4.9 provides a flow diagram of the essential components. The components include a waste transportation and handling system, storage for chemical reagents/binders, a mixer, typically a pug mill, and S/S waste storage area to allow the waste to stabilized.

Gel times for cement based process can vary from 1 to 48 hours depending on the solids content while cement/fly ash, lime/fly ash and kiln dust systems may take more than 48 hours. Use of silicates in cement systems can reduce the gel time to less than 1 hour (Conner, 1990). Examples of S/S include a full-scale fixation of lead and chrome in a metal finishing waste (F006, D002) using lime and sulfide. With lead content as high as 970 mg/kg in the waste, the lead in the leachate of the Toxicity Characteristics Leaching Procedure (TCLP) test was less than 0.1 mg/L (Shively and Crawford, 1986). Chromium in the leachate was less than 0.05 for a waste chromium concentration as high as 11,200 mg/L. Another example is the use of Portland cement to stabilize cadmium hydroxide sludge in a bench-scale test. Using a mix ratio of one part waste to 0.45 parts of cement, a TCLP concentration of 0.03 mg/L was achieved for a waste cadmium concentration of 15,340 mg/L (Butler et al., 1988).

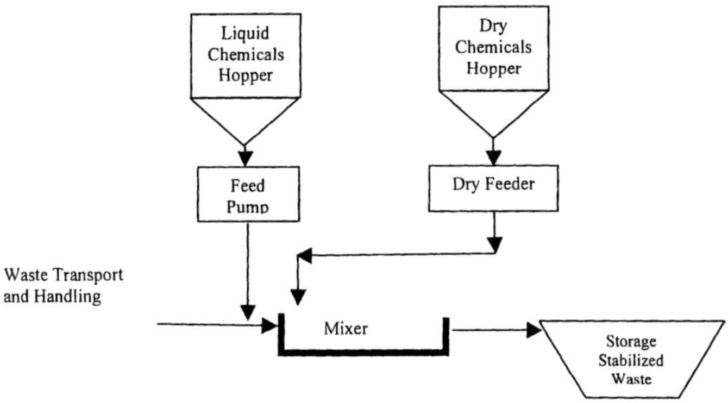

Figure 4.9 Schematic diagram of a typical stabilization/solidification system.

4.4.4 Testing

For regulatory purposes, all S/S wastes are tested for leachability of the hazardous constituents using the TCLP test. Other protocols to test the S/S waste are related to the final use of the stabilized materials. Examples of these tests include:

- moisture content - US EPA Method 9005-SW846 (Paint filter test) to determine the presence of free liquids;
- strength testing - usually ASTM D2166-85 or ASTM D1633-84 is used to test cohesive soil-like or cement-like material behavior under mechanical stress; and
- freeze-thaw durability (ASTM D4842) - to assess the resistance of the material to natural weathering stresses such as freezing and thawing.

4.5 Adsorption and Ion Exchange

Adsorption is described as the preferential accumulation of the pollutants at the surface of a solid phase or adsorbent. Adsorption is one of the more widely applied technologies for the treatment of hazardous wastes and contaminated groundwater, and can be used to remove a wide range of pollutants including synthetic organic chemicals such as pesticides and petroleum hydrocarbons, inorganic compounds such as heavy metals, and anions such as perchlorate. Common adsorbents used for environmental applications include activated carbon and synthetic ion exchange resins. Other natural and synthetic adsorbents include activated alumina, forage sponge and sorption clays.

4.5.1 Process Fundamentals

Adsorption is the accumulation of the pollutants (or solute) at the interface of two phases. Examples of adsorption include surface complexation reactions (surface hydrolysis, formation of coordinate bonds at the surface with metals and ligands), electrostatic interactions at the surface, and hydrophobic expulsion of hydrophilic compounds resulting in an accumulation at the interface. The pollutant (or solute) is usually called the adsorbate while the surface is called the adsorbent.

Absorption is the intraphase distribution of a solute where a solute is dissolved or absorbed into the absorbing phase. Figure 4.10 illustrates the adsorption and absorption processes. Examples of absorption are the dissolution of oxygen from air into water or gasoline mixed with water where some gasoline is dissolved in water and some water is dissolved in gasoline. The gasoline may be considered as dissolved in the absorbing phase (water), i.e., there is a distribution between the two phases. This phenomenon is sometimes referred to as partitioning. The extent of the solute dissolved in the absorbing phase is related to the solubility of the solute in the absorbing phase. In other words, the character of the solute and its interaction with the absorbing phase or solution play an important role in its distribution.

When both absorption and adsorption are occur simulataneously and cannot be distinguished, the term sorption is used. Example of sorption includes distribution of nonpolar compounds onto soil humic material. The structure of soil humic material consists of many different aromatic and aliphatic organic compounds forming a mesh-like structure. The nonpolar compounds may adsorb on the surface of one of the many functional groups or may reside or be absorbed/partitioned within the mesh-like structure.

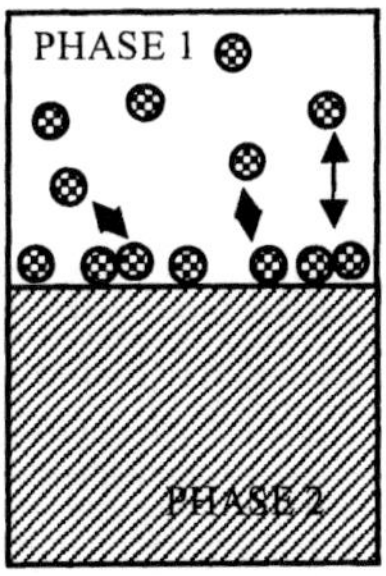

Adsorption – accumulation of solutes at the interface of two phases

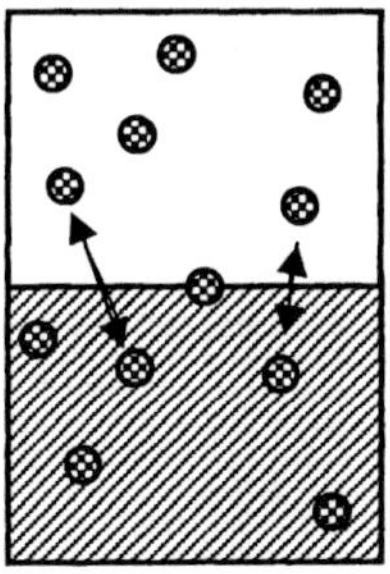

Absorption – distribution/penetration of solutes from one phase into another

Figure 4.10 Illustration of adsorption and absorption phenomena.

Ion exchange is a form of adsorption whereby an ion in the solid phase is replaced by another ion in a solution in contact with the solid. A more restrictive definition is the replacement of an adsorbed, readily exchangeable ion by another. Since replacement takes place at the interface – this process may be classified as adsorption.

Adsorption mechanisms can be categorized as chemical, electrostatic and physical adsorption. Chemical adsorption involves solute-sorbent interaction with the formation of a covalent bond by merging of electron clouds or formation of hydrogen bonds with bonding forces extending over short distances. It is characterized by large heats of interaction, in the range of 100 – 400 kJ/mole. Desorption of the adsorbed compound may result in a different compound. Chemical adsorption is sometimes termed chemisorption.

Electrostatic adsorption involves high energy Coulombic forces as opposed to formation of chemical bonds. These forces result in ion-ion interactions, as in ion exchange, or in dipole-ion interactions. Examples of electrostatic sorption include sorption of Ca^{2+} on clay surface, interactions of NH_3 through permanent dipoles or water with temporary dipoles with surface of soils. The heat of sorption for electrostatic adsorption may be as high as 200 kJ/mole.

Physical adsorption is the result of dipole forces or the action of van der Waal forces comprised of London dispersive forces between the adsorbate and the adsorbent. London dispersive forces are the result of interactions amongst rapidly fluctuating temporary dipoles and quadrapole movements associated with the motion of electrons within orbitals. Dipole-dipole interactions are the more important of all interactions, and the forces vary inversely with the sixth power of the distance between molecules. Typical values for heat of physical adsorption range from 5 to 10 kJ/mole.

Adsorption is often described in terms of isotherms. Isotherms show the equilibrium relationship between the bulk equilibrium aqueous phase activity (concentration) of the adsorbate and the amount adsorbed on the interface at constant temperature. Isotherms are based on empirical models such as the linear model, and the Freundlich model or models developed from first principles such as the Langmuir model, the Brunauer, Emmett and Teller (BET) model, and the Gibbs model.

4.5.2 Adsorbents

Activated carbon is most commonly used as an adsorbent for the removal of a wide range of toxic organic compounds from drinking water, industrial wastewater, contaminated groundwater and hazardous waste. Contaminants that can be removed by sorption to activated carbon include petroleum hydrocarbons, volatile organic compounds (VOCs), halogenated VOCs, metals, pesticides and explosives. Most adsorption systems are used in a polishing step following treatment such as solids removal and filtration. Carbon adsorption is effective in treating water with low contaminant concentrations (< 10 mg/L). Sorbents used to remediate groundwater contaminated with metals include activated alumina, which removes a variety of contaminants, including fluoride, arsenic, and selenium, and forage sponge, which is an open-celled cellulose sponge incorporating an amine-containing chelating polymer that selectively absorbs dissolved heavy metals.

Activated carbon comes in two forms: granular and powdered. Common sizes of granular activated carbon (GAC) are US standard mesh of 12 x 40 (1.68 to 0.42 mm) and 8 x 30 (2.38 to 0.59 mm) with a uniformity coefficient of 1.9. The surface areas of GAC range from several hundreds to more than 1,500 m^2/g. Powdered activated carbon (PAC) is commercially available with 65 to 90 % passing through 325 mesh (44 μm) sieve and with surface areas similar to that of GAC. GAC and PAC are made from wood, peat, lignite, bituminous coal, and coconut shells. The surface of GAC consists of functional groups such as OH-, and COO- that can adsorb a variety of compounds (Crittenden et al., 2005). For example, metal ions (M) bind to the hydroxyl functional groups as shown in equations 4.14 to 4.16 where S is the solid surface (Letterman, 1999):

$$S\text{-}OH + M^{z+} \leftrightarrow S\text{-}OM^{(z-1)}+ + H^+ \quad \text{(Eq. 4.14)}$$

$$2S\text{-}OH + M^{z+} \leftrightarrow (S\text{-}O)_2M^{(z-2)+} + 2H^+ \quad \text{(Eq. 4.15)}$$

$$\text{S-OH} + M^{z+} + H_2O \leftrightarrow \text{S-OMOH}^{(z-2)+} + 2\,H^+ \qquad \text{(Eq. 4.16)}$$

The earliest exchangers used in ion exchange systems were natural materials - natural zeolite and aluminosilicates. Today, synthetic ion exchange materials consisting of cross-linked polymer matrices containing charged functional groups attached by covalent bonding find wide use. The base material is polystyrene, which is cross-linked for structural stability with 3 to 8% divinylbenzene. Ion exchange materials can have different functional groups. For example, strong acid ion exchange resins have sulfonate (SO_3^-) groups while strong basic ion exchange resins have quaternary amine ($N(CH_3)_3^+$) groups. Both strong acid cation and strong base anion exchange resins ionize throughout the pH range and are regenerated by an acid and base, respectively.

4.5.3 Process Design

The reactor configuration typically used for carbon adsorption systems is the fixed bed system. Removal of suspended solids from waste streams is an important design consideration to prevent accumulation of solids in the GAC column and an increase in pressure drop. Fixed-bed GAC systems are backwashed regularly to remove accumulated solids. Backwashing results in downtime and may cause loss of carbon. In many treatment systems, two or more GAC columns are provided to allow for continuous treatment of contaminated water. Carbon adsorption systems can be deployed rapidly, and removal efficiencies of contaminants are high. However, the cost for operating the system may be high if there is a need to regenerate or dispose of the spent carbon. The spent GAC may be classified as hazardous waste and disposed of accordingly.

Various approaches are utilized in sizing GAC reactors. These methods include the use of pilot- and laboratory-scale column tests such as the rapid small-scale column test (RSSCT) (Crittenden et al., 2005), the bed depth service time (BDST) method (Droste, 1997) and the kinetic approach (Reynolds and Richards, 1996). Some parameters considered in a typical design include:

- Surface loading rate (SLR) = flow rate /column plan area
- Empty Bed Contact Time (EBCT) = depth of carbon bed/SLR
- GAC depth – selected based on required breakthrough
- Type of GAC – GAC made from different materials have different sorption capacities and characteristics

- Carbon usage rate (CUR) is the mass of carbon required per unit volume of water treated.

The minimum contact time needed can be estimated using the above mentioned design parametes and the carbon needed to provide the necessary removal for a given volume of water treated can be determined. Typical values for some design parameters used in GAC column design are summarized in Table 4.8.

Table 4.8 Typical GAC design parameters.

Properties	Typical	Range
SLR ($m^3/m^2/hr$)	8	1.9 – 30
EBCT (mins)	10	3 – 35 as high as 4 hr for high concentrations
GAC depth (m)	1	0.2 – 8

In the case of ion exchange, the ions to be removed determine whether cation or anion exchange resins or both are to be utilized. As in GAC systems, pretreatment of wastewater is essential for effective performance of the ion exchange system. Factors that affect the efficiency of ion exchangers include oil and grease that may clog the exchange resin, suspended solids greater than 10 mg/L may cause resin blinding, the pH of the water to be treated may affect selectivity of the ion exchange resin, and characteristics of the rinse-water generated during the regeneration step which may require additional treatment and disposal.

The volume of resin needed to treat a wastewater is determined from pilot studies or estimated based on data provided by resin suppliers (see Dow, 2002). The typical resin bed depth for co-current regeneration is 1.2 m and 2 m for counter-current packed bed systems. The vessel diameter is designed for a maximum pressure drop of 1 bar (15 lbf/ft^2) across the resin bed to allow for resin settling and accumulation of suspended solids during the service run. The maximum diameter for efficient flow distribution is 3.5 m. Surface loading rates typically range from 5 to 60 m/hr. Regenerant used for cation resins are typically strong acids such as sulfuric acid, nitric acid and hydrochloric acid while the regenerant for anion resins is sodium hydroxide.

Case Study

An application of GAC for removal of organic compounds is the pump-and treat-system at the Rocky Mountain Arsenal known as the North Boundary Containment System. The system captures contaminated groundwater flowing northward across the arsenal using a soil-bentonite slurry wall. The pollutants in the contaminated ground water are di-isopropylmethylphosphonate (DIMP), organochloropesticides, chloroform and N-nitrosodimethylamine (NDMA). Groundwater is extracted and pumped into an influent sump and then treated using a GAC column. The treated water is pumped into a system of 15 groundwater recharge trenches. The GAC groundwater treatment system is designed to treat an average flow of 480 gpm and a maximum flow of 720 gpm. Removal of DIMP by GAC was found to be effective with effluent DIMP typically less than 1 ppb, below the state treatment requirement of 8 ppb. However, NDMA concentrations in influent and effluent waters were 0.35 and 0.20 μg/L, respectively, suggesting that the existing GAC units were not effective in removing NDMA from the groundwater (Gunnison et al., 2000).

4.6 Electrochemical Processes

Pollutants can be destroyed using electrochemical process in a manner similar to chemical oxidation and reduction except that electricity is applied across electrodes to create the oxidation/reduction potential instead of using an external oxidizing agent. Reactions are generally conducted in an electrochemical cell and may be enhanced with addition of oxidizing chemicals. Better removal of specific pollutants can be achieved by using a separation membrane. The use of electrodes of special materials allows for selectivity in pollutants removed and may, at the same time, prevent the production of unwanted by-products. Advantages of electrochemical processes include (Juttner et al., 2000):

- versatility in treatment as it can be applied for small to large volumes and for a variety of pollutants in gases, liquids and solids;
- pollutants are oxidized and reduced directly or indirectly;
- in comparison to thermal processes, electrochemical processes are generally operated at lower temperature and pressure making them relatively energy efficient.
- process allows for precise control of the reactions with a certain level of selectivity; and

- the use of electricity to drive the reaction prevents the need for reactive chemicals, such as oxidizing or reducing agents thereby eliminating handling and disposal of hazardous materials.

A disadvantage of electrochemical processes is the in-situ solute mass transfer limitation due to the size of electrode area since reactions occur at the interface of the electrodes and are dependent on the conducting medium. Another disadvantage is the stability of the electrodes in contact with the waste and the loss of effectiveness of the electrochemical process due to electrode fouling.

4.6.1 Process Fundamentals

Electrochemical processes use charged electrodes to carry out chemical reactions for the destruction of waste constituents. By controlling the electrode potential and the environment at the surface of the electrodes, the conditions necessary to initiate and control chemical reactions are created. Electrochemical processes can be divided into the following: direct electrolysis (reactions at the cathode or the anode) and indirect electrolysis. Figure 4.11 illustrates the two processes.

Waste constituents capable of undergoing direct electrochemical oxidation or reduction at an electrode can, in principle, be transformed and/or removed aqueous solutions by the application of appropriate potentials in electrochemical reactors (see Figure 4.9). Oxidation or reduction processes occur directly on inert electrodes without the involvement of other substances (e.g., electron mediators, oxidizing agents). Direct electrolysis includes both cathodic processes and anodic processes. For example, many metal ions can be removed by cathodic deposition when a metal ion is reduced by accepting electrons at the cathode resulting in its deposition. The pollutants and electrode materials involved in the cathodic and anodic electrode reactions of an electrochemical cell may interact in an undesired manner resulting in side reactions that may decrease the effectiveness of the system. To prevent side reactions and to provide a certain level of selectivity, special separators like ion selective membranes which are thin polymeric materials may be placed within the electrochemical cell to selectively and directionally allow ions to migrate under the influence of the electric field. Direct processes are often limited by diffusion and thus low current densities need to be used if high efficiencies are desired. In addition, the high overpotential applied often necessitates noble metals or similarly expensive compounds to be used as electrodes.

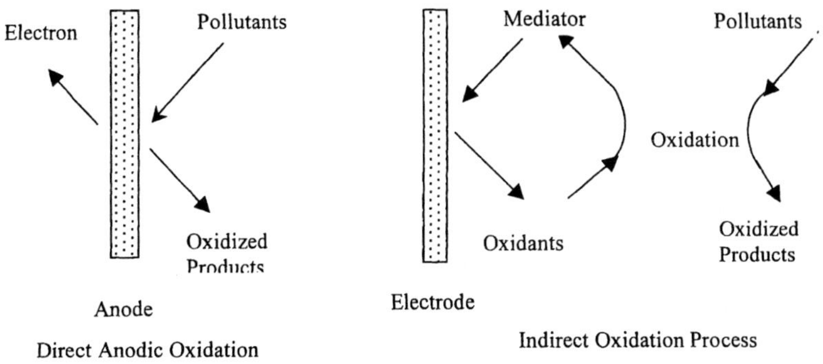

Figure 4.11 Direct and indirect electrolysis in electrochemical reactions.

As an alternative, cathodic or anodic redox mediated processes can be used to overcome some limitations of direct electrolysis by electrochemically generating redox reagents as a chemical reactant (or catalyst) to convert pollutants to less harmful products. The redox reagent acts as an intermediary for shuttling electrons between the pollutant substrate and the electrode. For example, it is possible to generate chlorine from the chloride in solution at the anod,e which in turn, can be used as an oxidizing agent to oxidize the pollutants.

Factors affecting the selection of an electrochemical reactor include (i) the electrode geometry and motion, (ii) energy usage, (iii) cell geometry or division and the electrolyte gap (iv) forms of reactants and products, and (v) batch or continuous operation (Walsh, 2001). Limitations for electrodes include (i) changes in activity and surface area due to catalysis, blockage, and potential-distribution, (ii) adsorption/desorption of reactant, products and contaminants, (iii) film formation/removal via passivation or polymerization, and (iv) phase transformation such as intercalation and dehydration. If membranes are included they can play a critical role in electrochemical reactors. The advantage with membranes include prevention of reactant or product loss at the counter-electrode, controlled migration of ions, use of chemically dissimilar electrolytes and protection of the counter-electrode from corrosion. However, since most membranes have a finite life time, performance may be time-dependent and may complicate cell design and construction, and increase power costs.

To improve the efficiency of electrochemical cells, different types of cell constructions have been developed in recent years. Electrochemical cell designs are generally focused on optimizing the space time yield of the system by improving mass transport of the desired ion; accommodation of large electrode areas, and a combination of improved mass transfer coefficients and enlarging specific electrode area. Examples of cells that improve solute mass transfer are the Pump cell, the Chemelec cell, and the Reno Cell. The multiple cathode cell and the Swiss-role cell were developed to accommodate large electrode areas in a small cell volume (Juttner et al., 2000).

4.6.2 Process Application

An example of the application electrochemical process for site remediation is the electrokinetics technology – a process that separates and extracts heavy metals, radionuclides, and organic contaminants from saturated or unsaturated soils, sludges, and sediments (US EPA, 1996). The process entails applying a low intensity direct current across electrode pairs implanted in the ground on each side of the contaminated soils. Figure 4.12 provides a schematic representation of a typical electrokinetic treatment system. The electrical current causes electroosmosis and ion migration, which move the aqueous phase contaminants in the subsurface towards an electrode. Depending on their electrical charge, contaminants in the aqueous phase or those desorbed from the soil surface are transported towards respective electrodes. The contaminants are deposited onto the electrode or extracted and processed using a recovery system. Under certain situations, surfactants and complexing agents may be added to increase the solubility and assist the movement of the contaminant, or introduced at the electrodes to enhance contaminant removal rates.

An improvement of the electrokinetics process is the Electrochemical Remediation Technologies (ECRTs) process which uses a proprietary AC/DC electrical signal to mineralize organic compounds and to mobilize and remove metal contaminants (Japp, 2000). The proprietary AC/DC converters produce a low-voltage, low-amperage electrical field that polarizes the soil or sediment such that the soil particles charge and discharge electricity causing redox reactions to occur at the interfaces within the soil–groundwater–contaminant–electrode system mineralizing organics and increasing mobilization of metals. Metals migrate to the electrodes where they are deposited and removed with the electrodes. The differences between ECRTs and traditional electrokinetics are that relatively low energy input is required to perform remediation with ECRTs and the ECRTs generally are effective within months, instead of years. Furthermore, metals generally migrate to and deposit at both electrodes in the ECRTs process, unlike classical electrokinetics techniques, in which metals migrate in the direction of only one electrode. ECRTs reaction rates are inversely proportional to grain size and ECRTs can remediate clays and silts faster than sands and gravels. ECRTs have been applied in commercial operations.

Electrochemical processes are generally useful for treating liquid wastes. Electrochemical processes can be used to remove certain metal ions from solution by controlled migration of ions through ion-exchange membranes. The metal may be precipitated as the hydroxide due to increasing pH near the cathode or in other processes, anodically generated iron (III) or aluminum (III) ions can be used to precipitate or flocculate metal hydroxides (Walsh, 2001).

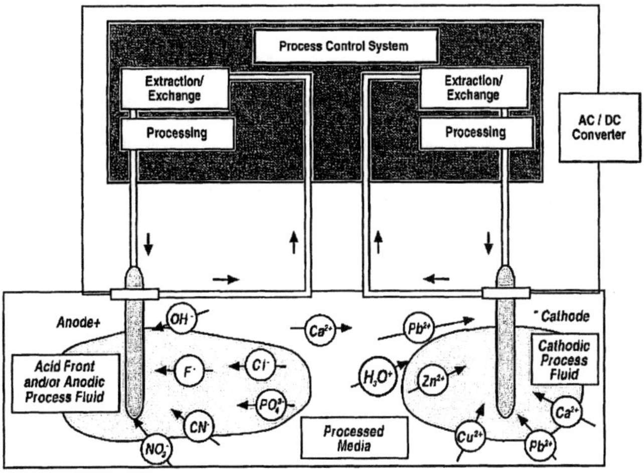

Figure 4.12 Schematic diagram of a typical in situ electrokinetic treatment system (US EPA, 1996).

While treating wastes containing polychlorinated biphenyls (PCBs), the CerOx process, a mediated electrochemical oxidation (MEO) or catalyzed electrochemical oxidation (CEO), was found to successfully destroy PCBs at a concentration of 2 mg/L in alcohol in a patented electrochemical cell (called the T-Cell). In this process, a cerium metal ion, oxidized to Ce^{4+}, is placed in contact with an organic compound, which reduces it to Ce^{3+}. The process operates at low temperature (90-95 ºC) and near atmospheric pressure. In the T-Cell, the anolyte (cerium reagent) and catholyte (nitric acid) are physically separated at all times by a fluoropolymer membrane. Ce^{4+} ions are produced in the cell and then mixed with the waste stream prior to passing into a liquid phase reactor. Excess reagent is maintained in the reactor by monitoring Ce^{3+}/Ce^{4+} ratios in the effluent and adjusting the organic waste input accordingly. The liquid effluent from the reactor is returned to the cell for regeneration of the cerium reagent. The cerium is recirculated and reoxidized to Ce^{4+} and the operation is repeated. The process can be used to treat wastes containing DDT, silvex and chlordane and pharmaceuticals wastewater and difficult chemical waste streams (Anonymous, 2000).

Another process that is similar to the CerOX process is the AEA Silver II process which utilizes Ag^{2-} ions in solution for the oxidation of organics. The oxidizing agent reacts with the organics to produce carbon dioxide, neutral salts and dilute acid solution. The process operates at low temperature (60-90°C) and atmospheric pressure. This technology has demonstrated high destruction efficiencies in trials with a range of chemicals, including some pesticides.

4.7 Chemical Leaching and Solvent Extraction

Chemical leaching and solvent extraction are ex-situ chemical processes for separating contaminants from excavated soils, sludges or sediments. Chemical leaching typically utilizes inorganic liquids such as acids for separating and recovering metals or salts from soils and sludges while solvent extraction makes use of non-aqueous solvents to separate organic contaminants from soils and sludges. Leaching or extraction may be combined in a soil washing process to reduce the volume of contaminated soils for disposal.

4.7.1 Process Fundamentals

Chemical leaching enhances desorption of metals from soils or sludges through dissolution of complexes or alteration of ionic states by adjusting the pH of the matrix. Acids are commonly used to leach metals and inorganic pollutants from soils. However, alkaline, carbonates and chelating agents are also used to selectively leach metals (US EPA, 1997b). Knowledge of the valence states, solubility and likely reaction products of all metals present is essential for proper application of a leaching agent. For example, cadmium, trivalent chromium and lead typically exist in the cationic state and are not mobile, however, they are more mobile under acidic conditions. Arsenic and hexavalent chromium typically exist in anionic forms in the environment. Mercury may transition between several valence states (US EPA, 1997c). Solvent extraction enhances dissolution, solubilization and desorption through preferential partitioning of the contaminants into the extraction solvent.

Given sufficient time for steady state conditions to occur, the theoretical contaminant removal efficiency can be calculated based on the distribution coefficient (K_d) of the contaminant between soil and chemical leaching solution (Sharma and Reddy, 2004):

The mass balance for a chemical leaching process can be expressed as:

$$C_{si}M_s = C_{sf}M_s + V_l C_l \quad \text{(Eq. 4.17)}$$

$$C_{sf} = K_d C_l \qquad \text{(Eq. 4.18)}$$

where, C_{si} is the initial contaminant concentration in soil (mg/kg), M_s is the dry mass of soil in kg, C_{sf} is the final contaminant concentration in soil (mg/kg), V_l is the volume of leaching solution in L, and C_l is the contaminant concentration in the leaching solution (mg/L).

$$\text{Removal efficiency} = \frac{C_{sf}}{C_{si}} \times 100\% = \frac{1}{1 + V_l / M_s K_d} \times 100\% \qquad \text{(Eq. 4.19)}$$

4.7.2 Remediation Uses

Chemical leaching is appropriate for separating heavy-metal contaminants from soils, sediments, and sludges from metal plating and finishing operations with the possibility of reclaiming the leached metals. Recoverable metals include chromium, copper, cadmium, cobalt, gold, lead, molybdenum, mercury, nickel, palladium, platinum, silver, tin, titanium, vanadium and zinc (NFESC, 2004a). The easiest separation to achieve is for soils with a single metal in a single valence state. Complex mixtures may be difficult to leach and/or complicate to treat producing a recoverable mixture of metals that needs further treatment before use or disposal (NFESC, 2004a). Meeting treatment goals for metals in the soils so that the soils can be returned to the excavation site is achievable but may be cost-prohibitive (US DOE 2004a). The particle size distributions of the soils or sludges have an impact on the successful application of chemical leaching. For example, soils with high clay content may require longer leaching times and the fine clay particles remaining in the leachate may make separation and treatment of the leachate difficult.

Solvent extraction is appropriate for organic contaminants such as VOCs, PCBs, halogenated solvents, petrochemicals, and explosive chemicals. However, safety issues arise while recycling solvent containing explosives if the solvent is recovered through distillation (US EPA, 1993a). Solvent extraction has been used for contamination from paint, rubber and pesticide/insecticide wastes, coal tar and petroleum refinery oily wastes but is least effective for very high-molecular weight organics and very hydrophilic compounds (US DOE, 2004a; NFESC, 2004d). This technology has been applied at Superfund sites containing multiple forms of contamination and at nuclear facilities (Hyman and Hladek, 2002). Solvent extraction is appropriate for fine- and coarse-grained soils, sediments and sludges (Braids, 2002). An important cost-saving advantage of solvent extraction over soil washing is the ability to recycle the solvent for multiple use cycles (US EPA, 1993b). The presence of inorganics typically does not interfere with solvent extraction of organics, and inorganics may be altered to a more easily extractable state by the solvent extraction process (NFESC, 2004d). Solvent extraction has been used prior to bioremediation to help increase the biodegradation of some polycyclic aromatic hydrocarbons (PAHs) (Lee et al., 2001). Treatment goals in the application of chemical leaching or solvent extraction may include meeting a contaminant concentration in the soil and achieving a significant volume reduction of the waste. Treatment requirements for RCRA listed metals wastes and contaminated media are in the Code of Federal Regulations, Part 40, Sections 268 and 269 (40 CFR 268 and 269), respectively. Sampling of leached soils or sludges should be done to ensure that treatment goals are met. Typically, either TCLP or total extractable metals method is used to assess whether treatment goals are met. Permits may also be required for the disposal of leachates and solvent-contaminated wastewaters at a municipal waste water treatment plant.

4.7.3 Process Design and Considerations

Chemical leaching and solvent extraction are considered short-term or medium-term treatment applications depending on the throughput of the soils treated (NFESC, 2004d). Examples of throughputs of commercial chemical leaching units treating excavated soil are in the range of 10-100 cubic yards per day (NFESC, 2004a) and for a mobile solvent extraction unit may range between 20-200 cubic yards per day (NFESC, 2004c). System design should include safe application of leaching solution or extraction solvents and for the recycling of fluids as much as possible to minimize costs. In addition, volatile emissions from leaching solutions during treatment and recycling must be contained.

The chemical leaching process produces treated leachate, recovered metals or salts, and treated soils or sludges which must be disposed of safely, while the solvent extraction process produces concentrated contaminants, separated solvent/water and treated sediment (US EPA, 1993b). Equipment needed include particle separation tank, chemical leaching or solvent extraction reactor, separation unit for leachate/solvent from soils or sludges, regeneration and recycling unit for leachate, treatment and disposal systems for leachate, and neutralization system for soils and sludges (US EPA, 2004).

In the process design of a chemical leaching or solvent extraction system, it is important to characterize the soils or sludges for particle size distribution, cation exchange capacity, organic content, moisture content, buffering capacities, pH, and the presence of organics such as VOCs (US DOE, 2004c). For extraction of metals, the valencies of metals to be removed must be determined. Site requirements including utilities and treatment area for staging of equipment, excavation and stockpiling of soils should be determined. Selection of leaching fluid or extraction solvent is important and is usually accomplished through bench-scale treatability studies. Acids are most often used to leach metals from soils and sludges. However, acids can alter soil silicates and will likely leach many species of metals at the same time. An alkaline extraction can be effective in leaching some metals, such as uranium, while leaving most other metals unleached from soils (Francis et al., 1999). The applicability of several leaching solutions for metals is shown in Table 4.9. Treatability studies are needed to test appropriateness of the leaching solution for the contaminated soils/sludges, and to determine the volume of leaching solution and contact time needed to attain cleanup goals. Testing should ensure that soils are not undesirably altered and that secondary contaminants are not formed (Francis et al., 2004). Pilot testing is essential to ensure that remediation meets the regulatory goals (Braids, 2002; US DOE, 2004c).

Table 4.9 Chemical leaching agents.

Leaching Agent	Used For	Reference
water	hexavalent chrome	US EPA, 1997b
hydrochloric acid, ethylenediamine tetraacetic acid (EDTA) (chelating agent)	lead, uranium, zinc, copper, cadmium, iron, chromium, arsenic, mercury	US EPA, 1997b; Peters, 1999, Kim et al., 2003
ethylenediaminedisuccinic acid (EDDS – a biodegradable chelating agent)	lead, zinc, copper	Tandy et al., 2004
polyethylenimine (PEI) (water soluble polymer)	lead	Sauer et al., 2004

Solvent extraction uses hydrocarbon solvents such as liquid propane, butane or triethylamine to remove contaminated halogenated and non-halogenated hydrocarbons (Hyman and Hladek, 2002; Sharma and Reddy, 2004). Extraction solvents should have high solubility with the contaminant but avoid extracting humic materials or dissolving the soils (NFESC, 2004d). Pressurized gases such as liquefied propane or butane or supercritical carbon dioxide fluid may also be used to separate polar organics (Braids, 2002). Typical components in chemical leaching systems are illustrated in Figure 4.13. Solvent extraction systems are similar to chemical leaching systems but with fewer reaction units as shown in Figure 4.14.

Oversized debris and large grit from excavated soils are removed using vibrating screens. The pretreatment can be designed to separate particulate metals before leaching or extracting by gravity separation to reduce contact time and reduce volume of leachate or solvent (NFESC, 2004a). Screened soils are mixed with the leaching solutions or solvents in chemical leaching units which are agitated to enhance desorption. The optimum soil/solvent or soil/leaching mixture for extraction is typically less than 20% solids and with particles less than ¼ inch in diameter (US EPA, 1993b). For solvent extraction, water in the soils may interfere with extraction. Therefore, the soil is often dewatered prior to extraction or kept dry during the pretreatment process (Hyman and Hladek, 2002).

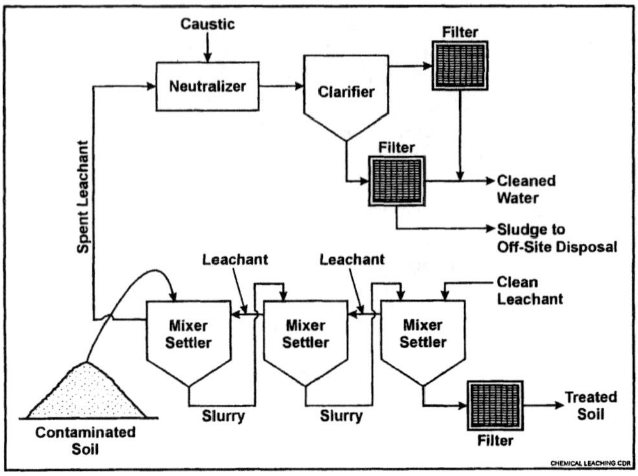

Figure 4.13 Schematic diagram of chemical leaching system (NFESC, 2004c).

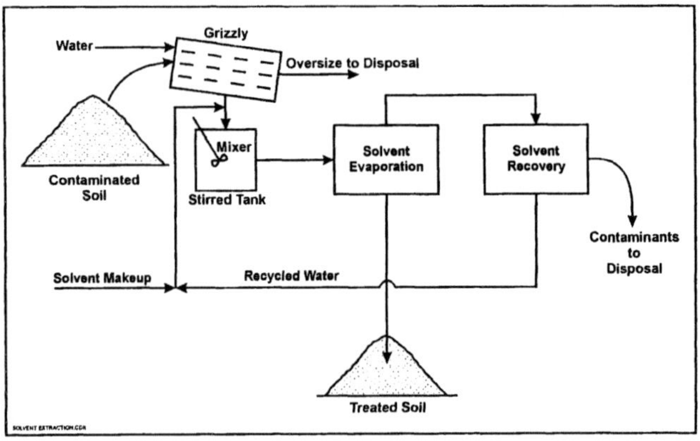

Figure 4.14 Schematic diagram of solvent extraction system (NFESC, 2004d).

For chemical leaching systems (See Figure 4.13), there can be more than one mixing units which also act as settling units. In the mixers, the leachate containing pollutants is skimmed off while the slurry is passed on to another mixer where more leaching solutions are added. Countercurrent flow of clean leachant to the contaminated soils assists in minimizing the amount of leachant used. The treated soils or sludges are then dewatered and disposed of accordingly. In the case of solvent extraction, an evaporator may be used to separate the solvent/contaminants from the treated soil (Figure 4.14).

The leachate generated is generally treated, regenerated and recycled for continued use on site. The leaching process may produce a salts, which can also be recovered, concentrated, and purified. Typical salts include copper chloride, copper ammonium chloride, and nickel carbonate. Products from catalyst leaching include nickel-copper-cobalt concentrate, alumina trihydrate, chromium oxide, molybdenum trioxide, and vanadium pentoxide (NFESC, 2004a). In some cases, metals are removed from the leachate through a process of precipitation, sedimentation and concentration. An example of an on site treatment process includes the addition of a precipitant such as lime to precipitate the metals (US EPA, 1997a; US DOE, 2004a). The final leaching solution can be recycled by supplementing with fresh leaching solution, or disposed at a wastewater treatment plant (NFESC, 2004a). In the case of solvent extraction, solvents are recovered and separated from the pollutants using distillation units (NFESC, 2004d) or the solvent extracts are disposed by incineration. Soils are often dewatered, and neutralized with lime and fertilizer if acid extraction was used, before returning them to the excavation (US DOE, 2004a). Examples of cases where chemical leaching and solvent extraction were used to treat contaminated soils or sludges are presented in Table 4.10.

Estimated costs for chemical leaching treatment are between $135 and $450 per cubic yard of soil and for solvent extraction in the range of $90 to $600 per cubic yard of soil treated. These costs include debris screening, unit mobilization and leasing, treatment pad installation, excavation, utilities, chemicals, unit operation, health and safety administration, sampling and analysis, on site disposal of leached soils, onsite leachate regeneration and treatment, offsite disposal of treated leachate to public wastewater treatment plant. Cost estimates do not include project design and management, regulatory fees, site characterization testing, treatability testing, bonding or contingencies (NFESC, 2004a, NFESC, 2004d; US EPA, 1993b).

4.8 Soil Flushing

In soil flushing, the soil is saturated with a flushing solution through irrigation or injection wells in order to solubilize and mobilize the contaminants. The contaminants and flushing solution are eventually extracted using down gradient wells or trenches and the extracted solution is treated above ground. The flushing solution can be recycled or treated and reinjected or disposed. Soil flushing is an in-situ adaptation of the pump-and-treat method for treating groundwater using a flushing solution. Soil flushing is often used to treat saturated soils and groundwater but has also been used to treat soils in the unsaturated and vadose zones. Soil flushing is also known as in-situ flushing, injection/recirculation or in-situ soil washing.

4.8.1 Process Fundamentals

Transport of contaminants during the flushing process is primarily by advection within the flushing solution. The rate of fluid flow is governed by the hydrogeology of the area, including porosity and hydraulic conductivity of contaminated soils, preferential flow channels, and the viscosity of the flushing solution (Sharma and Reddy, 2004). The flushing solution may react with contaminants or other soil components and create crystalline obstructions that reduce soil porosity, and therefore reduce the hydraulic conductivity of soils. Contaminant mobility also depends on the desorption of the contaminant from soil, solubility of the contaminant in water and the flushing solution, interfacial tension, the densities and viscosities of the contaminant plume(s) and the flushing solution. These factors are influenced by properties of soil and groundwater, including soil type, organic carbon content, pH, and cation exchange capacity (CH2M HILL, 1997; US EPA, 1997b). Cation exchange capacity is dependent on the amount of organic material and type of minerals present and as pH or CEC decreases, metals may become more mobile in soil.

Flushing solution additives include surfactants that can increase the mobility of the organic contaminants. Surfactants form hydrophilic micelles that enhance the mobility of hydrophobic contaminants in groundwater (Li et al., 2003). The surfactant(s) of choice have minimal losses to sorption or precipitation in soils, low toxicity and are readily biodegradable. Surfactants yield lower than expected recoveries if they are adsorbed onto the soil matrix, precipitated by cations in groundwater, or form macromolecules (liquid crystals) with the contaminant thereby reducing the effective soil porosity.

Table 4.10 Applications of chemical leaching, solvent extraction and soil flushing.

Location and Site Characteristics	Technology	Contaminant	Cost	Results	Ref
RMI Titanium plant, Ashtabula, OH. High clay content silt loams and clay loams; low organic material	Chemical extraction with bicarbonate solution, dewatering and ion exchange to remove uranium from liquid (64 tons of soils treated in 38 batches)	Uranium Most uranium present as U^{+6} Uranium levels in feed soil were 74 - 146 pCi/g with a hot spot with 587 pCi/g	Pilot plant cost \$638,670 including mobilization, treatment and preparatory work. Full-scale estimated at \$250-350/ton	Average 87-91% removal efficiency with residual 12-14 pCi/g of uranium 95% volume reduction; 5% off-site disposal	US DOE, 2000
Superfund site treatability study	Solvent extraction with hexane	Metals (As, Pb, barium), VOCs, SVOCs (PCB)	No estimate for full-scale costs	98% removal of organics.	Tillman, 1995
Hill Air Force Base, Salt Lake City UT. Sand and gravel 0-60 ft, silty clay 60 - 260 ft.	Soil flushing using: Cosolvent solubilization (70% ethanol, 0-10% pentanol, 20% water) Cosolvent mobilization (81-96% tert butanol, 0-15% hexanol, 4% water) Surfactant solubilization (5% Dowfax™, 95% water) Surfactant mobilization (0.4% AOT™ 100, 3% Tween™ 80, 96.6% water) Microemulsion (0-2.5% pentanol, 2.5% Brij™, 95-97.5% water) Macromolecular (10% hydroxy propyl-cyclodextrin)	LNAPLs including VOCs, BNAEs, pesticides, PCBs, dioxins, and furans. VOCs including 1,2-DCE, chlorobenzene, 1,4-dichloro benzene, 1,2-dichlorobenzene and 1,2,4-trichlorobenzene	No costs available	Percent removal of tested NAPLs: Cosolvent solubilization 94% Cosolvent mobilization 84-91% Surfactant solubilization 42% Surfactant mobilization 92% Microemulsion 93% Macromolecular 63%	US EPA, 1999

Location and Site Characteristics	Technology	Contaminant	Cost	Results	Ref
Dover Air Force Base, Dover, DE. 15 x 10 ft test cell with watertight walls Unconsolidated coastal plain sediments, average hydraulic conductivity of 1 x 10^{-3} cm/sec.	Soil flushing with 70% *n*-propanol and 30% saltwater. 3.2 L/min flooded over surface	80 mg/L averaged PCE concentration in groundwater	No estimate for full-scale	80% reduction estimated in 37 days	Wood and Falta, 2003
King of Prussia waste processing facility Superfund site, Winslow Township, NJ. 10-acres. Sandy. 22 months	Soil washing including screening, separation, froth flotation, sludge management. 25 tons per h.	Metals, primarily beryllium, chromium, copper, nickel, zinc.	$7.7 million including off-site disposal costs.	Reduced concentrations of 11 metals to treatment goals.	Sharma and Reddy, 2004
Borden, Canadian military base. Sand perched on clay aquatard with K > 10^{-3} cm/s. Contaminant in vadose and saturated zones. 3 x 3 x 3 m demonstration site. Five injection wells and five extraction wells.	Soil flushing. Water flush followed by surfactant: 1% nonylphenol polyethoyxylate (Witconol NP-100) and 1% phosphate ester of nonylphenol polyethoxylate (Rexophos 25-97). Air stripping removed PCE from flushing solution. 15 month testing duration.	Tetrachloro-ethylene (PCE) added to soil two months prior to flushing.	No estimate for full-scale	Decreased residual saturation in vadose zone from 10 to 4%. Reduced pool saturation in saturated zone from 20 to 3%. Reduced contaminant pool height from 50 cm to 2-3 cm.	Sharma and Reddy, 2004
Christi, TX Chlorocarbon plant. Clay (10 - 12 ft), clayey sand (10-30% clay K >10^{-3} cm/s)(12 - 14 ft), sand with 1-5% clay (16-24 ft). 25 x 35 x 12 ft. Groundwater was highly saline. Six injection wells, one central extraction well, several monitoring wells.	Soil flushing. Surfactant: 1% Witco 2722, Witco Corp. Air stripping removed CTET from solution, which was re-injected. Witco 2722 was heavily sorbed to tanks and wells, causing biofouling. Surfactant was replaced by Tergitol 15-S012, Union Carbide Corp.	Carbon tetrachloride (CTET)	No estimate of project cost	No reduction of initial CTET concentration > 2,000 mg/L (10 - 14 ft), reduced CTET from initial 574 -2674 mg/L to <10 mg/L (16 - 24 ft).	Sharma and Reddy, 2004

Surfactants can also be combined with co-solvents and water to form microemulsions, which increase the mobility and extractability of non-aqueous phase liquid (NAPL) contaminants (US EPA, 1999). Co-solvents such as isopropyl, tert-butyl alcohol and propanol (highly water-soluble alcohols) are miscible in water and organic contaminants. Less water-soluble alcohols such as butanol, pentanol, hexanol and heptanol can be added to improve miscibility with the contaminants. Organic acids and amines may also be used (US EPA, 1999). Co-solvents are added to flushing solutions with surfactants because they reduce the amount of surfactant needed by decreasing sorption of surfactants onto soils. Co-solvents increase the solubility of the contaminants in water, increase the mobility of the contaminants and/or increase desorption from soils (US EPA, 1999). For example, the removal of 1,1-bis(*p*-chlorophenyl)-2,2,2-trichloroethane (*p,p*-DDT) was improved from about 11% to 77% by adding ethanol and 1-propanol as co-solvents to contaminated soils (Smith et al., 2004). Co-solvents also prevent formation of surfactant/contaminant macromolecules (US EPA, 1999) and decrease the viscosity of surfactants allowing easier injection into the soil (Li et al., 2003).

The intentional formation of macromolecules can also improve contaminant mobility in porous soils. A macromolecule is formed from an injected seed compound that is highly mobile in water and allows for hydrophobic bonding of contaminants. For example, cyclodextrin may be used to enhance desorption of PAH-contaminated soils, and it is non-toxic and may also improve in-situ bioremediation (US EPA, 1999).

4.8.2 Applications to Remediation

Soil flushing can be applied to organic or water-soluble inorganic contaminants including volatile and semi-volatile organic compounds (VOCs and SVOCs), fuels, pesticides, radioactive compounds and some metals. Soil flushing is often used to enhance in-situ bioremediation (Testa and Jacobs, 2002). Successful flushing may require that organic contaminants have moderate to high solubility in water. Solubility may be improved by adding a co-solvent or a variety of surfactant to the flushing solution.

The most important considerations for applicability of soil flushing are: the type and extent of contaminants and soil characteristics (US EPA, 1997b). Soil flushing is more effective in soils with high hydraulic conductivity (Braids, 2002; Testa and Jacobs, 2002). The optimal hydraulic conductivity is greater than 1.0×10^{-3} cm/s, while less than 1.0×10^{-5} cm/s is considered poor (US EPA, 1997b). Low hydraulic conductivity results in slow flushing rates and likely accompanies strong contaminant sorption to clay or organic materials. Soils with organic carbon contents less than 1% are ideal, while those with more than 10% carbon are not good candidates since humic materials may interfere with flushing (Sharma and Reddy, 2004).

Advantages of soil flushing over pump-and-treat methods include decreased cost and reduced exposure of contaminants to workers. However, an important disadvantage is the possibility of spreading the contaminant plume beyond the recovery zone. To avoid plume migration, the flushing design must consider the effects of surfactants and co-solvents on flushing solution and plume density and viscosity. The design must control groundwater flow with restrictive barriers if necessary and avoid migration through preferential pathways in the aquifer (Commission on Geosciences, Environment and Resources, 1999). Likewise, contaminated areas may not receive adequate flushing if they are hidden behind clay and sand lenses (Sharma and Reddy, 2004). Using a non-toxic, biodegradable flushing solution is a possible response to plume migration issues.

Because of plume migration issues and the presence of potentially hazardous flushing solution residuals in soils, permit approval may be difficult (Testa and Jacobs, 2002). It is critical to understand the applicable policies since regulations may affect extraction and/or re-injection of the solution including the need for suitable containment and recovery of the contaminants and flushing solution (Li et al., 2003, US DOE, 2004b). Treatment standards for wastes are covered under the Resource Conservation and Recovery Act (RCRA) in the Code of Federal Regulations (CFR) - 40 CFR 268 for metals and 40 CFR 269 for contaminated media. Permits are also needed for discharging extracted groundwater, whether treated or untreated, into municipal sewers.

4.8.3 Process Design and Considerations

A very important factor in the successful implementation of a soil flushing technology is a thorough characterization of the site geochemistry and hydrogeology, soil porosity and permeability, particle size distribution, carbon content, moisture content, texture, cation exchange, buffering capacity and pH (US DOE, 2004b; US EPA, 1997b). A mapping of the site hydrogeology can be performed through soil cores and geological maps along with slug tests. The extent of contamination, contaminant concentration, solubility, partition coefficients in flushing solution and soils, oxidation-reduction potential (ORP) and complex stability constants should be properly estimated (US DOE, 2004b).

Based on the contaminant type and the geochemistry of the site, a flushing solution with the proper surfactants or co-solvents can be selected and tested. In some cases, geochemical models can be used to select a flushing solution for soil and contaminant conditions (Sharma and Reddy, 2004). Types of flushing agents suitable for different groups of contaminants are listed in Table 4.11. The advantages of different forms of soil flushing and application issues for surfactants and co-solvents in a soil flushing system are listed in Table 4.12. Inorganic contaminants may be flushed with water, organic or inorganic acids, and complexing or chelating agents. Inorganic flushing agents are listed in Table 4.9. The soil/solution ratio is also important, with more solution possibly resulting in more contaminant removal (Smith et al., 2004).

Table 4.11 Types of flushing agents (Sharma and Reddy, 2004).

Flushing Agent	Contaminants Targeted
Clean water	High-solubility organics; soluble inorganic salts
Surfactants	Low-solubility organics; petroleum products
Water/surfactants	Medium-solubility organics
Cosolvents	Hydrophobic contaminants
Acids	Basic organic contaminants, metals
Bases	Phenols, metals
Reductants/oxidants	Metals

Organic contaminants such as non-aqueous phase liquids (NAPLs) are often flushed using surfactants or co-solvents. The use of microemulsions allows extraction of NAPLs in a single, low-viscosity phase (US EPA, 1999). Some microemulsion surfactants are non-toxic and readily biodegradable (US EPA, 1999). Humic acids or complexing agents such as sodium citrate and EDTA may be an alternative to surfactants for PAH contamination (Lesage et al., 2001; Subramaniam et al., 2004). High-pressure steam injection can also be used to increase solubilization and volatilization of contaminants (Sharma and Reddy, 2004). Flushing fluids should be recycled as much as possible. Recovery of contaminants from flushing solution can be achieved by air stripping, liquid/liquid extraction, precipitation, filtration, or distillation (Strbak, 2000).

Batch studies and soil column testing must be done to select the flushing solution(s) and estimate the concentrations of the reagent(s). Pilot testing (field-scale demonstration) is also essential to accentuate potential heterogeneity of the site that may affect adequate cleaning to regulatory levels (Braids, 2002).

A typical soil flushing setup is shown in Figure 4.15. The equipment and processes included in a soil flushing train include flushing wells for saturated zone, irrigation for vadose zone, mixing equipment for solutions, piping and wastewater treatment to recycle flushing solution and treat flushing solution before disposal. The method of injection is gravity infiltration or pressure injection with pumps and wells for vadose or saturated zones, respectively. The number of injection wells or infiltration zones and recovery wells or trenches needed must be estimated. The well array design must consider the number and location of wells, and pumping rates necessary to intercept the contaminant flow (Li et al., 2003). Wells can be vertically or horizontally bored. Trenches are excavated and backfilled with porous material. Pumps may also be placed in excavated trenches where groundwater is shallow. Piping and pumps must be compatible with the solution used (Li et al., 2003). The treatment unit area should include flushing solution preparation, recycling and treatment system piping and containment, with considerations for chemical resistant materials. Extraction is critical to the recovery of the flushing solution. Multiple Phase Extraction (MPE) can be used to simultaneously extract groundwater, free product (NAPL) and VOCs. If needed, sampling wells downgradient of extraction point and in deeper aquifers can be installed.

Table 4.12 Cosolvent and surfactant applications (US EPA, 1999).

Technology	Remedial Fluid	Advantages	Design Issues
Cosolvent solubilization	Low molecular weight alcohols	Fast; easy to manage and operate; can achieve remedial objectives in reasonable time frames using food grade biodegradable materials	Potential unstable flow conditions; potential density override of remedial fluid; waste handling
Cosolvent mobilization	High molecular weight alcohols	Fast; easy to manage and operate; can achieve remedial objectives in a reasonable time frame	Potential loss of hydraulic control at DNAPL sites; unstable flow conditions; density override potential; waste handling
Surfactant solubilization	Hydrophilic surfactant	Faster than pump-and-treat; products used as surfactants can be food grade additives	Potential creation of liquid crystals at DNAPL sites; waste handling
Surfactant mobilization	Hydrophobic surfactants	Fast; can achieve remedial objectives in a reasonable time frame	Potential for the formation of macro molecules or liquid crystals; potential loss of hydraulic control at DNAPL sites; potential separation of remedial components; waste handling
Surfactant micro-emulsion	Surfactant and alcohol cosolvent	Fast; can achieve remedial objectives within a reasonable time frame	Waste handling
Macro molecule	Cyclodextrin	Reasonably fast, remedial fluid biodegradable	Biodegradable remedial fluid; some may be lost during reprocessing; preferentially removes smaller compounds; waste handling

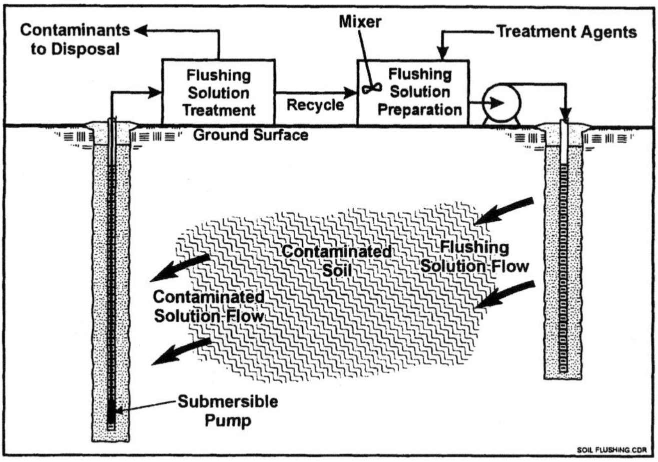

Figure 4.15 Typical soil flushing system (NFESC, 2004b).

Since there is very little control on the flow of the flushing solution after it enters the subsurface, extraction wells may have to be strategically placed around the contaminated plume to avoid accidental spreading of the contaminant plume. Sheet piling or slurry walls can be used to contain or direct the flow to ensure fluid recovery and avoid spreading of the plume. Soil flushing may result in sludges or solid wastes from the treatment and recycling of the flushing solution. Water recovered may be discharged to municipal wastewater treatment plants or reintroduced into the aquifer upgradient of the plume (Li et al., 2003).

The costs of soil flushing are estimated to be between \$75 and \$300 per cubic yard of contamination plume. This estimate includes injection and recovery well and pump installation, operating and maintenance labor, sampling well installation, utilities, flushing solution preparation system installation, chemicals, flushing solution treatment, system installation, site supervision, site quality assurance and health and safety support, sampling and analysis for process control, and off-site disposal of sludge residual from flushing solution treatment. The estimate does not include project design and management, regulatory fees, site characterization, treatability, site pretreatment and contingencies (NFESC, 2004b). Recycling of flushing fluids is a major cost. The largest variable cost factor is the amount and type of surfactant used (US DOE, 2004b).

4.9 Soil Washing

Soil washing is an ex-situ process that separates contaminated fine material (silt, clay and organic) from relatively contaminant-free coarse material (sand and gravel) in excavated soils. The volume reduction in contaminated soil significantly decreases the cost of treating the soils. Contaminants present in the fines are then removed or separated by leaching or solvent extraction, solidified or stabilized or simply disposed of in a secured landfill.

4.9.1 Process Fundamentals

The purpose of separating the coarse and fine materials is to isolate the soil fines which retain the bulk of the hydrophobic and inorganic contaminants. Separation of the fines typically results in clean coarse materials that can be safely disposed. The contaminants on the fines are then removed chemically by using acids, surfactants or solvents (by desorption, dissolution, solubilization and/or oxidation-reduction) and to a limited extent by physical agitation (attrition) during the wash (Sharma and Reddy, 2004). The pH of washing solutions may be adjusted (for metals) to improve solubilization and separation of contaminants from soils. In the case of PAHs and PCBs, which sorb strongly to soils, surfactants and co-solvents are often added to remove the contaminants. Surfactants form hydrophilic micelles that increase the solubility and, therefore, the mobility, of hydrophobic contaminants in the wash water. Co-solvents are miscible in water and the contaminant, and increase the solubility of the organic contaminant.

4.9.2 Remediation Uses

Soil washing is primarily used for VOCs and SVOCs, fuels and metals, radionuclides, inorganic chemicals, pesticides, PAHs, PCBs. Removal efficiencies for VOCs and SVOCs are generally 90 - 99% and 40 - 90%, respectively (US EPA, 1993b). Soil washing is often used as a pretreatment step in conjunction with other techniques, such as biological treatment, incineration, and solidification or stabilization (Dennis, 1999, Testa and Jacobs, 2002).

In general, high hydraulic conductivity soils (sand or gravel) with low preference for contaminant adsorption produce the highest removals. If soils contain 30-50% silt, clay or organic material the contaminants are more difficult to remove (Sharma and Reddy, 2004). The optimum particle size that can be washed is in the range of 0.24 to 2 mm (US DOE, 2004c). Advantages of soil washing include reduction of volume of contaminated material, improved consistency of material sent for further treatment, and good control of process parameters. Disadvantages include sensitivity to fines-content of soils (Dennis, 1999). Contaminants may be difficult to desorb from clays or soils with high humic-content. Such soils may require pretreatment (US DOE, 2004c). Complex mixtures such as metals and organics may be difficult to separate.

Treatment goals may include meeting a contaminant concentration and achieving a significant volume reduction of the waste. Treatment requirements for RCRA listed metal wastes and contaminated media are described in the Code of Federal Regulations, Part 40, Sections 268 and 269, respectively.

4.9.3 Process Design and Considerations

Soil washing units are often mobile and the systems may be proprietary. Typical unit processes for a soil washing system are illustrated in Figure 4.16. Pretreatment of the treated soil usually consists of vibrating screens or a trommel to separate oversize debris and large grit from excavated soils with vibrating screens. The soil is then mixed with an aqueous solution. The washing solution is water-based and may include pH adjustment and a chelating agent to leach metals, or a surfactant to remove organic contaminants (US DOE, 2004c). Various washing agents for soil washing applications are listed in Tables 4.9, 4.11 and 4.12. Scrubbing agents may be proprietary to the company providing services. Note that it may be difficult to find a washing solution suitable for soils with multiple forms of contamination, and multiple series of washing may be needed. Varying the chemical composition or soil/water ratios will change the performance of the unit. Chemicals in the washing solution may influence the available disposal routes and cost of disposal for the used solution (Dennis, 1999).

The slurry is agitated in a mixing tank to separate the fines and coarse particles and to dislodge fines attached to the coarse particles (US DOE, 2004a). The soils are then transferred to a size separation reactor such as a hydrocyclone to separate the coarse and fine materials. Contaminants and fines can also be removed from the coarse fraction by air flotation, high-pressure water jets or vibration (U SEPA, 1997a). The coarse materials are disposed of while the fines are further treated in a reactor using leaching chemicals, acid/bases and/or solvents. The treated fines are separated and the supernatant is treated and recycled back into the soil washing system. Most fines are removed from the wash solution through flocculation, or gravity sedimentation. Treated soils are often dewatered, and neutralized with lime and fertilizer if acid extraction was used, before returning them to the excavation (USDOE, 2004a). Units containing several separation processes, such as particle size separation, gravity separation or attrition scrubbing are more effective at treating soils with multiple types of contamination (Dennis, 1999). The washing solution can usually be used for multiple cycles of washing. The washing solution and rinse waters are regenerated with a precipitant such as lime used with a flocculent or chelating agent such as EDTA, citric acid or ammonium acetate to recover heavy metals (US EPA, 1997a; USDOE, 2004a).

The throughput of a mobile soil washing unit may range between 20 and 200 cubic yards per day (NFESC, 2004c, Sharma and Reddy, 2004). Typical costs for soil washing are between \$90 and \$400 per cubic yards of soil treated with reported costs closer to \$120-\$200 per ton including excavation (Dennis, 1999; Sharma and Reddy, 2004; US DOE 2004c). These costs include debris removal, unit and crew mobilization, treatment pad installation, excavation, solvent, operation and maintenance, utilities, sampling, disposal of sludge from onsite recycling of washing solution, off-site disposal of washing solution into municipal sewers, and returning soils to excavation. Costs do not include project design and management, regulatory fees, site characterization testing, treatability testing, bonding or contingencies (NFESC, 2004c; US EPA, 1993b) and are dependent on site conditions, waste quantity and concentration and treatment goals. The cost to dispose of the surfactant will likely be relatively high (Braids, 2002). Soil washing is probably only cost-effective on larger sites with at least 5,000 tons of contaminated soil or more (Braids, 2002; Sharma and Reddy, 2004).

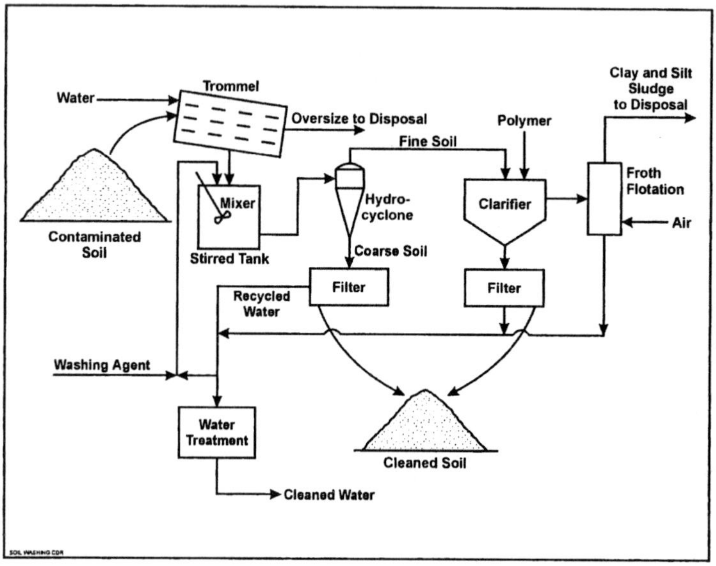

Figure 4.16 Typical soil washing system (NFESC, 2004c).

4.10 Conclusion

Chemical remediation technologies are an important group of technologies for the treatment of hazardous wastes. Depending on the technology and contaminant type, chemical technologies typically can treat a hazardous waste over a shorter period of time than bioremediation technologies. In addition, chemical technologies can completely destroy organic hazardous wastes to harmless and innocuous compounds or change the state of inorganic contaminants such as metals to less toxic state and/or bind them in a chemical matrix making them unavailable for exposure. However, chemical technologies are generally more expensive than physical and biological technologies requiring high initial capital costs for the reactors and subsequent operating costs of chemicals, labor and energy. Selection of chemical treatment technologies is dependent on the quantity of waste, concentrations of the hazardous constituents in the waste, the waste matrix, the physical-chemical properties of the hazardous constituents (recalcitrant nature, sorption affinity, etc), the final concentrations requirements, time constraints for remediation, capital and operating costs and on site constraints such as land availability, presence of obstructive structures, labor availability and labor skill levels. A thorough understanding of the physical principles and the chemical reactions within each treatment process is essential in the selection of the right treatment process for the treatment of hazardous wastes. By combining a series of treatment processes, most hazardous wastes can be treated to levels where it can be safely disposed or reuse on site.

References

Aieta, E.M., K. M. Reagan, J.S. Lang, L. McReynolds, J. Kang, and W. H. Glaze, (1988) Advanced oxidation processes for treating groundwater contaminated with TCE and PCE: Pilot-scale evaluations. *J AWWA*. 80(5):64 - 72.

Anonymous, (2000) Electrochemical process oxidizes PCBs. Chemical Engineering Progress. 96(12):17

Braids, Olin C., (2002) Soil. *in* Handbook of Complex Environmental Remediation Problems. J. Lehr, M. Hyman, T. E. Gass, W. J. Seevers (eds.). McGraw-Hill, New York, pp 2.1-2.18.

Bremner, D., (1990) Historical Introduction to Sonochemistry, *in* Advances in Sonochemistry. T.J. Mason (ed.). Vol 1. JAI Press Ltd., Greenwich, CT. p 1.

Butler, L.G., Cartledge, F.K., Chalasani, D., Eaton, H.C., Frey, F., Tittlebaum, M.E., Yang, S.L., (1988) Immobilization Mechanisms in Solidification/Stabilization Using Cement/silicate Fixing Agents, Louisiana State Univ., Baton Rouge, LA.

Cater, S.R., Bircher, K.G., and Stevens, R.D.S., (1990) Ray ox? A second generation enhanced oxidation process for groundwater remediation. *Proceedings of a Symposium on Advanced Oxidation Processes for the Treatment of Contaminated Water and Air*, June, Toronto, Canada.

Conner, J.R., (1990) Chemical Fixation and Solidification of Hazardous Wastes, Van Nostrand Reinhold, New York.

Copa, W.M., and Gitchel, W.B., (1988) Standard Handbook of Hazardous Waste Treatment and Disposal, H.M. Freeman (Ed.), McGraw-Hill, New York.

CH2M Hill, (1997) Technology Practices Manual for Surfactants and Cosolvents, US Department of Defense, Washington, DC.

Crittenden, J.C., Trussell, R. R., Hand, D.W., Howe, K.J., Tchobanoglous, G., (2005) Water Treatment Principles and Design, 2nd Edition, John Wiley and Sons, Inc., Hoboken, NJ.

Dennis, Ronald M. (1999) Preface. *Journal of Hazardous Materials*, 66 (1-2) 9-10.

Dow, (2002) Dowex Marathon C, Ion Exchange Resins: Engineering Information, Dow Chemical Company, Midland, MI.

Droste, R.L., (1997) Theory and Practice of Water and Wastewater Treatment, John Wiley & Sons, Inc., New York.

Commission on Geosciences, Environment and Resources. (1999) Groundwater and Soil Cleanup: Improving Management of Persistent Contaminants. National Academies Press: Washington DC.

Dussert, B.W., (1997) Advanced Oxidation. *Industrial Wastewater*. November/December, pp.29 - 34.

Farhataziz, P.C., and Ross, A.B., (1977) Selective specific rates of reactions of transients in water and aqueous solutions. P-III. Hydroxyl radical and perhydroxyl radical and their radical ions, Nat'l Standard Reference Data Service 59, US National Bureau and Standards, Washington, DC.

Francis, C.W., Timpson M. E., and Wilson J. H. (1999) Bench- and pilot-scale studies relating to the removal of uranium from uranium-contaminated soils using carbonate and citrate lixiviants. *Journal of Hazardous Materials*. 66 (1-2) 67-87.

Glaze, W.H., (1987) Drinking water treatment with ozone, *Environ. Sci. & Technol.* 21(3):224-230.

Gloyna, E. F. and Li, L. (1995) Progress in supercritical water oxidation: research and development, Fifth Intl. Chem. Oxidation Symposium and Principles and Practices Workshop, Nashville, TN.

Gunnison, D., Zappi, M.E., Teeter, C., Pennington, J.C., Bajpai, R., (2000) Attenuation mechanisms of *N*-nitrosodimethylamine at an operating intercept and treat groundwater remediation system, *Journal of Hazardous Materials* 73: 179–197.

Haas, C.N., and Vamos, R.J., (1995) Hazardous and Industrial Waste Treatment, Prentice Hall, Englewoods Cliffs, NJ.

Hoigne, J., and Bader, H., (1976) The role of hydroxyl radical reactions in ozonation processes in aqueous solutions, *Water Res.* 10:377.

Hoigne, J., and Bader, H., (1983) Rate constants of reactions of ozone with organic and inorganic compounds in water: 1. Non-dissociating organic compounds, *Water Res.* 17:173 - 183.

Hyman, M. and Hladek, K., (2002) Management of Wastes from Nuclear Facilities *in* Handbook of Complex Environmental Remediation Problems, J. Lehr, M. Hyman, T. E. Gass, W. J. Seevers (eds.) McGraw-Hill, New York, pp 6.1-6.39.

Hua, I., and Hoffmann, M. R. (1996) Kinetics and mechanism of the sonolytic degradation of CCl_4 : Intermediates and byproducts. *Environ. Sci. Technol.,* 30:864.

Japp, J.M., (2002) Electrochemical remediation technology remove mercury in soils, *US EPA Technology News and Trends*, 2 (September):5.

Juttner, K., Galla, U., Schmieder, H., (2000) Electrochemical approaches to environmental problems in the process industry, *Electrochimica Acta,* 45:2575–2594.

Killilea, W., (1998) Plan for supercritical water oxidation for VX/NaOH hydrolysate, United States Army Alternative Technologies and Approaches Program, Aberdeen, MD, USA. http://www.clu-in.org/acwaatap/

Kim, C.S., Lee, Y., Ong, S. K., (2003) Factors affecting EDTA extraction of lead from lead-contaminated soils, *Chemosphere*, 51:845-853.

Koszalka, D. P., Soodma, J. F., Bever, R. A., (1992) Ultrasonic decontamination of groundwater. AIChE Summer National Meeting, Minneapolis, MN, August, AICHE, New York.

Kruger, K., Schulze, Th.-L., Peters, D., (1999) Sonochemical treatment of natural ground water at different high frequencies preliminary results. *Ultrason. Sonochem.*, 6:123.

Lee, P.H., Ong, S.K., Golchin, J., and Nelson, G.L. (2001) Use of solvents to enhance PAH biodegradation of coal-tar-contaminated soils, *Water Research,* 35(16):3941-3949.

Lesage, S., Brown, S., Millar, K., and Novakowski, K., (2001) Humic acids enhanced removal of aromatic hydrocarbons from contaminated aquifers: developing a sustainable technology. *J. of Environmental Science and Health, Part A*, 36(8):1515-1533.

Letterman, R.D., (1999) Water Quality and Treatment, McGraw-Hill, Inc., New York.

Li, T., Patel, R.U., Ramsden, D.K., and Greene, J., (2003) Ground Water Recovery and Treatment *in* MTBE Remediation Handbook, E. E. Moyer and P. T. Kostecki (eds.). Amherst Scientific Publishers: Amherst, MA, pp 289-327.

Lin, J., Chang, C., Wu, J., (1996) Decomposition of 2-chlorophenol in aqueous solution by ultrasound/H_2O_2 process, Water Sci. and Tech., 33(6):75-81.

Mason, T., and Lorimer, J., (1988) Sonochemistry: Theory, Applications, and Uses of Ultrasound in Chemistry, Ellis Norwood, NY.

Modell, M., (1989) Supercritical Water Oxidation, Standard Handbook of Hazardous Waste Treatment and Disposal, H. M. Freeman (ed.), McGraw-Hill, NY. pp 8.153 - 8.168.

NCMS, (1994) Pollution Prevention and Control Technologies for Plating Operations, http://www.nmfrc.org/bluebook/tocmain.htm, Ann Arbor, MI.

NEESA, 1993. Precipitation of Metals from Ground Water. NEESA Document Number 20.2-051.6, Naval Energy and Environmental Support Activity, Port Hueneme, CA.

NFESC (Naval Facilities Engineering Service Center) (2004a) "Chemical Leaching". (http://enviro.nfesc.navy.mil/erb/restoration/technologies/remed/phys_chem/phc-05.asp)

NFESC (Naval Facilities Engineering Service Center) (2004b) "Soil Flushing". (http://enviro.nfesc.navy.mil/erb/restoration/technologies/remed/phys_chem/phc-24.asp)

NFESC (Naval Facilities Engineering Service Center) (2004c) "Soil Washing". (http://enviro.nfesc.navy.mil/erb/restoration/technologies/remed/phys_chem/phc-27.asp)

NFESC (Naval Facilities Engineering Service Center) (2004d) "Solvent Extraction". (http://enviro.nfesc.navy.mil/erb/restoration/technologies/remed/phys_chem/phc-31.asp)

Peters, R. W., (1999) Chelant extraction of heavy metals from contaminated soils. *J. of Hazardous Materials,* 66(1-2):151-210.

Raven, P.H. and Berg, L.R. (2006) Environment, 5th Edition, John Wiley and Sons, Hoboken, NJ.

Reynolds, T. D., and Richards, P.A., (1996) Unit Operations and Processes in Environmental Engineering, PWS Publishing Company, Boston, MA.

Rice, R.G., and Netzer, A., (1982) Handbook of Ozone technology and Applications, Vol. 1, Ann Arbor Sci. Publications, Ann Arbor, MI.

Ruppert, G., Bauer, R., and Heisler, G.J., (1993) The Photo-Fenton reaction-an effective photochemical wastewater treatment process. *J. Photochemistry and Photobiology A:Chemistry.* 73:75-78.

Sauer, N. N., Deborah S. E., and Brandy L. D., (2004) Lead extraction from contaminated soil using water-soluble polymers. *J. Environmental Engineering,* 130(5):585-588.

Sharma, H. D. and Reddy K. R., (2004). Geoenvironmental Engineering: Site Remediation, Waste Containment, and Emerging Waste Management Technologies. John Wiley & Sons, Hoboken, NJ, pp 413-418, 524-538.

Shih, T.C., Wangpaichitr, M., Suffet, M., (2003) Evaluation of granular activated carbon technology for the removal of methyl tertiary butyl ether (MTBE) from drinking water. *Water Res.* 37(2):375-85

Shively, W.E., and Crawford, M.A., (1986) EP Toxicity and TCLP extractions of Industrial and Solidified Hazardous Waste,CH2M Hill, Boston, MA.

Smith, E., Smith, J., Naidu, R., and Juhasz, A.L., (2004) Desorption of DDT from a contaminated soil using cosolvent and surfactant washing in batch experiments. *Water, Air, and Soil Pollution,* 151(1-4):71-86.

Staehelin, J. and Hoigne, J., (1985) Decomposition of ozone in water in the presence of organic solutes acting as promoters and initiators of radical chain reactions, *Environ. Sci. and Technol.*, 19(12):1206.

Strbak, L., (2000) In-Situ Flushing with Surfactants and Cosolvents, US EPA Technology Innovation Office.
(http://cluin.org/download/studentpapers/strbak_flushing.pdf)

Subramaniam, K., Stepp, C., and Pignatello, J. J., (2004) Enhancement of polynuclear aromatic hydrocarbon desorption by complexing agents in weathered soil. *Environ. Eng. Sci.,* 21(4):515-523.

Suslick, K.S., (1990) Sonochemistry, Science, 247:1439.

Suslick, K.S., and Hammerton, D.A., (1986) The site of sonochemical reactions, IEEE Trans. *Ultrasonic Ferroelec. Freq. Contr.,* 33:143.

Suslick, K.S., Cline, Jr., R.E., and Hammerton, D.A., (1986) The sonochemical hot spot, *J. Amer. Chem. Soc.,* 108:5641.

Tandy, S., Bossart, K., Mueller, R., Ritschel, J., Hauser, L., Schulin, R., and Nowack, B., (2004) Extraction of heavy metals from soils using biodegradable chelating agents, *Environ. Sci. and Technol.*, 38(3):937-944.

Testa, S. M., and Jacobs, J.A., (2002) Oil Spills and Leaks *in* Handbook of Complex Environmental Remediation Problems, J. Lehr, M. Hyman, T. E. Gass, W. J. Seevers (eds.). McGraw-Hill: New York, pp 9.45-9.46.

Tillman, J., (1995) Removal of PCBs from Contaminated Soil Using the CF Systems Solvent Extraction Process: a Treatability Study: Project Summary, EP

1.89/2:600/SR-95/505, Risk Reduction Engineering Laboratory, U.S. Environmental Protection Agency, Cincinnati, OH,.

US DOE, (2000) Cost and Performance Report: Chemical Extraction for Uranium Contaminated Soil at the RMI Titanium Company Extrusion Plant, Ashtabula, Ohio. (http://costperformance.org/profile.cfm?ID=231&CaseID=231)

US DOE, (2004a) Remediation Technologies Screening Matrix and Reference Guide, version 4.0. Section 4.17 Chemical Extraction. (http://www.frtr.gov/matrix2/section4/4_17.html)

US DOE, (2004b) Remediation Technologies Screening Matrix and Reference Guide, version 4.0. Section 4.8 Soil Flushing. (http://www.frtr.gov/matrix2/top_page.html)

US DOE, (2004c) Remediation Technologies Screening Matrix and Reference Guide, version 4.0. Section 4.21 Soil Washing. (http://www.frtr.gov/matrix2/top_page.html)

US EPA, (1998) Advanced Photochemical Oxidation Processes, EPA 16251R-981004, Office of Research and Development, Environmental Protection Agency, Washington DC.

US EPA, (1993a) Approaches for the Remediation of Federal Facility Sites Contaminated with Explosive of Radioactive Wastes, EP 1.8: R 28/4, Office of Research and Development, US EPA, Cincinnati, OH.

US EPA, (1993b) Selecting Remediation Techniques for Contaminated Sediment, EP 1.8/R28/3, Office of Research and Development , US EPA, Washington, DC.

US EPA, (1997a) Best Management Practices (BMPs) for Soil Treatment Technologies: Suggested Operational Guidelines to Prevent Cross-Media Transfer of Contaminants During Clean-up Activities, EP 1.8M 31/4, Office of Solid Waste, US EPA, Washington, DC.

US EPA, (1996) Recent Development for In Situ Treatment of Metal Contaminated Soil, EPA Office of Solid Waste & Emergency Response. Technology Innovative Office. Washington, DC.

US EPA, (1997b) Recent Developments for In Situ Treatment of Metal Contaminated Soils, EPA 542-R-97-004, US EPA, Washington, DC.

US EPA, (1997c) Technology Alternatives for the Remediation of Soils Contaminated with As, Cd, Cr, Hg, and Pb, EP 1.3/3:SO3, Office of Research and Development , US EPA, Washington, DC.

U.S. EPA (1998) Chemical Hazard Data Availability Study, US EPA Office of Pollution Prevention and Toxics, Washington, DC.

US EPA, (1999) In Situ Enhanced Source Removal, EPA/600/C-99/002, US EPA, Washington, DC.

US EPA, (2000) EPA Superfund Record of Decision Amendment: New Hanover City Airport Burn Pit, Wilmington, NC, EPA/AMD/R04-00/142, US EPA, Region 4, Atlanta, GA.

US EPA, (2004) Remediation and Characterization Innovative Technologies database (EPA REACH IT). (http://www.frtr.gov/matrix2/appd_a/vendor.html#soil_ex_chem) .

US EPA, (2000) Solidification/Stabilization Use at Superfund Sites, EPA-542-R-00-010, Office of Solid Waste and Emergency Response, Cincinnati, OH

US EPA. (1997) Innovative Site Remediation Technology, Solidification/Stabilization. Design & Application, Volume 4. (EPA 542-B-97-007), US EPA, Cincinnati, OH.

US EPA, (1995) In Situ Remediation Technology: Electrokinetics, EPA542-K-94-007, US EPA, Office of Solid Waste and Emergency Response Technology Innovation Office, Washington, DC.

US EPA, (1983) Development Document for Effluent Limitations Guidelines and Standards for the Metal Finishing Point Source Category, EPA 440/1-83-091, US EPA, Washington, DC.

Walsh, F. C., (2001) Electrochemical technology for environmental treatment and clean energy conversion, *Pure Appl. Chem.*, 73(12):1819–1837.

Wood, A. L. and Falta, R., (2003) SERDP and NRMRL Sponsor Field Test of Cosolvent-Enhanced DNAPL Removal" Technology News and Trends, US EPA, Washington, DC. (http://www.cluin.org/products/newsltrs/tnandt/view.cfm?issue=0503.cfm)

Zepp, R.G., Faust, B.C., and Hoigne, J. (1992) The Hydroxyl radical formation in aqueous reactions (pH 3-8) of iron with hydrogen peroxide: The Photo-Fenton Reaction. *Environ. Sci. & Technol.* 26:313-319.

CHAPTER 5

Redox and Precipitation

Irene M. C. Lo and Keith C. K. Lai

5.1 Introduction

Redox reaction and precipitation are of paramount importance in both natural water as well as water and wastewater treatment processes. The behavior of compounds containing carbon, nitrogen, sulfur, iron and manganese etc in natural water and the treatment processes is largely influenced by redox reaction (Snoeyink and Jenkins 1980). Therefore, understanding the redox condition in polluted areas is often a prerequisite for selecting appropriate treatment or remediation approaches (Christensen et al. 2000). For example, knowledge of redox condition of contaminant plume in subsurface allows evaluation of the plume development and risks to downgradient groundwater resources as well as assessment of natural attenuation as a remediation option.

Precipitation and dissolution are the basic principles of many water and wastewater treatment processes. Dissolution of minerals is a prime factor in determining the chemical composition of natural water. On the other hand, precipitation of minerals from supersaturated solution is a key step in altering the natural water chemical composition (Stumm and Morgan 1996). The treatment processes, including lime-soda softening, iron removal, coagulation with hydrolyzed metal salts and phosphate precipitation, are all based upon the precipitation phenomenon (Snoeyink and Jenkins 1980).

In this chapter, the mechanisms involving removals of hexavalent chromium [Cr(VI)], uranium, nitrate (NO_3^-) and nitrite (NO_2^-), chlorinated aliphatic hydrocarbons (CAHs) as well as chlorinated organic micropollutants from aqueous environment via either redox process or the processes of redox and precipitation are the main scope to be emphasized. However, to help readers to readily get familiar with various removal mechanisms, some basic theories and definitions relating to redox process and precipitation will be briefly introduced in advance. Afterwards, some case studies with regard to the redox process or redox and precipitation processes commonly occurred in natural environment will then be focused so as to emphasize their efficacies and potentials for remediating hazardous wastes.

5.2 Redox Reaction

5.2.1 Basic Theory

The term redox is a shortened form of reduction and oxidation. By definition, oxidation is a reaction in which a substance loses or donates electrons, whereas a substance gains or accepts electrons in a reduction reaction. Since redox reaction involves electron transfer from one atom to another, the oxidation state of reactants and products after the reaction are changed. The oxidation state, sometimes referred to as an oxidation number, represents a hypothetical charge that an atom would have if the ion or molecule were to dissociate (Freeze and Cherry 1979). Table 5.1 summarizes the rules used to deduce the oxidation number from the formula of a substance.

As an example of redox reaction, the reaction between dichromate ion ($Cr_2O_7^{2-}$) and ferrous iron (Fe^{2+}) is considered, which is shown in Eq. (5.1).

$$Cr_2O_7^{2-} + 6Fe^{2+} + 14H^+ \Leftrightarrow 2Cr^{3+} + 6Fe^{3+} + 7H_2O_{(l)} \qquad \text{(Eq. 5.1)}$$

$$6Fe^{2+} \Leftrightarrow 6Fe^{3+} + 6e^- \qquad \text{(Eq. 5.2)}$$

$$Cr_2O_7^{2-} + 14H^+ + 6e^- \Leftrightarrow 2Cr^{3+} + 7H_2O_{(l)} \qquad \text{(Eq. 5.3)}$$

Totally six electrons are transferred from Fe^{2+} to reduce Cr(VI) in $Cr_2O_7^{2-}$ to trivalent chromium [Cr(III)]. Since the Fe^{2+} donates electrons for the reduction of $Cr_2O_7^{2-}$, it is called reducing agent or reductant. Conversely, $Cr_2O_7^{2-}$ is said to be an oxidizing agent or oxidant because it accepts the electrons released from the oxidation of Fe^{2+} (Snoeyink and Jenkins 1980). For the sake of balancing chemical equation, redox reaction is usually separated into two half reactions for balancing the number of element and electroneutrality. Eqs. (5.2) and (5.3) show the balanced half reactions for the redox reaction between $Cr_2O_7^{2-}$ and Fe^{2+} [i.e., Eq. (5.1)]. The general rules used for balancing redox reactions are summarized in Table 5.2. Although free electrons are used to balance the electroneutrality of half reactions, electrons can only be exchanged and are not found in free state. As a result of the conservation of the electrons, a reduction reaction and an oxidation reaction must always be coupled (Appelo and Postma 1993).

Table 5.1 Rules for assigning oxidation number to the atoms in ions or molecules.

[a]Rule		[a]Oxidation Number
1.	The sum of the oxidation numbers of all the atoms in the species is equal to its total charge	
2.	For atoms in their elemental form	0
3.	For atoms of Group I For atoms of Group II For atoms of Group III (except boron) For atoms of Group IV (except carbon and silicon)	+1 +2 +3 (EX_3), +1 (EX) +4 (EX_4), +2 (EX_2)
4.	For hydrogen	+1 in combination with nonmetals -1 in combination with metals
5.	For fluorine	-1 in all its compounds
6.	For oxygen	-2 unless combined with fluorine -1 in peroxides (O_2^{2-}) -½ in superoxides (O_2^-) -⅓ in ozonides (O_3^-)
7.	Halogens	-1 in most compounds, unless the other elements include oxygen or more electronegative halogens

[a, b]Examples

Carbon compound		Sulfur compound		Nitrogen compound		Iron compound	
Substance	Oxidation number	Substance	Oxidation number	Substance	Oxidation number	Substance	Oxidation number
HCO_3^-	+4	S	0	N_2	0	Fe	0
CO_3^{2-}	+4	H_2S	-2	SCN^-	-3	FeO	+2
CO_2	+4	HS^-	-2	N_2O	-3	$Fe(OH)_2$	+2
CH_2O	0	FeS_2	-1	NH_4^+	-3	$FeCO_3$	+2
$C_6H_{12}O_6$	0	FeS	-2	NO_2^-	+3	Fe_2O_3	+3
CH_4	-4	SO_3^{2-}	+4	NO_3^-	+5	$Fe(OH)_3$	+3
CH_3OH	-2	SO_4^{2-}	+6	HCN	-3	FeOOH	+3

Note: To determine an oxidation number, work through the rules in the order given. Stop as soon as the oxidation number has been assigned.

[a]Shriver and Atkins (1999).

[b]Freeze and Cherry (1979) as well as Stumm and Morgan (1996).

Table 5.2 Guides for balancing redox reactions.

Step	Description
1.	For each half reaction, write the oxidized and reduced species into the equation and balance the elements at left and right, except hydrogen and oxygen
2.	Balance the number of oxygen atoms by adding H_2O
3.	Balance the number of protons by adding H^+
4.	Balance electroneutrality by adding electrons
5.	Subtract the two half reactions to obtain the complete redox reaction

5.2.2 Redox Equilibria

In terms of Gibbs free energy, the redox reaction between $Cr_2O_7^{2-}$ and Fe^{2+} [i.e., Eq. (5.1)] can be represented by Eq. (5.4). Water does not appear in the equation since by definition, it is defined having unit activity.

$$\Delta G_r = \Delta G_r{}^0 + RT\ln\left(\frac{\{Cr^{3+}\}^2\{Fe^{3+}\}^6}{\{Cr_2O_7{}^{2-}\}\{Fe^{2+}\}^6\{H^+\}^{14}}\right) \quad \text{(Eq. 5.4)}$$

This Gibbs free energy of a reaction (ΔG_r) can be related to the voltage developed by the redox reaction in an electrochemical cell by Eq. (5.5).

$$\Delta G = -nFE \quad \text{(Eq. 5.5)}$$

Substitution of Eq. (5.5) into Eq. (5.4) results in Nernst equation [i.e., Eq. (5.6)] in which E^0 is the sum of the standard potentials for the oxidation half reaction ($E^0{}_{oxd}$) and reduction half reaction ($E^0{}_{red}$) [i.e., Eq. (5.7)].

$$E = E^0 - \frac{RT}{nF}\ln\left(\frac{\{Cr^{3+}\}^2\{Fe^{3+}\}^6}{\{Cr_2O_7{}^{2-}\}\{Fe^{2+}\}^6\{H^+\}^{14}}\right) \quad \text{(Eq. 5.6)}$$

$$E^0 = E^0{}_{ox} + E^0{}_{red} \quad \text{(Eq. 5.7)}$$

The standard potential for various half reactions, like the Gibbs free energy of formation, cannot be determined absolutely. They are conventionally measured with reference to a particular half reaction, which is assigned a standard potential of zero at 25 ^{0}C and 1 atm. This reaction is the reduction of hydrogen ion (H^+) to hydrogen gas (H_2) [i.e., Eq. (5.8)].

$H^+ + e^- \Leftrightarrow \frac{1}{2} H_{2(g)}$ (Eq. 5.8)

The physical setup, which defines the standard potential of a half reaction is illustrated in Fig. 5.1 (Snoeyink and Jenkins 1980). At the left side of the electrochemical cell, a hydrogen gas electrode or standard hydrogen electrode is shown, which consists of a platinum electrode over which hydrogen gas is bubbled in a solution of pH = 0, thereby fulfilling standard state conditions. In the right compartment, an inert platinum electrode is immersed into the solution containing ferrous and ferric irons. The two electrodes are connected to a voltmeter and the electrical circuit is closed by a salt bridge, which also helps maintain the electroneutrality in the compartments.

The Nernst equation for the redox reaction shown in Fig. 5.1 is listed as follows.

$$E = E^0 - \frac{RT}{F}\ln\left(\frac{\{Fe^{2+}\}\{H^+\}}{\{Fe^{3+}\}(P_{H_2})^{1/2}}\right) \quad \text{(Eq. 5.9)}$$

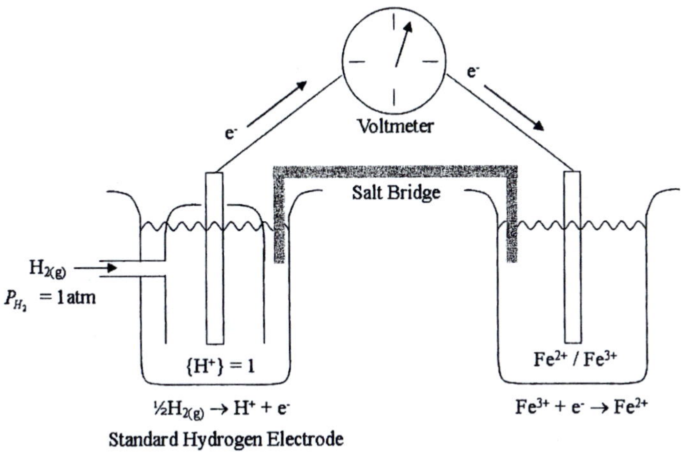

Figure 5.1 Electrochemical cell for the definition of standard potential of a half reaction.

Since both partial pressure of hydrogen gas (P_{H_2}) and $\{H^+\}$ are in unity in the standard hydrogen electrode, they are usually omitted from Eq. (5.9) and indicated instead by adding the postscript *h* to *E* [i.e., Eq. (5.10)].

$$Eh = E^{\circ} - \frac{RT}{F}\ln\left(\frac{\{Fe^{2+}\}}{\{Fe^{3+}\}}\right) \qquad \text{(Eq. 5.10)}$$

Note once more that both redox potential (*Eh*) and E^0 for any half reactions are expressed as the potential with respect to the standard state H^+/H_2 reaction. The value of the standard potential of a half reaction indicates the tendency to release or to accept electrons so that listing of the standard potentials for various half reactions are useful to obtain a first overview over possible reactions. Table 5.3 lists the standard potentials for some of half reactions (Sillen and Martell 1964; Stumm and Morgan 1970; Snoeyink and Jerkins 1980; Dean 1985).

As seen in Table 5.3, the standard potentials for Eq. (5.2) and Eq. (5.3) are -0.77 and +1.33 V, respectively. Therefore, the standard potential for the redox reaction between $Cr_2O_7^{2-}$ and Fe^{2+}, based upon Eq. (5.7), is equal to 0.56 V corresponding to ΔG_r^0 of -324 kJ. Positive value of the standard potential or negative value of ΔG_r^0 indicates that Eq. (5.1) proceeds spontaneously to right if all the activities are in unity.

In a similar fashion as pH, which manifests the distribution of all acid-base equilibria, *Eh* determines the distribution of all redox equilibria in aqueous environment. However, in contrast to pH, *Eh* is difficult to be measured unambiguously in most of natural water. As aforementioned, *Eh* measurement is made with an inert platinum electrode against the standard hydrogen electrode, which unfortunately is impractical to set up in field. Therefore, a reference electrode of known potential rather than the standard hydrogen electrode is usually applied instead and the measured potential ($E_{measured}$) is then accordingly corrected with reference to the standard hydrogen electrode using Eq. (5.11). For instance, calomel reference electrode ($KCl_{(sat)}$, $Hg_2Cl_{2(s)}$:$Hg_{(l)}$), which has a potential ($E_{reference}$) of 0.244 V at 25 ^{0}C, is one of the most common reference electrodes used for *Eh* measurement.

$$Eh = E_{measured} - E_{reference} \qquad \text{(Eq. 5.11)}$$

5.2.3 *Electron Activity and pε Concept*

The concept of electron activity is a theoretical treatment of redox reaction, which considerably simplifies the algebra of redox reaction. In lieu of expressing the half reaction in terms of Gibbs free energy, the law of mass action is applied. Considering the half reaction of the oxidation of Fe^{2+} to ferric iron (Fe^{3+}), it can be expressed as Eq. (5.12).

Table 5.3 Standard potentials for some of half reactions at 25 ^{0}C.

Half reaction	E^0 **(V)**
$H^+ + e^- \Leftrightarrow \frac{1}{2}H_{2(g)}$	0
$Na^+ + e^- \Leftrightarrow Na_{(s)}$	-2.72
$Mg^{2+} + 2e^- \Leftrightarrow Mg_{(s)}$	-2.37
$Cr_2O_7^{2-} + 14H^+ + 6e^- \Leftrightarrow 2Cr^{3+} + 7H_2O_{(l)}$	+1.33
$Cr^{3+} + e^- \Leftrightarrow Cr^{2+}$	-0.41
$MnO_4^- + 2H_2O_{(l)} + 3e^- \Leftrightarrow MnO_{2(s)} + 4OH^-$	+0.59
$MnO_4^- + 8H^+ + 5e^- \Leftrightarrow Mn^{2+} + 4H_2O_{(l)}$	+1.51
$Mn^{4+} + e^- \Leftrightarrow Mn^{3+}$	+1.65
$MnO_{2(s)} + 4H^+ + 2e^- \Leftrightarrow Mn^{2+} + 2H_2O_{(l)}$	+1.23
$Fe^{3+} + e^- \Leftrightarrow Fe^{2+}$	+0.77
$Fe^{2+} + 2e^- \Leftrightarrow Fe_{(s)}$	-0.44
$Fe(OH)_{3(s)} + 3H^+ + e^- \Leftrightarrow Fe^{2+} + 3H_2O_{(l)}$	+1.06
$3Fe^{3+} + 4H_2O_{(l)} + e^- \Leftrightarrow Fe_3O_{4(s)} + 8H^+$	+0.55
$Cu^{2+} + e^- \Leftrightarrow Cu^+$	+0.16
$Zn^{2+} + 2e^- \Leftrightarrow Zn_{(s)}$	-0.76
$CO_{2(g)} + 8H^+ + 8e^- \Leftrightarrow CH_{4(g)} + 2H_2O_{(l)}$	+0.17
$CO_{2(g)} + 3H^+ + 4e^- \Leftrightarrow CH_3COO^-$ (Acetate)	-0.06
$NO_3^- + 2H^+ + 2e^- \Leftrightarrow NO_2^- + H_2O_{(l)}$	+0.84
$NO_3^- + 10H^+ + 8e^- \Leftrightarrow NH_4^+ + 3H_2O_{(l)}$	+0.88
$2NO_3^- + 12H^+ + 10e^- \Leftrightarrow N_{2(g)} + 6H_2O_{(l)}$	+1.24
$NO_2^- + 8H^+ + 6e^- \Leftrightarrow NH_4^+ + 2H_2O_{(l)}$	+0.89
$2NO_2^- + 8H^+ + 6e^- \Leftrightarrow N_{2(g)} + 4H_2O_{(l)}$	+1.53
$SO_4^{2-} + 3H^+ + 2e^- \Leftrightarrow HSO_3^- + H_2O_{(l)}$	+0.11
$S_{(s)} + 2e^- \Leftrightarrow S^{2-}$	-0.44
$SO_4^{2-} + 10H^+ + 8e^- \Leftrightarrow H_2S_{(g)} + 4H_2O_{(l)}$	+0.34
$SO_4^{2-} + 9H^+ + 8e^- \Leftrightarrow HS^- + 4H_2O_{(l)}$	+0.24

$$K_1 = \frac{\{Fe^{3+}\}\{e^-\}}{\{Fe^{2+}\}} \quad \text{(Eq. 5.12)}$$

In marked contrast to the Nernst equation, the electron activity appears explicitly in the activity product. It is worth to note that the electron activity should not be interpreted as the concentration of electrons since it can only be exchanged and cannot exist in free state. However, electron activity can manifest the tendency to release or accept electrons. In analogy to pH, Eq. (5.13) indicates the definition of pε.

$$p\varepsilon = -\log \{e^-\} \qquad \text{(Eq. 5.13)}$$

Rewriting Eq. (5.12) in logarithmic form yields Eq. (5.14).

$$\log K_1 = \log \{Fe^{3+}\} - p\varepsilon - \log \{Fe^{2+}\} \qquad \text{(Eq. 5.14)}$$

Similarly, the reduction half reaction of $Cr_2O_7^{2-}$ shown in Eq. (5.3) can be written as Eqs. (5.15) and (5.16) using the law of mass action.

$$K_2 = \frac{\{Cr^{3+}\}^2}{\{Cr_2O_7^{2-}\}\{H^+\}^{14}\{e^-\}^6} \qquad \text{(Eq. 5.15)}$$

$$\log K_2 = 2\log \{Cr^{3+}\} - \log \{Cr_2O_7^{2-}\} + 14\,pH + 6\,p\varepsilon \qquad \text{(Eq. 5.16)}$$

By applying the pε concept, redox speciation in the redox reaction between $Cr_2O_7^{2-}$ and Fe^{2+} can be simply determined from Eqs. (5.14) and (5.16). This is because the two half reactions in the same solution have equal value of pε if they are in equilibrium.

Both Nernst equation and pε concept are commonly used in the literature for the description of redox equilibria. Latter is extremely useful in coping with the redox equilibria in the solution involving both redox and other equilibria, such as acid-base and complexation. The reason is that pε concept lets the algebra of redox reaction become similar to other mass action expressions, thereby allowing the same algorithm to be used in the computation procedure. However, the disadvantage of using pε concept is that it is non-measurable quantity. When the problems with topics, such as analytical methods and corrosion involving electrochemical cells, are concerned, Nernst equation is more preferable since it can be related to voltage measurements directly. Certainly, there must be a simple relationship between *Eh* and pε, which is illustrated in Eq. (5.17).

$$Eh = \frac{2.303RT}{F} p\varepsilon \qquad \text{(Eq. 5.17)}$$

5.2.4 Common Redox Processes in Aquifers

Redox processes, which commonly occur in subsurface environment, are illustrated in Table 5.4 in which their half reactions for reduction and oxidation as well as the corresponding standard pε for the half reactions at pH 7 [i.e., $p\varepsilon^0(W)$] are also included (Christensen et al. 2000). From a thermodynamical point of view, a reduction half reaction (upper part of Table 5.4) can combine with any oxidation reaction if the $p\varepsilon^0(W)$ of the reduction half reaction is higher than the $p\varepsilon^0(W)$ of the oxidation half reaction (lower part of Table 5.4). Besides, sequences of half reactions can be constructed which range from highly oxidized conditions to highly reduced conditions. Therefore, according to Table 5.4, oxygen gas reduces prior to nitrate, which again is followed by the reduction of manganese oxides and then thereupon by the reduction of iron oxyhydroxides (Appelo and Postma 1993).

Table 5.4 Redox processes commonly occurred in aquifers as well as the corresponding half reactions and standard electron activities [i.e., $p\varepsilon^0(W)$] at pH 7

Type of reaction	Half Reaction	$p\varepsilon^0$ (W)
O_2 reduction	$\frac{1}{4} O_{2(g)} + H^+ + e^- \Leftrightarrow \frac{1}{2} H_2O_{(l)}$	+13.75
Denitrification	$\frac{1}{5} NO_3^- + \frac{6}{5} H^+ + e^- \Leftrightarrow \frac{1}{10} N_{2(g)} + \frac{3}{5} H_2O_{(l)}$	+12.62
Mn(IV) reduction	$\frac{1}{2} MnO_{2(s)} + \frac{1}{2} HCO_3^- (10^{-3}) + \frac{3}{2} H^+ + e^- \Leftrightarrow \frac{1}{2} MnCO_{3(s)} + H_2O_{(l)}$	+8.9[a]
Fe(III) reduction	$FeOOH_{(s)} + HCO_3^- (10^{-3}) + 2H^+ + e^- \Leftrightarrow FeCO_{3(s)} + 2H_2O_{(l)}$	-0.8[a]
Organic-C reduction	$\frac{1}{2} CH_2O + H^+ + e^- \Leftrightarrow \frac{1}{2} CH_3OH$	-3.01
SO_4^{2-} reduction	$\frac{1}{8} SO_4^{2-} + \frac{9}{8} H^+ + e^- \Leftrightarrow \frac{1}{8} HS^- + \frac{1}{2} H_2O_{(l)}$	-3.75
CO_2 reduction	$\frac{1}{8} CO_{2(g)} + H^+ + e^- \Leftrightarrow \frac{1}{8} CH_{4(g)} + \frac{1}{4} H_2O_{(l)}$	-4.13
Organic-C → CO_2	$\frac{1}{4} CH_2O + \frac{1}{4} H_2O_{(l)} \Leftrightarrow \frac{1}{4} CO_{2(g)} + H^+ + e^-$	-8.20
Organic-C→VFA	$\frac{1}{2} CH_2O + \frac{1}{2} H_2O_{(l)} \Leftrightarrow \frac{1}{2} HCOO^- + \frac{3}{2} H^+ + e^-$	-7.68
Sulfide oxidation	$\frac{1}{8} HS^- + \frac{1}{2} H_2O_{(l)} \Leftrightarrow \frac{1}{8} SO_4^{2-} + \frac{9}{8} H^+ + e^-$	-3.75
Fe(II) oxidation	$FeCO_{3(s)} + 2H_2O_{(l)} \Leftrightarrow FeOOH_{(s)} + HCO_3^- (10^{-3}) + 2H^+ + e^-$	-0.8[a]
CH_4 oxidation	$\frac{1}{2} CH_{4(g)} + \frac{1}{2} H_2O_{(l)} \Leftrightarrow \frac{1}{2} CH_3OH + H^+ + e^-$	+2.88
Nitrification	$\frac{1}{8} NH_4^+ + \frac{3}{8} H_2O_{(l)} \Leftrightarrow \frac{1}{8} NO_3^- + \frac{5}{4} H^+ + e^-$	+6.16
Mn(II) oxidation	$\frac{1}{2} MnCO_{3(s)} + H_2O_{(l)} \Leftrightarrow \frac{1}{2} MnO_{2(s)} + \frac{1}{2} HCO_3^- (10^{-3}) + \frac{3}{2} H^+ + e^-$	+8.9[a]

Note: VFA refers to volatile fatty acid and the source of the data is from Christensen et al. 2000.

[a]These data correspond to bicarbonate ion (HCO_3^-) concentration of 10^{-3} M.

5.3 Definition – Precipitation

5.3.1 Steps for Precipitation

Nucleation, crystal growth as well as aging, ripening and agglomeration of solids are the main steps involved in precipitation (Walton 1967; Nielson 1964). Basically, precipitation can only occur in a supersaturated solution. However, it is possible for a solution, which is only slightly supersaturated, to be stable indefinitely with respect to a solid phase unless there is either continuous increase in the degree of supersaturation or an addition of fine particle of a substance into this solution.

Nucleation is the initial phase of precipitation, which involves the formation of nuclei from clusters of molecules or ion pairs of the component ions of precipitates, or from fine particles chemically unrelated to precipitates but with some similarities in crystal lattice structure (Snoeyink and Jerkins 1980). Spontaneous precipitation of solid phase can only take place on the surface of nuclei. Precipitation from homogeneous solution (i.e., solution contains no solid phase) requires formation of nuclei from ions in solution. This initial phase of precipitation is called homogeneous nucleation, whereas it is defined as heterogeneous nucleation if foreign particles form the nuclei for subsequent precipitation. In natural water, most of nucleation is heterogeneous since aqueous solution usually contains fine particles of various types.

Because of the creation of an organized structure with defined surfaces from a random arrangement of solution constituents, the formation of nuclei from precipitate ions is an energy-consuming process. Stable nuclei can only be formed when an activation energy barrier is surmounted. To overcome this energy barrier, supersaturated solution or solution, which has a concentration greater than that predicted by the equilibrium with precipitates, is required. Therefore, homogeneous nucleation possesses higher activation energy barrier or requires higher degree of supersaturation than heterogeneous nucleation. Besides, in a similar fashion as catalysts used in chemical reaction, foreign particles can catalyze nucleation by reducing the activation energy barrier, thereby requiring lower degree of supersaturation of solution for the heterogeneous nucleation compared to homogeneous nucleation. Fig. 5.2a illustrates the catalytic influence from foreign particles on the nucleation in which the activation energy barrier of heterogeneous nucleation ($\Delta G_{heterogeneous}$) is many times smaller than that of homogeneous nucleation ($\Delta G_{homogeneous}$) (Stumm 1992). The activation energy for nucleation under different saturation states (Ω) is illustrated in Fig. 5.2b (Appelo and Postma 1993). As is evident, the activation energy decreases with increasing saturation state or degree of supersaturation.

The volume of nuclei showing the activation energy barriers of nucleation (i.e., $\Delta G_{homogeneous}$, $\Delta G_{heterogeneous}$, $\Delta G_{\Omega 1}$ and $\Delta G_{\Omega 2}$) is known as the critical nuclei, and is dependent upon the saturation state of solution. Once the nucleus size passes the size of the critical nucleus, the solids enter the domain of crystal growth in which there is continuous deposition of precipitate constituent ions onto the nuclei (Appelo and Postma 1993). Generally, it is observed that high saturation state yields poor crystals. This is because the activation energy barrier is low under this circumstance, thereby allowing many nuclei to form but few grow larger afterwards. At low saturation state, the activation energy barrier is high so that crystal growth dominates over nucleation, thereby resulting in larger crystals.

The crystal structure of the initially formed precipitates may change to more stable phase since the initial solids formed by the precipitation may not thermodynamically be the most stable solids. This change in crystal structure over time is often called aging. Because the more stable solids usually have comparatively low solubility than the initially formed solids, this change may be consequently accompanied by the additional precipitation. Except for the change of crystal structure, a phenomenon called ripening may also take place whereby the crystal size of the precipitates increases. Small particles usually have higher solubility than large particles. Therefore, the solution concentration in equilibrium with small particles is higher than that in equilibrium with large particles. Accordingly, large particles continue to grow since the solution is still supersaturated with respect to it. However, the continuous decrease in solution concentration owing to the growth of large particles consequently creates an undersaturated condition with respect to small particles, thereby leading the dissolution of small particles. Besides, agglomeration of particles to form larger particles also enhances the conversion of small particles to large particles.

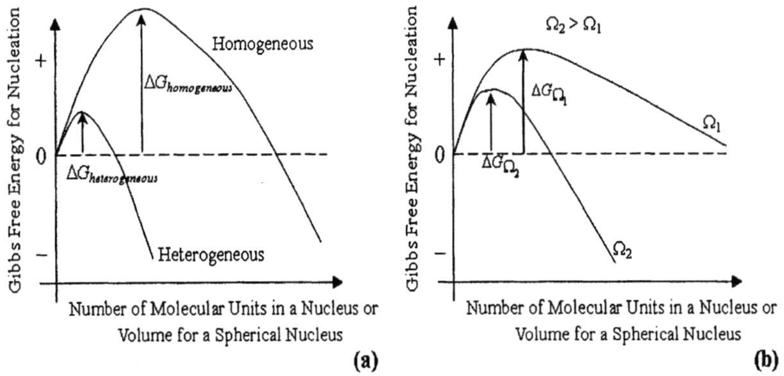

Figure 5.2 (a) Comparison of the activation energy of homogeneous and heterogeneous nucleation at a specific saturation state (Ω) and (b) comparison of the activation energy of nucleation under different saturation states.

5.3.2 Solubility Product and Saturation Index

Considering the dissolution of gypsum [see Eq. (5.18)], the equilibrium constant for this dissolution reaction, based upon the law of mass action, is shown in Eq. (5.19). Solubility product (K_{sp}) is the name given to the equilibrium constant, which delineates the reaction relating to the dissolution of a precipitate in pure water to form its constituent ions.

$$CaSO_{4(s)} \Leftrightarrow Ca^{2+} + SO_4^{2-} \quad \text{(Eq. 5.18)}$$

$$\text{Equilibrium Constant, } K_{sp} = \{Ca^{2+}\}\{SO_4^{2-}\} \quad \text{(Eq. 5.19)}$$

As stated previously, supersaturation is one of the prerequisites for the precipitation. Therefore, understanding the Ω of solution can indicate the possibility of precipitation and also the quality of crystal growth. To calculate the Ω or saturation index, actual ion activity product (IAP) in water sample is compared to the activity product at equilibrium. As an example, the saturation index for the precipitation of gypsum can be determined from Eq. (5.21).

$$\Omega = \frac{\{Ca^{2+}\}_{actual}\{SO_4^{\ 2-}\}_{actual}}{\{Ca^{2+}\}_{equilibrium}\{SO_4^{\ 2-}\}_{equilibrium}} = \frac{IAP}{K_{sp}} \qquad \text{(Eq. 5.21)}$$

For unit Ω, the solution is known in equilibrium. Besides, it is under supersaturation if the Ω is larger than one, whereas it is said to be under subsaturation if the Ω is smaller than one (Appelo and Postma 1993).

5.3.3 Effect of Ionic Strength and Common-Ion Effect

In pure water, an ion, such as calcium, is surrounded by a shield of water molecule dipoles. Under this circumstance, all the calcium ions or concentrations can participate in calcium precipitation and dissolution so that calcium activity or effective concentration of calcium for the precipitation is exactly equal to its molality and, based upon Eq. (5.22), a unit activity coefficient ($\gamma_{Ca}{}^{2+}$) is achieved. However, in the presence of charged solutes, which are not the constituent ions for calcium precipitation, additional electrostatic shielding of the calcium ions by these charged solutes occurs. Only certain portion of calcium ions is surrounded by water molecule dipoles for the precipitation and dissolution. Therefore, the calcium activity or its effective concentration for the precipitation is always less than its molality, thereby resulting in $\gamma_{Ca}{}^{2+}$ of less than unity. Since the presence of the charged solutes reduces the activity of calcium ions for the precipitation, more calcium ions can dissolve into the solution containing higher ionic strength and consequently result in higher solubility than in pure water. This phenomenon is known as ionic strength effect.

$$\{Ca^{2+}\} = \gamma_{Ca^{2+}}(Ca^{2+}) \qquad \text{(Eq. 5.22)}$$

Addition of electrolyte, which does not contain precipitate constituent ions, into the solution saturated with the constituent ions can increase the solubility of the constituent ions because of ionic strength effect. Contrarily, if the electrolyte containing precipitate constituent ions is added into the solution saturated with the constituent ions, the solubility of the ions decreases and precipitation occurs eventually. This is because the activity product [e.g., $\{Ca^{2+}\}\{SO_4^{2-}\}$ for gypsum] must adjust to attain a value equal to its K_{sp}. This process is known as common-ion effect (Freeze and Cherry 1979).

5.4 Contaminants Amenable to Redox Processes and Precipitation

5.4.1 Chromium

Except for lead, chromium is the most common inorganic pollutant detected in groundwater at hazardous sites in United States (National Research Council 1994). Its prevalence in contaminated sites is due to the wide application in industries, such as steel production, leather tanning, electroplating, pigment and chemical manufacturing as well as corrosion prevention, in the past several decades (Buerge and Hug 1997; Sedlak and Chan 1997). In natural environment, chromium usually exists in trivalent and hexavalent forms in which former is dominant under anoxic or suboxic conditions, while latter dominates under oxic condition. Cr(III) basically is essential in human glucidic metabolism. It is comparatively immobile in aqueous environment since its oxides and hydroxides are characterized by very low K_{sp} (Loyaux-Lawniczak et al. 2000). According to Fig. 5.3, at pH less than 3.6, chromium ion (Cr^{3+}) is predominant. At pH above 3.6, trivalent form of chromium is dominated by chromium hydroxyl species, including $Cr(OH)^{2+}$, $Cr(OH)_2^+$, $Cr(OH)_3$ and $Cr(OH)_4^-$ (Richard and Bourg 1991; Palmer and Wittbrodt 1991; USEPA 2000).

Conversely, Cr(VI) is highly toxic since it is a strong oxidant and also a potential carcinogen (Fendorf and Li 1996). Cr(VI) oxyanions, such as chromate ($HCrO_4^-$), bichromate ions (CrO_4^{2-}) and $Cr_2O_7^{2-}$, are more soluble than Cr(III) hydroxyl species and thereby spread easily in aquifers (Nriagu and Nieboer 1988). Since the toxicity and mobility of Cr(VI) are in marked contrast to Cr(III), reduction of Cr(VI) to Cr(III) followed by Cr(III) precipitation is the major principles applied for the remediation of Cr(VI) contaminated groundwater (Powell et al. 1995; Blowes et al. 1997; Lo et al. 2003a). This remedial principle is also applied for remediating a series of inorganic contaminants, including uranium, selenium and technetium (Powell et al. 1998).

Reduction of Cr(VI) thermodynamically requires the presence of electron donors or reductants. Fe^{2+} in minerals and solution, reduced sulfur compounds and soil organic carbon are the reductants commonly found in natural aquifers. Ferrous iron-bearing minerals include carbonate- $[Fe_4^{2+}Fe_2^{3+}(OH)_{12}]^{2+}[CO_3{\bullet}2H_2O]^{2-}$, sulfate- $[Fe_4^{2+}Fe_2^{3+}(OH)_{12}]^{2+}[SO_4{\bullet}nH_2O]^{2-}$ and chloride-containing green rusts $[Fe_3^{2+}Fe^{3+}(OH)_8]^+[Cl]^-$ as well as magnetite (Fe_3O_4) (Loyaux-Lawniczak et al. 2000; Peterson et al. 1997; Williams and Scherer 2001). Eq. (5.23) shows the possible chemical equation involving in the redox reaction between CrO_4^{2-} and sulfate-containing green rust (Loyaux-Lawniczak et al. 2000). As evident, CrO_4^{2-} is reduced to Cr(III) substituted ferrihydrite (i.e., $Fe_{45/11}Cr_{10/11}HO_8$) with concomitant production of hydroxide ion (OH^-), which subsequently provides thermodynamically favorable condition for further precipitation of Cr(III) (Palmer and Wittbrodt 1991).

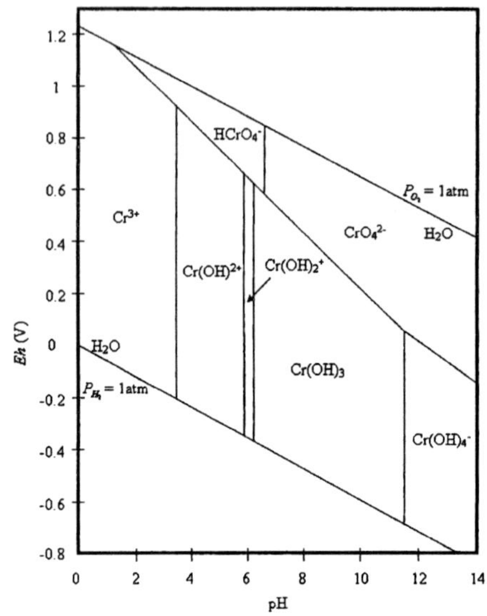

Figure 5.3 Pourbaix diagram for chromium.

$$15Fe_4^{2+}Fe_2^{3+}(OH)_{12}SO_{4(s)}+20CrO_4^{2-} \Leftrightarrow 22Fe_{45/11}Cr_{10/11}HO_{8(s)}+15SO_4^{2-}+10OH^-+74H_2O_{(l)} \quad \text{(Eq. 5.23)}$$

The redox reaction between CrO_4^{2-} and Fe_3O_4 is illustrated in Eq. (5.24) (Peterson et al. 1997). After the reduction, the Cr(III) subsequently precipitates as $Cr(OH)_3$ via Eq. (5.25).

$$6Fe_3O_{4(s)} + 2CrO_4^{2-} + 5H^+ \Leftrightarrow 9Fe_2O_{3(s)} + 2Cr^{3+} + 5OH^- \quad \text{(Eq. 5.24)}$$

$$2Cr^{3+} + 5OH^- \Leftrightarrow Cr(OH)_{3(s)} + Cr(OH)_2^+ \quad \text{(Eq. 5.25)}$$

The efficacy of aqueous Fe^{2+} on reducing Cr(VI) has been well recognized (Fendorf and Li 1996). Virtually, in treating industrially generated chromium wastes, aqueous Fe^{2+} is one of the dominant reductants of Cr(VI) (Eary and Rai 1988). Eq. (5.26) expresses the general chemical equation for the redox reaction between Cr(VI) and aqueous Fe^{2+}.

$$H_xCrO_4^{x-2} + 3Fe^{2+} + (y+3z)H_2O_{(l)} \Leftrightarrow Cr(OH)_y^{3-y} + 3Fe(OH)_z^{3-z} + (x+y+3z)H^+ \quad \text{(Eq. 5.26)}$$

Reduced sulfur compounds, including sulfide (S^{-2}) and sulfite (SO_3^{2-}), are also effective reductants with respect to Cr(VI) (Palmer and Wittbrodt 1991). In the presence of excess SO_3^{2-}, the reduction of Cr(VI) follows Eq. (5.27). In the presence of excess Cr(VI), however, the redox reaction follows Eq. (5.28). Although S^{2-} can thermodynamically reduce Cr(VI) (see Table 5.3), studies indicate that Fe^{2+} must be present in order to obtain both Cr(VI) reduction and precipitation [i.e., Eq. (5.29)] (Palmer and Wittbrodt 1991; Simon et al. 2002).

$$2HCrO_4^- + 6H^+ + 4HSO_3^- \Leftrightarrow 2Cr^{3+} + 2SO_4^{2-} + S_2O_6^{2-} + 6H_2O_{(l)} \quad \text{(Eq. 5.27)}$$

$$2HCrO_4^- + 5H^+ + 3HSO_3^- \Leftrightarrow 2Cr^{3+} + 3SO_4^{2-} + 5H_2O_{(l)} \quad \text{(Eq. 5.28)}$$

$$Cr_2O_7^{2-} + 2FeS_{(s)} + 7H_2O_{(l)} \Leftrightarrow 2Cr(OH)_{3(s)} + 2Fe(OH)_{3(s)} + 2S_{(s)} + 2OH^- \quad \text{(Eq. 5.29)}$$

Apart from Fe^{2+} and reduced sulfur, another important soil constituent, which can contribute to the Cr(VI) reduction, is soil organic carbon. In practice, the amount of $Cr_2O_7^{2-}$ being reduced by soil is a recognized method often used for measuring the content of soil organic carbon (Nelson and Sommers 1982). Eq. (5.30) depicts the idealized chemical equation involved.

$$2Cr_2O_7^{2-} + 3C + 16H^+ \Leftrightarrow 4Cr^{3+} + 3CO_{2(g)} + 8H_2O_{(l)} \quad \text{(Eq. 5.30)}$$

In many types of soil, it is found that soil organic carbon mainly exists in the form of humic substances, including humic acid, fulvic acid and humin. It is well known that the functional groups, such as carboxylic acids, alcohols and phenols, in these humic substances act as the main reductants for the reduction of Cr(VI) (Wiberg 1965; Hayes 1985). Besides, Stollenwerk and Grove (1985) found that the addition of fulvic acid to groundwater samples spiked with Cr(VI) showed significant reduction in Cr(VI) concentration, particularly in acidified samples. In addition, Bloomfield and Pruden (1980) reported that water soluble soil organic matter is effective in reducing Cr(VI) at pH less than 4 but not effective at pH greater than 5.

Zero-valent metal, for example metallic iron (Fe^0), is another thermodynamically favorable reductant for Cr(VI). It is commercially available (Powell et al. 1995), effective under both anoxic and oxic conditions (Cantrell et al. 1995) as well as more reactive compared with most of the natural reductants in aquifers (Rai and Zachara 1988; Henderson 1994; Cantrell et al. 1995; Fendorf and Li 1996; Sedlak and Chan 1997). Theoretically, Fe^0 donates electrons to reduce Cr(VI) to Cr(III) and simultaneously it is oxidized to ferric iron (Fe^{3+}) as shown in Eq. (5.31). The reduced Cr(III) is then removed from aqueous solution through the precipitation of chromium hydroxide [i.e., Eq. (5.25)], or co-precipitation of mixed chromium-iron hydroxide solids as delineated in Eq. (5.32) (Powell et al. 1995; Blowes et al. 1997) or mixed chromium-iron oxyhydroxide solids as depicted in Eq. (5.33) (Eary and Rai 1988; Schwertmann et al. 1989). Since heterogeneous nucleation usually has lower activation energy barrier than homogeneous nucleation, the reduced Cr(III) usually forms precipitates on the surface of Fe^0 or reductant solids rather than in the bulk solution (Appelo and Postma 1993).

$$CrO_4^{2-} + Fe^0_{(s)} + 4H_2O_{(l)} \Leftrightarrow Cr^{3+} + Fe^{3+} + 8OH^- \quad \text{(Eq. 5.31)}$$

$$(1-x)Fe^{3+} + xCr^{3+} + 3H_2O_{(l)} \Leftrightarrow Cr_xFe_{1-x}(OH)_{3(s)} + 3H^+ \quad \text{(Eq. 5.32)}$$

$$(1-x)Fe^{3+} + xCr^{3+} + 2H_2O_{(l)} \Leftrightarrow Cr_xFe_{1-x}OOH_{(s)} + 3H^+ \quad \text{(Eq. 5.33)}$$

Generally, the decrease in Cr(VI) concentration owing to the reduction and precipitation processes can be described by a pseudo first-order kinetic model, which is shown in Eq. (5.34) (Cantrell et al. 1995; Alowitz and Scherer 2002). Moreover, the rate of Cr(VI) reduction was found directly proportional to the ratio of the surface area of reductant to the solution volume or the surface area concentration (λ_s) (Williams and Scherer 2001). Therefore, the rate of Cr(VI) reduction can also be described by Eq. (5.35).

$$\frac{d[C]}{dt} = -k_{obs}[C] \quad \text{(Eq. 5.34)}$$

$$\frac{d[C]}{dt} = -k_{sa}\lambda_s[C] \quad \text{(Eq. 5.35)}$$

To compare the efficiency of different reductants on Cr(VI) reduction, the observed pseudo first-order rate constants (k_{obs}) or half-lives ($t_{1/2}$) of Cr(VI) reductions using various types of reductant are summarized in Table 5.6. As seen, Fe^0 and carbonate-containing green rust have the highest reactivity on reducing Cr(VI) in which the $t_{1/2}$ range from 2.4 to 8.7 min. Besides, Lam (2004) also found that each gram of Fe^0 could remove approximately 2.3 to 4.2 mg of Cr(VI) from groundwater under 23 ^{0}C. Fe^{2+} in aqueous solution and soil has comparatively low reactivity. The aquifer materials in Odessa, Texas containing Fe^{2+} and soil organic carbon possess the lowest reactivity in which the corresponding $t_{1/2}$ is about 2.5 yrs (Henderson 1994). As shown in Fig. 5.4, Fe^0 has the lowest standard potential for the reduction half reaction. From a thermodynamical point of view, Fe^0 has the highest reducing power for Cr(VI) reduction or is the strongest reductant among Fe^{2+}, Fe_3O_4, reduced sulfur and soil organic carbon (Palmer and Wittbrodt 1991; Snoeyink and Jenkins 1980). Therefore, in the latter part of this chapter, the mechanisms of the redox reactions and the precipitation between contaminants and Fe^0 will primarily be discussed.

Table 5.6 Observed pseudo first-order rate constants (k_{obs}) and half-lives ($t_{1/2}$) for Cr(VI) reduction

Reactive Material	Pseudo First-Order Rate Constant (k_{obs})	Half-Life ($t_{1/2}$)	Experiment	Literature Cited	Remarks
Trinity Sandy Aquifer in Odessa, TX	3.2×10^{-5} hr^{-1}	2.5 yrs	Field experiment	Henderson (1994)	Nearly neutral in pH in the groundwater
Ferrous iron in soil	7×10^{-4} hr^{-1} 1.1×10^{-3} hr^{-1}	990 hrs 630 hrs	Batch experiment	Rai and Zachara (1988)	Laboratory experiments
Aqueous ferrous iron	[a]2.0×10^{-6} s^{-1}	96.3 hrs	Batch experiment	Sedlak and Chan (1997)	Initial Fe^{2+} conc of 0.9 μM (50.2 μg/L) and at pH 7.2
	[b]2.1×10^{-4} min^{-1} [b]2.4×10^{-4} min^{-1}	55.2 hrs 48.4 hrs	Initial rate and stopped-flow methods	Fendorf and Li (1996)	Initial Fe^{2+} conc of 100 μM (5580 μg/L) at pH 6.67
Carbonate containing green rust	2.6×10^{-3} s^{-1} 4.0×10^{-3} s^{-1}	4.4 min 2.9 min	Batch experiment	Williams and Scherer (2001)	In the presence of 0.5 g/L of carbonate containing green rust
Zero-valent iron (Fe^0)	[c]6.5 hr^{-1} *(0.13 hr^{-1})	6.4 min *(5.2 hrs)	Batch experiment	Cantrell et al. (1995)	Initial Cr(VI) conc of 500 μg/L and at pH 8.4
	[c]7.7 hr^{-1} *(0.16 hr^{-1})	5.4 min *(4.4 hrs)			Initial Cr(VI) conc of 10,000 μg/L and at pH 8.4
	[d]4.8 to 17.2 hr^{-1} *(0.25 to 0.9 hr^{-1})	2.4 to 8.7 min *(0.77 to 2.7 hrs)	Batch experiment	Alowitz and Scherer (2002)	Initial Cr(VI) conc of 10000μg/L and at pH 7.0
	[e]2.3 to 4.2 mg Cr(VI)/g Fe^0		Column experiment	Lam (2004)	Initial Cr(VI) conc of 25,000 μg/L and at pH 7.1

*The figures listed in parentheses are the k_{obs} and $t_{1/2}$ at 1 m^2 L^{-1} of the surface area concentration (λ_s)

[a] $\frac{d[Cr(VI)]}{dt} = -k[Fe^{2+}][Cr(VI)] = -k_{obs}[Cr(VI)]$; [b] $\frac{d[Cr(VI)]}{dt} = -k'[Fe^{2+}]^{0.6}[Cr(VI)] = -k_{obs}[Cr(VI)]$

[c]2.43 m^2/g of Fe^0 specific surface area and 48.6 m^2/L of the surface area concentration (λ_s); [d]2.30 m^2/g of Fe^0 specific surface area and 19 m^2/L of the surface area concentration (λ_s); [e]Cr(VI) removal capacity (mg of Cr(VI) removed per gram of Fe^0). 1.8 m^2/g of Fe^0 specific surface area and 7716 m^2/L of the surface area concentration (λ_s)

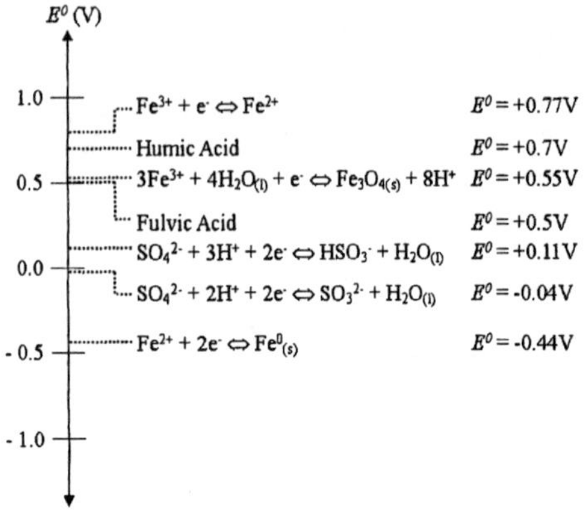

Figure 5.4 Standard potentials for various reduction half reactions.

5.4.2 Uranium

Uranium is the heaviest naturally occurring element. All of its isotopes (i.e., ^{238}uranium and ^{235}uranium) are radioactive in nature with a decay $t_{1/2}$ between 10^8 and 10^9 yrs (Simon et al. 2002). Groundwater within mine waste piles and leachate derived from mine wastes customarily contain high concentration of uranium (Dubrovsky et al. 1984; Olsen et al. 1986; Morin et al. 1988). It is dangerous to human not just because of its radioactivity but also due to its toxicity as a heavy metal. The maximum contaminant level (MCL) of uranium in water, based upon the radiation limit, is 300 μg/L. However, with regard to its toxicity, the level is down to 20 μg/L (Simon et al. 2002).

Naturally, uranium mainly exists in the oxidation states +4 [i.e., U(IV)] and +6 [i.e., U(VI)]. In analogy to chromium, U(VI), for example uranyl ion (UO_2^{2+}), is more mobile than U(IV) and the solubility of U(IV) oxides, such as uraninite (UO_2), is of the order of 10^{-3} mg/L in a pH between 4 and 14. The substantial decrease in the uranium solubility after reducing from U(VI) to U(IV) indicates that reduction and precipitation is one of the possible measures for the remediation of U(VI) contaminated water and groundwater. Actually, good results of U(VI) removal through reduction and precipitation have been reported in which, following the pseudo first-order kinetics, nearly 100% of U(VI) at an initial concentration up to 18000 mg/L can be removed through the redox reaction and precipitation with Fe^0. As you can see from Table 5.7, Gu et al. (1998) reported that in a glass vial containing Fe^0 and U(VI) spiked solution, half of 1000 mg/L of U(VI) concentration can be removed from the aqueous solution within 5.7 min or less. Although adsorption of UO_2^{2+} onto the corrosion products of Fe^0, such as iron oxides and oxyhydroxides, is known to take place particularly at a high pH (Langmuir 1978; Hsi and Langmuir 1985; Ho and Miller 1986), Gu et al. (1998) found that less than 4% of UO_2^{2+} is adsorbed by the Fe^0 corrosion products and the reduction and precipitation with Fe^0 is the major pathway for the removal of UO_2^{2+}. According to Eq. (5.36), by using the electrons released from Fe^0 oxidation, UO_2^{2+} is reduced to UO_2, which exists as precipitates on the Fe^0 surface (Powell et al. 1998; Ott 2000). Since the standard potential of U(VI)/U(IV) is always higher than that of Fe^{2+}/Fe^0, the reduction of U(VI) by Fe^0 can occur spontaneously (Simon et al. 2002). However, it is possible that the reduced U(IV) species on the Fe^0 surface reoxidize and consequently remobilize when the reduced system becomes more oxidized.

$$Fe^0_{(s)} + UO_2^{2+} \Leftrightarrow Fe^{2+} + UO_{2(s)} \qquad \text{(Eq. 5.36)}$$

Table 5.7 Pseudo first-order rate constants (k_{obs}) and the corresponding half-lives ($t_{1/2}$) of U(VI) reduction using Fe^0 as a reductant

Fe^0 type	Fe^0 Specific Surface Area (m^2/g)	Observed Pseudo First-Order Rate Constant, k_{obs} (min^{-1})	Half-Life, $t_{1/2}$, (min)
[a]Master-builder Fe^0	0.98	0.39 (0.002)	1.79 (348)
[a]Peerless Fe^0, medium	0.10	0.29 (0.015)	2.41 (47.8)
[a]Peerless Fe^0, coarse	0.08	0.12 (0.008)	5.68 (92.4)
[a]Cercona cast Fe^0	0.02	0.19 (0.048)	3.61 (14.6)
[a]Cercona Fe^0-palladized	na	0.31	2.22
[b]Fe^0 in natural GW	na	(0.023)	(30.1)
[b]Fe^0 in deionized water	na	(0.035)	(19.8)
[b]Fe^0 in 0.4 M NaCl soln.	na	(0.079)	(8.77)
[b]Fe^0 in 0.4 M $NaNO_3$ soln.	na	(0.0072)	(96.3)

Note: na refers to not available. The figures shown in parentheses are the k_{obs} and $t_{1/2}$ at 1 m^2/L of λ_s.
[a]The initial uranyl ion (UO_2^{2+}) concentration is 1000 mg/L. All the results are obtained from batch experiments containing 2 g of Fe^0 and 10 mL of solution. The surface area concentration (λ_s) for Master-builder Fe^0, medium size Peerless Fe^0, coarse size Peerless Fe^0 and Cercona cast Fe^0 are 196, 20, 16 and 4 m^2/L, respectively. The source of the data is from Gu et al. (1998).
[b]Iron coupons punched from mild steel plate measuring 1.43 cm in diameter and 0.15 cm in thickeness were used as Fe^0. The pH of the natural groundwater is 6.84, and its calcium, magnesium and carbonate concentrations are 52, 5.9 and 96 mg/L, respectively. The source of the data is from Farrell et al. (1999).

5.4.3 Nitrate and Nitrite

Nitrate (NO_3^-) pollution of water and groundwater has become environmental problems in many parts of the world. Anthropogenic sources, such as nitrogen fertilizers, nitrogen pesticides, animal wastes, and septic systems, account for most NO_3^- contamination of groundwater (Choe et al. 2000). Besides, storm and irrigation runoffs from farmlands always bring high concentration of NO_3^- to aquifers (Huang et al. 1998). Although excess amount of NO_3^- can cause eutrophication in water bodies, NO_3^- itself is relatively non-toxic to human. However, nitrite (NO_2^-), derived by microbially mediated reduction of NO_3^-, can cause human health problems, such as methemoglobinemia, liver damage and even cancers (Cabel et al. 1982; Choe et al. 2004). In United States, 18% of private wells were found containing NO_3^- level above the drinking water standard of 10 mg/L nitrate-nitrogen (NO_3^--N). Another 37% of the wells have the levels greater than 3 mg/L NO_3^--N, thereby indicating the necessity of the remediation of NO_3^- contaminated groundwater (Till et al. 1998).

Currently, ion exchange, reverse osmosis, biological denitrification and chemical reduction are the technologies commonly utilized for the removal of NO_3^-. Ion exchange and reverse osmosis require frequent regeneration of the media and generate secondary brine wastes. Biological denitrification requires intensive maintenance and a constant supply of organic substrate. In comparing with the approach of chemical reduction, microbial process is relatively slow and incomplete (Choe et al. 2000).

In the reduction processes by Fe^0, NO_3^- may reduce to NO_2^- (5.37), ammonium ion (5.38) or nitrogen gas (5.39) and simultaneously NO_2^- may also reduce to ammonium ion (5.40) or nitrogen gas (5.41) depending upon the reaction conditions. In analogy to chromium and uranium, the NO_3^- and NO_2^- reduction by Fe^0 is in pseudo first-order with respect to NO_3^- or NO_2^- concentrations (Choe et al. 2000; Cheng et al. 1997; Alowitz and Scherer 2002). According to Table 5.3, it is readily to observe that the standard potentials for the NO_3^- and NO_2^- reduction half reactions (i.e., between +0.84 and +1.53 V) are much higher than that of Fe^{2+}/Fe^0 (i.e., -0.44 V) (Dean 1985). Therefore, Eqs. (5.37) to (5.41) are thermodynamically favorable and the reduction of NO_3^- and NO_2^- by Fe^0 can occur spontaneously. Finally, ammonium ion, which is one of the end products produced from the redox reactions [i.e., (5.38) and (5.40)], can be removed from aqueous environment by raising the solution pH, thereby releasing ammonia gas (NH_3) from the remediated solution (Cheng et al. 1997).

$$Fe^0_{(s)} + NO_3^- + 2H^+ \Leftrightarrow Fe^{2+} + NO_2^- + H_2O_{(l)} \quad \text{(Eq. 5.37)}$$

$$4Fe^0_{(s)} + NO_3^- + 10H^+ \Leftrightarrow 4Fe^{2+} + NH_4^+ + 3H_2O_{(l)} \quad \text{(Eq. 5.38)}$$

$$5Fe^0_{(s)} + 2NO_3^- + 12H^+ \Leftrightarrow 5Fe^{2+} + N_{2(g)} + 6H_2O_{(l)} \quad \text{(Eq. 5.39)}$$

$$3Fe^0_{(s)} + NO_2^- + 8H^+ \Leftrightarrow 3Fe^{2+} + NH_4^+ + 2H_2O_{(l)} \qquad \text{(Eq. 5.40)}$$

$$3Fe^0_{(s)} + 2NO_2^- + 8H^+ \Leftrightarrow 3Fe^{2+} + N_{2(g)} + 4H_2O_{(l)} \qquad \text{(Eq. 5.41)}$$

Huang et al. (1998) studied the temporal change of the concentration of NO_3^-, NH_3 and Fe^{2+} in a batch containing 0.3 g of Fe^0 and 15 mL of solution spiked with 500 mg/L of NO_3^-. The observation of a simultaneous decrease in NO_3^- concentration as well as increase in the concentration of NH_3 and Fe^{2+} conclusively indicates the reduction of NO_3^- to NH_3 and oxidation of Fe^0 to Fe^{2+}. The similarity of the maximum molarity between NO_3^- and NH_3 shows that NH_3 is the main end product of the redox reaction under those experimental conditions. Since reduction of 1 mole of NO_3^- to NH_3 stochiometrically requires 4 mole of Fe^0 [see Eq. (5.38)], the molarity of Fe^{2+}, at the end of the experiment, should be at least 4 times higher than the initial molarity of NH_3.

One of the most important factors affecting the rate of the NO_3^- and NO_2^- reductions is pH since all the redox reactions between Fe^0 as well as NO_3^- and NO_2^- involve the consumption of H^+ (Choe et al. 2004). Table 5.8 summarizes some of the k_{obs} of the NO_3^- and NO_2^- reductions under various solution pH (Alowitz and Scherer 2002; Cheng et al. 1997). As evident, the rate of the NO_3^- and NO_2^- reductions decreases with increasing solution pH.

Cheng et al. (1997) also reported 73% decrease in the k_{obs} of the NO_3^- reduction with increasing the solution pH from 5.0 to 7.0. In addition, no reduction of NO_3^- by Fe^0 was observed in an unbuffered solution. Therefore, it is expected that acidic environment or addition of H^+ is the prerequisite for the reduction of NO_3^- and NO_2^- by Fe^0.

Microbially mediated reduction of NO_3^- or biological denitrification is another approach for removing NO_3^- from water and groundwater. Chemoautotrophs using the H_2 gas produced from the anaerobic oxidation of Fe^0 with water [i.e., Eq. (5.42)] can denitrify NO_3^- to nitrogen gas [Eq. (5.43)] (Till et al. 1998). Microbially catalyzed NO_3^--dependent Fe^{2+} oxidation [Eq. (5.44)] involves oxidation of solid phase Fe^{2+} compounds, such as α-FeOOH and $FeCO_3$, into $Fe(OH)_3$ with concomitant denitrification of NO_3^- (Weber et al. 2001). Recently, public concerns focus on the enhancement from anaerobic ammonium-oxidizing bacteria (AMMONOX) to the abiotic NO_3^- reduction by Fe^0. As aforesaid, NH_3 is a primary product of the abiotic reductions of NO_3^- and NO_2^- by Fe^0 [Eqs. (5.38) and (5.40)], which has an adverse aesthetic impact on drinking water (Jafvert and Valentine 1992). According to Eq. (5.45), AMMONOX can oxidize ammonium ion to nitrogen gas by utilizing NO_3^- as an electron acceptor. Therefore, the growth of AMMONOX on the Fe^0 surface does not only help reduce the NO_3^- concentration; it can also help remove ammonium, thereby allowing elimination of the subsequent process for the treatment of NH_3 in remediated solution (Westerhoff and James 2003).

Table 5.8 Pseudo first-order rate constants (k_{obs}) and the corresponding half-lives ($t_{1/2}$) of nitrate (NO_3^-) and nitrite (NO_2^-) reductions using Fe^0 as a reductant.

Fe^0 Type	pH	Observed Pseudo First-Order Rate Constant, k_{obs} (hr^{-1})	Half-Life, $t_{1/2}$, (hr)
Nitrate			
	5.5	0.95±0.027 (0.0099±0.0003)	0.73±0.02 (70.1±2.0)
	6.0	0.58±0.010 (0.0060±0.0001)	1.20±0.02 (114.7±2.0)
	6.5	0.20±0.086 (0.0021±0.0009)	4.25±1.83 (408.1±175.5)
	7.0	0.13±0.024 (0.0014±0.0003)	5.52±1.02 (529.8±97.8)
[a]Fisher Fe^0	7.5	0.15±0.057 (0.0016±0.0006)	5.40±2.05 (518.4±197.0)
	8.0	0.13±0.042 (0.0014±0.0004)	5.95±1.92 (571.4±184.6)
	8.5	0.025±0.0008 (0.0003±0.00001)	27.75±0.89 (2663.8±85.2)
	9.0	0.008±0.0001 (0.0001±0.000001)	90.03±1.52 (8642.5±145.9)
	2.75	0.0571±0.0027 (0.0026±0.0001)	12.2±0.6 (267.6±12.7)
	2.89	0.0913±0.0028 (0.0042±0.0001)	7.6±0.23 (167.1±5.1)
[b]Aldrich Fe^0	3.19	0.0256±0.0033 (0.0012±0.0002)	27.3±3.5 (605.6±78.1)
	3.36	0.0592±0.0117 (0.0027±0.0005)	12.0±2.4 (268.0±53.0)
	6.16	0.00904±0.00148 (0.0004±0.0001)	77.7±6.3 (1733.0±283.7)
	8.00	0.00886±0.00082 (0.0004±0.00004)	78.5±7.2 (1735.6±160.6)
Nitrite			
	5.5	2.6±0.22 (0.137±0.012)	0.27±0.02 (5.1±0.4)
	6.0	2.0±0.37 (0.105±0.019)	0.36±0.07 (6.8±1.3)
	6.5	1.0±0.07 (0.053±0.004)	0.70±0.05 (13.2±0.9)
[c]Fisher Fe^0	7.0	0.61±0.12 (0.032±0.006)	1.2±0.23 (22.5±4.4)
	7.5	0.23±0.023 (0.012±0.001)	3.0±0.30 (57.8±5.8)
	8.0	0.10±0.020 (0.005±0.001)	7.2±1.44 (137.2±27.4)
	8.5	0.035±0.003 (0.002±0.0002)	20.0±1.77 (379.2±33.6)

Note: All the studies are conducted in batch experiments. The figures shown in parentheses are the k_{obs} and $t_{1/2}$ at 1 m^2/L of the surface area concentration (λ_s).
[a]Initial nitrate concentration 50 mg/L, 2.30±0.18 m^2/g of the Fe^0 specific surface area and 96 m^2/L of the λ_s (Alowitz and Scherer 2002).
[b]Initial nitrate concentration 20 mg/L NO_3^--N, 0.55±0.05 m^2/g of the Fe^0 specific surface area and 22 m^2/L of the λ_s (Su and Puls 2004).
[c]Initial nitrite concentration 40 mg/L, 2.30±0.18 m^2/g of the Fe^0 specific surface area and 19 m^2/L of the λ_s (Alowitz and Scherer 2002).

$$Fe^0_{(s)} + 2H_2O_{(l)} \Leftrightarrow H_{2(g)} + Fe^{2+} + 2OH^- \quad \text{(Eq. 5.42)}$$

$$2NO_3^- + 5H_{2(g)} \Leftrightarrow N_{2(g)} + 4H_2O_{(l)} + 2OH^- \quad \text{(Eq. 5.43)}$$

$$10Fe^{2+} + 2NO_3^- + 24H_2O_{(l)} \Leftrightarrow 10Fe(OH)_{3(s)} + N_{2(g)} + 18H^+ \quad \text{(Eq. 5.44)}$$

$$5NH_4^+ + 3NO_3^- \Leftrightarrow 4N_{2(g)} + 9H_2O_{(l)} + 2H^+ \quad \text{(Eq. 5.45)}$$

5.4.4 Chlorinated Aliphatic Hydrocarbons

Of the 25 most frequently detected groundwater contaminants at hazardous waste sites, 10 are CAHs, such as 1,1,1-trichloroethane (1,1,1-TCA), tetrachloroethylene (PCE), trichloroethylene (TCE) and dichloroethylene (DCE) isomers (National Research Council 1994). The prevalence of CAHs in groundwater is due to their extensive applications as dry cleaning and metal degreasing agents, refrigerants as well as propellants over the past 50 years (Orth 1992; Fetter 1999; Montgomery 2000; Lai and Lo 2002). CAH contaminated groundwater has drawn a great concern from public or environmentalists for the necessity of remediation because of the persistence of CAHs in hydrogeological environment and also their toxicity to human body (Montgomery 2000). For instance, the $t_{1/2}$ of the natural attenuation of CAHs may range from 128 to 2310 days (Rügge et al. 1999) and epoxidation of CAHs by liver oxidation enzyme was proven producing carcinogenic compounds, thereby resulting in drinking water limits in the order of a few micrograms per litre (Gotpagar et al. 1997; Montgomery 2000).

Reduction or reductive dechlorination of CAHs by reductants, such as Fe^0, is one of the methods for removing CAHs from water and groundwater (Boronina et al. 1995; Warren et al. 1995; Arnold and Roberts 1998; Boronina et al. 1998; Fennelly and Roberts 1998; Lo and Lai 2002; Lo et al. 2003b; Andrea et al. 2005). Oxidation of Fe^0 under both anaerobic [Eq. (5.42)] and aerobic conditions [Eqs. (5.46) and (5.47)] provides electron sources and thermodynamically favorable condition for the reduction of CAHs. Hydrogenolysis [Eq. (5.48)], reductive-β-elimination [Eq. (5.49)] and direct catalytic hydrogenolysis by H_2 [Eq. (5.50)] are the main pathways involved (Matheson and Tratnyek 1994; Roberts et al. 1996; Farrell et al. 2000; Liang et al. 2000; Lai and Lo 2003; Lai et al. 2004).

$$2Fe^0_{(s)} + O_{2(g)} + 2H_2O_{(l)} \Leftrightarrow 2Fe^{2+} + 4OH^- \quad \text{(Eq. 5.46)}$$

$$4Fe^{2+} + 4H^+ + O_{2(g)} \Leftrightarrow 4Fe^{3+} + 2H_2O_{(l)} \quad \text{(Eq. 5.47)}$$

$$RHCl{=}RHCl + H^+ + 2e^- \Leftrightarrow RH_2{=}RHCl + Cl^- \quad \text{(Eq. 5.48)}$$

$$RHCl{=}RHCl + 2e^- \Leftrightarrow RH{\equiv}RH + 2Cl^- \quad \text{(Eq. 5.49)}$$

$$RHCl{=}RHCl + H_{2(g)} \xrightarrow{Catalyst} RH_2{=}RHCl + H^+ + Cl^- \qquad \text{(Eq. 5.50)}$$

Hydrogenolysis is the replacement of a chlorine atom with a hydrogen atom, while reductive-β-elimination involves the removal of either two chlorine atoms or a chlorine and a hydrogen atom with concomitant formation of a triple bond (Roberts et al. 1996; Farrell et al. 2000). Direct catalytic hydrogenolysis by H_2 requires the presence of catalyst, such as Fe_3O_4, in which H_2 released from Eq. (5.42) is utilized as an electron donor (Matheson and Tratnyek 1994; Ritter et al. 2002).

Completely reductive dechlorination of CAHs by Fe^0 is not a one-step process. Instead, by following the pseudo first-order kinetics [i.e., Eqs. (5.34) or (5.35)], CAHs are sequentially reduced to less chlorinated forms and finally transformed to non-chlorinated end products. As illustrated in Fig. 5.5, through the pathways of hydrogenolysis, reductive-β-elimination and hydrogenation, chlorinated ethenes, such as PCE and TCE, are sequentially dechlorinated into DCE isomers, vinyl chloride (VC), dichloroacetylene, chloroacetylene, acetylene and finally reduced to ethene and ethane (Arnold and Roberts 1998). Since the standard potentials for various reductive dechlorination reactions (see Table 5.9) are higher than that of Fe^{2+}/Fe^0 (i.e., –0.44 V), reductive dechlorination of CAHs by Fe^0 can occur spontaneously.

The chlorinated intermediates generated from the pathways of hydrogenolysis, reductive-β-elimination and hydrogenation, such as cis-DCE and VC, are also toxic to human. Therefore, incomplete reduction of CAHs by Fe^0 may potentially generate more toxic contaminants. To prevent the production of highly toxic daughter products, the reaction time for the reductive dechlorination should be carefully determined based upon the k_{obs} of different CAHs so that the reaction time is long enough to allow Fe^0 to completely dechlorinate CAHs. Table 5.10 summarizes the representative kinetic data for the reductive dechlorination by Fe^0.

Apart from Fe^0, reductive dechlorination of chlorinated ethenes by ferrous iron- or sulfur-containing minerals has also been reported (Butler and Hayes 1999; Lee and Batchelor 2002). Table 5.11 illustrates the typical k_{obs} of the reductive dechlorination of PCE, TCE, cis-DCE and VC by green rust and mackinawite (FeS). Lee and Batchelor (2002) reported that dichloroacetylene, chloroacetylene, acetylene and ethene are the main dechlorination products detected in the dechlorination process of PCE and TCE by green rust, thereby proposing the domination of reductive-β-elimination over hydrogenolysis as a major pathway of the reductive dechlorination. However, the dechlorination products of both hydrogenolysis and reductive-β-elimination were also detected in the dechlorination process of PCE and TCE by FeS (Butler and Hayes 1999).

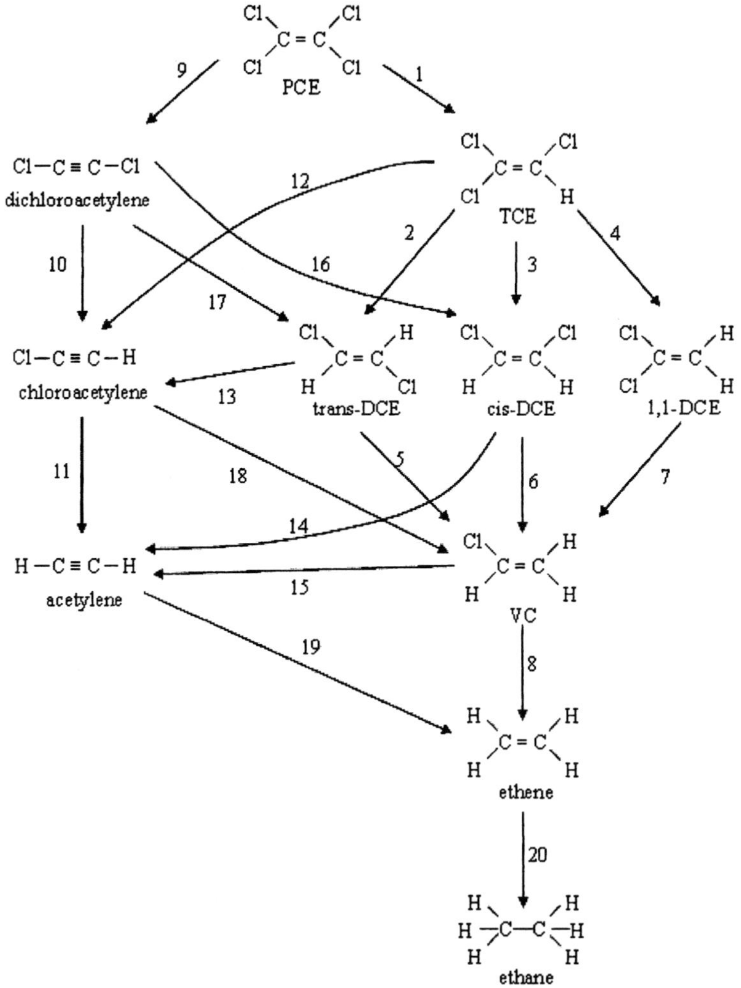

Figure 5.5 Hypothesized reaction pathways for chlorinated ethenes and intermediates during reductive dechlorination by Fe^0. Reactions 1–8, 10 and 11 correspond to hydrogenolysis pathway, while reactions 9, 12–15 are pathways of reductive-β-elimination. Reactions 16–20 are hydrogenation reactions (Blowes et al. 1999).

Table 5.9 Standard potentials for various reductive dechlorination reactions of chlorinated aliphatic hydrocarbons (CAHs) by Fe^0

Contaminants	Pathway	Reaction	E^0 (W) (V)
PCE	[a]Hydrogenolysis	PCE → TCE	0.592
TCE	[a]Hydrogenolysis	TCE → cis-DCE	0.530
TCE	[a]Hydrogenolysis	TCE → 1,1-DCE	0.513
TCE	[a]Hydrogenolysis	TCE → trans-DCE	0.509
1,1-DCE	[a]Hydrogenolysis	1,1-DCE → VC	0.423
cis-DCE	[a]Hydrogenolysis	cis-DCE → VC	0.407
trans-DCE	[a]Hydrogenolysis	trans-DCE → VC	0.428
VC	[a]Hydrogenolysis	VC → ethene	0.481
Dichloroacetylene	[a]Hydrogenolysis	Dichloroacetylene → Chloroacetylene	0.560
Chloroacetylene	[a]Hydrogenolysis	Chloroacetylene → acetylene	0.500
PCE	[b]Reductive-β-elimination	PCE → dichloroacetylene	0.631
TCE	[b]Reductive-β-elimination	TCE → chloroacetylene	0.599
cis-DCE	[b]Reductive-β-elimination	cis-DCE → acetylene	0.568
trans-DCE	[b]Reductive-β-elimination	trans-DCE → acetylene	0.589
Dichloroacetylene	[c]Hydrogenation	Dichloroacetylene → cis-DCE	0.490
Dichloroacetylene	[c]Hydrogenation	Dichloroacetylene → trans-DCE	0.470
Chloroacetylene	[c]Hydrogenation	Chloroacetylene →VC	0.370

Source: Roberts et al. (1996)

[a]Standard potential corresponding to hydrogenolysis [Eq. (5.48)]: $RHCl{=}RHCl + H^+ + 2e^- \Leftrightarrow RH_2{=}RHCl + Cl^-$.

[b]Standard potential corresponding to reductive-β-elimination [Eq. (5.49)]: $RHCl{=}RHCl + 2e^- \Leftrightarrow RH{\equiv}RH + 2Cl^-$.

[c]Standard potential corresponding to reduction of triple bond to double bond (hydrogenation).

5.4.5 Polychlorinated Biphenyls

Polychlorinated biphenyls (PCBs) is a kind of chlorinated organic micropollutants recently drawing more public concern because of their environmental persistence and bioconcentration in the food chain (Hester and Harrison 1996). Although its production has been ceased since 1970s, heavy uses of PCBs as organic diluents, plasticizers, pesticide extenders, adhesives, dust-reducing agents, cutting oils, flame retardants, heat transfer fluids as well as dielectric fluids for transformers and capacitors since 1930s result in a direct or indirect release of PCBs into water and soil environments (Kim et al. 2004). It is expected that 31% of the world PCB production has been released into the environment and another 65% are currently in use or has been disposed in landfills and hazardous waste dumps, thereby having a possibility of releasing to environment (Hester and Harrison 1996).

Table 5.10 Representative kinetic data for the reductive dechlorination of chlorinated aliphatic hydrocarbons (CAHs) by Fe^0.

CAH	Observed pseudo first-order rate constant, k_{obs} (hr^{-1})	Half-life, $t_{1/2}$, (hr)
Tetrachloroethylene	[a]2.1, [b]0.104 – 0.779	[a]0.33, [b]0.89 – 6.64
Trichloroethylene	[a]0.39, [b]0.104 – 0.786	[a]1.78, [b]0.882 – 6.69
cis-Dichloroethylene	[a]0.041, [b]0.011 – 0.022	[a]16.9, [b]31.8 – 63.6
trans-Dichloroethylene	[a]0.12, [b]0.374 – 0.419	[a]5.78, [b]1.65 – 1.85
1,1-Dichloroethane	[b]0.015 – 0.262	[b]2.65 – 46.8
1,1-Dichloroethylene	[a]0.064, [b]0.033 – 0.466	[a]10.8, [b]1.49 – 20.7
Vinyl Chloride	[a]0.05, [b]0.026 – 0.074	[a]13.9, [b]9.37 – 26.2
Hexachloroethane	[a]31	[a]0.022
1,1,2,2-Tetrachloroethane	[a]13	[a]0.053
1,1,1,2-Tetrachloroethane	[a]14	[a]0.050
1,1,1-Trichloroethane	[a]11, [b]0.087 – 0.579	[a]0.063, [b]1.20 – 7.95
Chloroform	[a]0.92, [b]0.071 – 0.364	[a]0.753, [b]1.91 – 9.78

[a]All k_{obs} and $t_{1/2}$ are determined based upon 1000 m^2/L of the surface area concentration (λ_s). The source of the data is from Johnson et al. (1996).
[b]All k_{obs} and $t_{1/2}$ are determined based upon 7790 m^2/L of λ_s. The source of the data is from Lai et al. (2006).

Table 5.11 Kinetic data for the reductive dechlorination of chlorinated aliphatic hydrocarbons (CAHs) by green rust and iron sulfide

CAH	[a]Reduced by Green Rust		[b]Reduced by Iron Sulfide	
	Observed Pseudo First-Order Rate Constant, k_{obs} (hr^{-1})	Half-Life, $t_{1/2}$, (hr)	Observed Pseudo First-Order Rate Constant, k_{obs} (hr^{-1})	Half-Life, $t_{1/2}$, (hr)
PCE	0.0978 (0.000162)	7.09 (4278)	0.00057 (0.00114)	1216 (607.9)
TCE	0.0516 (0.0000854)	13.4 (8115)	0.00149 (0.00298)	465 (232.6)
cis-DCE	0.0313 (0.0000518)	22.1 (13378)		
VC	0.0469 (0.0000776)	14.8 (8925)		

Note: The figures shown in parentheses are the k_{obs} and $t_{1/2}$ at 1 m^2/L of the surface area concentration (λ_s).
[a]The experimental results are obtained from batch kinetic experiments conducted at 23°C and pH 7 with 604 m^2/L of the λ_s (Lee and Batchelor 2002).
[b]The experimental results are obtained from batch kinetic experiments conducted at 25°C and pH 8.3 with 0.5 m^2/L of the λ_s (Butler and Hayes 1999).

PCBs consist of the structures having 1 to 10 chlorine atoms bounding to a biphenyl molecule (Figs. 5.6a and 5.6b). Of the 209 possible congeners, only about 130 of the structures are energetically favorable in the manufacturing processes. Like many other aromatic hydrocarbons, PCBs are highly lipophilic and chemically stable. Accumulation of PCBs in the human body through food chain may cause neurodevelopmental and neurobehavioral deficits in children and cancer (Hester and Harrison 1996). Nowadays, incineration, biodegradation and reductively chemical dechlorination are the technologies commonly used to destruct PCBs in water and soil (Grittini et al. 1995). Incineration of PCB-contaminated soil is effective but it may produce another undesirable products, such as dioxin (Kim et al. 2004). Aerobic and anaerobic biodegradation of PCBs may take several months for complete decomposition. Reductively dechlorination of PCBs using Fe^0 as reductant was found only happening at 400 ^{0}C, whereas there was no dechlorination at room temperature (Chuang et al. 1995). However, rapidly reductive dechlorination was observed using bimetallic materials, such as palladized Fe^0 (Pd/Fe), as a reductant (Grittini et al. 1995; Korte et al. 2002; Doyle et al. 1998).

In analogy to the dechlorination of CAHs, reductive dechlorination of PCBs is also a stepwise process in which PCBs are sequentially dechlorinated to less chlorinated biphenyls and finally reduced to biphenyl. As depicted in Eq. (5.51), by using the electrons released from the Fe^0 oxidation, the PCB containing x+y number of chlorine atoms is sequentially reduced into biphenyl with concomitant release of chloride ions and Fe^{2+}. Palladium in this redox reaction serves as a catalyst rather than taking part into the reaction. Fig. 5.7 illustrates the possible reduction pathways of 4,4',5,5'-tetrachlorobiphenyl (4,4',5,5'-TeCB) on the Pd/Fe surface. It is interesting to note that 4,4',5,5'-TeCB is first reduced to trichlorobiphenyls (TCBs) and subsequently to dichlorobiphenyls (DCBs). Afterwards, DCBs is reductively dechlorinated to monochlorophenyls (MCBs) and finally to biphenyl. Each dechlorination step certainly involves direct transfer of two electrons and the release of a chlorine atom (Yak et al. 2000).

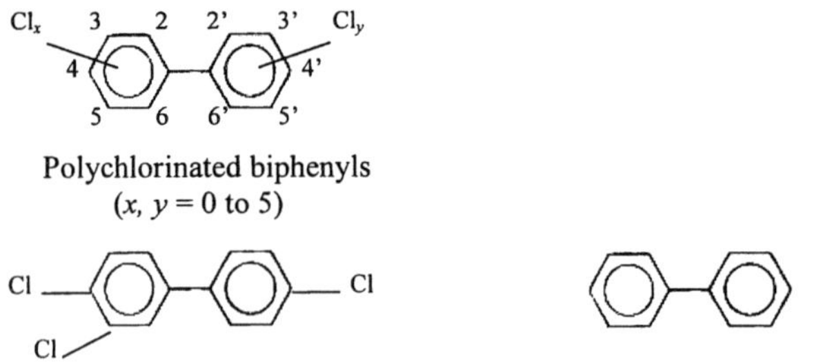

4,4',5-Trichlorobiphenyl **(a)** Biphenyl **(b)**

Figure 5.6 Basic structure and nomenclature of (a) polychlorinated biphenyls (PCBs) and (b) biphenyl.

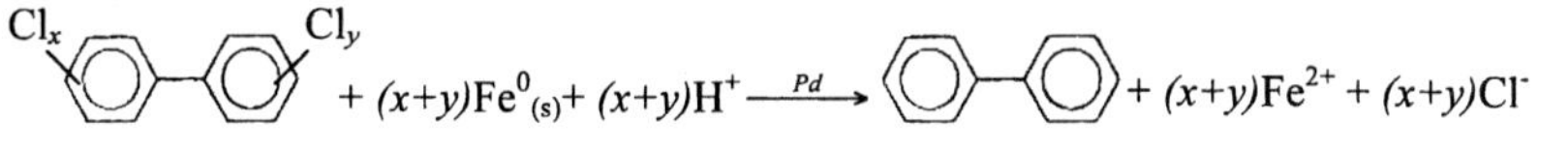

(Eq. 5.51)

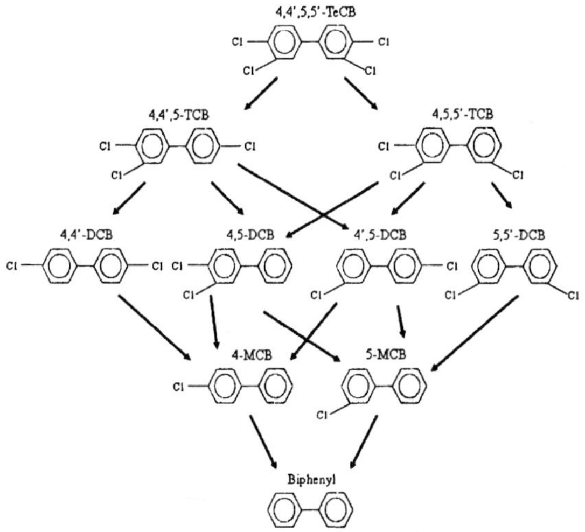

Figure 5.7 Possible dechlorination pathways of 4,4',5,5'-tetrachlorobiphenyl (4,4',5,5'-TeCB) on the Pd/Fe surface.

With regard to the kinetics of the reductive dechlorination of PCBs, Kim et al. (2004) reported that the reductive dechlorination by Pd/Fe was a pseudo first-order reaction in which the rate of dechlorination were in order of 4-MCB > 3-MCB > 2-MCB. According to Table 5.12, the $t_{1/2}$ of the dechlorination of 4-MCB by Pd/Fe is about 41.8 hrs, whereas it is 70.7 and 87.7 hrs for 3-MCB and 2-MCB, respectively. Similar trend of variation in the dechlorination rate constant of MCBs could also be observed for palladized zinc (Pd/Zn). In addition, it was also found that PCBs with chlorine atom at carbon 4 would be preferentially dechlorinated and PCBs with chlorine atom at positions 3 or 5 would be followed. Finally, PCBs with chlorine atom at positions 2 or 6 had the least priority to be dechlorinated by bimetallic materials (Kim et al. 2004). Besides, based upon the $t_{1/2}$ at 1 m^2/L of the λ_s listed in Table 5.12, it is readily to observe that DCBs, including 3,4-DCB, 2,4-DCB and 2,3-DCB, are comparatively more recalcitrant to be dechlorinated by Pd/Fe compared to MCBs.

Table 5.12 Kinetic data for the reductive dechlorination of polychlorinated biphenyls (PCBs) by Pd/Fe and Pd/Zn

	[a]Reduced by Pd/Fe			[b]Reduced by Pd/Zn	
PCB	**Observed Pseudo First-Order Rate Constant, k_{obs} (hr^{-1})**	**Half-Life, $t_{1/2}$, (hr)**	**PCB**	**Observed Pseudo First-Order Rate Constant, k_{obs} (hr^{-1})**	**Half-Life, $t_{1/2}$, (hr)**
4-MCB	0.0166 (0.0060)	41.8 (115.5)	4-MCB	0.144 (0.048)	4.8 (14.6)
3-MCB	0.0098 (0.0045)	70.7 (154.0)	3-MCB	0.08 (0.028)	8.7 (25.2)
2-MCB	0.0079 (0.0040)	87.7 (173.3)	2-MCB	0.069 (0.024)	10.0 (28.9)
3,4-DCB	0.0118 (0.0020)	58.7 (355.4)			
2,4-DCB	0.0094 (0.0016)	73.7 (447.1)			
2,3-DCB	0.0086 (0.0015)	80.6 (477.9)			

Note: The experimental results were obtained from batch reactors containing 20 mL of solution and 1 to 2 g of bimetallic materials. The figures shown in parentheses are the k_{obs} and $t_{1/2}$ at 1 m^2/L of the surface area concentration (λ_s).
[a]Pd/Fe specific surface area 0.062±0.002 m^2/g.
[b]Pd/Zn specific surface area 0.059±0.002 m^2/g.

In comparing the performance of Pd/Fe with Pd/Zn on the reductive dechlorination, latter shows higher reactivity in reducing MCBs (Table 5.12), most likely owing to the more negative of the standard potential of Zn (i.e., - 0.76 V) compared to the Fe^0 (i.e., -0.44 V). Therefore, Zn can thermodynamically provide relatively high reducing force for the MCB reduction. Doyle et al. (1998) also reported that palladized sodium (Pd/Na) and magnesium (Pd/Mg) may possess higher reactivity on PCB dechlorination than the Pd/Fe because both sodium and magnesium have more negative standard potential, which are -2.72 and -2.37 V, respectively (see Table 5.3). According to Eq. (5.51), reductive dechlorination of PCBs requires H^+ and Chuang et al. (1995) found that water was the dominant hydrogen donor for this redox reaction. Unfortunately, extractions of waste or soil for PCBs usually do not employ water. Korte et al. (2002) mentioned that solvent with lower dielectric constant (i.e., less polar) exhibited greater inhibition to the palladium-catalyzed reductive dechlorination reaction. Therefore, reductive dechlorination of PCBs by Pd/Fe in ethanol was observed to be better than that in isopropanol since the former possesses a higher dielectric constant.

5.5 Redox and Precipitation in Natural Systems

5.5.1 Natural Reduction of Hexavalent Chromium

One of the examples showing natural reduction and precipitation of Cr(VI) is at the Chromium I Superfund site, which is located in Odessa, Texas (Henderson 1994). Between 1969 and 1978, several chrome plating factories were located there. During most of this period, wastewater and rinse water containing primarily dissolved Cr(VI) were reportedly discharged directly into an aquifer, thereby resulting in 72 mg/L of maximum Cr(VI) concentration in groundwater. The unconfined aquifer at Odessa, Texas ranging in the thickness from 55 to 70 ft extended to a depth of approximately 140 ft below the ground surface (Fig. 5.8). Besides, the aquifer is a Cretaceous Trinity Formation (Fm) in which the sands were moderately well-sorted, ferruginous to calcareous and were also considered to be hydraulically continuous.

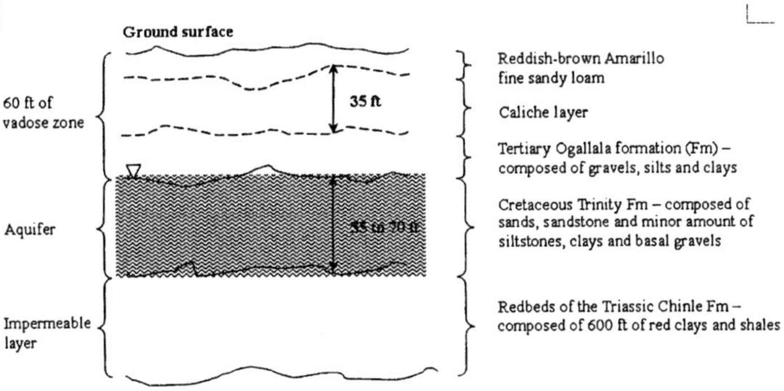

Figure 5.8 Hydrogeologic conditions of the Trinity Sand Aquifer at Odessa, TX.

Investigation of Cr(VI) contamination in the Trinity Sand aquifer conducted from 1986 to 1991 showed significantly temporal decrease in Cr(VI) concentration in the groundwater. The maximum Cr(VI) concentration decreased from 72 mg/L in 1986 to 5.3 mg/L in 1991. The relatively constant of the overall areal extent of the plume between 1986 and 1991 indicated the insignificance of the influence of groundwater advection and dispersion on the disappearance of Cr(VI). After taking account of both dissolved and adsorbed masses of Cr(VI) in the aquifer, Henderson (1994) determined that only 30% of the total Cr(VI) present in the aquifer in 1986 remained in 1991. Fitting pseudo first-order kinetic model to the mass data of total Cr(VI) shown in Figure 5.9 resulted in $t_{1/2}$ of approximately 2.5 years, with a corresponding k_{obs} of 3.2×10^{-5} hr^{-1}.

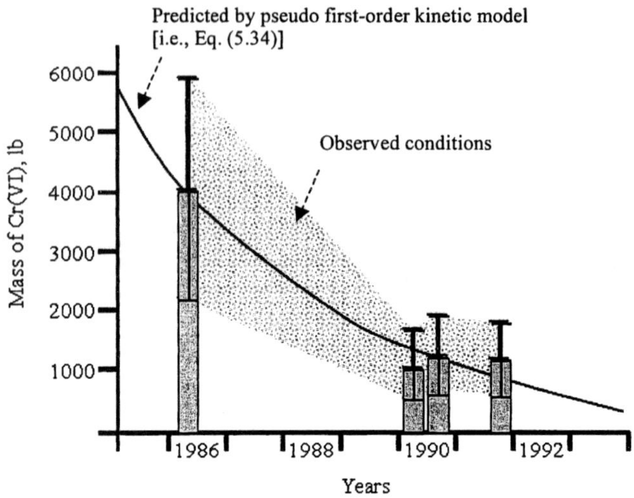

Figure 5.9 Temporal variation of the Cr(VI) mass in the trinity sand aquifer between 1986 and 1991.

Analyses of the sediment samples collected from the Trinity Sand aquifer showed that each kilogram of the soils contained approximately 67 mg of total chromium [i.e., Cr(III) and Cr(VI)] in which about 60 to 90% of the total chromium onto the soils was inferred existing as Cr(III) solids. Therefore, reduction of Cr(VI) to Cr(III) followed by heterogeneous precipitation was believed to be the main underlying culprit leading the 70% loss of Cr(VI) concentration in the groundwater between 1986 and 1991. According to Fig. 5.3, the pH of 6.7 to 7.4 and the *Eh* of 0.26 to 0.53 V measured in the Trinity Sand groundwater further indicated that Cr(III) is the thermodynamically stable valence state of chromium in the Trinity Sand aquifer in which Cr(III) mainly existed as $Cr(OH)_3$ solids. As aforesaid, the sands of the Trinity Formation were observed to be ferruginous. In addition, dissolved organic carbon concentration as high as 8 mg/L had also been measured in both groundwater and the sediment of the aquifer. Therefore, Fe^{2+} and dissolved organic carbon present in the Trinity Sand aquifer were believed acting as the major reductants for the Cr(VI) reduction as depicted in Eqs. (5.26) and (5.30), respectively. The abundance of iron oxyhydroxides in the Trinity Sand sediments also provided favorable conditions for the co-precipitation of mixed chromium-iron hydroxide [Eq. (5.32)] or oxyhydroxide solids [Eq. (5.33)]. By relying on the natural reduction and precipitation of Cr(VI), the Cr(VI) concentration in the

Trinity Sand aquifer could drop below the MCL of 0.1 mg/L for chromium within 10 years without requiring any engineering interventions or remediation.

5.5.2 Naturally Reductive Dechlorination of Chlorinated Solvents

Anaerobic transformation of TCE was observed at the superfund site in St. Joseph, Michigan (Semprini et al. 1995). Most likely owing to the disposal of wastewater from an automotive brake manufacturer into an unlined lagoon close to the site from 1950s to 1970s, TCE concentration as high as 100 mg/L was detected in the groundwater at St. Joseph, Michigan. The geologic formation of the interest is an unconfined aquifer consisting of a layer of unconsolidated fine sand with some silt and overlaying a lacustrine clay unit. The water table at the superfund site varied from 7 to 16 m below ground surface. The thickness of the whole sand layer (i.e., unsaturated and saturated layers) ranged from 12 to 32 m (Semprini et al. 1995).

Analyses of groundwater samples collected in 1991 showed high concentration of TCE, DCE isomers, VC and ethene within 20 m of the center of the plume. The detection of high concentration of the hydrogenolysis products of TCE indicated the occurrence of the reductive dechlorination of TCE in the aquifer (National Research Council 2000). The maximum TCE concentration appeared at a depth of about 65 ft where methane concentration was nearly at the minimum. Reductive dechlorination of TCE to cis-DCE was observed at a depth where intermediate increase in methane concentration and decrease in sulfate concentration were observed, thereby indicating fast transformation of TCE to cis-DCE in a transition zone between the zone of sulfate-reducing and the zone of methanogenesis. Complete disappearance of TCE and cis-DCE achieved at a depth of 75 ft in which maximum concentration of VC, ethene and methane and minimum concentration of sulfate were detected. This phenomenon showed further reduction of TCE and cis-DCE to VC and ethene required more reducing conditions (i.e., methanogenic conditions) rather than sulfate-reducing conditions.

Further verification of the reduction of TCE to DCE isomers in less reducing conditions of sulfate reduction and reductive dechlorination of DCE isomers to VC and ethene under the redox condition of methanogenesis could also be obtained from the two-dimensional chemical concentration contour over the aquifer. It is particularly interesting to note that the zone of high TCE concentration was associated with an area, which was deficient in methane or absent of highly active methanogenic condition. Raised cis-DCE level occurred in the transition zone from low to high methane concentration, thereby showing the dechlorination of TCE to cis-DCE under less reducing conditions. Besides, decrease in cis-DCE concentration was associated with low sulfate values. Furthermore, elevated VC and ethene concentrations were observed associating with the region of elevated methane concentration and thereby showed the necessity of the methanogenic or more reducing conditions for the further dechlorination of TCE and cis-DCE to VC and ethene.

Anaerobic transformation of TCE observed at the superfund site in St. Joseph, Michigan involved reductive dechlorination, which certainly had to be coupled with oxidation reactions. The organic matter leaching from the disposal lagoon containing chemical oxygen demand (COD) as high as 400 mg/L was the main electron donor for the reductive dechlorination of TCE, sulfate reduction and methane production (National Research Council 2000). Semprini et al (1995) reported that 8 to 25% of the TCE in the groundwater were finally converted to ethene and up to 15% of the diminution in the COD in the aquifer was associated with the reductive dechlorination of TCE. As shown in Fig. 5.10, microorganisms in the aquifers first converted COD or organic electron donors to sugars, amino acids and organic acids, which subsequently were fermented to alcohols and fatty acids for energy. Then another type of microbes oxidized the alcohols and fatty acids and thereby produced acetate and H_2. Finally, the other set of microbes utilized acetate and H_2 as electron donors and simultaneously used TCE, sulfate and carbons dioxide as electron acceptors, thereby resulting in the reductive dechlorination of TCE in the aquifer (McCarty 1997; McCarty 1999).

5.6 Conclusion

The natural reduction of hexavalent chromium in the Trinity Sand aquifer at Odessa, Texas and the naturally reductive dechlorination of CAHs at the superfund site in St. Joseph, Michigan showed the potential of the processes of redox and precipitation for hazardous waste remediation. Remediation of hazardous wastes, such as Cr(VI) and U(VI), can be attained by first reducing to immobile forms. Then they were removed from aqueous environments by precipitating as oxide, hydroxide or oxyhydroxide solids. Remediation of NO_3^- or NO_2^- contaminated groundwater can be achieved through the reduction of NO_3^- or NO_2^- to ammonium and nitrogen gas. Stepwise reductive dechlorination of CAHs and PCBs into less chlorinated or non-chlorinated forms is the main approach used to remediate chlorinated aliphatic and aromatic organics from groundwater. By selecting the reductants or reactive materials possessing lower standard potentials than the contaminants, redox reactions between the contaminants and reductants can thermodynamically occur. Since reactive materials are required for triggering the redox reactions, the precipitations following the reduction usually occur onto the surface of the reactive materials rather than in the bulk solution.

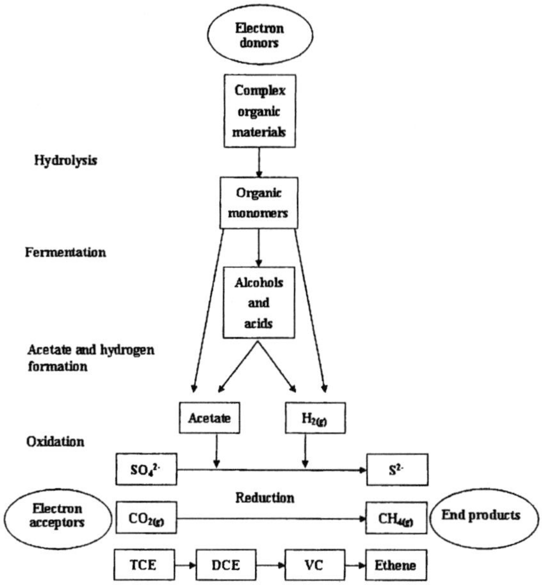

Figure 5.10 Possible steps involved in the reductive dechlorination of TCE in the aquifer at St. Joseph, Michigan.

Notation

The following symbols have been used in this chapter:

$[C]$	=	concentration of contaminant or hazardous waste (mg L^{-1} or M)
E	=	potential developing the electromotive force (emf) of the electrochemical cell (V)
E^0	=	standard potential or the potential with all species present at unit activity at 25 ^{0}C and 1 atm (V)
E^0 (W)	=	standard potential in neutral water (V)
Eh	=	redox potential (V)
$E_{measured}$	=	potential measured by reference electrode (V)
E^0_{ox}	=	standard potential for the oxidation half reaction (V)
E^0_{re}	=	standard potential for the reduction half reaction (V)

$E_{reference}$	=	potential of reference electrode with respect to standard hydrogen electrode (V)
F	=	Faraday's constant (96.42 kJ V^{-1} mol^{-1} or 23.06 kcal V^{-1} mol^{-1})
ΔG	=	change in Gibbs free energy (kJ mol^{-1})
$\Delta G_{heterogeneous}$	=	activation energy of heterogeneous nucleation (kJ mol^{-1})
$\Delta G_{homogeneous}$	=	activation energy of homogeneous nucleation (kJ mol^{-1})
ΔG_r	=	change in Gibbs free energy of the reaction (kJ mol^{-1})
ΔG_r^0	=	change in standard Gibbs free energy of the reaction (kJ mol^{-1}), which is equal to ΔG_r when products or reactants are present at unit activity at a specified standard state (i.e., 25 ^{0}C and 1 atm)
ΔG_{Ω_1}	=	activation energy of nucleation at saturation state Ω_1 (kJ mol^{-1})
ΔG_{Ω_2}	=	activation energy of nucleation at saturation state Ω_2 (kJ mol^{-1})
ΔH_r^0	=	enthalpy change of chemical reaction (kJ mol^{-1})
IAP	=	ion activity product
(i)	=	molality of species i (mol kg^{-1})
$\{i\}$	=	activity of species i
$\{i\}_{actual}$	=	activity of species i in actual water sample
$\{i\}_{equilibrium}$	=	activity of species i at equilibrium
K_1, K_2	=	equilibrium constants
K_{sp}	=	solubility product
K_{sp,T_1}	=	solubility product at temperature T_1
K_{sp,T_2}	=	solubility product at temperature T_2
K_W	=	equilibrium constant for water dissociation
k	=	rate constant (M^{-1} s^{-1})
k'	=	rate constant ($M^{-0.6}$ min^{-1})
k_{obs}	=	observed pseudo first-order rate constant (s^{-1}, min^{-1}, hr^{-1} or yr^{-1})
k_{sa}	=	k_{obs} normalized to the surface area concentration (L m^{-2} hr^{-1})
n	=	number of electron transferred in the reaction
n_H	=	number of moles of protons exchanged per mole of electrons
$p\varepsilon^0$	=	standard pε at 25 ^{0}C and 1 atm
$p\varepsilon^0$ (W)	=	$p\varepsilon^0$ in neutral water
	=	$p\varepsilon^0 + \frac{n_H}{2}\log K_w$
P_{H_2}	=	partial pressure of hydrogen gas (atm)
R	=	ideal gas constant (8.314×10^{-3} kJ mol^{-1} K^{-1})
T, T_1, T_2	=	absolute temperature (K)
$t_{1/2}$	=	half-life (sec, min, hr or yr)
γ_i	=	activity coefficient of species i (kg mol^{-1})
λ_s	=	surface area concentration (m^2 L^{-1})
Ω	=	saturation state

References

Alowitz, M. J., and Scherer, M. M. (2002). "Kinetics of nitrate, nitrite, and Cr(VI) reduction by iron metal." *Environ. Sci. Technol.*, 36(3), 299-306.

Andrea, P. D., Lai, K. C.-K., Kjeldsen, P., and Lo, I. M.-C. (2005). "Effect of groundwater inorganics on the reductive dechlorination of TCE by zero-valent iron." *Water, Air, and Soil Pollut.*, 162, 401-420.

Appelo, C. A. J., and Postma, D. (1993). *Geochemistry, groundwater and pollution*, Rotterdam, Brookfield, Vt., USA.

Arnold, W. A., and Roberts, A. L. (1998). "Pathways of chlorinated ethylene and chlorinated acetylene reaction with Zn(0)." *Environ. Sci. Technol.*, 32(19), 3017-3025.

Bloomfield, C., and Pruden, G. (1980). "The behavior of Cr(VI) in soil under aerobic and anaerobic conditions." *Environ. Pollut.*, 23, 103-114.

Blowes, D. W., Ptacek, C. J., and Jambor, J. L. (1997). "*In-situ* remediation of Cr(VI)-contaminated groundwater using permeable reactive walls: Laboratory studies." *Environ. Sci. Technol.*, 31(12), 3348-3357.

Blowes, D. W., Puls, R. W., Gillham, R. W., Ptacek, C. J., Bennett, T. A., Bain, J. G., Hanton-Fong, C. J., and Paul, C. J. (1999). *An in situ permeable reactive barrier for the treatment of hexavalent chromium and trichloroethylene in ground water: Volume 2 performance monitoring*, EPA/600/R-99/095b, Office of Research and Development, United States Environmental Protection Agency, Washington, D. C. 20460, USA.

Boronina, T., Klabunde, K. J., and Sergeev, G. (1995). "Destruction of organohalides in water using metal particles: Carbon tetrachloride/water reactions with magnesium, tin and zinc." *Environ. Sci. Technol.*, 29(6), 1511-1517.

Boronina, T., Lagadic, I., and Klabunde, K. J. (1998). "Activated and nonactivated forms of zinc posers: Reactivity toward chlorocarbons in water and AFM studies of surface morphologies." *Environ. Sci. Technol.*, 32(17), 2614-2622.

Buerge, I. J., and Hug, S. J. (1997). "Kinetics and pH dependence of chromium(VI) reduction by iron(II)." *Environ. Sci. Technol.*, 31(5), 1426-1432.

Butler, E. C., and Hayes, K. F. (1999). "Kinetics of the transformation of trichloroethylene and tetrachloroethylene by iron sulfide." *Environ. Sci. Technol.*, 33(12), 2021-2027.

Cabel, B., Kozicki, R., Lahl, U., Podbielski, A., Stachel, B., and Struss, S. (1982). "Pollution of drinking water with nitrate." *Chemosphere*, 11(11), 1147-1154.

Cantrell, K. J., Kaplan, D. I., and Wietsma, T. W. (1995). "Zero-valent iron for the in situ remediation of selected metals in groundwater." *J. Hazard. Mater.*, 42(2), 201-212.

Cheng, I. F., Muftikian, R., Fernando, Q., and Korte, N. (1997). "Reduction of nitrate to ammonia by zero-valent iron." *Chemosphere*, 35(11), 2689-2695.

Choe, S., Chang, Y. Y., Hwang, K. Y., and Khim, J. (2000). "Kinetics of reductive denitrification by nanoscale zero-valent iron." *Chemsophere*, 41(8), 1307-1311.

Choe, S., Liljestrand, H. M. and Khim, J. (2004). "Nitrate reduction by zero-valent iron under different pH regimes." *Applied Geochemistry*, 19(3), 335-342.

Christensen, T. H., Bjerg, P. L., Banwart, S. A., Jakobsen, R., Heron, G., and Albrechtsen, H. J. (2000). "Characterization of redox conditions in groundwater contaminant plumes." *J. Contam. Hydrol.*, 45(3-4), 165-241.

Chuang, F. W., Larson, R. A., and Wessman, M. S. (1995). "Zero-valent iron-promoted dechlorination of polychlorinated biphenyls." *Environ. Sci. Technol.*, 29(9), 2460 – 2463.

Dean, J. A. (1985). *Lange's handbook of chemistry*, 13th Ed., McGraw-Hill, New York.

Doyle, J. G., Miles, T. A., Parker, E., and Cheng, I. F. (1998). "Quantification of total polychlorinated biphenyl by dechlorination to biphenyl by Pd/Fe and Pd/Mg bimetallic particles." *Microchem. J.*, 60(3), 290-295.

Dubrovsky, N. M., Morin, K. A., Cherry, J. A., and Smyth, D. J. A. (1984) "Uranium tailings acidification and subsurface contaminant migration in a sand aquifer." *Water Poll. Res. J. Canada*, 19, 55-89.

Eary, L. E., and Rai, D. (1988). "Chromate removal from aqueous wastes by reduction with ferrous ion." *Environ. Sci. Technol.*, 22(8), 972-977.

Farrell, J., Kason, M., Melitas, N., and Li, T. (2000). "Investigation of the long-term performance of zero-valent iron for reductive dechlorination of trichloroethylene." *Environ. Sci. Technol.*, 34(3), 514-521.

Fendorf, S. E., and Li, G. C. (1996). "Kinetics of chromate reduction by ferrous iron." *Environ. Sci. Technol.*, 30(5), 1614-1617.

Fennelly, J. P., and Roberts, A. L. (1998). "Reaction of 1,1,1-trichloroethane with zero-valent metals and bimetallic reductants." *Environ. Sci. Technol.*, 32(13), 1980-1988.

Fetter, C. W. (1999). *Contaminant hydrogeology*, 2nd Ed., Prentice-Hall, Inc., Simon & Schuster/A Viacom Company, Upper Saddle River, New Jersey 07458, USA.

Freeze, R. A., and Cherry, J. A. (1979). *Groundwater*, Prentice-Hall, Inc., Englewood Cliffs, New Jersey 07632.

Grittini, C., Malcomson, M., Fernando, Q., and Korte, N. (1995). "Rapid dechlorination of polychlorinated biphenyls on the surface of a Pd/Fe bimetallic system." *Environ. Sci. Technol.*, 29(11), 2898-2156.

Gotpagar, J., Grulke, E., Tsang, T., and Bhattacharyya, D. (1997). "Reductive dehalogenation of trichloroethylene using zero-valent iron." *Environ. Prog.*, 16(2), 137-143.

Gu, B., Liang, L., Dickey, M. J., Yin, X., and Dai, S. (1998). "Reductive precipitation of uranium(VI) by zero-valent iron." *Environ. Sci. Technol.*, 32(21), 3366-3373.

Hayes, M. H. B. (1985). *Humic substances in soil, sediment, and water*, John Wiley and Sons, New York.

Henderson, T. (1994). "Geochemical reduction of hexavalent chromium in the Trinity Sand Aquifer." *Ground Water*, 32(3), 477-486.

Hester, R. E., and Harrison, R. M. (1996). *Chlorinated organic micropollutants*, Royal Society of Chemistry, Cambridge, UK.

Ho, C. H., and Miller, N. H. (1986). "Adsorption of uranyl species from bicarbonate solution onto hematite particles." *J. Colloid Interface Sci.*, 110(1), 165-171.

Hsi, C. K. D., and Langmuir, D. (1985). "Adsorption of uranyl onto ferric oxyhydroxides: Application of the surface complexation site-binding model." *Geochim. Cosmochim. Acta.*, 49(9), 1931-1941.

Huang, C. P., Wang, H. W., and Chiu, P. C. (1998). "Nitrate reduction by metallic iron." *Wat. Res.*, 32(8), 2257-2264.

Jafvert, C. T., and Valentine, R. L. (1992). "Reaction scheme for the chlorination of ammoniacal water." *Environ. Sci. Technol.*, 26(3), 577-586.

Johnson, T. L., Scherer, M. M., and Tratnyek, P. G. (1996). "Kinetics of halogenated organic compound degradation by iron metal." *Environ. Sci. Technol.*, 30(8), 2634 – 2640.

Kiely, G. (1998). *Environmental engineering*, International Eds., Irwin McGraw-Hill.

Kim, Y. H., Shin, W. S., and Ko, S. O. (2004). "Reductive dechlorination of chlorinated biphenyls by palladized zero-valent metals." *Journal of Environmental Science & Health Part A-Toxic/Hazardous Substances & Environmental Engineering*, 39(5), 1177-1188.

Korte, N. E., West, O. R., Liang, L., Gu, B., Zutman, J. L., and Fernando, Q. (2002). "The effect of solvent concentration on the use of palladized-iron for the step-wise dechlorination of polychlorinated biphenyls in soil extracts." *Waste Management*, 22(3), 343-349.

Lai, K. C.-K., and Lo, I. M.-C. (2002). "Bench-scale study of the effects of seepage velocity on the dechlorination of TCE and PCE by zero-valent iron." *The 6th International Symposium on Environmental Geotechnology*, July 2-5, Seoul, Korea.

Lai, K. C.-K., and Lo, I. M.-C. (2003). "Field monitoring of the performance of a PRB at the Vapokon site, Denmark." *Permeable Reactive Barriers (PRBs) Action Team Meeting*, RTDF, October 15-16, Niagara Falls, New York.

Lai, K. C.-K., Lo, I. M.-C., and Kjeldsen, P. (2004). "Remediation of chlorinated aliphatic hydrocarbons in groundwater using Fe^0 PRB." *1st International Symposium on Permeable Reactive Barriers*, PRB-Net, March 14-16, Belfast, Northern Ireland.

Lam, S. C. (2004). *Geochemistry effect and competitive effect of TCE on the reactivity of zero-valent iron for chromium removal*, MPhil Thesis, The Hong Kong University of Science and Technology.

Langmuir, D. (1978). "Uranium solution-mineral equilibria at low temperatures with applications to sedimentary ore deposits." *Geochim. Cosmochim. Acta.*, 42(6), 547-569.

Lee, W. J., and Batchelor, B. (2002). "Abiotic reductive dechlorination of chlorinated ethylenes by iron-bearing soil minerals. 2. Green rust." *Environ. Sci. Technol.*, 36(24), 5348-5354.

Liang, L. Y., Korte, N., Gu, B., Puls, R., and Reeter, C. (2000). "Geochemical and microbial reactions affecting the long-term performance of in situ "iron barriers." *Adv. Environ. Res.*, 4(4), 273-286.
Liang, L. Y., Korte, N., Gu, B., Puls, R., and Reeter, C. (2000). "Geochemical and microbial reactions affecting the long-term performance of in situ "iron barriers." *Adv. Environ. Res.*, 4(4), 273-286.
Lo, I. M.-C., and Lai, K. C.-K. (2002). "Preliminary design of permeable reactive wall based on column studies." *The 6th International Congress on Environmental Geotechnics*, ISEG, August 11-15, Rio de Janeiro, Brazil.

Lo, I. M. C., Lam, S. C., and Chan, K. K. (2003a). "Effects of groundwater characteristics on chromate removal by permeable reactive barriers." *International Conference on Contaminated Land,* 12-16 May, Gent, Belgium.

Lo, I. M.-C., Lai, K. C.-K., and Kjeldsen, P. (2003b). "Performance of permeable reactive barrier on remedying aliphatic chlorinated organic contaminated groundwater." *International Conference on Soil and Groundwater Contamination & Cleanup in Arid Countries*, ICSGCC, January 20-23, Muscat, Sultanate of Oman.

Loyaux-Lawniczak, S., Refait, P., Ehrhardt, J. J., Lecomte, P., and Génin, J. M. R. (2000). "Trapping of Cr by formation of ferrihydrite during the reduction of chromate ions by Fe(II)-Fe(III) hydroxysalt green rusts." *Environ. Sci. Technol.*, 34(3), 438-443.

Matheson, L. J., and Tratnyek, P. G. (1994). "Reductive dehalogenation of chlorinated methanes by iron metal." *Environ. Sci. Technol.*, 28(12), 2045-2053.

McCarty, P. L. (1997). "Breathing with chlorinated solvents." *Science*, 276(5318), 1521-1522.

McCarty, P. L. (1999). *Chlorinated organics*, American Academy of Environmental Engineers, Annapolis, Md.

Montgomery, J. H. (2000). *Groundwater chemicals desk reference*, 3rd Ed., Lewis Publishers, CRC Press LLC.

Morin, K. A., Cherry, J. A., Dave, N. K., Lim, T. P., and Vivyurka, A. J. (1988) "Migration of acidic groundwater seepage from uranium-tailings impoundments, 1. Field study and conceptual hydrogeochemical model." *J. Contam. Hydrol.*, 2(4), 271-303.

National Research Council (1994). *Alternatives for ground water cleanup*, National Academy Press, Washington, D. C., USA.

National Research Council (2000). *Natural attenuation for groundwater remediation*, Committee on Intrinsic Remediation, Water Science and Technology Board, Board on Radioactive Waste Management, as well as Commission on Geosciences, Environment, and Resources, National Academy Press, Washington, D. C.

Nelson, D. W., and Sommers, L. E. (1982). *Methods of soil analysis, part 2*, American Society of Agronomy, Inc. and Soil Science Society of America, Inc., Madison WI.

Nielson, A. E. (1964). *Kinetics of precipitation*, Macmillan.

Nriagu, J. O., and Nieboer, E. (1988). *Chromium in the natural and human environments*, A Wiley-Interscience Publication, John Wiley and Sons, New York.

Olsen, C. R., Lowry, P. D., Lee, S. Y., Larsen, I. L., and Cutshall, N. H. (1986). "Geochemical and environmental processes affecting radionuclide migration from a formerly used seepage trench." *Geochim. Cosmochim. Acta*, 50(4), 593-607.

Orth, W. S. (1992). *Mass balance of the degradation of trichloroethylene in the presence of iron filings*, M. Sc. Thesis, Department of Earth Sciences, University of Waterloo, Ontario, Canada.

Ott, N. (2000). *Permeable reactive barriers for inorganics*, National Network of Environmental Management Studies Fellow, Office of Solid Waste and Emergency Response, United States Environmental Protection Agency, Washington, D. C. 20460, USA.

Palmer, C. D., and Wittbrodt, P. R. (1991) "Processes affecting the remediation of chromium-contaminated sites." *Environmental Health Perspectives*, 92, 25-40.

Peterson, M. L., White, A. F., Brown, G. E., and Parks, G. A. (1997). "Surface passivation of magnetite by reaction with aqueous Cr(VI): XAFS and TEM results." *Environ. Sci. Technol.*, 31(5), 1573-1576.

Powell, R. M., Puls, R. W., Hightower, S. K., and Sabatini, D. A. (1995). "Coupled iron corrosion and chromate reduction: Mechanisms for subsurface remediation." *Environ. Sci. Technol.*, 29(8), 1913-1922.

Powell, R. M., Blowes, D. W., Gillham, R. W., Schultz, D., Sivavec, T., Puls, R. W., Vogan, J. L., Powell, P. D., and Landis, R. (1998). *Permeable reactive barrier technologies for contaminant remediation*, EPA/600/R-98/125, Office of Research and Development, United States Environmental Protection Agency, Washington, D. C. 20460, USA.

Rai, D., and Zachara, J. M. (1988). "Chromium reactions in geologic materials." *Electric Power Research Institute*, EA-5741.

Richard, F. C., and Bourg, A .C. M. (1991). "Aqueous geochemistry of chromium: A review" *Wat. Res.*, 25(7), 807-816.

Ritter, K., Odziemkowski, M. S., and Gillham, R. W. (2002). "An in situ study of the role of surface films on granular iron in the permeable iron wall technology." *J. Contam. Hydrol.*, 55(1-2), 87-111.

Roberts, A. L., Totten, L. A., Arnold, W. A., Burris, D. R., and Campbell, T. J. (1996). "Reductive elimination of chlorinated ethylenes by zero-valent metals." *Environ. Sci. Technol.*, 30(8), 2654-2659.

Rügge, K., Bjerg, P. L., Pedersen, J. K., Mosbæk, H., and Christensen, T. H. (1999). "An anaerobic field injection experiment in a landfill leachate plume, Grindsted, Denmark 1. Experimental setup, tracer movement, and fate of aromatic and chlorinated compounds." *Water Resour. Res.*, 35(4), 1231-1246.

Schwertmann, U., Gasser, U., and Sticher, H. (1989). "Chromium-for-iron substitution in synthetic goethites," *Geochim. Cosmochim. Acta*, 53(6), 1293-1297.

Seidell, A. (1958). *Solubilities*, 1, 4th Eds., American Chemical Society, D. van Nostrand Co., Princeton, NJ.

Sedlak, D. L., and Chan, P. G. (1997). "Reduction of hexavalent chromium by ferrous iron." *Geochim. Cosmochim. Acta*, 61(11), 2185-2192.

Semprini, L., Kitanidis, P. K., Kampbell, D. H., and Wilson, J. T. (1995). "Anaerobic transformation of chlorinated aliphatic hydrocarbons in a sand aquifer based on spatial chemical distribution." *Water Resour. Res.*, 31(4), 1051-1062.

Shriver, D., and Atkins, P. (1999). *Inorganic chemistry*, 3rd Ed., W. H. Freeman and Company, New York.

Sillen, L. G., and Martell, A. E. (1964). *Stability constants of metal ion complexes*, Special Publication No. 16, The Chemical Society, London.

Simon, F. G., Meggyes, T., and McDonald, C. (2002). *Advanced groundwater remediation: Active and passive technologies*, Thomas Telford.

Snoeyink, V. L., and Jenkins, D. (1980). *Water chemistry*, John Wiley & Sons, New York.

Stollenwerk, K. G., and Grove, D. B. (1985). "Reduction of hexavalent chromium in water samples acidified for preservation." *J. Environ. Qual.*, 14(3), 395-399.

Stumm, W. (1992). *Chemistry of the solid-water interface: Processes at the mineral-water and particle-water interface in natural systems*, Wiley, New York.

Stumm, W., and Morgan, J. J. (1970). *Aquatic chemistry*, Wiley-Interscience, New York.

Stumm, W., and Morgan, J. J. (1996). *Aquatic chemistry: Chemical equilibria and rates in natural waters*, 3rd Ed., John Wiley & Sons, New York.

Su, C., and Puls, R. W. (2004). "Nitrate reduction by zerovalent iron: Effects of formate, oxalalte, citrate, chloride, sulfate, borate, and phosphate." *Environ. Sci. Technol.*, 38(9), 2715-2720.

Till, B. A., Weathers, L. J., and Alvarez, P. J. J. (1998). "Fe(0)-supported autotrophic denitrification." *Environ. Sci. Technol.*, 32(5), 634-639.

United States Environmental Protection Agency (2000). *In site treatment of soil and groundwater contaminated with chromium, technical resource guide*, EPA-625-R-00-005, Office of Research and Development, Washington, D. C. 20460, USA.

Walton, A. G. (1967). *The formation and properties of precipitates*, Wiley-Interscience, New York.

Warren, K. D., Arnolod, R. G., Bishop, T. L., Lindholm, L. C., and Betterton, E. A. (1995). "Kinetics and mechanism of reductive dehalogenation of carbon tetrachloride using zero-valence metals." *J. Hazard. Mater.*, 41(2-3), 217-227.

Weber, K. A., Picardal, F. W., and Roden, E. E. (2001). "Microbially catalyzed nitrate-dependent oxidation of biogenic solid-phase Fe(II) compounds." *Environ. Sci. Technol.*, 35(8), 1644-1650.

Westerhoff, P. and James, J. (2003). "Nitrate removal in zero-valent iron packed columns." *Wat. Res.*, 37(8), 1818-1830.

Wiberg, K. B. (1965). *Oxidation in Organic Chemistry*, Academic Press, New York.

Williams, A. G. B., and Scherer, M. M. (2001). "Kinetics of Cr(VI) reduction by carbonate green rust." *Environ. Sci. Technol.*, 35(17), 3488-3494.

Yak, H. K., Lang, Q., and Wai, C. M. (2000). "Relative resistance of positional isomers of polychlorinated biphenyls toward reductive dechlorination by zerovalent iron in subcritical water." *Environ. Sci. Technol.*, 34(13), 2792 – 2798.

CHAPTER 6

Chemical Reactive Barriers

Irene M. C. Lo and Keith C. K. Lai

6.1 Introduction

Reactive barrier technology was first proposed for groundwater remediation in the 1970s when an *in situ* barrier containing limestone was used for the attenuation of acidity and dissolved metals in acidic mine drainage (Pearson and McDonnell 1975; Hedin et al. 1994; Naftz et al. 2002). In the early 1990s, recognition of the efficacy of zero-valent iron (Fe^0) in the abiotic reduction of chlorinated aliphatic hydrocarbons (CAHs) led to the first Fe^0-based reactive barrier installation at the CAH contaminated Canadian Forces Base (CFB) Borden site at Ontario, Canada. The success of the reactive barrier at the Borden site subsequently initiated a period of active installations resulting in the construction of more than 40 reactive barriers between 1993 and 2001 to treat CAHs, metals, nutrients and radionuclides in groundwater (Lo and Lai 2002; Lo et al. 2005; Naftz et al. 2002; USEPA 2002).

Unlike a physical barrier containing contaminated groundwater and preventing flow of groundwater into a zone of contamination (Lo 2003), a reactive barrier provides a high permeability reactive zone for contaminated groundwater flow. Powell and Puls (1997) stated that a reactive barrier involves an emplacement of reactive materials in the subsurface to intercept a contaminant plume. As contaminated groundwater passes through the reactive zone, contaminants are either immobilized or transformed into more desirable states via the processes of abiotic reduction, reductive precipitation or adsorption, thereby attaining remediation concentration goals downgradient of the barrier (Bedient et al. 1999; Powell et al. 1998). Although the chemical reactive barrier technology are known by various names, such as a limestone drain (Hedin et al. 1994), geochemical barrier (Longmire et al. 1991) or an *in situ* treatment curtain (Blowes et al. 1995), permeable reactive barrier (PRB) is the preferred terminology (Naftz et al. 2002).

6.2 Mechanisms of Groundwater Remediation by PRBs

Removal mechanisms used by PRBs for groundwater remediation depend on the types of contaminants to be treated. For instance, the PRB at the Denver Federal Center, Colorado, utilizes abiotic reduction for the remediation of CAH contaminated groundwater (McMahon et al. 1999) while the PRB installed at United States Coast Guard (USCG) Support Center near Elizabeth City, North Carolina relies on the principle of reductive precipitation for the removal of hexavalent chromium [Cr(VI)] from the groundwater (Puls et al. 1999). The PRB at a former industrial site in Brunnam Gebirge, Austria utilizes adsorption as a mechanism for the remediation of aromatic hydrocarbons polluted groundwater (USEPA 2002).

6.2.1 Abiotic Reduction

Abiotic reduction of contaminants, such as CAHs, polychlorinated biphenyls (PCBs), nitrate, chlorate and bromate requires the release of electrons from oxidation reactions (Gillham and O'Hannesin 1994; Westerhoff 2003; Kim et al. 2004). These reactions can only proceed in the presence of reductants or electron donors. From a thermodynamic perspective, abiotic reduction of a contaminant occurs if the selected reductant possesses a lower standard potential (E^0) of half reaction than that of the contaminant (Christensen et al. 2000). Among various reductants, such as ferrous iron (Fe^{2+}) in minerals and solution, reduced sulfur compounds, soil organic carbon, hydrogen gas and zero-valent metals (Eary and Rai 1988; Palmer and Wittbrodt 1991; Matheson and Tratnyek 1994; Arnold and Roberts 1998; Arnold and Roberts 2000; Williams and Scherer 2001), Fe^0 is the most commonly used reactive material in PRBs because of its low cost, high reactivity and long-term stability (Hardy and Gillham 1996; Lai and Lo 2002; Scherer et al. 2000; USEPA 2002).

Through oxidation reactions with water [Eqs. (6.1) and (6.2)], electrons released from Fe^0 are directly transferred from the Fe^0 surface to the contaminants for the abiotic reduction. The Fe^{2+} released from the Fe^0 oxidation reactions provides another source for indirect transfer of electrons to the contaminants [Eq. (6.3)] (Powell et al. 1998). Besides, Matheson and Tratnyek (1994) mentioned another pathway of indirect electron transfer involving hydrogen gas released from the anaerobic oxidation of Fe^0 with water [i.e. Eq (6.1)] for the abiotic reduction of CAHs [Eq. (6.4)]. However, this redox reaction can only occur in the presence of a catalyst (House 1972).

$$Fe^0_{(s)} + 2H_2O_{(l)} \Leftrightarrow H_{2(g)} + Fe^{2+} + 2OH^- \qquad \text{(Eq. 6.1)}$$

$$2Fe^0_{(s)} + O_{2(aq)} + 2H_2O_{(l)} \Leftrightarrow 2Fe^{2+} + 4OH^- \qquad \text{(Eq. 6.2)}$$

$$4Fe^{2+} + 4H^+ + O_{2(aq)} \Leftrightarrow 4Fe^{3+} + 2H_2O_{(l)} \qquad \text{(Eq. 6.3)}$$

$$RHCl{=}RHCl + H_{2(g)} \xrightarrow{Catalyst} RH_2{=}RHCl + H^+ + Cl^- \quad \text{(Eq. 6.4)}$$

6.2.1.1 Reduction of Chlorinated Aliphatic Hydrocarbons

Chlorinated aliphatic hydrocarbons (CAHs), such as tetrachloroethylene (PCE), trichloroethylene (TCE) and 1,1,1-trichloroethane (1,1,1-TCA), are frequently detected as groundwater contaminants at hazardous sites because of their extensive usage as dry cleaning and metal degreasing agents (Orth 1992; National Research Council 1994; Fetter 1999; Montgomery 2000; Lai and Lo 2002). As described by Arnold and Roberts (2000), hydrogenolysis [Eq. (6.5)] and reductive-β-elimination [Eq. (6.6)] are the main pathways for the abiotic reduction of CAHs using Fe^0 as a reductant.

$$RHCl{=}RHCl + H^+ + 2e^- \rightarrow RH_2{=}RHCl + Cl^- \quad \text{(Eq. 6.5)}$$

$$RHCl{=}RHCl + 2e^- \rightarrow RH{\equiv}RH + 2Cl^- \quad \text{(Eq. 6.6)}$$

Hydrogenolysis involves the replacement of chlorine atom with hydrogen atom, whereas reductive-β-elimination involves release of two chlorine atoms or one chlorine and one hydrogen atom with concomitant formation of a triple bond (Su and Puls 1999). Through these pathways, CAHs are sequentially reduced to less chlorinated and non-chlorinated byproducts, which subsequently can be reduced to ethene and ethane via hydrogenation reaction (Ritter et al. 2002).

Abiotic reduction of CAHs using Fe^0 requires direct electron transfer from the Fe^0 surface to contaminants necessitating close contact of Fe^0 and the CAHs (Weber 1996). Therefore, the rate of CAH degradation by Fe^0 is mainly dependent upon the rates of mass transfer of CAHs between bulk solution and Fe^0 surface via the stagnant fluid layer surrounding the Fe^0 (i.e., boundary layer) and the electron transfer from Fe^0 to CAHs (Fogler 1992). The mass transfer rate of CAHs is directly proportional to the diffusion coefficient of CAHs in aqueous environment and inversely proportional to the thickness of the boundary layer (Fogler 1992). Groundwater velocity can affect the thickness of the boundary layer potentially influencing the mass transfer coefficient of CAHs, while temperature is a major factor affecting the diffusion coefficient of CAHs and the rate constant for the electron transfer from Fe^0 to CAHs (Fogler 1992; Su and Puls 1999).

Laboratory studies of the influence of groundwater velocity and temperature on the rate of CAH degradation conducted by Lai and Lo (2002) and Lai (2004) indicated that within a range between 31 and 1884 m/yr, groundwater velocity exerted positive effect on the rate of PCE and TCE degradation by Fe^0 (see Figures 6.1a and 6.1b).

Further, raising temperature from 10 to 23 ^{0}C caused an approximately 2-fold increase in the TCE and PCE degradation rate constants. Activation energies of 70.3 kJ/mol for TCE and 38.6 kJ/mol for PCE degradation indicated domination of the electron transfer process over the mass transfer process as a major rate-limiting step in the abiotic reduction of TCE and PCE by Fe^0 (Su and Puls 1999; Lai 2004).

6.2.1.2 Reduction of Polychlorinated Biphenyls

Polychlorinated biphenyls (PCBs) are chlorinated organic compounds that are commonly used as organic diluents, plasticizers and pesticide extenders (Kim et al. 2004). They are recalcitrant in the natural environment and can accumulate in the food chain (Hester and Harrison 1996). Abiotic reduction of PCBs is also a stepwise process in which PCBs are sequentially reduced to less chlorinated and non-chlorinated biphenyls. As illustrated in Eq. (6.7), by using the electrons released from the Fe^0 oxidation, the PCB containing x+y number of chlorine atoms is sequentially reduced into biphenyl. Palladium serves as a catalyst in this reaction.

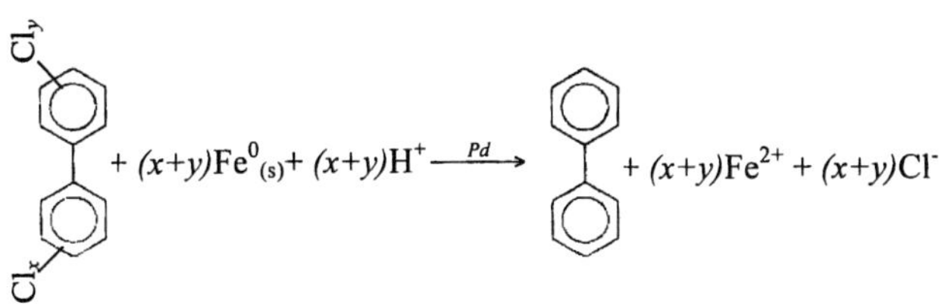

(Eq. 6.7)

6.2.1.3 Reduction of Nitrate, Chlorate and Bromate

Oxo-anions including nitrate (NO_3^-), chlorate (ClO_3^-) and bromate (BrO_3^-) can pose potential health risks when present in drinking water. Nitrate is a contaminant from agriculture and wastewater (Eichler and Schulz 1998), whereas chlorate is a byproduct of chlorine dioxide which is added as a disinfectant during water treatment (Siddiqui 1996). Bromate is formed during ozonation in water treatment plants (Siddiqui et al. 1995; Westerhoff et al. 1998). In the presence of Fe^0, nitrate is first reduced to nitrite [Eq. (6.8)] and subsequently reduced to ammonium [Eq. (6.9)]. Similarly, chloride and bromide are the main byproducts of the abiotic reduction of chlorate [Eq. (6.10)] and bromate [Eq. (6.11)], respectively (Westerhoff 2003).

$$Fe^0_{(s)} + NO_3^- + 2H^+ \Leftrightarrow Fe^{2+} + NO_2^- + H_2O_{(l)} \qquad \text{(Eq. 6.8)}$$

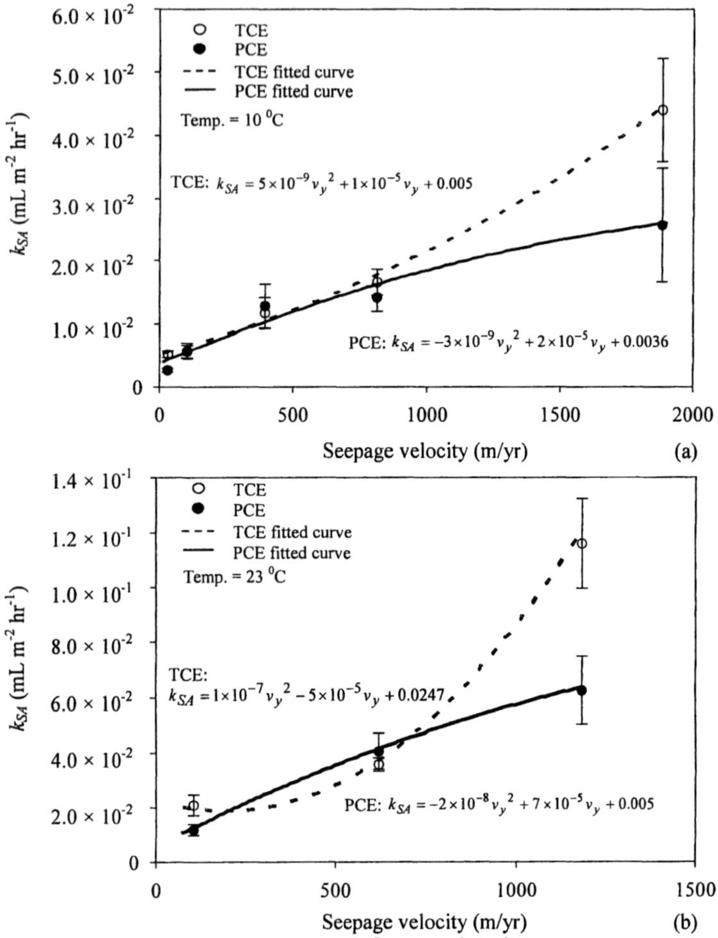

Figure 6.1 Variation of TCE and PCE degradation rate constants with seepage velocity at (a) 10 and (b) 23 ^{0}C. k_{SA} refers to the first-order degradation rate constant normalized to 1 m^2 Fe^0 surface area per millilitre of solution and v_y represents the longitudinal groundwater velocity. The quadratic equations show the mathematical equations of the TCE or PCE fitted curves. (Lai 2004).

$$3Fe^0_{(s)} + NO_2^- + 8H^+ \Leftrightarrow 3Fe^{2+} + NH_4^+ + 2H_2O_{(l)} \quad \text{(Eq. 6.9)}$$

$$ClO_3^- + 6H^+ + 3Fe^0_{(s)} \Leftrightarrow Cl^- + 3H_2O_{(l)} + 3Fe^{2+} \quad \text{(Eq. 6.10)}$$

$$BrO_3^- + 6H^+ + 3Fe^0_{(s)} \Leftrightarrow Br^- + 3H_2O_{(l)} + 3Fe^{2+} \quad \text{(Eq. 6.11)}$$

6.2.2 Reductive Precipitation

Reductive precipitation is another major mechanism utilized for groundwater remediation using PRBs (Powell et al. 1998). It involves first an abiotic reduction of a contaminant into a lower oxidation state and subsequent precipitation of the reduced contaminant on the surface of reactive materials (e.g., Fe^0). Contaminants including chromium (Lo et al. 2003a; Blowes et al. 1995; Powell et al. 1995), uranium (Gu et al. 1998), selenium (McRae et al. 1997) and technetium (Clausen et al. 1995) in groundwater can be effectively removed through this mechanism.

6.2.1.3 Reduction of Nitrate, Chlorate and Bromate

Oxo-anions including nitrate (NO_3^-), chlorate (ClO_3^-) and bromate (BrO_3^-) can pose potential health risks when present in drinking water. Nitrate is a contaminant from agriculture and wastewater (Eichler and Schulz 1998), whereas chlorate is a byproduct of chlorine dioxide which is added as a disinfectant during water treatment (Siddiqui 1996). Bromate is formed during ozonation in water treatment plants (Siddiqui et al. 1995; Westerhoff et al. 1998). In the presence of Fe^0, nitrate is first reduced to nitrite [Eq. (6.8)] and subsequently reduced to ammonium [Eq. (6.9)]. Similarly, chloride and bromide are the main byproducts of the abiotic reduction of chlorate [Eq. (6.10)] and bromate [Eq. (6.11)], respectively (Westerhoff 2003).

6.2.2 Reductive Precipitation

Reductive precipitation is another major mechanism utilized for groundwater remediation using PRBs (Powell et al. 1998). It involves first an abiotic reduction of a contaminant into a lower oxidation state and subsequent precipitation of the reduced contaminant on the surface of reactive materials (e.g., Fe^0). Contaminants including chromium (Lo et al. 2003a; Blowes et al. 1995; Powell et al. 1995), uranium (Gu et al. 1998), selenium (McRae et al. 1997) and technetium (Clausen et al. 1995) in groundwater can be effectively removed through this mechanism.

6.2.2.1 Reductive Precipitation of Chromium

In addition to lead, chromium is the most common inorganic groundwater contaminant detected at hazardous waste sites in United States (National Research Council 1994). Its prevalence is due to the wide application in industries, such as steel production, leather tanning and electroplating over the past several decades (Buerge and Hug 1997; Sedlak and Chan 1997). In natural environment, chromium mainly exists in hexavalent and trivalent forms. Hexavalent chromium [Cr(VI)] existing as bichromate ($HCrO_4^-$), chromate (CrO_4^{2-}) and dichromate ions ($Cr_2O_7^{2-}$) is relatively more toxic and soluble than trivalent chromium [Cr(III)], such as chromium ion (Cr^{3+}) and chromium hydroxyl species [i.e., $Cr(OH)^{2+}$, $Cr(OH)_2^+$, $Cr(OH)_3$ and $Cr(OH)_4^-$] (Nriagu and Nieboer 1988; Richard and Bourg 1991; Palmer and Wittbrodt 1991; USEPA 2000). Therefore, to remove Cr(VI) from groundwater, it is first reduced to trivalent forms using Fe^0 as a reductant [Eqs. (6.12) and (6.13)]. Then the reduced Cr(III) is removed from aqueous solution through the precipitation of chromium hydroxide [Eq. (6.14)], or co-precipitation of mixed chromium-iron hydroxide solids [Eq. (6.15)] (Powell et al. 1995; Blowes et al. 1997) or mixed chromium-iron oxyhydroxide solids [Eq. (6.16)] (Eary and Rai 1988; Schwertmann et al. 1989; Lo et al. 2005) on Fe^0 surface.

$$CrO_4^{2-} + Fe^0_{(s)} + 4H_2O_{(l)} \Leftrightarrow Cr^{3+} + Fe^{3+} + 8OH^- \quad \text{(Eq. 6.12)}$$

$$Cr_2O_7^{2-} + 2Fe^0_{(s)} + 7H_2O_{(l)} \Leftrightarrow 2Cr^{3+} + 14OH^- + 2Fe^{3+} \quad \text{(Eq. 6.13)}$$

$$2Cr^{3+} + 5OH^- \Leftrightarrow Cr(OH)_{3(s)} + Cr(OH)_2^+ \quad \text{(Eq. 6.14)}$$

$$(1-x)Fe^{3+} + xCr^{3+} + 3H_2O_{(l)} \Leftrightarrow Cr_xFe_{1-x}(OH)_{3(s)} + 3H^+ \quad \text{(Eq. 6.15)}$$

$$(1-x)Fe^{3+} + xCr^{3+} + 2H_2O_{(l)} \Leftrightarrow Cr_xFe_{1-x}OOH_{(s)} + 3H^+ \quad \text{(Eq. 6.16)}$$

As mentioned above, reductive precipitation of Cr(VI) by Fe^0 involves first abiotic reduction of Cr(VI) to Cr(III). Therefore, the presence of other contaminants, such as CAHs, nitrate, bromate and chlorate, theoretically can affect the amount of Cr(VI) to be removed by Fe^0 by competing with Cr(VI) for electrons released from the Fe^0 oxidation. Column experiments conducted by Lo et al. (2005) indicated that in the presence of TCE, the Cr(VI) removal capacities of Fe^0 were decreased by about 40% when compared with their respective Cr(VI) removal capacities with identical groundwater geochemistry but without TCE (see Table 6.1). By comparing the results from columns B0 and B1 listed in Table 6.1, the degradation rate constant of TCE was observed to decrease by about 50% when Cr(VI) was singly applied. The degradation rate constant of TCE was further reduced by 75% in the presence of both Cr(VI) and carbonate (compare columns B0 and B2). The findings reported by Lo et al. (2005) conclusively showed the competition between TCE and Cr(VI) for the Fe^0.

Table 6.1 Competitive effects of TCE on Cr(VI) removal by zero-valent iron (from Lo et al. 2005).

Columns	Cr(VI)	TCE	Ca^{2+}	Mg^{2+}	CO_3^{2-}	Cr(VI) Removal Capacity	Normalized First-Order Degradation Rate Constant
	mg/L	mg/L	mg/L	mg/L	mg/L	mg Cr(VI)/gram Fe^0	L hr^{-1} m^{-2}
[b]A1	25	0	0	0	0	4.1	-
[c]A2	25	0	0	0	120	4.2	-
[d]A3	25	0	40	24	120	2.3	-
[a]B0	0	200	0	0	0	-	0.040
[b]B1	25	200	0	0	0	2.2	0.021
[c]B2	25	200	0	0	120	2.5	0.010
[d]B3	25	200	40	24	120	1.4	0.038

Note: The pH of the influent solution for the column experiments (A1 to A3 and B0 to B3) was adjusted to 7.10 using 0.5 N NaOH and 0.5 N HCl.

[a]Control column for TCE [i.e., type B columns]

[b]Columns for investigating the competitive effect in solution without hardness and alkalinity

[c]Columns for investigating the competitive effect in solution spiked with 120 mg/L of carbonate

[d]Columns for investigating the competitive effect in solution spiked with 40, 24 and 120 mg/L of calcium, magnesium and carbonate, respectively.

6.2.2.2 Reductive Precipitation of Uranium

High concentration of uranium is often detected in the groundwater within mine waste piles and leachate derived from mine wastes (Dubrovsky et al. 1984; Olsen et al. 1986; Morin et al. 1988). Uranium exists naturally in the oxidation states +4 [i.e., U(IV)] and +6 [i.e., U(VI)] in which U(VI), for example uranyl ion (UO_2^{2+}), is more mobile than U(IV) such as uraninite (UO_2). According to Eq. (6.17), by using the electrons released from the Fe^0 oxidation, UO_2^{2+} can be first reduced to UO_2, and subsequently precipitated on the Fe^0 surface (Powell et al. 1998; Ott 2000).

$$Fe^0_{(s)} + UO_2^{2+} \Leftrightarrow Fe^{2+} + UO_{2(s)} \qquad \text{(Eq. 6.17)}$$

6.2.3 Adsorption

PRBs can also be designed with materials which can adsorb groundwater contaminants. These materials include metal oxides, zeolite, granular activated carbon and synthetic polymers and resins (Ott 2000). Adsorption of phosphorous by metal oxides, such as ferric oxide (Fe_2O_3) and aluminum oxide (Al_2O_3) has been widely studied by Baker et al. (1996, 1997). As shown in Figure 6.2, Fe_2O_3 or Al_2O_3 usually forms a hydroxylated surface with hydroxide ion and phosphorus adsorption only occurs on the hydroxylated mineral surface. The chemical binding of aqueous phosphate anions to this mineral surface takes place by a specific adsorption mechanism, which results in stable, covalently bonded surface complexes (Goldberg and Sposito 1985).

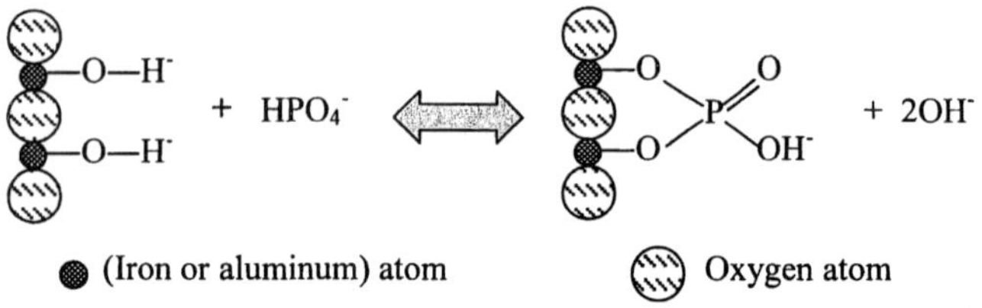

Figure 6.2 Schematic diagram showing specific adsorption of phosphate on the hydroxylated surfaces of metal oxides.

Surfactant modified zeolite (SMZ) is another material used in a PRB for adsorption of organic contaminants as well as anionic (e.g., CrO_4^{2-} and $Cr_2O_7^{2-}$) and cationic metal contaminants. Zeolites are naturally occurring aluminosilicates and their surfaces are usually negatively charged. They have been applied for the removal of cationic contaminants from aqueous solution. Addition of high molecular weight amines, such as hexadecyltrimethylammonium (HDTMA), can modify zeolite's surface. HDTMA can irreversibly exchange with cations on the zeolite's surface but it is too large to access the internal site (see Figure 6.3). Therefore, cationic contaminants can still be adsorbed to the zeolite's internal sites. The organic tail of HDTMA can adsorb organic contaminants by hydrophobic partitioning, while oxyanionic contaminants are attracted to its positive head by electrostatic interaction (Ott 2000).

6.3 Geochemical and Hydraulic Behaviors of PRBs

The oxidation of Fe^0 in groundwater results in chemically reduced conditions and elevated pH, which can affect a number of complexation and mineral reactions. Carbon, oxygen and hydrogen in groundwater can be transformed to lower oxidation states by redox reactions induced by the reducing conditions. Elevated pH can decrease the solubility of various secondary minerals, resulting in precipitation on Fe^0 surface. Moreover, surface complexation of cations on iron oxides can be enhanced at high pH (Yabusaki et al. 2001). As a result, across a Fe^0 PRB, there is significant variation in groundwater geochemical characteristics, which potentially affects the reactivity (Andrea et al. 2005) and hydraulic performance of an iron barrier for long-term remediation of contaminated groundwater (Lai et al. 2004b).

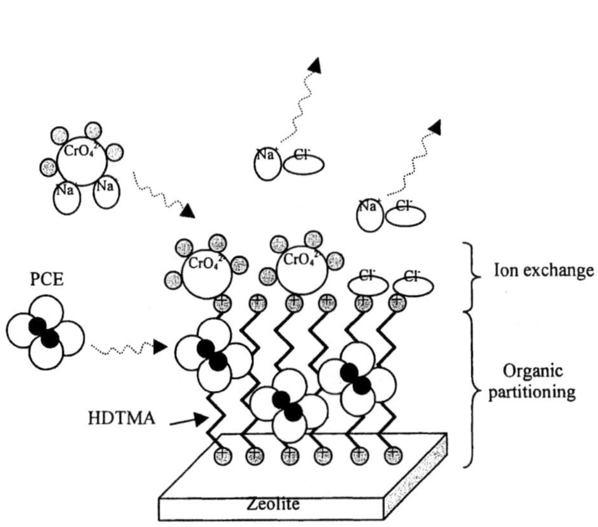

Figure 6.3 Schematic diagram showing the principles of the adsorption of organic and anionic contaminants by surfactant modified zeolite (SMZ) (Ott 2000).

6.3.1 Geochemical Behaviors of PRB

As illustrated in Eqs. (6.1) and (6.2), the hydroxide ions released from Fe^0 oxidation cause an increase in groundwater pH across a Fe^0 PRB. Field monitoring of the Fe^0 PRB at Vapokon site, Denmark revealed that after passing through the barrier, groundwater pH increased by 2.5 to 3.3 units (see Figure 6.4a) (Lo et al. 2003b; Lai et al. 2004a). Similar observations were obtained from the Fe^0 PRB installed at the USCG Support Center in North Carolina for TCE and Cr(VI) remediation in which the pH increased from 6.7 to 11.0 (Blowes et al. 1999b).

Electrons released from Fe^0 oxidation lead to a sharp decrease in redox potential (*Eh*) of groundwater across a Fe^0 PRB. Lai et al (2004a) reported an approximately 250 mV drop (from 105 to -150 mV) in *Eh* across the Fe^0 PRB at Vapokon site which also matched with the results observed from the PRB at USCG Support Center (see Figure 6.5).

By comparing the standard half-reaction potential of dissolved oxygen (DO) (E^0 = 1.27 V) with other groundwater contaminants, such as Cr(VI) (E^0 = 1.33 to 1.51 V) and CAHs (E^0 = 0.37 to 0.63 V), it can be seen that DO in groundwater is also a favorable electron acceptor with respect to Fe^0 (E^0 = - 0.44 V) (Roberts et al. 1996;

favorable electron acceptor with respect to Fe^0 (E^0 = - 0.44 V) (Roberts et al. 1996; Snoeyink and Jenkins 1980; Benjamin 2002). Therefore, DO can also be reduced within an iron barrier [Eq. (6.18)] and quick consumption of dissolved oxygen usually occurs at the entrance of a Fe^0 PRB.

$$O_{2(aq)} + 4H^+ + 4e \Leftrightarrow 2H_2O_{(l)} \quad \text{(Eq 6.18)}$$

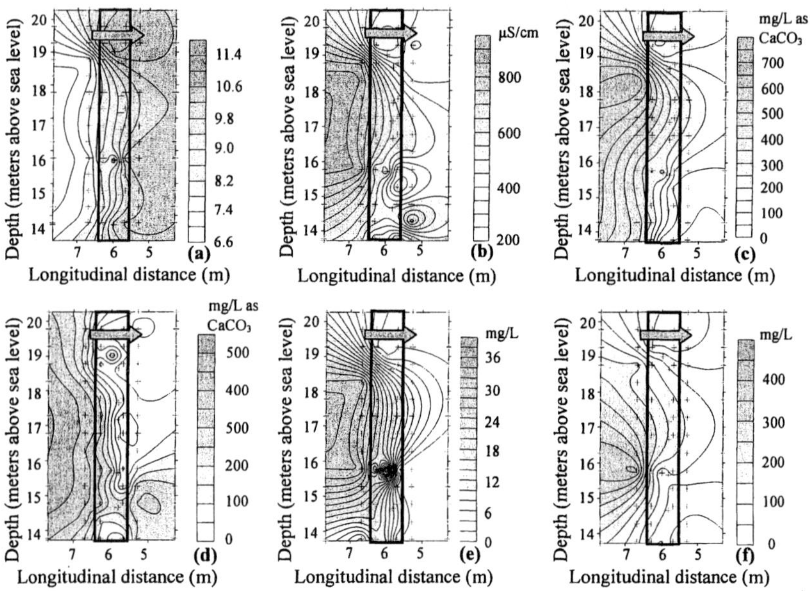

Figure 6.4 Side distributions of (a) pH and (b) groundwater specific conductivity, and concentrations of (c) calcium, (d) total alkalinity, (e) total dissolved iron and (f) sulfate across the permeable reactive barrier at Vapokon site in Sep. 2000 (□ represents the Fe^0 PRB, + refers to monitoring screens and | indicates groundwater flow direction).

Decrease in solubility or increase in stability of minerals under high pH and low redox potential within a Fe^0 PRB encourages precipitation of minerals, thereby resulting in the continuous decrease in concentrations of some common groundwater constituents along an iron barrier. As illustrated in Figures 6.4b to 6.4f, across the Fe^0 PRB at Vapokon site, there were observable decreases in the concentration of calcium ion

(Ca^{2+}), total alkalinity (TALK), total dissolved iron (Fe_T) and sulfate ion (SO_4^{2-}), which resulted in reduced groundwater specific conductivity (Lo et al. 2003b; Lai et al. 2004a). Reduction in the concentration of Ca^{2+}, TALK and Fe_T resulted from the precipitation of aragonite ($CaCO_3$) and siderite ($FeCO_3$), (Eqs. (6.19) and (6.20), respectively) (Naftz et al. 2002; Lo et al. 2005). Sulfur-bearing minerals, such as pyrite (FeS_2) and mackinawite (FeS), can also be formed within a Fe^0 PRB in which sulfide is first formed through biologically mediated reduction of sulfate by sulfate-reducing bacteria [Eqs. (6.21) and (6.22)] (Butler and Hayes 1998; Butler and Hayes 1999; Scherer et al. 2000; Gu et al. 1999; Gu et al. 2002; Wilkin and Puls 2003).

$$Ca^{2+} + CO_3^{2-} \Leftrightarrow CaCO_{3(s)} \qquad \text{(Eq. 6.19)}$$

$$Fe^{2+} + CO_3^{2-} \Leftrightarrow FeCO_{3(s)} \qquad \text{(Eq. 6.20)}$$

$$SO_4^{2-} + 2CH_2O \Leftrightarrow 2HCO_3^- + HS^- + H^+ \qquad \text{(Eq. 6.21)}$$

$$SO_4^{2-} + 4H_{2(g)} \Leftrightarrow S^{2-} + 4H_2O_{(l)} \qquad \text{(Eq. 6.22)}$$

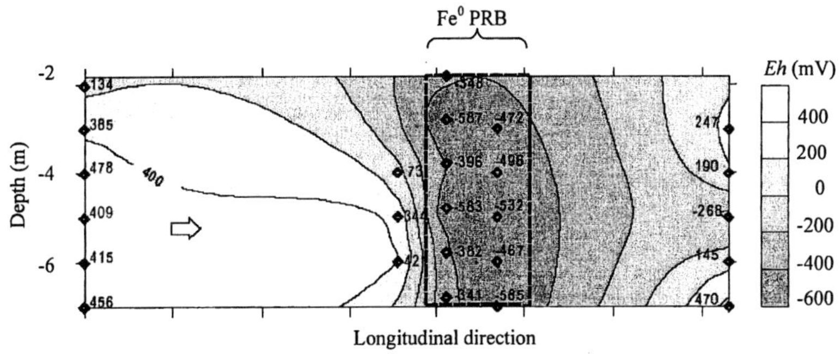

Figure 6.5 Side distribution of redox potential across the permeable reactive barrier at United States Coast Guard (USCG) Support Center in Elizabeth City, North Carolina in Nov. 1996 (Blowes et al. 1999b).

6.3.2 Precipitation of Minerals inside a PRB

Treatment of contaminated groundwater containing high DO encourages the precipitation of ferric hydroxides, oxides and oxyhydroxides, such as hematite (α-Fe_2O_3) and goethite (α-FeOOH). This is because Fe^0 can be oxidized into ferric iron (Fe^{3+}) in the presence of oxygen [Eqs. (6.2) and (6.3)]. These orange, amorphous and gelatinous grains together, thereby effectively blocking a significant fraction of the pore spaces inside a Fe^0 packed medium (Appelo and Postma 1996; Mackenzie et al. 1999; Liang et al. 2000).

Dissolved oxygen is usually consumed quickly at the entrance of an iron barrier. Beyond the entrance zone of the Fe^0 packed medium, the groundwater rapidly become anaerobic. Anaerobic oxidation of Fe^0 [Eq. (6.1)] is predominant and ferrous hydroxide [$Fe(OH)_2$] and magnetite (Fe_3O_4) are the major iron hydroxides or oxides to be precipitated (Lo et al. 2005) in the PRB. Eqs. (6.23) and (6.24) illustrate the chemical equations involved in the precipitation of ferrous hydroxide and magnetite, respectively (Liang et al. 2000; Mackenzie et al. 1999; Ritter et al. 2002).

$$Fe^{2+} + 2OH^- \Leftrightarrow Fe(OH)_{2(s)} \qquad \text{(Eq. 6.23)}$$

$$3Fe^{2+} + 4H_2O_{(l)} \Leftrightarrow Fe_3O_{4(s)} + 6H^+ + H_{2(g)} \qquad \text{(Eq. 6.24)}$$

Since ferrous hydroxide is thermodynamically unstable, it may be further transformed to magnetite through the Schikorr disproportionation reaction [Eq. (6.25)] or to green rust (Reardon et al. 1995; Roh et al. 2000; Ritter et al. 2002). Depending upon groundwater characteristics, carbonate- ($[Fe_4^{2+}Fe_2^{3+}(OH)_{12}]^{2+}[CO_3{\bullet}2H_2O]^{2-}$), chloride- ($[Fe_3^{2+}Fe^{3+}(OH)_8]^+[Cl]^-$) and sulfate-containing green rust ($[Fe_4^{2+}Fe_2^{3+}(OH)_{12}]^{2+}$ $[SO_4{\bullet}2H_2O]^{2-}$) are the types of green rusts formed on Fe^0 surface [Eqs. (6.26) to (6.28)]. Figure 6.6 shows photomicrographs of various minerals found in Fe^0 PRBs (Naftz et al. 2002; Roh et al. 2000).

$$3Fe(OH)_{2(s)} \Leftrightarrow Fe_3O_{4(s)} + H_{2(g)} + 2H_2O_{(l)} \qquad \text{(Eq. 6.25)}$$

$$6Fe(OH)_{2(s)} + CO_3^{2-} + 2H_2O_{(l)} \Leftrightarrow [Fe_4^{2+}\,Fe_2^{3+}(OH)_{12}]^{2+}[CO_3{\bullet}2H_2O]^{2-}{}_{(s)} + 2e^- \qquad \text{(Eq. 6.26)}$$

$$4Fe(OH)_{2(s)} + Cl^- \Leftrightarrow [Fe_3^{2+}\,Fe^{3+}(OH)_8]^+[Cl]^-{}_{(s)} + e^- \qquad \text{(Eq. 6.27)}$$

$$6Fe(OH)_{2(s)} + SO_4^{2-} + 2H_2O_{(l)} \Leftrightarrow [Fe_4^{2+}\,Fe_2^{3+}(OH)_{12}]^{2+}[SO_4{\bullet}2H_2O]^{2-}{}_{(s)} + 2e^- \qquad \text{(Eq. 6.28)}$$

Precipitation of minerals, such as aragonite, goethite, hematite and siderite, on Fe^0 surface may passivate the Fe^0 in a PRB by masking the redox active sites where the exchange of electrons between the Fe^0 surface and contaminants takes place (Reardon 1995; Ritter et al. 2002). However, magnetite and green rusts do not inhibit the Fe^0 reactivity in groundwater remediation. This is because electron transfer between the Fe^0 and contaminants can still occur via the thin coatings of magnetite and green rust on the Fe^0 surface (Ritter et al. 2002). In marked contrast to aragonite, goethite, hematite and siderite, precipitation of sulfur-bearing minerals, such as mackinawite, pyrite and sulfate-containing green rust, on the Fe^0 surface may enhance the Fe^0 reactivity because they can act as electron donors and mediators as well as possess more reactive sites per unit surface area than the Fe^0 (Kriegman-King and Reinhard 1994; Butler and Hayes 1998; Butler and Hayes 1999; Lee and Batchelor 2002a; Lee and Batchelor 2002b; Butler and Hayes 2001; Wilkin et al. 2003).

Figure 6.6 Photomicrographs of (a) aragonite ($CaCO_3$) crystal, (b) calcite ($CaCO_3$) in distorted rhomb shape, (c) cubed-shaped siderite ($FeCO_3$), (d) rounded or bytrodial amorphous iron sulfide, (e) goethite (α-FeOOH) and (f) green rust found in iron reactive barriers.

6.3.3. Hydraulic Behavior of PRBs

The influence of mineral precipitates on the Fe^0 reactivity is based upon the type of minerals formed inside a Fe^0 PRB, which, in turn, depends upon the groundwater characteristics. However, mineral precipitation can reduce the void space within a Fe^0 packed medium, thereby potentially deteriorating the barrier's hydraulic performance. Figure 7.7 illustrates the formation of precipitates on the surface of Fe^0 used in the PRB at the USCG Support Center (Wilkin et al. 2002). Approximately, 0.88 and 2.19% losses in porosity per year as a result of mineral precipitation were determined in the PRBs installed at Vapokon site and USCG Support Center, respectively (Lai et al. 2004a; Blowes et al. 1999a).

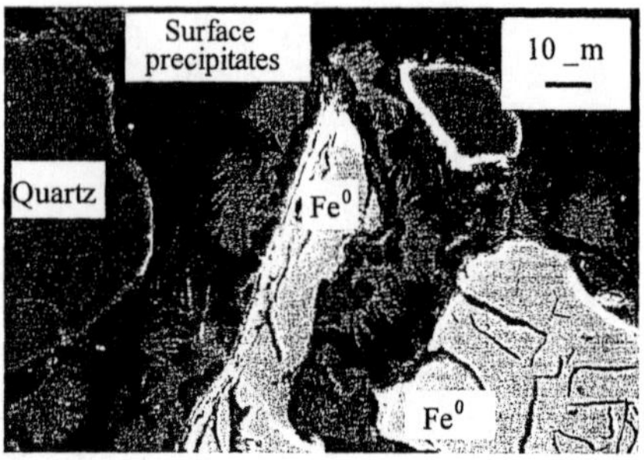

Figure 6.7 Minerals precipitated around the granular Fe^0 used in the permeable reactive barrier at United States Coast Guard (USCG) Support Center in Elizabeth City, North Carolina (Wilkin et al. 2002).

In addition to mineral precipitation, hydrogen gas released from the anaerobic oxidation of Fe^0 [Eq. (6.1)] and excessive microbial growth within a Fe^0 PRB can also lead to significant reductions in the iron barrier's permeability (Mackenzie et al. 1999; Gu et al. 2002). The released hydrogen gas may become entrapped within the void space of a Fe^0 PRB, forming a thin film of hydrogen gas on the Fe^0 surface. Excessive microbial growth can lead to a significant accumulation of biofilm on the Fe^0 surface. These decreases in the void space inside a Fe^0 PRB can deteriorate the PRB's permeability and induce bypass or even reversed flow of contaminated groundwater, that prevents the contaminants from flowing into a Fe^0 PRB (Ebert et al. 2003). As reported by Lai et al (2004b), after 4 years of operation, there was approximately 50% loss of permeability of the iron barrier at the Vapokon site. Changes in the perpendicular flow patterns of groundwater were also observed after one year of operation (see Figure 6.8). Since minerals precipitate locally rather than uniformly inside the Fe^0 PRB, there was preferential flow of contaminated groundwater across the reactive barrier (see Figure 6.9).

6.4 Conclusions

Chemical reactive barriers known as PRBs have been widely used for groundwater remediation since 1990s. PRBs are constructed by the placement of reactive materials in the subsurface designed to intercept a contaminant plume, provide a flow path through the reactive medium, and transform the contaminants into environmentally acceptable forms in the downgradient of the reactive barrier. Groundwater contaminants, such as CAHs, metals, nutrients and radionuclides, can be effectively removed from groundwater using a PRB through the processes of abiotic reduction, reductive precipitation and adsorption.

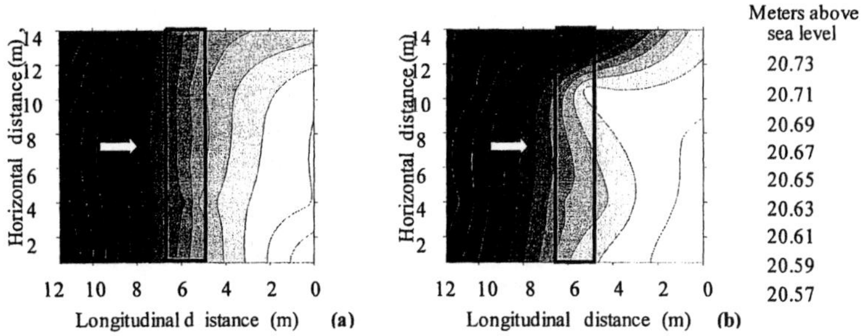

Figure 6.8 Water table contours across the permeable reactive barrier at Vapokon site measured after operating for (a) half year and (b) 3 years (□ represents the Fe^0 PRB and ➪ indicates groundwater flow direction).

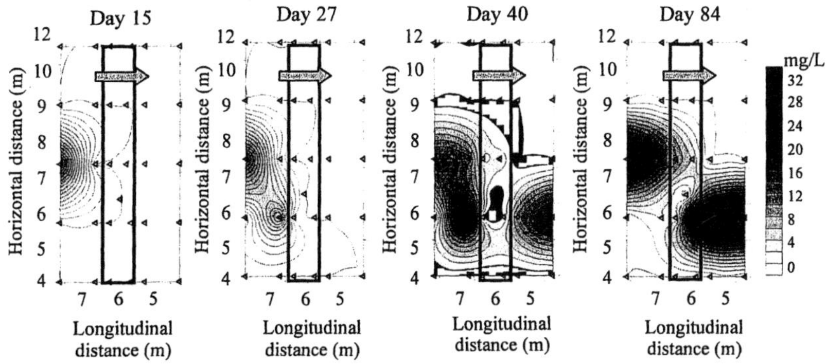

Figure 6.9 Preferential flow of conservative tracer (lithium) across the permeable reactive barrier at Vapokon site (□ represents the Fe^0 PRB, Δ refers to monitoring screens and ➪ indicates groundwater flow direction).

Fe^0 PRBs are commonly used in the remediation of contaminated groundwater because of their relatively low cost and high reactivity. However, it causes high pH and low redox potential conditions within the Fe^0 packed medium and thereby significantly enhances mineral precipitation and complexation reaction within the Fe^0 PRB. Basically, across an iron barrier, there are observable decreases in groundwater specific conductivity as well as reduction in the concentrations of dissolved oxygen, hardness and alkalinity. The precipitation of minerals within an iron barrier deteriorates its hydraulic conductivity and can potentially lead to preferential or bypassing flow of contaminated groundwater through the Fe^0 PRB.

References

Andrea, P. D., Lai K. C.-K., Kjeldsen, P., and Lo, I. M.-C. (2005). "Effect of groundwater inorganics on the reductive dechlorination of TCE by zero-valent iron." *Water, Air, and Soil Pollut.*, 162, 401-420.

Appelo, C., and Postma, D. (1996). *Geochemistry, groundwater and pollution*, A.A. Balkema, Rotterdam, Netherlands.

Arnold, W. A., and Roberts, A. L. (1998). "Pathways of chlorinated ethylene and chlorinated acetylene reaction with Zn(0)." *Environ. Sci. Technol.*, 32(19), 3017-3025.

Arnold, W. A., and Roberts, A. L. (2000). "Pathways and kinetics of chlorinated ethylene and chlorinated acetylene reaction with Fe(0) particles." *Environ. Sci. Technol.*, 34(9), 1794-1805.

Baker, M. J., Blowes, D. W., and Ptacek, C. J. (1996). "Development of a reactive mixture to remove phosphorous from on-site wastewater disposal systems." *Water Centre for Groundwater Research Annual Septic System Conference, Disposal Trenches, Pre-treatment and Re-use of Wastewater*, May 13, Waterloo, Ontario

Baker, M. J., Blowes, D. W., and Ptacek, C. J. (1997). "Phosphorous adsorption and precipitation in a permeable reactive wall: Applications for wastewater disposals systems." *International Containment Technology Conference and Exhibition*, Feb. 9-12, St. Petersburg, Florida.

Bedient, P. B., Rifai, H. S., and Newell, C. J. (1999). *Ground water contamination transport and remediation*, 2nd Ed., Prentice Hall PTR, Upper Saddle River, NJ 07458, USA.

Benjamin, M. M. (2002). *Water chemistry*, McGraw-Hill International Edition, Civil Engineering Series.

Blowes, D. W., Ptacek, C. J., Hanton-Fong, C. J., and Jambor, J. L. (1995). "In situ remediation of chromium contaminated groundwater using zero-valent iron." *American Chemical Society, 209th National Meeting*, April 2-7, Anaheim, California.

Blowes, D. W., Puls, R. W., Bennett, T. A., Gillham, R. W., Hanton-Fong, C. J., and Ptacek, C. J. (1997). "In situ porous reactive wall for treatment of Cr(VI) and trichloroethylene in groundwater." *International Containment Conference*, February 9-12, St. Petersburg, FL, 851-857.

Blowes, D. W., Gillham, R. W., Ptacek, C. J., Puls, R. W., Bennett, T. A., O'Hannesin, S. F., Hanton-Fong, C. J., and Bain, J. G. (1999a). *An in situ permeable reactive barrier for the treatment of hexavalent chromium and trichloroethylene in ground water Volume 1 design and installation*, EPA/600/R-99/095a, Office of Research and Development, United States Environmental Protection Agency, Washington, D. C. 20460, USA.

Blowes, D. W., Puls, R. W., Gillham, R. W., Ptacek, C. J., Bennett, T. A., Bain, J. G., Hanton-Fong, C. J., and Paul, C. J. (1999b). *An in situ permeable reactive barrier for the treatment of hexavalent chromium and trichloroethylene in ground water: Volume 2 performance monitoring*, EPA/600/R-99/095b, Office of Research and Development, United States Environmental Protection Agency, Washington, D. C. 20460, USA.

Buerge, I. J., and Hug, S. J. (1997). "Kinetics and pH dependence of chromium(VI) reduction by iron(II)." *Environ. Sci. Technol.*, 31(5), 1426-1432.

Butler, E. C., and Hayes, K. F. (1998). Effects of solution composition and pH on the reductive dechlorination of hexachloroethane by iron sulfide." *Environ. Sci. Technol.*, 32(9), 1276-1284.

Butler, E. C., and Hayes, K. F. (1999). Kinetics of the transformation of trichloroethylene and tetrachloroethylene by iron sulfide." *Environ. Sci. Technol.*, 33(12), 2021-2027.

Butler, E. C., and Hayes, K. F. (2001). "Factors influencing rates and products in the transformation of trichloroethylene by iron sulfide and iron metal." *Environ. Sci. Technol.*, 35(19), 3884-3891.

Christensen, T. H., Bjerg, P. L., Banwart, S. A., Jakobsen, R., Heron, G., and Albrechtsen, H. J. (2000). "Characterization of redox conditions in groundwater contaminant plumes." *J. Contam. Hydrol.*, 45(3-4), 165-241.

Clausen, J. L., Richards, W. L., Korte, N. E., and Liang, L. (1995). "ORNL/MMES research into remedial applications of zero-valence metals. 3: Removal of TCE, cis-1,2-DCE, vinyl chloride, and technetium." *209th ACS National Meeting* , American Chemical Society, April 2-7, Anaheim, California.

Dubrovsky, N. M., Morin, K. A., Cherry, J. A., and Smyth, D. J. A. (1984) "Uranium tailings acidification and subsurface contaminant migration in a sand aquifer." *Water Poll. Res. J. Canada*, 19, 55-89.

Eary, L. E., and Rai, D. (1988). "Chromate removal from aqueous wastes by reduction with ferrous ion." *Environ. Sci. Technol.*, 22(8), 972-977.

Ebert, M., Köber, R., Parbs, A., Schäfer, D., and Dahmke, A. (2003). "Ionic tracers for the reactivity assessment of Fe^0 PRB's." *Proceedings of the 8th International FZK/TNO Conference on Contaminated Soil, ConSoil 2003*, Gent, Belgium, 1582-1590.

Eichler, F., and Schulz, D. (1998). "The nitrogen reduction programme in the Federal Republic of Germany." *Environ. Pollut.*, 102(S1), 609-617.

Fetter, C. W. (1999). *Contaminant hydrogeology*, 2nd Ed., Prentice-Hall, Inc., Simon & Schuster/A Viacom Company, Upper Saddle River, New Jersey 07458, USA.

Fogler, H. S. (1992). *Elements of chemical reaction engineering*, 2nd Ed., Prentice-Hall Inc., Englewood Cliffs, New Jersey.

Gillham, R. W., and O'Hannesin, S. F. (1994). "Enhanced degradation of halogenated aliphatics by zero-valent iron." *Ground Water*, 32, 958-967.

Goldberg, S., and Sposito, G. (1985). "On the mechanism of specific phosphate adsorption by hydroxylated mineral surfaces: A Review." *Soil Sci. Plant Anal.*, 16(8), 801-821.

Gu, B., Liang, L., Dickey, M. J., Yin, X., and Dai, S. (1998). "Reductive precipitation of uranium(VI) by zero-valent iron." *Environ. Sci. Technol.*, 32(21), 3366-3373.

Gu, B., Phelps, T. J., Liang, L., Dickey, M. J., Roh, Y., Kinsall, B. L., Palumbo, A. V., and Jacobs, G. K. (1999). "Biogeochemical dynamics in zero-valent iron columns: Implication for permeable reactive barriers." *Environ. Sci. Technol.*, 33(13), 2170-2177.

Gu, B., Watson, D. B., Wu, L. Y., Phillips, D. H., White, D. C., and Zhou, J. Z. (2002). Microbiological characteristics in a zero-valent iron reactive barrier." *Environ. Monit. Assess.*, 77, 293-309.

Hardy, L. I., and Gillham, R. W. (1996). Formation of hydrocarbons from the reduction of aqueous CO_2 by zero-valent iron." *Environ. Sci. Technol.*, 30(1), 57-65.

Hedin, R. S., Watzlaf, G. R., and Nairn, R. W. (1994). "Passive treatment of acid mine drainage with limestone." *J. Environ. Qual.*, 23, 1338-1345.

Hester, R. E., and Harrison, R. M. (1996). *Chlorinated organic micropollutants*, Royal Society of Chemistry, Cambridge, UK.

House, H. O. (1972). *Modern synthetic reactions*, 2nd Ed., W. A. Benjamin, Menlo Park, California.

Kim, Y. H., Shin, W. S., and Ko, S. O. (2004). "Reductive dechlorination of chlorinated biphenyls by palladized zero-valent metals." *Journal of Environmental Science & Health Part A-Toxic/Hazardous Substances & Environmental Engineering*, 39(5), 1177-1188.

Kriegman-King, M. R., and Reinhard, M. (1994). "Transformation of carbon tetrachloride by pyrite in aqueous solution." *Environ. Sci. Technol.*, 28(4), 692-700.

Lai, K. C.-K., and Lo, I. M.-C. (2002). "Bench-scale study of the effects of seepage velocity on the dechlorination of TCE and PCE by zero-valent iron." *The 6th International Symposium on Environmental Geotechnology*, July 2-5, Seoul, Korea.

Lai, K. C.-K. (2004). *Laboratory and full-scale studies of a permeable reactive barrier on the dechlorination of chlorinated aliphatic hydrocarbons*, Ph.D. Thesis, Civil Engineering Department, The Hong Kong University of Science and Technology.

Lai, K. C.-K., Lo, I. M.-C., and Kjeldsen, P. (2004a). "Remediation of chlorinated aliphatic hydrocarbons in groundwater using Fe^0 PRB." *1st International Symposium on Permeable Reactive Barriers*, PRB-Net, March 14-16, Belfast, Northern Ireland.

Lai, K. C.-K., Kjeldsen, P., and Lo, I. M.-C. (2004b). "Evaluation of the hydraulic performance of a full scale permeable reactive barrier by a tracer test." *1st*

International Symposium on Permeable Reactive Barriers, PRB-Net, March 14-16, Belfast, Northern Ireland.

Lee, W. J., and Batchelor, B. (2002a). "Abiotic reductive dechlorination of chlorinated ethylenes by iron-bearing soil minerals. 1. Pyrite and magnetite." *Environ. Sci. Technol.*, 36(23), 5147-5154.

Lee, W. J., and Batchelor, B. (2002b). "Abiotic reductive dechlorination of chlorinated ethylenes by iron-bearing soil minerals. 2. Green rust." *Environ. Sci. Technol.*, 36(24), 5348-5354.

Liang, L. Y., Korte, N., Gu, B., Puls, R., and Reeter, C. (2000). "Geochemical and microbial reactions affecting the long-term performance of in situ iron barriers." *Adv. Environ. Res.*, 4(4), 273-286.

Lo, I. M.-C., and Lai, K. C.-K. (2002). "Preliminary design of permeable reactive wall based on column studies." *The 6th international Congress on Environmental Geotechnics*, ISEG, August 11-15, Rio de Janeiro, Brazil.

Lo, I. M.-C. (2003). "Innovative waste containment barriers for subsurface pollution control." *ASCE Practice Periodical of Hazardous, Toxic and Radioactive Waste Management*, 7(1), 37-45.

Lo I. M.-C., Lam, S. C., and Chan, K. K. (2003a). "Effects of groundwater characteristics on chromate removal by permeable reactive barriers." *International Conference on Contaminated Land*, 12-16 May, Gent, Belgium.

Lo, I. M.-C., Lai, K. C.-K., and Kjeldsen, P. (2003b). "Performance of permeable reactive barrier on remedying aliphatic chlorinated organic containminated groundwater." *International Conference on Soil and Groundwater Contamination & Cleanup in Arid Countries*, ICSGCC, January 20-23, Muscat, Sultanate of Oman.

Lo, I. M.-C., Lam, S. C., and Lai, K. C. K. (2005). "Competitive effects of TCE on Cr(VI) removal by zero-valent iron." *ASCE J. Environ. Eng.* (accepted for publication).

Longmire, P. A., Brookins, D. G., Thomson, B. M., and Eller, P. G. (1991). "Application of sphagnum peat, calcium carbonate, and hydrated lime for immobilizing U tailings leachate." *Scientific Basis for Nuclear Waste Management XIV*, Vol. 212, 623-631.

Mackenzie, P. D., Horney, D. P., and Sivavec, T. M. (1999). "Mineral precipitation and porosity losses in granular iron columns." *J. Hazard. Mater.*, 68(1-2), 1-17.

Matheson, L. J., and Tratnyek, P. G. (1994). "Reductive dehalogenation of chlorinated methanes by iron metal." *Environ. Sci. Technol.*, 28(12), 2045-2053.

McMahon, P. B., Dennehy, K. F., and Sandstrom, M. W. (1999). "Hydraulic and geochemical performance of a permeable reactive barrier containing zero-valent iron, Denver Federal Center." *Ground Water*, 37(3), 396-404.

McRae, C. W., Blowes, D. W., and Ptacek, C. (1997). "Laboratory-scale investigation of remediation of As and Se using iron oxides." *6th Symposium and Exhibition on Groundwater and Soil Remediation*, March 18-21, Montreal, Quebec, Canada.

Montgomery, J. H. (2000). *Groundwater chemicals desk reference*, 3rd Ed., Lewis Publishers, CRC Press LLC.

Morin, K. A., Cherry, J. A., Dave, N. K., Lim, T. P., and Vivyurka, A. J. (1988) "Migration of acidic groundwater seepage from uranium-tailings impoundments, 1. Field study and conceptual hydrogeochemical model." *J. Contam. Hydrol.*, 2(4), 271-303.

Naftz, D. L., Morrison, S. J., Davis, J. A., and Fuller, C. C. (2002). *Handbook of groundwater remediation using permeable reactive barriers: Applications to radionuclides, trace metals, and nutrients*, Amsterdam, Academic Press.

National Research Council (1994). *Alternatives for ground water cleanup*, National Academy Press, Washington, D. C., USA.

Nriagu, J. O., and Nieboer, E. (1988). *Chromium in the natural and human environments*, A Wiley-Interscience Publication, John Wiley and Sons, New York.

Olsen, C. R., Lowry, P. D., Lee, S. Y., Larsen, I. L., and Cutshall, N. H. (1986). "Geochemical and environmental processes affecting radionuclide migration from a formerly used seepage trench." *Geochim. Cosmochim. Acta*, 50(4), 593-607.

Orth, W. S. (1992). *Mass balance of the degradation of trichloroethylene in the presence of iron filings*, M.Sc. Thesis, Department of Earth Sciences, University of Waterloo, Ontario, Canada.

Ott, N. (2000). *Permeable reactive barriers for inorganics*, National Network of Environmental Management Studies Fellow, Office of Solid Waste and Emergency Response, United States Environmental Protection Agency, Washington, D. C. 20460, USA.

Palmer, C. D., and Wittbrodt, P. R. (1991). "Processes affecting the remediation of chromium-contaminated sites." *Environmental Health Perspectives*, 92, 25-40.

Pearson, F. H., and McDonnell, A. J. (1975). "Limestone barriers to neutralize acidic streams." *J. Environ. Engin. Div., EE3*, 425-441.

Powell, R. M., Puls, R. W., Hightower, S. K., and Sabatini, D. A. (1995). "Coupled iron corrosion and chromate reduction: Mechanisms for subsurface remediation." *Environ. Sci. Technol.*, 29(8), 1913-1922.

Powell, R. M., and Puls, R. W. (1997). *Permeable reactive subsurface barriers for the interception and remediation of chlorinated hydrocarbon and chromium (VI) plumes in ground water*, EPA/600/F-97/008, U. S. EPA Remedial Technology Fact Sheet, United States Environmental Protection Agency, Washington, D. C. 20460, USA.

Powell, R. M., Blowes, D. W., Gillham, R. W., Schultz, D., Sivavec, T., Puls, R. W., Vogan, J. L., Powell, P. D., and Landis, R. (1998). *Permeable reactive barrier technologies for contaminant remediation*, EPA/600/R-98/125, Office of Research and Development, United States Environmental Protection Agency, Washington, D. C. 20460, USA.

Puls, R. W., Blowes, D. W., and Gillham, R. W. (1999). "Long-term performance monitoring for a permeable reactive barrier at the U. S. Coast Guard Support Center, Elizabeth City, North Carolina." *J. Hazard. Mater.*, 68, 109-124.

Reardon, E. J. (1995). "Anaerobic corrosion of granular iron: Measurement and interpretation of hydrogen evolution rates." *Environ. Sci. Technol.*, 29(12), 2936-2945.

Richard, F. C., and Bourg, A .C. M. (1991). "Aqueous geochemistry of chromium: A review." *Wat. Res.*, 25(7), 807-816.

Ritter, K., Odziemkowski, M. S., and Gillham, R. W. (2002). "An in situ study of the role of surface films on granular iron in the permeable iron wall technology." *J. Contam. Hydrol.*, 55(1-2), 87-111.

Roberts, A. L., Totten, L. A., Arnold, W. A., Burris, D, R., and Campbell, T. J. (1996). "Reductive elimination of chlorinated ethylenes by zero-valent metals." *Environ. Sci. Technol.*, 30(8), 2654-2659.

Roh, Y., Lee, S. Y., and Elless, M. P. (2000). "Characterization of corrosion products in the permeable reactive barriers." *Environ. Geol.*, 40(1-2), 184-194.

Scherer, M. M., Richter, S., Valentine, R. L., and Alvarez, PJ. J. (2000). Chemistry and microbiology of permeable reactive barriers for *in situ* groundwater clean up." *Crit. Rev. Environ. Sci. Technol.*, 30(3), 363-411.

Schwertmann, U., Gasser, U., and Sticher, H. (1989). "Chromium-for-iron substitution in synthetic goethites." *Geochim. Cosmochim. Acta*, 53(6), 1293-1297.

Sedlak, D. L., and Chan, P. G. (1997). "Reduction of hexavalent chromium by ferrous iron." *Geochim. Cosmochim. Acta*, 61(11), 2185-2192.

Siddiqui, M. S., Army, G., and Rice, R. G. (1995). "Bromate ion formation – A critical review." *J. Am. Water Works Assoc.*, 87(10), 58-70.

Siddiqui, M. S. (1996). "Chlorine-ozone interactions: Formation of chlorate." *Water Res.*, 30(0), 2160-2170.

Snoeyink, V. L., and Jenkins, D. (1980). *Water chemistry*, John Wiley & Sons, New York.

Su, C., and Puls, R. W. (1999). "Kinetics of trichloroethene reduction by zero-valent iron and tin: Pretreatment effect, apparent activation energy, and intermediate products." *Environ. Sci. Technol.*, 33(1), 163-168.

United States Environmental Protection Agency (2000). *In site treatment of soil and groundwater contaminated with chromium, technical resource guide*, EPA/625/R-00/005, Office of Research and Development, Washington, D. C. 20460, USA.

United States Environmental Protection Agency (2002). *Field applications of in situ remediation technologies: Permeable reactive barriers*, EPA/68/W-00/084, Technology Innovation Office, Office of Solid Waste and Emergency Response, Washington, D. C. 20460, USA.

Weber, E. J. (1996). "Iron-mediated reductive transformations: Investigation of reaction mechanism." *Environ. Sci. Technol.*, 30(2), 716-719.

Westerhoff, P., Song, R., Army, G., and Minear, R. (1998). "NOM's role in bromine and bromate formation during ozonation." *J. Am. Water Works Assoc.*, 90(2), 82-94.

Westerhoff, P. (2003). "Reduction of nitrate, bromate, and chlorate by zero-valent iron (Fe^0)." *J. of Environ. Eng.*, 129(1), 10-16.

Wilkin, R. T., Puls, R. W., and Sewell, G. W. (2002). *Environmental research brief - long-term performance of permeable reactive barriers using zero-valent iron: An evaluation at two sites*, EPA/600/S-02/001, Office of Research and Development, United States Environmental Protection Agency, Washington, D. C. 20460, USA.

Wilkin, R. T., and Puls, R. W. (2003). *Capstone report on the application, monitoring, and performance of permeable reactive barriers for ground-water remediation: Volume 1 – performance evaluations at two sites*, EPA/600/R-03/045a, Office of Research and Development, United States Environmental Protection Agency, Cincinnati, OH 45268, USA.

Wilkin, R. T., Puls, R. W., and Sewell, G. W. (2003). "Long-term performance of permeable reactive barriers using zero-valent iron: Geochemical and microbiological effects." *Ground Water*, 41(4), 493-503.

Williams, A. G. B., and Scherer, M. M. (2001). "Kinetics of Cr(VI) reduction by carbonate green rust." *Environ. Sci. Technol.*, 35(17), 3488-3494.

Yabusaki, S., Cantrell, K., Sass, B., and Steefel, C. (2001). "Multicomponent reactive transport in an in situ zero-valent iron cell." *Envir. Sci. Technol.*, 35(7), 1493-1503.

CHAPTER 7

Biological Degradation

Kshipra Misra, S. Bala Subramanian, Satinder K. Brar, R. D. Tyagi and Rao.Y. Surampalli

7.1 Introduction

Toxic, carcinogenic and mutagenic chemicals in the environment have the potential to cause major health disorders. Such health effects include acute, chronic, local and systemic allergic reactions, which in turn affect liver, skin, kidney and the nervous, respiratory, circulatory and reproductive systems of human beings and other organisms (Steven et al., 2000).

The primary sources of toxic pollutants in the environment include accidental releases and industrial or sanitary wastewater discharges (Burrows, 1982). Various industries responsible for releasing the waste either in the form of waste water or in the form of hazardous waste are: emerging and synthetic chemicals production plants, unproductive and disorganized agricultural practices, pharmaceuticals, boiler and cooling water treatment plant, aircrafts and vehicle wash racks, motor pools, laboratories, laundries, coal pile runoff, paint stripping, metal plating and munitions manufacturing and processing (Stephen, 2002). Besides, modern scientific technology and its applications play its own active role in producing and releasing toxic pollutants into the environment (Petrovic et al., 2004).

Toxic pollutants can be broadly categorized as organic, inorganic and biological agents. Organic compounds such as aliphatic and aromatic hydrocarbons, chlorinated biphenyls (PCBs), and polycyclic aromatic hydrocarbons (PAHs) are well established carcinogens and mutagens. Inorganic pollutants include gases, minerals, metals and their compounds (Lajoie et al., 1993).

These pollutants are mostly xenobiotic in nature and, in some cases, resemble natural compounds. The physical removal of toxic pollutants is usually a temporary solution and chemical treatments traditionally employed have the potential to further contaminate the environment. These methods often 'relocate' the contaminants instead of eliminating them. The toxicity and persistence of these compounds has necessitated the need to develop innovative biological processes to eliminate them from the environment.

Various studies have reported efficient degradation of pollutants by microorganisms due to metabolic pathways and degradative genes responsible for enzyme production, biosorption, co-metabolism, bioaccumulation, and biotransformation (Ju, 1997 and Davis, 2002). Microorganisms can degrade low molecular weight compounds intracellularly while a variety of macromolecules, can also be transformed by means of extra cellular enzymes (Evans, 2003). For example, the *Pseudomonas* species is highly versatile in metabolic functions and has a specific role in bioremediation and biodegradation of toxic pollutants. Many aromatic compounds such as benzoate, *p*-hydroxybenzoate, maldelate, tryptophan, phthalate, salicylate have been transformed into a common intermediate beta-ketoadipate by the catalytic action of 1,2-dioxygenase produced by *Pseudomonas* species (Cafaro et al., 2004). Even recalcitrant compounds like PAHs, can be transformed by microbes and used in the synthesis of cellular components and byproducts such as biosurfactants (Cooper et al., 1984).

Aliphatic hydrocarbons are assimilated by a wide variety of microorganisms, but not all species are capable of utilizing them as a growth substrate. Whilst many details of microbial hydrocarbon metabolism have already been elucidated (Ratledge, 1984) including enzymology (Providenti, 1993), regulation and genetics (Witholt et al., 1990), less attention has been paid to the primary interaction of microorganisms and the hydrocarbons involved.

Various microorganisms such as bacteria, fungi and algae are actively involved in removal of heavy metal pollution from the environment via bioassimilation. Microbes adsorb the toxic metals on their cell wall and recover metal compounds from polluted sites (Nandakumar et al., 1995). *Chlamydomonas reinhardtii* grown in laboratory media were found to fractionate radioisotope selenium (Thomas, 2004). Microorganisms are known to produce metalothione, a protein secreted by bacterial species to adsorb metal ions from the surrounding sites through biosorption. Similarly various plant species are also known to remove toxic metals from the soil and water through biosorption and bioassilmilation (Hong et al., 2000 and Zumriye Aksu, 2005).

Plants can also mediate the degradation of toxic environmental pollutants through several processes jointly known as phytoremediation processes. Phytoremediation technologies can employ selected plants to clean-up sites contaminated with a variety of pollutants such as heavy metals, radionuclides, chlorinated solvents, petroleum hydrocarbons, polychlorinated biphenyls, polynuclear aromatic hydrocarbons, chlorinated pesticides, organophosphate pesticides, and explosives (Lasat, 2002; Singh et al. 2003). The main advantage of phytoremediation is its low cost in comparison to conventional engineering techniques (Kassel et al. 2002).

This chapter discusses various types of pollutants, their degradative pathways, microorganisms responsible for assimilation and degradation of toxic compounds, and factors affecting and inducing biodegradation rates and bioavailability of toxic pollutants in soil and groundwater systems. Case studies describing degradation procedures with special reference to removal of explosives and petroleum hydrocarbons have been henceforth presented.

7.2 Definitions

This section describes several important processes related to biodegradation and biotransformation of contaminants in soil and groundwater systems.

Biodegradation

Biodegradation is a process mediated by microbial/enzymatic reactions that results in the *in vitro* or *in vivo* transformation of toxic chemicals usually into non-toxic or less harmful products. It can also be defined as a process of degradation in which contaminants such as oil, metals PAHs, and PCBs are used as an energy source for microorganisms. Biodegradation processes are regulated by factors such as temperature, nutrients and oxygen availability. These processes can mediate the primary alteration of the chemical structure of a substance resulting in loss of an undesirable property of that substance. Biodegradation processes can also cause the complete breakdown of a compound to oxidized or reduced simple molecules such as CO_2, CH_4, NO_3^-, N_2 and H_2O (Martin et al., 2003).

Mineralization

Mineralization is a biochemical process by which natural or synthetic organic compounds are completely converted into inorganic derivatives such as CO_2, NO_3^-, N_2 and H_2O. This process is also known as biomineralization (Schwartz et al., 1997).

Biotransformation

Biotransformation is the process whereby a substance is altered by a biochemical reaction such as oxidation, reduction, hydrolysis, conjugation, alkylation, dealkylation, deamination and decarboxylation within or outside the cell (Ornston et al., 1990). Biotransformtion processes are mediated by enzymes.

Bioaccumulation

Bioaccumulation or bioassimilation is a process resulting in the cellular uptake of the contaminant. It incorporates the concepts of bioconcentration and biomagnification (Hong et al., 2000).

Bioavailability

Bioavailability is typically defined as a measure of the rate and extent at which a contaminant is available to a micro or macroorganism for use as a growth nutrient. It is an important parameter to determine the quantity of chemical compounds absorbed or degraded by microbial or ecological receptors at a particular contaminated site (Davis et al., 2002).

7.3 Degradation Pathways and Mechanisms

Biodegradation of pollutants is mediated by microbial metabolism via enzymatic conversion of toxic substrates into byproducts that are usually less toxic to the organism. Bacterial metabolic pathways are complex; they are regulated by several catalytic enzymes and controlled by degradative genes. Recent advances in biological degradation of selected pollutants, mechanisms of degradative pathways and microbes responsible for degradation are summarized in Table 7.1.

This section discusses the mechanisms and pathways of biodegradation of common soil and groundwater pollutants including aromatic hydrocarbons, PAHs, and polychlorinated biphenyls (PCBs).

7.3.1 Aromatic Hydrocarbons

Catabolic transformation of aromatic hydrocarbons can occur via "ortho" and "meta" cleavage of the dioxygenate ring structure (Dagely, 1986). The genes that encode enzymes for ortho-cleavage pathways are present in the chromosomal DNA while those for meta-cleavage pathways are generally located on plasmids. These two alternative approaches of ring cleavage are illustrated in Figure 7.1.

Prokaryotic (bacteria) and eukaryotic (fungi) metabolic preferences reveal that ortho-catabolism of proto-catechuate by the former proceeds via lactonization of β-carboxy-cis, cis muconate into γ-carboxy-muconolactone in the former but into β-carboxymuconolactone in the latter (Figure 7.2). Both pathways eventually lead to the common metabolic intermediate β-ketoadipate (Ornston, 1982).

Table 7.1 Types of pollutant and microbes responsible for their biodegradation in different environments.

Toxic Pollutants	Degrading Organisms	Reference
Aliphatic Hydrocarbons	*Acinetobacter, Aeromonas eutrophus, Alcalilgenes, Bacillus subtilis, Corynebacterium Micrococcus, Moraxella, Mycobacterium, Psedomonas mendocina, Pseudomonas pickettii, Pseudomonas putida, Pseudomonas stutzeri*	Witholt et al., 1990.
Chlorinated aromatic compounds	**Bacteria** *Enterobacteriaceae, Ralstonia, Flavobacterium, Gluonbacter, Halobacterium, Hyphomicrobium sp., Micrococcus, Moraxella Nocardia* **Algae** *Anabaena, Calothrix, Cyanobacteria (Blue green algae), Diatoms, Nostoc.* **Yeast and Fungi** *Aspergillus aculeatus, Aspergillus niger, Coriolus versicolor, Emericella nidulans, Frankia, Ganoderma sp., Geotrichum candidum, Laminaria digitata, Penicillium miczynskii, Phanerochaete chrysosporium, Rhizopus oryzae, Torulopsis bombicola, Trametes versicolor,*	Robert and Robert, 2001; Davis et al, 2002;. Xiujin et al., 2004.
Aromatic and polycyclic aromatic hydrocarbons	*Xanthomonas, Zoogloea spp, Ralstonia, Nocardia, Psedomonas mendocina, Pseudomonas pickettii, Pseudomonas putida, Pseudomonas stutzeri, Pseudomonas testosterone*	Canter, 1997; Ronald, 1987; Autheunisse, et al., 1987; Kappesser, et al., 1989; Cafaro et al., 2004
Inorganic nitrogen compounds	Anabaena, Moraxella, and Mycobacterium, *Nitrosomonas and Nitrobacter SPP, Nostoc, Rhizobium, Bradyrhizobium, Rhodopseudomonas, Spirillum, Thiobacillus*	Brady, 1979; Martin, 2003; Stephen, 2004.
Inorganic heavy metals	Bacteria, fungi and micro algae are involved in removal of heavy metals through biosorption. *Rhizopus arrhizus, Mycobacterium chlorophenolicium, Bacillus subtilis, Penicillium miczynskii, Emercilla nidulans, aeromonas species, Rhizopus oryzae, Asperigillus niger, Candida lipolytica,*	Toshiyuki et al., 1998; Zumriye, 2005.

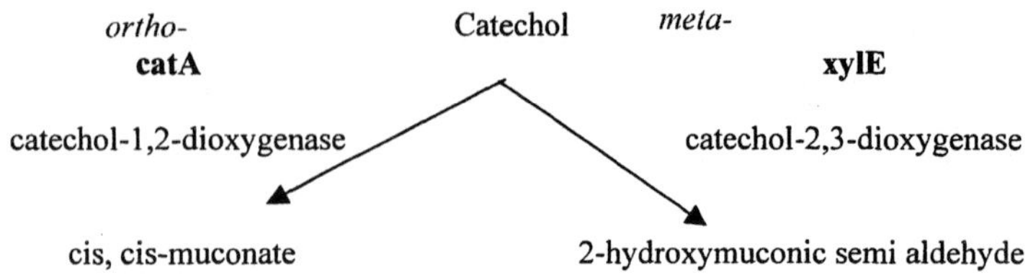

Figure 7.1 Mechanism for degradation pathway of catechol (Dagely, 1986).

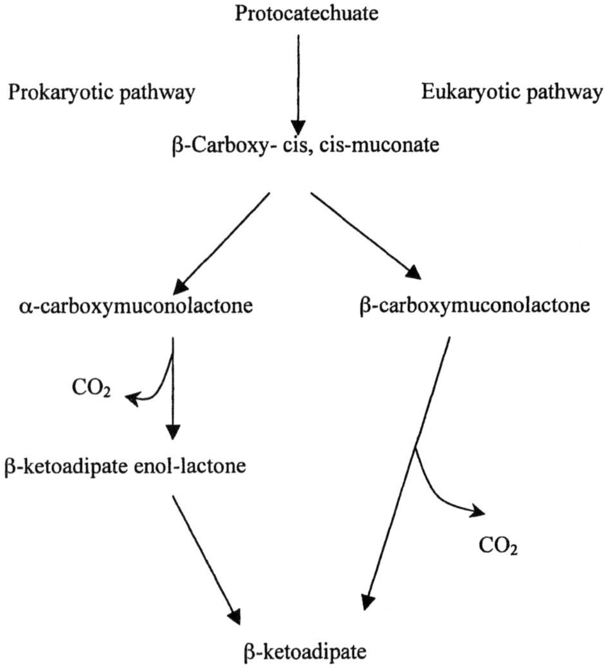

Figure 7.2 Catabolism of protocatechute via prokaryotic and eukaryotic pathways leads to the common intermediate β-ketoadipate (Ornston, 1982).

7.3.1.1 Central Metabolic Pathways

Metabolic pathways are very important to understand biodegradation of pollutants. For example, catechol and protocatechuate are considered to be the principal catabolites of aromatic hydrocarbons. The central metabolic process in any organism is not unidirectional but can proceed via many different routes. These pathways are cycles of variable enzymatic activities, designed to shuffle metabolic intermediates where needed and to supplement or replenish nutrients when necessary. In the aerobic biotransformation of aromatic hydrocarbons, the ultimate destination of the metabolic pathways is the Krebs' cycle (Figure 7.3) (Neidhardt, 1990).

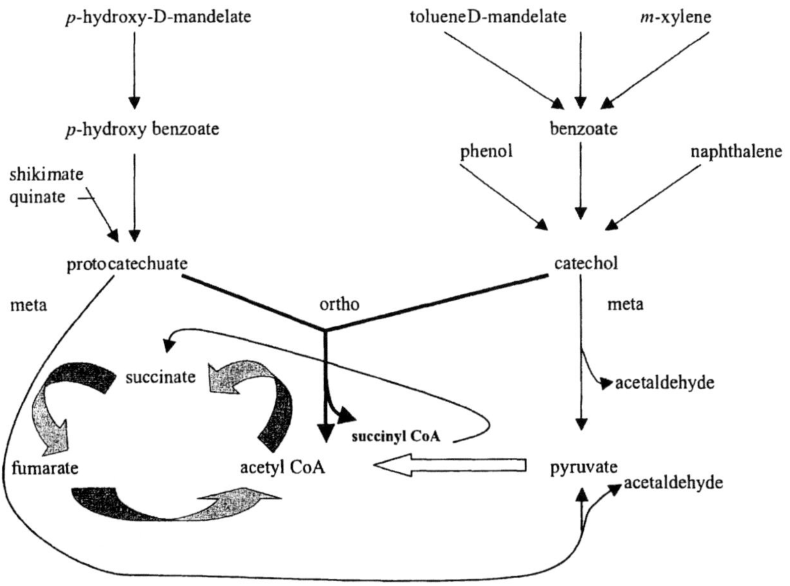

Figure 7.3 Central metabolic pathways for aromatic hydrocarbons (Harayama et al., 1989).

The principle of enzyme regulation was first conceptualized in the 1940's by relating the synthesis of enzymes to induction by substrate molecules (Stainer, 1947).

Most metabolic processes have been found to converge upon a few compounds, such as catechol, protocatechuate, or benzoate. The principal catabolic intermediates provide alternative routes or metabolic options, whose availability depends on the concentration of each principal catabolite and the level of enzymatic activity in pathways that might permit their further degradation. For example, in some *Pseudomonas* species, the major metabolic route for degradation of toluene and alkyl-substituted toluenes contrasts with that of the beta-ketoadipate pathway. It involves the meta-fission of catechol by introducing oxygen into the ring structure via a catechol 2,3 dioxygenase rather than by a catechol 1,2-dioxygenase (Burlage et al., 1989).

7.3.2 PAHs

Biological degradation of PAHs with up to five rings has been well studied (Dagely, 1986). Lower molecular weight PAHs are degraded easily while high molecular weight PAHs are difficult to degrade by normal enzymatic processes (Ronald, 1987). Microorganisms utilize molecular oxygen for the initial hydroxylation of PAHs, which is catalyzed by two types of enzymes – mono-oxygenases and dioxygenases (Gibson, 1982).

Biodegradation of the two-ring PAH naphthalene in soil was first reported in the 1920's and, since then, has been studied for the ability of bacteria to utilize the compound as the sole source of carbon and energy (Ferrero, 2002). The *Pseudomonas* species has been reported to degrade naphthalene using naphthalene dioxygenase. The metabolic pathway incorporates both atoms of molecular oxygen to form cis-naphthalene dihydrodiol, which is further converted to 1,2-dihydroxy naphthalene and ultimately to pyruvate and salicylate (Figure 7.4) (Cerniglia and Heitkamp, 1989). Oxidation of salicylate by salicylate hydroxylase yields catechol, which undergoes either ortho or meta fission depending on the bacterial species.

Cunninghamela elegans and a variety of fungi have also been reported to transform naphthalene to metabolites that are similar to those produced by mammalian enzymes and laboratory animals. Like prokaryotic organisms, fungi do not utilize naphthalene as a sole source of carbon and energy (Skubal et al, 2001) but utilize cytochrome P-450 monooxygenase to form naphthalene 1,2-oxide. The enzyme epoxide hydrolase converts naphthalene 1,2-oxide to trans-naphthalene dihydrodiol.

Blue green algae can also degrade naphthalene through oxidation producing cis-naphthalene dihydrodiol as a minor metabolite. Efficient and complete biodegradation of naphthalene can be achieved by cloning and genetic improvement of microorganisms through biotechnology and genetic engineering tools (Thomas, 2004).

naphthalene
↓
(+)-cis-1,2-dihydroxy-1,2-dihydroxynaphthalene
↓
1-2-dihydroxynaphthalene
↓
cis-*o*-hydroxybenzal pyruvic acid
↓
salicylaldehyde
↓
salicylic acid
↓
catechol
↙ ↓ ↘
meta pathway ↓ ortho pathway
ring fission pathways

Figure 7.4 Steps involved in aerobic bacterial degradation of naphthalene (Cerniglia, and Heitkamp, 1989).

7.3.3 PCBs

Polychlorinated biphenyls (PCBs) are a series of technical mixtures of many isomers and compounds that vary from mobile oily liquids to white crystalline solids and hard non-crystalline resins. PCBs are among the major toxic substances manufactured in several countries. These are used in a variety of applications such as transformer oils, capacitor dielectrics and heat transfer fluids and are found in large quantities in the environment (D'Silva, 2003).

PCB degrading bacteria are ubiquitous and mostly aerobic in nature. Gram negative soil bacteria include *Pseudomonas, Acinetobacter, Achromobacter, Alcaligenes, Moraxella* and *Acetobacter* while Gram positive bacteria include *Arthrobacter* and *Cornyebacterium* species. These strains utilize PCBs as a sole source of carbon and energy and co-metabolize them into a number of components like chlorobenzoic acids via ring-dioxygenation and meta-cleavage (Kas, 1997).

The degradation rate of PCBs decreases with increase in chlorine substitution. PCBs containing two chlorines in the ortho- position of a single ring and each ring show a striking resistance to degradation. PCBs containing all chlorines on a single ring are generally degraded faster than those containing the same number on both rings. PCBs having two chlorines at the 2 and 3 positions of one ring are susceptible to microbial attack by *Alcaligenes and Acinetobacter* (Xiujin, 2004).

The oxidative degradation of PCBs into chloro-benzoic acid by microbes involves four enzymes such as biphenyl dioxygenase, dihydrodiol dehydrogenase, 2,3-dihydroxybiphenyl dioxygenase and hydrolase produced from the bphA, bphB, bphC and bphD genes, respectively, in *Alcaligenes and Acinetobacter* (Furukawa et al., 1982). A schematic diagram showing possible mechanism is illustrated in Figure 5. Degradative pathways for PCBs are well studied. However, cloning of bph operon into other soil bacterial species using molecular biological techniques is yet to be established in the laboratory.

7.3.4 Metals

Biotransformation of metals and understanding their metabolic pathways is important particularly for the removal of toxic metals from contaminated soil and groundwater environments. Lead, cadmium, chromium, zinc, arsenic, mercury and copper are major toxic metal pollutants of environmental concern (Furukawa et al., 1982).

Metal compounds are often assimilated by microorganisms through the action of extra cellular enzymes and cofactors. Algae can accumulate toxic metals on cell wall surfaces. The biological assimilation of inorganic compounds and heavy metals is defined using several terms such as bioassimilation, biodegradation, biosorption, biomagnification, bioaccumulation, biotransformation and biovolatilization (Zumriye, 2005). These metabolic processes, however, have not been fully studied at the molecular level for heavy metals and radioactive isotopes.

7.3.5 Nitrogen Compounds

Plants can store high levels of nitrate, or translocate it from tissue without deleterious effect. However, if livestock and humans consume plant material rich in nitrates, they may suffer methemoglobinemia. In this disease, liver reduces nitrate to nitrite, which combines with hemoglobin and renders it unable to bind oxygen. In contrast to nitrate, high levels of ammonium are toxic to both plants and animals (Lea, 1992).

biphenyl

↓ biphenyl dioxygenase (bphA gene)

2, 3-dihydroxy-4-phenylhexa-2,4-diene

↓ dihydrodiol dehydrogenase (bphB gene)

2, 3-dihydroxybiphenyl

↓ 2, 3-dihydroxybiphenyl dioxygenase (bphC)

2-hydroxy-6-oxo-6-phenylhexa-2, 4-dienoic acid

↓ hydrolase (bphD)

benzoic acid

Figure 7.5 Degradative pathways of PCBs (Furukawa et al., 1982).

These nitrogenous compounds are assimilated and degraded by natural transformation process such as, ammonification, volatilization, nitrification, denitrification, plant and microbial uptake, deposition, and adsorption. These are already well studied and reported by several authors (Campbell, 1996; Marschner, 1995).

Biochemical pathways and microbial metabolism can also be used for removal of another class of nitrogenous toxic compounds-cyanides (Stephen, 2004). Cyanide has played an important role in the evolution of life on earth (Oro, 1981) and remains an important form of nitrogen for microorganisms, fungi and plants. Major source of cyanide discharges include petrochemical refining, synthesis of organic chemicals and plastics, electroplating, aluminum, metal mining and processing industries. The release of cyanide from these industries has been estimated to be >14 million kg yr^{-1} (ATSDR, 1997).

Some microorganisms are capable of synthesizing and biodegrading cyanide compounds. Cyanide degradation takes place via hydrolytic, oxidative, reductive, and substitution / transfer reactions. Several reviewers have discussed these pathways by various microorganisms (Raybuck, 1992; Dubey, 1995; Barclay, 2002). Nevertheless, additional organisms with the capacity for cyanide biodegradation are still being studied (Adjei et al., 1999; Kwon et al 2002; Yanese et al., 2000). As shown in Figures 7.6, 7.7, 7.8 and 7.9, more than one pathway can be utilized for cyanide biodegradation in some organisms (Ezzi-Mufaddal, 2002). The catalytic pathways are controlled by external factors such as availability of oxygen, pH and cyanide concentration. Besides, cyanide bio-availability and solubility in soil-water systems are also determining factors (Aronstein et al., 1994).

cyanide hydratase (fungal enzyme)

cyanide ⟹ formamide

cyanidase (bacterial enzyme)

cyanide ⟹ formate + ammonia

Figure 7.6 Hydrolytic reactions for microbial degradation of cyanide.

cyanide monooxygenase　　cyanase

cyanide ⟹ cyanate ⟹ ammonia + CO_2

Figure 7.7 Oxidative reactions for microbial degradation of cyanide.

methanogens (anaerobic biodegradation)

cyanide ⟹ methane + ammonia

Figure 7.8 Reductive reactions for microbial degradation of cyanide.

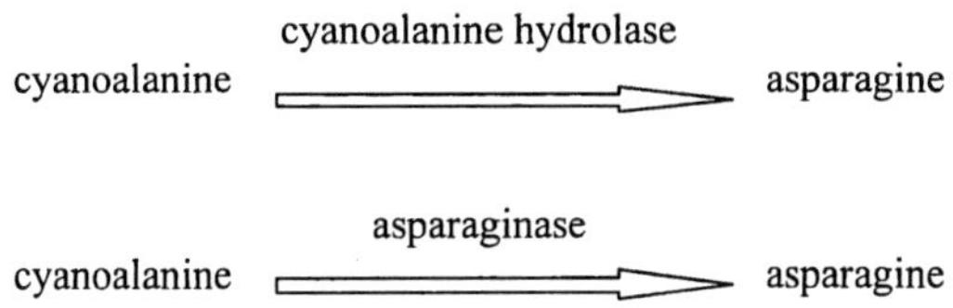

Figure 7.9 Substitution/transfer reactions for microbial degradation of cyanoalanine.

7.4 Rates of Transformation

The fundamental prerequisites for microbial degradation include: 1) the necessary metabolic capacity within a microorganism to bring about biodegradation; 2) the presence of such biodegrading organisms in the contaminated environment; 3) accessibility or bioavailability of the contaminants to the microbes; and 4) conducive environmental conditions for the proliferation of biodegrading microorganisms.

A lag or acclimatization period is often seen prior to the onset of biodegradation of any organic compound. Little or no contaminant degradation is observed during this period. The lag phase is caused by the time needed by a microorganism for induction of genes followed by the synthesis of enzymes needed for biodegradation.

The concentration of a contaminant in soil or groundwater also affects its rate of degradation. A minimum concentration of compound is required for the growth of microorganisms and if the compound concentration is lower than the required level, biodegradation may not occur. For example, the threshold concentration of organic compounds to sustain microbial growth is between 0.1 and 10 mg/L of water, or mg/kg of soil Alexander (1999). Very high contaminant concentration may cause toxicity to most bacteria. In such cases, significant degradation does not occur until rare species capable of surviving such concentrations are able to produce sufficient biomass (Chapelle, 2001).

7.4.1 Factors Affecting Biodegradation of Contaminants

Oxygen

Availability of oxygen can increase the growth rate and yield of aerobic organisms. Aerobes have mono- and dioxygenases, which are uniquely effective in the oxidation of hydrocarbons. However, the presence of O_2 can suppress degradation of halogenated pollutants through inhibition of reductive dechlorination. In the case of hydrocarbon degradation, oxygen is rapidly depleted at heavily contaminated sites resulting in anaerobic conditions (Donald, 1991).

Inorganic Nutrients

Availability of inorganic nutrients such as nitrogen and phosphorus may limit the biomass production in the presence of excess organic carbon which in turn can reduce the rate of degradation for organic pollutants (Johns, 1991). On the other hand, addition of inorganic nutrients such as nitrogen/phosphorus is also used to stimulate degradation of oil spills and soil/subsurface petroleum hydrocarbon degradation (Bitsch, 2002).

Sorption

Sorption of toxic pollutant on solid-surfaces decreases availability to microbes. Biodegradation of sorbed compounds requires organisms that are able to either facilitate the mass-transfer of such compounds into the aqueous phase or utilize extracellular enzymes and cofactors to help transform sorbed chemicals. Some examples exhibiting such mechanisms include selenium uptake by *Chlamydomonas* species and removal of basic blue, acid blue and congo red dyes from aqueous solution by biosorption on *Asperigillus niger* (Fu, 2002).

Solubility

Cellular intake of contaminants usually requires that the contaminants should be present in the aqueous phase. Thus, solubility of organic contaminants and metals can significantly impact their ability to be removed by bacteria (Ryan, 1988). Organic contaminants such as PAHs and PCBs are highly insoluble in water and, therefore, difficult to degrade in aqueous environments

Non-Aqueous Phase Liquid

Contaminants that exist as non-aqueous phase liquids (NAPLs) are difficult to biodegrade due to mass-transfer rate (solubilization) limitations. The reduced biodegradation potential is due to lack of bioavailability of the pollutant in the nonaqueous phase.

Contaminant Structure

The presence and location of halogen, amine, methoxy, and phenoxy groups in substituted aliphatic, aromatic and polynuclear aromatic hydrocarbons can significantly reduce or completely inhibit the ability of enzymes to initiate an oxidative attack on the contaminant (Harvey, 1991).

Bioavailability

Bioavailability, in the context of soil and groundwater contamination, is described as the ability of a chemical pollutant to interact with ecological receptors such as bacteria and higher organisms. Bioavailability assays are routinely used to assess remediation end points at contaminated sites.

Miscellaneous Factors

Microbial numbers (concentration of cells), co-metabolism/synergism, plasmid-born degradation and reductive dechlorination etc and some additional factors such as pH, suspended solids, soil texture and permeability, soil moisture, free ammonia, biological oxygen demand (BOD), chemical oxygen demand (COD), dissolved oxygen (DO), total nitrogen, temperature, acidity, and alkalinity can also affect the rate of biodegradation (Dushenkov, 1997).

7.5 Case Studies

Selected examples of effective bioremediation of toxic contaminants using microorganisms and plants are discussed in this section.

Case 1

Kas et al. (1997) studied the biodegradation of alkanes and PCBs in the laboratory using bioreactors containing mixtures of indigenous soil bacteria. Implementation of bioremediation protocols resulted in the removal of oil within a three-month period. The microbiological, biochemical and genetic characteristics of bacterial strain *Pseudomonas* C12B degrading petroleum hydrocarbons and alkanes were studied. Bacterial strain was isolated originally for its ability to utilize alkylsulfonates as the sole source of carbon and energy demonstrating the metabolic ability of bacteria to degrade various chemical pollutants.

Biodegradation of PCBs was carried out using two biological models, bacterial co-cultures and plant cells cultivated *in vitro*. An industrial mixture of PCBs was used to study the bacterial enzymatic degradation with the decrease in initial PCB concentration to 20% after 45 days. Degradation using plant cells involved selection of various plant species differing in their growth parameters and morphology (amorphous, differentiated shoot forming or "hairy root"). Differentiated or hairy root cultures exhibited better degradative abilities than undifferentiated amorphous cultures.

Case 2

Skubal et al., (2001) studied natural biotransformation of petroleum hydrocarbons to evaluate biogeochemical characterization in the field and laboratory microcosm. They assessed the biotransformation potential of trichloroethylene (TCE) and toluene in a plume containing petroleum hydrocarbons and chlorinated solvents at the former Wurtsmith Air Force base in Oscoda, MI.

The study evaluated the impact of terminal electron accepting process (TEAPs), microbial phylogeny and contaminant composition on hydrocarbon degradation. Biotransformation was assessed and indicated the presence of reduced electron acceptors, relevant microbial communities and occurrence of metabolic byproducts like dissolved methane and carbon dioxide. The accumulation of metabolic products of benzene, toluene, ethyl benzene and xylene (BTEX) and dechlorination products further confirmed the biodegradation of BTEX in the methanogenic zone by reductive dechlorination.

Mineralization of TCE and toluene was found to be less in the microcosms in the initial days of incubation, without added electron acceptors and nutrient compounds. Evidence of methanogenesis appeared after 8 months and CH_4 and CO_2 were found during degradation of contaminants other than toluene. Cis-dichloroethylene was found only in one methanogenic microcosm after 500 days of incubation. The study revealed that dynamic redox potential zonation prevented the reductive dehalogenation of highly chlorinated solvents during the course of an year. The predominat TEAP at the highly contaminated water shifted from methanogenesis to iron- and sulfate-reduction.

Case 3

Fu and Viraraghavan (2002) investigated the removal of dyes such as basic blue 9 (cationic), acid Blue 29 (anionic), congo red (anionic) and disperse red 1 (nonionic) dyes from aqueous solution by biosorption process using dead *Asperigillus niger* fungus. They found that *A. niger* removes dyes from aqueous environment utilizing three major functional groups-carboxyl, amino and phosphate. The lipid fraction in the biomass of fungus also played an important role in the biosorption.

Similarly, Aksu (2001) also investigated biosorption of two reactive dyes onto dried activated sludge containing bacteria and protozoa. He inferred that activated sludge has an extensive uptake capacity of organic pollutants due to acidic polysaccharides, lipids, amino acids and other cellular components available on the cell wall of bacteria.

Case 4

Hong et al. (2000) studied the biosorption of 1,2,3,4 tetrachlorodibenzo-p-dioxin (1,2,3,4-TCDD) and some pesticides like polychlorinated dibenzofurans (PCDFs) by *Bacillus pumilus.* The results indicated that the removal of these compounds was more efficient when dead organisms were used rather than the live bacterial strains. The authors suggested that in addition to attachment to microorganisms, extracellular polymeric substances might have been involved in the biosorption process.

Biosorption of the pesticide lindane was studied by Ju et al., (1997) under pHs ranging between 2.9 to 6.9 using *E. coli, Z. ramigera, Bacillus megaterium,* and *B. subtilis.* It was proposed that the repulsive electrostatic force for the adsorption of organic halide on the bacterial cell surface decreased when the lower pH generated less negative charges on the cell wall. As the cell and lindane molecules approached each other due to low repulsive electrostatic force, the Van der Waals force was intensified and biosorption was enhanced.

Case 5

Stephen (2004) studied the removal of cyanide compounds from the waste products of a number of industrial processes. Biodegradation of cyanide compounds occurred via four reaction modes including hydrolytic, oxidative, reductive and substitution. He suggested that biodegradation of cyanide using bacteria and plants can be successfully used for the removal of cyanide and cyanide compounds from contaminated waters.

Case 6

Contamination of soil and groundwater by explosives and their transformation products is a common problem at explosives manufacturing, processing, storage and disposal facilities around the world and its need to be mitigated. Of nearly twenty different energetic compounds used in conventional munitions by the military today, hexahydro-1,3,5-trinitro-1,3,5-triazine (RDX) and octahydro-1,3,5,7- tetranitro-1,3,5,7-tetrazocine (HMX) are the most powerful and commonly used. These compounds are highly toxic in nature and have resulted in severe soil and groundwater contamination due to their constant release in the environment (Cataldo, 1990).

Khan et al. (2004) conducted laboratory scale studies on phytoremediation for removal of RDX and HMX by black grass *(Ophiophogon blackei),* tomato

(Lycopersicon Sps.) and basil (*Ocimum sanctum*). Two sets of pots were used for the study. These pots were planted with seeds and cultivated in the greenhouse with additional lights of 3200-3400 Lumens (at 220/240V). Temperature was maintained at 15°C during night and 25-27°C during the day. One pot per species was prepared with non-contaminated soil as a control for uptake studies and one more pot per contamination level with no plants cultivated in the soil was prepared as control. Results of this study revealed the phytoavailability of RDX and HMX and their possibility of removal from the soil plants. *Ophiophogon* sps. (Black grass) is a versatile plant species that can withstand higher concentrations of RDX and HMX upto 301 mg/Kg. *Lycopersicon* Sps. (tomato) showed a high removal of almost 96 % at a concentration of 25 mg/Kg and significant removal rates at other concentrations of 37.5 and 50 mg/Kg. *Ocimum* Sps. was very sensitive to the concentrations of both RDX and HMX.

Stephen et al. (2002) studied the degradation of explosives from pink water generated from military wastewater discharge. This hazardous pink-water contained dissolved trinitrotoluene (TNT) and cyclo trimethylene trinitrosamine (RDX), as well as some by-products. They studied anaerobic treatment of explosives by fluidized bed reactor (FBR) using granular activated carbon (GAC). They have demonstrated bench scale batch studies using an anaerobic bacterial consortium, which fed ethanol as a sole source of carbon and energy and converted the TNT into triaminotoluene (TAT), which was consequently degraded into undetectable end products.

RDX was also sequentially degraded into nitrioso-, dinitroso-, trinitroso- and hydroxylaminodinitroso-RDX followed by triazine ring cleavage and formed methanol and formaldehyde as major end products. The anaerobic bacterial consortia were isolated from the sludge digesters of municipal waste water treatment plants. The same procedure was studied in field conditions to treat large volume of samples. They concluded that FBR-GAC was cost effective and more efficient method of treatment for explosives compared to the only GAC based system. Thus, this method could be used and was recommended for the treatment of pink water.

Kyung et al. (2002) designed a bench scale anoxic membrane bioreactor (MBR) system for treatment of explosives. It consisted of a bioreactor coupled with ceramic cross flow ultrafiltration module. The system was evaluated to treat synthetic waste water containing alkaline hydrolysis byproducts (hydrolysates) of RDX. The MBR system removed 95% of the carbon sources, 93.3% amount of nitrate and 55% of nitrite.

Case 7

Marine pollution by hydrocarbons of crude and refined petroleum products is known to cause serious environmental problems. Several petroleum hydrocarbons are known to be toxic and natural remediation of these substances is known to take months to years. These toxic hydrocarbons often enter food chains causing long term effects in marine organisms, fishes and birds leading to permanent inheritable genetic changes.

The rate of bioremediation at oil spill sites can be enhanced by using efficient oil degrading bacteria, slow release nutrients and bioemulsifiers. Bioemulsifiers are structurally diverse group of surface-active molecules of biological origin. Like chemical surfactants, biosurfactants are amphiphilic molecules having hydrophilic and hydrophobic moieties. These properties confer them an excellent ability to reduce surface and interfacial tension in both aqueous solutions and hydrocarbon mixtures.

Susan et al (2004) studied degradation of petroleum hydrocarbon using biosurfactant produced from consortium of oil degrading bacteria. The bioemulsifier used in this study was produced using a consortium of five different marine bacteria having the ability to degrade and utilize hydrocarbons of high speed diesel.

The bioemulsifier was found to reduce the surface tension of distilled water from 72 mN/m to 36 mN/m at a concentration of 0.25% (w/v). This material was found to show reduction in surface tension of seawater and synthetic seawater to almost same extent as observed in distilled water. The surface-active agents reduced surface tension and helped in the entry of bacteria into oil layers. They also increase the bioavailability of hydrocarbons to microorganisms and enhance dissolved oxygen concentrations for the growth and multiplication of aerobic bacteria.

The bioemulsifier was sprayed along with oil degrading bacteria in oil polluted site in sea water and resulted in the emulsification of oil in to small droplets, which were readily taken up by bacteria as an energy source. Nutrient compounds were also added in hydrophobic phase which helped in the fast removal of oil without any adverse impacts on the environment. The study confirmed very effective bioremediation of floating oil within a period of 10 days in the area of $25 m^2$ in seawater.

7.6 Conclusion

Traditionally, physical and chemical methods have been adopted as means of detoxification and degradation of these contaminants. Currently, these technologies are loosing ground due to drawbacks such as recompartmentalization of pollutants, which transfers contaminants from one medium to another without destroying them. Biological methods seem to offer a respite as they can treat and handle these pollutants in a natural context. Many biologically mediated pollutant removal techniques have been successfully employed in the field. Extensive studies will continue evaluating the use of novel biological technologies and improvement of pre-existing ones to enhance contaminant removal. Genetics in biological treatment is being studied more extensively to understand molecular changes during various biologically facilitated mechanisms and obtain solutions that can offer plenty of advantages.

References

Adjei MD, Ohta Y. 1999. Isolation and characterization of a cyanideutilizing Burkholderia cepacia strain. World J Microbiol Biotechnol: 15:699-704

Aksu Z. 2001. Biosorption of reactive dyes by dried activated sludge: equilibrium and kinetic modeling. Biochem Eng J. 7: 79-84.

Ankley G T, Burkhard L P. 1992. Environ Toxicol Chem 11:1235.

Aronstein B N, Maka A, Srivastava VJ: 1994. Chemical and biological removal of cyanides from aqueous and soil-containing systems. Appl Biochem Microbiol. 41:700-707.

ATSDR: 1997. Toxicological Profile for Cyanide. Atlanta, GA: US Department of Health Human Services, Public Health Service.

Barclay M, Day J. C., Thompson I. P., Knowles C. J., and Bailey M. J. 2002. Substrate-regulated cyanide hydratase (chy) gene expression in Fusarium solani: the potential of a transcription-based assay for monitoring the biotransformation of cyanide complexes. Environ Microbiol, 4:183-189.

Bitsch N, Dudas C, Koerner W, Failing K, Biselli S, Rimkus G, and Brunn H. 2002. Arch Environ Contam Toxicol. 43:257.

Burlage et al., 1989. The TOL (pWWO) catabolic plasmid. Appl. Environ. Microbiol. 55:1323-1328.

Burrows. W.D.1982. Tertiary treatment of effluent from Holston Army Ammunition Plant, US Army Armament R&D Command, Dover, NJ, Report No. 8207.

Cafaro V., V. Izzo, R. Scognamiglio, E. Notomista, P. Capasso, A. Casbarra, P. Pucci, and A. L. Donato. 2004. Phenol hydroxylase and toluene/o-xylene monooxygenase from *Pseudomonas stutzeri* OX1: Interplay between two enzymes. Appl. Env. Microbiol. 70(4): 221-2219.

Campbell, W. H. 1996. Nitrate reductase biochemistry comes of age. *Plant Physiol.* 111:355-361.

Cerniglia, C. E., and M. A. Heitkamp. 1989. Microbial biodegradation of polycyclic aromatic hydrocarbons in aquatic environment. In metabolism of *Polycyslic Aromatic Hydrocarbons in the aquatic Environment.* Ed. U. Varanasi. Boca Raton, FL; CRC Press, Inc. 41-68.

Cooper, D. G. and Paddock, D. A. 1984. Production of a biosurfactants from *Torulopsis bombicola.* Appl. Env. Microbiol. 47: 173-176.

Dagely, S. 1986. Biochemistry of aromatic hydrocarbon degradation in *Pseudomonads.* In the Bacteria, X. Ed. J. R. Sokatch, L. N. Ornston. New York: Academic press. 527-555.

Daughton CG, Ternes TA (1999) Environ Health Perspect 107:907

Davis, J. W., Odom, J. M., DeWeed, K. A., Stahl, D. A., Fishbain, S. S., West, R. J. 2002. Natural attenuation of chlorinated solvents at area 6, Dover Air force base: characterization of microbial community structure . Jounal of contaminant hydrology, 57(1-2), 41-59.

Donald, J., and P. E. Freeman. 1991. Application of the powdered activated Carbon/ Activated Sludge (PACT) and wet air oxidation process to the treatment of explosives contaminated waste water, Proceedings of the ninth International Symposium on Compatibility of Plastics and other Materials with Explosives, Propellants, San Diego.

D'Silva, K., Fernandes, A., White, S., Rose, M. 2003. Analysis of polybrominated diphenyl ethers in foodstuffs - extension of an extraction method for PCDD/Fs and PCBs to incorporate PBDEs, Organohalogen Compounds, 61: 179-182.

D'Silva, K., Fernandes, A., Rose, M. 2004. Brominated Organic Micropollutants - Igniting the flame retardant issue, Critical Reviews in Environmental Science and Technology, 34: 141-207.

Dubey S K., and Holmes D. S. 1995. Biological cyanide destruction mediated by microorganisms. World J Microbiol Biotechnol, 11:257.

Dushenkov, S., D. Vasudev, Y. Kapulnik, D.Gleba,D.Fleisher, K.C. Ting, and B. Ensley. 1997. Removal of Uranium from water using terrestrial plants, Environ. Sci. Technol. 31: 3468-3474.

Evans, G. M. and J. C. Furlong. 2003. Environmental Biotechnology: theory and applications. West Sussex, UK, John Wiley and Sons.

Ezzi-Mufaddal I, Lynch James M. 2002. Cyanide catabolizing enzymes in Trichoderma spp. Enzyme Microb Technol, 31:1042-1047.

FAO, Food and Agriculture Organization of the United Nations, 2003. 'International Code of Conduct on the Distribution and Use of Pesticides', FAO, Rome,

Ferrero, M., E. L. Brossa, J. Lalucat, E. G. Valdes, R. R. Mora. and R. Bosch. 2002. Coexistence of two distinct copies of naphthalene degradation genes in Pseudomonas strains isolated from the western Mediterranean region. Appl. Environ. Microbiol. 68(2): 957-962.

Fu Y., and Viraraghavan, T. 2002. Removal of Congo Red from an aqueous solution by fungus *Asperhillus niger.*Advances Environ Res. 7: 239-247.

Furukawa, K., and A. M. Chakrabarty. 1982. Involvement of plasmids in total degradation of chlorinated biphenyls. Appl. Environ. Microbiol. 44: 619-626.

Gibson, D. T. 1982. Microbial degradation of hydrocarbons. Toxicol. Environ. Chem. 5: 237-250.

Gibson, D. T. 1988. microbial metabolism of aromatic hydrocarbons and the carbon cycle. In microbial metabolism and the carbon cycle. Ed. S. R. Hagedorn, R. S. Hanson, D. A. kunz. Chur, Switzerland. Harwood Academic Publishers. 33-58

Harayama, S., and K. N. Timmis. 1989. Catabolism of aromatic hydrocarbons by Pseudomonas. In Genetics of bacterial Diversity. Ed. D. A. Hopwood, K. F. Chater, London; Academic press.

Harvey, S. D. 1991. Fate of the explosive hexahydro-1,3,5-trinitro-1,3,5 triazine (RDX) in soil and bioaccumulation in bush bean hydroponic plants, Environmental Toxicology and Chemistry,10: 845-855.

Hong H. B., Hwang S. H., and Chang Y. S. 2000. Biosorption of 1,2,3,4-tetrachlorodibenzo-p-dioxin and polychlorinated dibenzofurans by *Bacillus pumilus*. Water Res. 34: 349-353.

Johns, H.O., Bragg, J.R., Dash, L. C. and Owens, E.H. 1991. Proceedings of the 1991 International oil spill conference. March 4-7,1991, San Deigo, C.A. pp 167-176.

Ju, Y. H., Chen T. and Liu, J. C. 1997. A study on the biosorption of lindane. Colloids Surf B. 9: 187-196.

Kas, J., J. Burkhard, K. Demnerova, J. Kostal, T. Macek, M. Mackova, and J. Pazlarova. 1997. Perspectives in biodegradation of alkanes and PCBs. *Pure &Appl. Chern.,* Vol. 69, No. 11, pp. 2357-2369.

Kassel A. G., et al. 2002. Phytoremediation of trichloroethylene using Hybrid Poplar. Physiol. Mol. Biol. Plants 8(1): 3-10.

Kostal, J., Suchanek, M., Klierova, H., Kralova, B., Demnerova, K., and McBeth, D. 1998. *Pseudomonas* C12-b, an SDS degrading strain , harbours a plasmid coding for degradation of medium chain length n-alkanes. International Biodeterioration and Biodegradation . 42:221-228.

Kwon HK, Woo SH, Park JM. 2002. Thiocyanate degradation by Acremonium strictum and inhibition by secondary toxicants. Biotechnol Lett. 24:1347-1351.

Kyung D. Z. and M. K. Stenstrom. 2002. Application of a membrane bioreactor for treating explosives process wastewater. Water Research. 36: 1018-1024.

Lajoie C. A, Zylstra G. J, DeFlaun M. F. and P. F. Strom. 1993. Development of field application vectors for bioremediation of soils contaminated with polychlorinated biphenyls. Appl. Environ. Microbiol. 59: 1735-1741.

Lasat, M. M. 2002. Phytoextraction of toxic metals: a review of biological mechanisms. *J. Environ. Qual.* 31, 109-20

Lea, P. J., Blackwell Raymond, D., and Joy Kenneth, W. 1992. Ammonia assimilation in higher plants. *In: "Nitrogen Metabolism of Plants" Proc Phytochem Soc Europe* 33:153-186.

Marschner, H. 1995. *Mineral Nutrition of Higher Plants*. 2nd ed. Academic Press, London.

Martin Burger and Lousie E. Jackson. 2003. Microbial immobilization of ammonium and nitrate in relation to ammonification and nitrification rates in organic and conventional cropping systems. Soil biology and biochemistry. 35: 29-36.

Nandakumar, B. A., V. Dushenkov, H. Motto, I. Raskinse. 1995. Phytoextraction: Use of plants to remove heavy metals from soils, Environ. Sci. Technol. 29(5): 1232-1235.

Neidhardt, et al., 1990. Physiology of the bacterial cell. New York: Sinauer.

Norberg King T J, Durhan E J and Ankley G T. 1991. Environ Toxicol Chem 10:891.

Ornston, L. N., and W. K. Yeh. 1982. Recurring themes and repeated sequences in metabolic evolution., In A. M. Chakrabarty (ed.), Biodegradation and Detoxification of environmental pollutants, CRC press, Miami.

Ornston, L. N., J. E. Houghton, E. L. Neidle, and L. A. Gree. 1990. subtle selection and novel mutation during divergence of the β-ketoadipate pathway. In *Pseudomonas*. Biotransformations, pathogenesis, and evovlving biotechnology. Ed. R. Silver, A. M. Chakrabarty. Ed. Washington D.C.: ASM publications. 207-225.

Oro′ J, Lazcano-Araujo A. 1981. The role of HCN and its derivatives in prebiotic evolution. In Cyanide in Biology. Edited by Vennesland B, Conn EE, Knowles CJ, Westley J, Wissing F: London: Academic Press:517-541.

Petrovic, M., E. Eljarrat, M. J. Lopez de Alda ·D. Barcelo. 2004. Endocrine disrupting compounds and other emerging contaminants in the environment: A survey on new monitoring strategies and occurrence data. Anal Bioanal Chem 378 : 549-562.

Providenti, M.A., Lee, H. and Trevors J. C. 1993. Selected factors limiting the microbial degradation of recalcitrant compounds. J.Industrial Microbiol. 12,379-395.

Ratledge, C. 1984. Microbial conversion of alkanes and fatty acids. JAOCS 61,447-453.

Raybuck S. A. 1992. Microbes and microbial enzymes for cyanide degradation. Biodegradation, 3:3-18.

Robert C. Ahlet and Robert W. peters. 2001. Treatment of PCB-contaminated soils: I. evalution of in situ Reductive Dechlorination of PCBs. Environmental Progress. 108-116.

Ronald Eisler. 1987. Polycyclic aromatic hydrocarbon hazards to fish, wildlife, and invertebrates: a synoptic review. Contaminant Hazard Reviews. Biological Report 85(1.11): 1-55.

Ryan, A. J., R. M. Bell, J. M. Davidson, G. A. and O'Conner. 1988. Plant uptake of non-ionic organic chemicals from soils, Chemosphere. 2299-2323.

Sarvar Khan, Satinder K. Brar and Kshipra Misra. 2004. Phytoremediation of RDX and HMX by Ophiophogon Blackei and Lycopersicon sps., Chapter 6 in Book Titled` Bio-Technological Applications in Environmental Management' editor; Dr. R.K. Trivedy.

Schwartz, O. J., and Jones, L.W. 1997. Chapter 14:Bioaccumulation of xenobiotic organic chemicals by terrestrial plants,In W. Wang, J.W. Gorsuch, and J.S. Hughes (eds.) Plants for Environmental Studies. 418-449.

Shore LS, Gurevitz M, and Shemesh M . 1993. Bull Environ ContamToxicol 51:361

Singh, O. V, and S. labana. 2003. Phytoremediation: an overview of metallic ion decontamination from soil. *Appli. Microbiol. Biotechnol.* 61, 405-12.

Skubal, K. L, M. J. Barcelona and P. Adriaens. 2001. An assessment of natural biotransformation of petroleum hydrocarbons and chlorinated solvents at an aquifer plume transect. Journal of contaminant hydrology. 49: 151-169.

Stainer, R. Y. 1947. Simultaneous adaptation: A new technique for the study of metabolic pathways. J. Bacterial. 54: 339-348

Stephen W. M., N. R. Adrian, R. F. Hickey and R. L. Heine. 2002. Anaerobic treatment of pinkwater in a fluidized bed reactor containing GAC. Journal of Hazardous Matterials 92: 77-88.

Stephen Ebbs. 2004. Biological degradation of cyanide compounds. Current Opinion in biotechnology. 15: 231-236.

Titus, S.; Kumar, P. and P.C. Deb. Bioemulsifier produced by hydrocarbon utilizing bacteria for the bioremediation of pollutant oil. Paper presented in *World Conference on Disaster Management and Infrastructure Control System* held in Hydrerabad during 10-12 November 2003.

Thomas M. Johnson. 2004. A review of mass-dependent fractionation of selenium isotopes and implications for other heavy stable isotopes. Chemical geogology, 204: 201-214.

Thompson, P.L., L.A. Ramer, and J.L. Schnoor. 1998. Uptake and transformation of TNT by a poplar hybrid. Environ. Sci. Tech. 32(7): 975-80.

Thornton, F. S., M. I. Bright, D. N. Lerner and J. H. Tellam. 2000. Attenuation of landfill leachate by UK Triassic sandstone aquifer materials. 2. Sorption and degradation of organic pollutants in laboratory columns. Journal of contaminant Hydrology 43: 355-383.

Toshiyuki M., M. Ueda, T. Kawaguchi, M. Arai and Atsuo Tanaka. 1998. Assimilation of cellooligosaccharides by a cell surface-engineered yeast expressing β-glucosidase and carboxymethylcellulase from *Aspergillus aculeatus*. Appl. Env. Microbiol. 64 (12): 4857-4861.

Williams Forth, D. Henschler, W. Rummel, K. Starke, 1996. 'Allgemeine und spezielle Pharmakologie und Toxikologie', Spektrum Akademischer Verlag, Heidelberg, , p. 981.

Witholt, B., de Smet, M.J., Kingma, J., van Beilen, J.B., Kok, M., Lageveen, R.G. and Eggink, G. 1990. Bioconversion of aliphatic compounds by Pseudomonas putida in multiphase bioreactors; Background and economic potential. Trends Biotechnol. 8, 46-52

Xiujin Qiu, Melinda K. Stock and john W. Davis. 2004. Remediation options for chlorinated volatile organics in a partially anaerobic aquifer. Remediation Autumn. 39-47.

Yanese H, Sakamoto A,Okamoto K, Kita K, Sato Y. 2000. Degradation of the metal-cyano complex tetracyanonickelate (II) by Fusarium oxysporum N-10. Appl Biochem Microbiol, 53:328-334.

Zumriye Aksu. 2005. Application of biosorption for the removal of organic pollutants: a review. Process Biochemistry 40: 997–1026.

CHAPTER 8

Bioremediation with Bacteria and Enzymes

Kshipra Misra, S. Bala Subramanian, Satinder K. Brar, R. D. Tyagi and R.Y. Surampalli

8.1 Introduction

In the past decades, significant amounts of chemicals have been released into the environment by agricultural, industrial, commercial and other human activities. These chemicals have caused harm to the ecosystem and human health. These pollutants are mainly released into the environment as metal, non-metal, metalloid, inorganic and organic compounds (Cluis, 2004). Organic contaminants include aliphatic, alicyclic, aromatic and polycyclic aromatic hydrocarbons comprising halogenated and non-halogenated compounds, pesticides and explosives. Inorganic pollutants may be metals such as Ag, Al, As, Be, Cd, Cr, Cu, Hg, Fe, Ni, Pb, Sb, Se, Zn, and radioactive elements (Meagher, 2000).

Most remediation techniques involving physical and chemical methods are expensive and produce secondary pollutants in the environment. In order to overcome these problems, biological degradation of pollutants is favored because of its cost effective and eco–friendly approach (Hattan et al., 2003). Bioremediation is a process in which degradation of toxic compounds results in their conversion into non–toxic substances such as CO_2 and H_2O. This process can be facilitated either at contaminated sites (*in-situ* bioremediation) or in bioreactors (*ex-situ* bioremediation) using microorganisms to achieve complete detoxification of toxic compounds (Hwang and Cutright, 2002).

Microorganisms are ubiquitous in nature and have tremendous metabolic ability to degrade and utilize most toxic compounds as sources of energy and growth. They possess characteristic degradative enzymes for biodegradation of respective contaminants through aerobic or anaerobic processes. Bacteria can be classified as aerobic and anaerobic based on their requirement for oxygen for growth. Aerobic bioremediation has 10 to 100 times higher degradation efficiency than anaerobic processes and is, therefore, a commonly used practice (Ahlet, et al., 2001). The rate of degradation can be further enhanced by dispersing adapted bacteria at contaminated sites through a process know as bioaugmentation (Quan, et al., 2004) or by adding required nutrients to stimulate the growth of indigenous microorganisms through biostimulation (Trindade et al. 2005).

In general, biodegradation follows microbial metabolic pathways such as aerobic respiration, anaerobic respiration, fermentation and co-metabolism. Bioremediation of toxic compounds also depends upon the bioavailability of contaminant to microbes, environmental factors and site conditions such as temperature, pH, nutrients, electron acceptor(s), redox potential, water activity, osmotic pressure and concentration of contaminants (Evans 2003; Thakur 2004).

This chapter gives descriptions and presents case studies associated with common soil and groundwater bioremediation techniques. These techniques include biosparging, bioventing, biostimulation, bioaugmentation, bioleaching, anaerobic and aerobic biotransformation, biological fixation, enzyme-catalyzed treatment, biological reactors and natural attenuation.

8.2 Biosparging

Biosparging is an *in-situ* remediation technology that utilizes naturally occurring microorganisms to degrade organic contaminants of concern (COCs) within the saturated zone. The rate of bioremediation is enhanced by inducing air (or oxygen) flow using air injection wells, and if necessary, by the addition of air to the saturated zone (Weston 1988).

A schematic of a biosparging system is illustrated in Figure 8.1. The process is similar to an *in-situ* air sparging system (IAS), except that a lower air flow rate is used to enhance biotransformation and minimize volatilization (primary mechanism of IAS). The air flow rate is controlled by the metabolic demand of microorganisms to successfully remediate the saturated zone. Biosparging has proven most effective in reducing petroleum products at leaky underground storage tank sites (Norris, et al. 1994). Although constituents adsorbed to aquifer material can also be treated to a certain extent, the technique is not suitably effective for highly volatile contaminants.

8.2.1 Factors Affecting Biosparging Processes

There are various factors which are responsible for effective *in-situ* bioremediation of ground water contaminated with petroleum hydrocarbons (EPA-542-R-00-008). These can be divided into two categories: site characteristics and constituent characteristics.

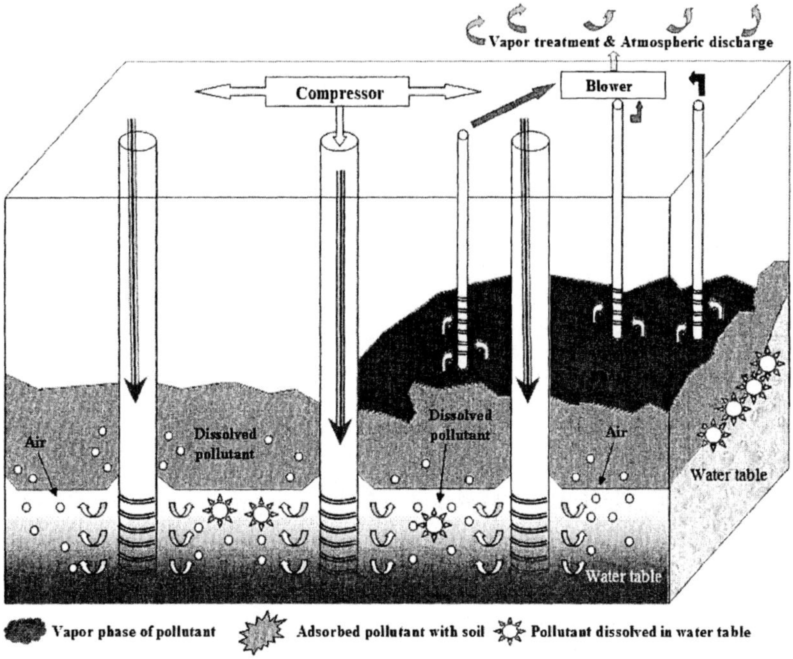

Figure 8.1 Schematic representation of *in-situ* bioremediation of organic contaminant of concern (COCs) using biosparging system combined with soil vapor extraction for ground water treatment.

8.2.1.1 Site Characteristics

Various soil characteristics play a pivotal role in controlling biosparging processes. These include intrinsic soil permeability, soil structure and stratification, temperature, pH, oxygen concentration, nutrients, microbial density and iron content.

Intrinsic Soil Permeability

The intrinsic permeability of a soil is a measure of its ability to transmit fluids which determines the rate at which oxygen can be supplied to the hydrocarbon degrading bacteria. At least 3 to 3 ½ pounds of oxygen is generally needed to degrade one pound of petroleum hydrocarbons. The coarse grained soils (sands) have greater intrinsic permeability compared to fine grained soils (clays and silts) and should be greater than 10^{-9} for effective bioremediation (EPA 542-R-00-006).

The intrinsic permeability of a saturated zone can be calculated using the following equation (Aelion et al. 1995):

$$k = K\,(\mu/\rho g) \qquad \text{(Eq. 8.1)}$$

where k = intrinsic permeability (cm^2)
K = hydraulic conductivity (cm/sec)
μ = water viscosity (g/cm · sec)
ρ = water density (g/cm^3)
g = acceleration due to gravity (cm/sec^2)

Soil Structure and Stratification

These characteristics include the type of soil present and its micro and macro structure. Soil structure and stratification can control the biosparging pressure and distribution of oxygen and nutrients in the saturated zone (Clayton et al. 1995).

Temperature

The optimal bacterial growth is found to be in the range of 10 to 45° C. The rate of microbial activity typically doubles for every 10° C rise in temperature within this range. The rate of degradation decreases below and above the mentioned range of temperature (Filler 1997; Webb and Phelan 1997). In most areas of the U.S., the average groundwater temperature is about 13°C, but groundwater temperatures may be somewhat lower or higher in the extreme northern and southern states. In most cases, subsurface microbial activity has been found to decrease significantly at temperatures below 10°C and distinctly bring to an end below 5°C. Biosparging is an *in-situ* technology, the bacteria are likely to experience stable groundwater temperatures with only slight seasonal variations.

pH

Values of pH between 6 and 8 are most suitable for bacterial growth. The pH of groundwater can be adjusted prior to and or during biosparging process (McCray and Falta 1997). However, pH adjustment is expensive approach due to natural buffering capacity of the groundwater system which requires continuous adjustment and monitoring throughout the operation. In addition, during pH adjustment, it may lead to rapid changes in pH and lead to unfavourable conditions for microbial activity.

Oxygen Concentration

The rate of biodegradation greatly depends on the availability and supply of oxygen at a contaminated site. Oxygen serves as a terminal electron acceptor in aerobic metabolic processes. In the absence or low availability of dissolved oxygen microbes can utilize other electron acceptors (such as nitrate or sulfate) for degradation of contaminants (McCray and Falta 1996). However, this occurs at significantly reduced rates of transformation.

Nutrients

Nutrients play a major role in bacterial growth and metabolism. Frequent addition of nutrients is necessary to maintain the required bacterial populations at contaminated sites. However, over addition of nutrients at the polluted sites may inhibit the rate of metabolism (Norris, et al. 1993). Nitrogen addition can lower pH, depending on the amount and type of nitrogen added.

Microbial Density

Microbial density is an important factor for effective biosparging with a typical range from 10^4 to 10^7 CFU/g of soil. The minimum plate count of heterotrophic bacteria in a biosparging zone should not be less than 10^3 CFU/g (Sleep 1998). Otherwise, the rate of remediation will be too slow.

Iron

Ferrous iron [Fe^{+2}] present in soil precipitates to iron oxide [Fe^{+3}] by oxidation reaction. Ferric iron precipitates can block the soil pore spaces and reduce soil permeability (Mesania and Jennings 2000). Hence, care should be exercised in such soils or groundwater systems.

8.2.1.2 Constituent Characteristics

Success of biosparging process also depends on certain features of chemical constituents present:

Chemical Structures

It is a chief rate determining parameter of biodegradation in biosparging processes. Low molecular weight (nine atoms or less) aliphatic and mono aromatic compounds biodegrade faster than higher molecular weight and complex compounds as summarized in table 6.1 (Nakhla and Niaz 2002).

Table 8.1 Types of constituents and their rates of biodegradation.

	Chemical Contaminants	Sources	Rate of Degradation
Alkanes and mono-aromatics	n-butane, l-pentane, n-octane, nonane, methyl butane, dimethylpentenes, methyloctanes, benzene, toluene, ethylbenzene, xylenes, propylbenzenes, decanes, dodecanes, tridecanes and tetradecanes.	Gasoline, diesel, kerosene, heating fuels, heating oil lubricating oils	Easy for microbial degradation (faster rate)
Poly aromatics	Naphthalenes, fluoranthenes, pyrenes, acenaphthenes	Diesel, kerosene, heating oil, lubricating oils	Complex microbial degradation (slow rate)

Concentration and Toxicity

The presence of very high concentration of petroleum compounds (> 50,000 ppm) or soluble heavy metals (> 2,500 ppm) at contaminated sites can be toxic and tend to retard the growth and reproduction of bacteria responsible for biodegradation. Very low concentrations of contaminants also diminish bacterial activity towards initiation of biodegradation processes. Therefore, an optimum level of pollutant concentration is required (Chapelle 1999). Pollutant concentration below 0.1 ppm is not generally treatable using biological process. Similarly > 95 % degradation of total petroleum hydrocarbons (TPH) is also very difficult to biodegrade due to presence of recalcitrant or non-biodegradable petroleum hydrocarbons.

Vapor Pressure

Vapor pressure plays an important role in evaluating bioremediation rates. Constituents with higher vapor pressures are generally volatilized and not biodegraded. Typically, constituents with vapor pressures > 0.5 mm Hg are likely to be volatilized by induced air stream and those with vapor pressures < 0.5mm Hg undergo *in-situ* biodegradation mediated by soil bacteria (Widdowson et al. 1997).

Product Composition and Boiling Point

Both of these parameters control constituent volatility. Compounds of higher molecular weight and higher boiling points require longer duration for microbial degradation (Leahy and Colwell 1990). Petroleum products are often classified by their boiling point (rather than vapor pressures) and generally all petroleum-derived organic compounds are biodegradable. Products which have boiling points of ≤ 250°C to 300°C will volatilize to some extent and can be removed by a combination of volatilization and biodegradation in a biosparging system. For example in biosparging, biodegradation of petroleum hydrocarbons such as gasoline (40-225°C), kerosene (180-300°C), diesel fuel (200-338°C) and heating oil (> 275°C) requires lesser time than the lubricating oil which is non volatile.

Henry's Law Constant

As has been already discussed in chapter 2, Henry's Law constant is used for the quantitative measurement of volatility of a constituent. It is an important factor that quantifies the relative tendency of a dissolved constituent to convert into vapor phase.

8.2.2 System Design

Laboratory treatability experiments followed by field pilot scale studies are often carried out for successful evaluation of the potential and effectiveness of biosparging for a given contaminated site (Aelion et al. 1996). Commonly, microbial screening and biodegradation studies at the laboratory level and biosparging treatability tests at the field level are conducted to determine, verify and quantify the potential effectiveness of the approach and provide necessary data to design a system.

The essential goals in designing an air sparging system are to configure the wells and monitoring points in order to optimize influence of air on the plume for maximum removal of toxic contaminants. There is also a need to provide optimum monitoring and vapor extraction points ensuring minimal migration of the vapor plume (Johnson et al 1993). The placement and number of air sparge points can affect the sparging pressure and distribution of air in the saturated zone. These air sparging points are required to aerate the dissolved phase plume determined primarily by permeability and structure of soil. The bubble radius primarily depends on hydraulic conductivity of the aquifer material in which sparging takes place and should be determined based on the pilot scale studies. Other factors which affect sparging are soil heterogeneities and differences between lateral and vertical permeability of the soils (Flathman and Jerger 1994). General guidelines for developing a biosparging pilot test plan are summarized in Table 8.2.

8.2.3 Advantages and Disadvantages

Biosparging is an enhanced *in-situ* bioremediation technology widely used for degradation of organic pollutants and petroleum hydrocarbons at contaminated sites in ground water. Various advantages and disadvantages are presented in Table 8.3.

Table 8.2 Summary of guidelines for developing a biosparging pilot test plan (EPA-600-R-92-173).

Major Requirements	Details
Site diagram	i. Horizontal and vertical delineation of the plumes ii. Utilities, surface seals, and potential receptors (workers, air intake systems, buildings, sewer systems or other sub surface confined spaces) iii. Test well network consisting of dedicated air sparging wells (ASWs) and observation wells.
Design details for biosparging pilot test	Dedicated ASWs are required to effectively implement the pilot test and its design i. Number of ASW within the contaminated area (at least one) ii. Location of ASWs (below the plume). iii. Distribution of multiple ASWs (based upon the distribution of plume and geologic heterogeneity) iv. Construction
Observation well design	i. Number of observation wells. ii. Proper design. iii. Location (in the radial pattern). iv. Construction.
Monitoring proposal for the system	i. Pressure reading of ASWs and observation wells. ii. Water elevation in the observation wells. iii. Visual observations like bubbles. iv. Dissolved oxygen and CO_2 measured in the observation wells v. CO_2 levels in the exhaust vapors. vi. Sparging rate at the compressor discharge flow gauge vii. Radius of influence viii. Sparging vapor concentrations of COCs, in the observation wells.

8.2.4 Applications

Biosparging can be successfully applied at sites contaminated with mid-weight petroleum products such as diesel fuel, jet fuel and also lighter petroleum products including gasoline which tend to volatilize readily and is removed more rapidly using air sparging. However, heavier products such as lubricating oils, which generally take longer time to biodegrade, can also be removed to certain extent using biosparging (Riser-Roberts 1992). The bioventing technology is presented in Figure 8.2.

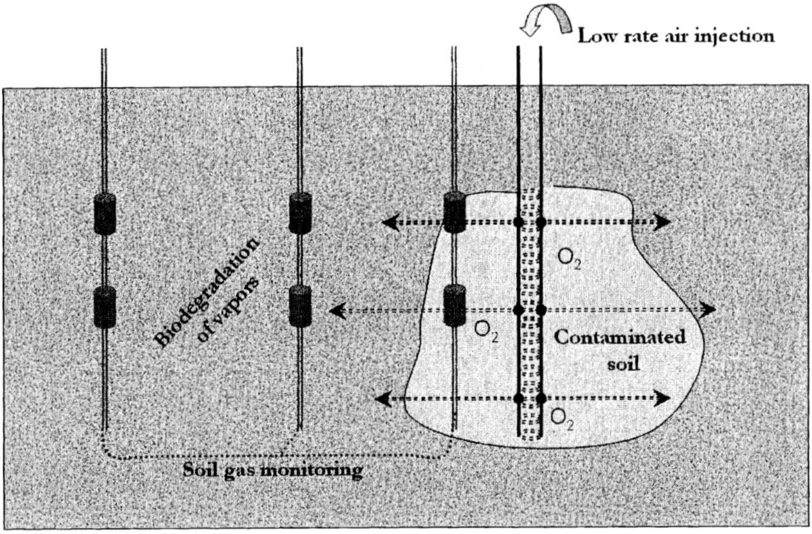

Figure 8.2 Schematic diagram of Bioventing system used for biodegradation of the pollutants from ground soil (note: figure not to scale).

Table 8.3 Advantages/disadvantages of biosparging processes (Boersma et al., 1995).

Advantages	Limitations
Readily available equipment and easy installation	Only useful where air sparging is suitable, can not be applied ; i. in confined aquifers, ii. in the presence of free products, and iii. in heterogeneous soils
Implemented with minimal disturbance to site operations	Interactions among chemical, physical and biological process are not well understood.
Short treatment times, 6 months – 2 years under optimal conditions	Lack of field and laboratory data to optimize and support design considerations.
Enhances the effectiveness of air sparging for treating a wider range of petroleum hydrocarbons	Cannot be used at nearby basements and sewers; creates ground water mounding
Requires no pretreatment, storage, or discharge of ground water	May induce the migration of free products and constituents to spread the contaminants.
Low air injection rates minimize potential need for vapor capture and treatment	Lack of pH control

8.3 Bioventing

Bioventing is an *in-situ* remediation technology that uses indigenous microorganisms to biodegrade organic constituents adsorbed to soils in the unsaturated (vadose) zone. Soils in the capillary fringe and saturated zone are not affected. The activity of indigenous bacteria is enhanced by inducing air/ oxygen flow into the unsaturated zone (using extraction or injection wells) and, if necessary, by adding nutrients (Norris et al. 1994).

When extraction wells are used for bioventing, the process is similar to soil vapor extraction (SVE). However, while SVE removes constituents primarily through volatilization, bioventing systems promote biodegradation of constituents and minimize volatilization (generally by using lower air flow rates than for SVE) (Kirtland and Aelion 2000).

All aerobically biodegradable constituents such as petroleum products including gasoline, jet fuels, kerosene, and diesel fuel can be treated very effectively using bioventing processes. Bioventing is most often used at sites with mid-weight petroleum products (i.e., diesel fuel and jet fuel), because lighter products (i.e., gasoline) tend to volatilize readily and can be removed more rapidly using SVE. Heavier products (e.g., lubricating oils) generally take longer to biodegrade (Nakhla and Niaz 2002).

Co-metabolic bioventing is applicable in anaerobic biodegradation of contaminants such as trichloroethylene (TCE), trichloro ethane (TCA), ethylene dibromide and dichloroethylene (DCE). Bioremediation cannot degrade inorganic contaminants but can be used to change the valency state of inorganics leading to microbial adsorption, uptake, accumulation and concentration (Mihopoulos et al 2002). The advantages and disadvantages of bioventing are summarized in Table 8.4.

8.3.1 Factors Affecting Bioventing Processes

Key factors affecting the bioventing process are similar to those impacting biosparging. These processes are summarized in Table 8.5. Besides, certain limiting factors such as the presence of water table within several feet of the surface, other factors such as the presence of saturated, oil lenses, lower soil permeability, vapor build-up in basements within the radius of influence of air injection wells, low moisture content of soil and lower temperatures may also slow down the remediation process (Bachman et al 2001).

Table 8.4 Advantages and disadvantages of bioventing.

Advantages	Limitations
Ready availability of equipment Easy installation.	Higher contaminant concentrations may be toxic to microorganisms and inhibit growth.
Can be operated with minimal disturbance to the site and in inaccessible areas (e.g., under buildings).	Cannot be used in areas with low soil permeability, high clay content, insufficient delineation of subsurface conditions.
Requires short treatment times (usually 6 months to 2 years under optimal conditions)	Difficult to remove very low concentration of contaminants.
Cost competitive ($45-140/ton of contaminated soil)	Prior permission required for nutrient and air injection at the site.
Easily combined with other technologies like air sparging and groundwater extraction	Aerobic biodegradation of many chlorinated compounds may not be effective unless there is a primary-metabolite present, or an anaerobic cycle.
May not require costly off gas treatment	Slow rate of remediation at low temperatures

Table 8.5. Key factors affecting bioventing process.

Site Characteristics	Constituent Characteristics
Intrinsic permeability,	Chemical structure
Soil structure and stratification	Concentration and toxicity
Microbial presence	Vapor pressure
Soil pH	Product composition and boiling point
Moisture content	Henry's law constant
Soil temperature	
Nutrient concentrations	
Depth to groundwater	

8.3.2 Case Studies

Bioventing is a commonly used bioremediation approach due to easy availability of hardware components. Bioventing has received increased public acceptance, particularly for its use in conjunction with soil vapor extraction (SVE) (Mihopoulos et al 2000). The time required to clean up contaminated sites using bioventing mainly depends upon specific soil and pollutant properties. Several examples of the application of this technique are summarized in this section.

Air Force Center for Environmental Excellence (AFCEE), US Air Force has been continuously using bioventing process for clean up of oil from different sites at a cost of \$ 10 - \$ 50 per cubic yard (http://www.afcee.brooks.af.mil). Through the efforts of the AFCEE Bioventing Initiative and the US EPA Bioremediation Field Initiative, bioventing has been implemented at more than 150 sites (through out US) and has emerged as one of the most cost-effective, safe and efficient technologies currently available for vadose zone (i.e., unsaturated soils) remediation of petroleum-contaminated sites. Bioventing is particularly effective in quickly treating the most toxic, soluble, and mobile constituents in fuels (i.e., benzene, toluene, ethyl benzene, and xylenes). Results are typically seen within less than one year. The technology has been applied at waste oil, heating oil, diesel, jet fuel, and gasoline contaminated sites.

Kirtland and Aelion (2000) conducted bioventing studies for the removal of petroleum hydrocarbons from soil and ground water. The main objective of the study was to evaluate and quantify the mass removal rate of petroleum contaminants in low permeability soil sediments using air sparging and soil vapor extraction in pulsed and continuous operation. Samples were collected from exhaust gas and vadose zone air and were analyzed for O_2, CO_2, BTEX (benzene, toluene, ethyl benzene, xylene) and total combustible hydrocarbon (TCH) concentrations using portable hand meters and gas chromatograph. Based on the data obtained, a conclusion was drawn that pulsed operation (mean mass removal rate 17.6 kg day^{-1}) was preferable to continuous operation (mean mass removal rate 14.3 kg day^{-1}) because of increased mass removal and decreased energy consumption.

Filler et al. (2001) carried out successful bioremediation of petroleum hydrocarbons using combination of bioventing with active warming, fertilization and power cycling at Arctic region. The bioremediation of petroleum-contaminated soils in the Arctic was carried out using innovative engineering and environmental manipulation to enhance microbial activity beyond the natural effective season. Temperature and substrate availability were used as important parameters to extend the beneficial activity of microbes in Arctic biopiles. A multidisciplinary team of engineers, microbiologists and electricians has designed and installed a thermally enhanced biopile at a diesel-contaminated gravel pad in Prudhoe Bay, AK. The combination of bioventing with active warming, fertilization and power cycling was used to enhance remediation at this site. They found that the components which mainly rendered the success of this process were thermal insulation system (TIS) design, microbiological monitoring plan, and power optimization. Researchers successfully developed an integrated approach for bioremediation of diesel-contaminated soils at an Arctic site.

Mihopoulos et al. (2002) studied bioventing processes under anaerobic conditions to understand the oxygen exclusion patterns in vadose zone. Anaerobic condition was maintained by supplying nitrogen and oxygen exclusion was performed by using venting flows at the contaminated sites. It was important in designing an anaerobic bioventing (ABV) process for bioremediation of soil contaminated with chlorinated solvents. This study helped in designing an effective ABV soil remediation system and it was particularly attractive for *in-situ* bioremediation process, by minimizing *ex -situ* treatment costs.

8.4 Biostimulation

Biostimulation refers to the addition of specialized nutrients in the contaminated sites so that naturally occurring microbes are present in sufficient numbers and types to break down the waste material effectively. It is a remediation process whereby the microbial activity is enhanced to a great extent by timely addition of appropriate nutrients. This accelerates the naturally occurring biodegradation process under favorable environmental conditions such as oxygen / air, temperature, pH, addition / presence of suitable microbial populations, water content and proper mixing (Trindade et al. 2005).

Biostimulation can be effectively employed using essential nutrients such as carbon, nitrogen, phosphorus, potassium, calcium and magnesium for rapid biodegradation of contaminants by increasing the bacterial density (Evans et al. 2004). This process is mainly used to clean up oil spillage as well as other organic compounds such as trichloroethylene (TCE) from soil and wetlands (Eguchi et al. 2001). Biostimulation can be successfully carried out *in-situ* as well as *ex-situ* utilizing aerobic or anaerobic bacteria, and can be combined with other bioremediation processes.

8.4.1 Factors Affecting Biostimulation Processes

The following important factors have to be considered for effective and complete biostimulation at contaminated sites (William 1997):

- Physical characterization of soil and chemical constituents
- Chemical characterization of soil and contaminants
- Characterization of microbial populations

The soil structure plays an important role in the dispersion of nutrients and the bio-availability of toxic pollutants to bacteria and, therefore, is required to be characterized properly before initiating the biostimulation process. Physical characterization of the organic pollutant is also conducted in terms of concentration and their positioning below and above the ground level (Suzuki 1996, 1998).

Chemical nature of the contaminated sites also determines the rate of biodegradation and microbial activity for successful bioremediation (William 1997). In addition to actual contaminants to be treated, the nature of the soil in terms of its chemical composition helps in deciding the extent of biostimulation. This includes: total organic carbon, chemical oxygen demand, calcium, copper, chloride, iron, magnesium, nitrate, dissolved oxygen, potassium, sodium, sulfate, total phosphate and pH which can influence the rate of biodegradation. Additionally, microbial characterization in terms of the types of bacteria, aerobic and anaerobic bacterial populations, biodegradation ability of microorganisms and the rate of bioremediation have to be considered to promote the biostimulation process (Eguchi et al 1998, 2000).

8.4.2 Case Studies

Lin and Mendelssohn (1998) studied the combined effects of biostimulation and phytoremediation at oil-spill habitat using marsh sods of *Spartina alterniflora* and *Spartina patens*. Oil degradation rate was compared with or without addition of fertilizers. Addition of fertilizer such as N–P–K at rates of 666 kg N ha^{-1} (NH_4-N), 272 kg P ha^{-1} (P_2O_5) and 514 kg K ha^{-1} (K_2O) significantly increased the rate of biodegradation in soil. Conclusively, the addition of nutrients increased biomass of plants and in turn enhanced rate of phytoremediation process.

Eguchi et al. (2001) conducted degradation of organic pollutant, especially trichloroethylene (TCE) using anaerobic microbial populations. The degradation study of TCE was conducted with methane injection. Methane, oxygen, nitrate and phosphate were introduced into the groundwater contaminated with TCE. After a week of biostimulation, reduction in methane concentrations was observed and methane utilizing bacteria increased from 10 to 10^4 CFU/mL. The study indicated that biostimulation using methane and nutrients was useful for the removal of TCE from the contaminated site.

Evans et al. (2004) studied the removal of oil contamination by biostimulation process (nitrogen, phosphate and sulfur addition) utilizing diversity of a bacterial community of an acidic Cambisol Atlantic forest in sub-surface site. The experiment was based on enumeration of bacterial populations and hydrocarbon degraders in microcosms using plating techniques and molecular fingerprinting. Polymerase chain reaction (PCR) followed by denaturing gradient gel electrophoresis (DGGE) was used to generate microbial community fingerprints by employing 16S rRNA gene as a molecular marker. Biostimulation increased the soil pH and the levels of phosphorous, K, Ca, and Mg. Oil contamination caused increase in soil organic carbon level (170-190% higher than the control). At control site, bacterial numbers remained stable and there was little change in oil contaminated sites. On the other hand, at biostimulated sites, an increased bacterial number and decrease in the number of DGGE bands were observed. The addition of inorganic compounds to soil had greater impact on the development of bacterial community compared to oil contamination.

Trindade et al. (2005) carried out a comparative study of bioremediation for weathered oil and a soil recently contaminated with oil. The application of optimized conditions stimulated biodegradation of both weathered and recently oil-contaminated soils leading to the increase in oil mineralization and total petroleum hydrocarbon (TPH) removal. Although the weathering process was responsible for slower rate of biodegradation and TPH removal obtained after 41 days of experiments, the well adapted native microbial consortium was responsible for better decontamination. This study illustrated the important role of the weathering phenomenon in biodegradation of oil.

8.5 Bioaugmentation

Bioaugmentation is a significant *in- situ* and *ex-situ* treatment process in which naturally occurring microbes are added to contaminated sites in order to eliminate toxic contaminants. It is considered an extremely cost-effective technique in comparison to other existing bioremediation technologies (Head and Oleszkiewich 2004).

Further, bioaugmentation can be employed to degrade a wide range of chemical contaminants such as ammonia, hydrogen sulfide, pesticides, petroleum compounds and a growing number of toxic organic chemicals present in soil and water (Vogel 1996). The advantages of this technique include:

i. Total degradation of toxic compounds;
ii. This changes the species composition of existing biomass resulting in improved waste water / soil treatment;
iii. Bacteria isolation can be attempted from the contaminated soils and water of sludge storage areas and lagoons.

8.5.1 Methodology for Isolation of Bacteria

Criteria for isolation and selection of bacterial strain used for degradation of desired pollutant of interest are given below (Christopher and Plowman 1997):

i. Selection of contaminant of interest;
ii. Selection of bacterial strain which survives and utilizes the contaminant for growth;
iii. Selection of strains which consume oxygen and oxidize the target compound thereby converting into non-toxic gases;
iv. Removal of minor strains to increase overall efficiency of the process;
v. Selection of strain which degrades higher concentration of target compounds;

vi. Selection of microbes which degrade variety of toxic compounds and must be non-pathogenic.

8.5.2 Case Studies

Bioaugmentation is a highly cost-effective bioremediation process and utilizes naturally occurring micro-organisms for removal of toxic pollutants. It clearly provides certain advantages over biostimulation, especially, when pollutants are toxic in nature and the contaminated site does not contain appropriate microorganisms. Determination of the potential success of bioaugmentation requires an understanding of bioavailability of the pollutant, survival and activity of added microorganisms and general environmental conditions which control the rate of bioremediation (Vogel 1996).

Hwang and Cutright (2002) conducted a study to evaluate biodegradability of aged pyrene (PYR) and phenenthrene (PHE) in a natural soil. A stock solution of each compound was prepared in hexane and spiked (in kg soil) to yield 100 mg PYR, 100 mg PHE, and (50 mg PYR + 50 mg PHE). PAHs at (50 + 50) mg/kg was spiked to evaluate cometabolic substrate utilization in the presence of more easily biodegradable co-solute and to compare the biodegradation results in terms of the same concentration level (i.e., total PAH = 100 mg/kg). Bioaugmentation and biostimulation was performed by inoculating an enriched bacterial population and the addition of nutrients, respectively. Aging resulted in higher concentrations of both compounds and smaller bacterial activity in the solution phase. Surprisingly, total biodegradation rate was increased in the aged soil system than fresh soil. They concluded that biostimulation was not appropriate where as bioaugmentation achieved an effective biodegradation in this study.

Jianlong et al. (2002) used bioaugmentation process to enhance the removal of quinoline compounds which were poorly removed by aerobic – anoxic – oxic (A1-A2-O) reactor system from a waste water plant. Quinoline degrading bacterium identified as *Burkholderia pickettii* was used as a bioaugmentation microorganism. The removal rate of quinoline was increased by combining bacteria with the oxic reactor. This study concluded that bioaugmentation can be used as an efficient method for removal of recalcitrant organic compounds from waste water.

Quan et al (2004) carried out bioaugmentation experiments using bacteria for the removal of toxic and recalcitrant organic substances intermittently appearing in waste water. 2,4 Dichlorophenol (DCP) was chosen as a target compound and 2,4-DCP degrading special mixed culture was used as bioaugmentation product. Conventional activated sludge (CAS) system was used to study effectiveness of bioaugmentation process. There was not only greater removal of 2,4-DCP but other chlorophenol compounds also degraded. A separate bioaugmentation bioreactor was combined with original CAS system at a different location. This study concluded that bioaugmentation could be used to improve CAS process facing shock loads of toxic pollutants.

Head and Oleszkiewicz (2004) studied bioaugmentation using nitrifying bacteria for waste water treatment plants in cold climates. The objective was to study the impact of sudden decrease in temperature on nitrification rate. Experiments were carried out in sequencing batch reactors (SBRs) at 10° C by seeding nitrifying bacteria acclimated at 20° C, resulting in complete removal of ammonia.

8.6 Anaerobic Biotransformation

Anaerobic biotransformation is used for the degradation of organic compounds such as chlorinated hydrocarbons, polychlorinated phenols and nitro-aromatics which are resistant to aerobic treatment. Anaerobic treatment processes produce much smaller quantities of biomass (Cheng et al. 1996).

Moreover, anaerobic biotransformation of polychlorinated hydrocarbons, such as carbon tetrachloride (CT), tetrachloroethylene (PCE) and 1,1,1-trichloroethane (TCA) is of particular interest in bioremediation of contaminated groundwater. Dechlorination rates are generally faster under these conditions (Doong and Wu 1995). Substrate concentration, substrate availability, temperature, pH, co-factor supply, concentration of cells, viability of cells and type of operation (batch or continuous) are some of the major factors responsible for successful anaerobic biotransformation processes (Marschner et al. 2001).

8.6.1 Case Studies

Doong and Wu (1995) studied effects of different substrates on anaerobic biotransformation of heavily chlorinated hydrocarbons to assess the bioavailability of substrate on *in-situ* bioremediation of contaminated ground waters in anaerobic fed batch process. Feeding of different substrates (glucose, methanol, acetates and humic acids) into the microcosms improved the removal efficiency by increasing rate of biotransformation. The substrate concentrations in fed batch were increased from 10-30 mg/L. The corresponding removal efficiencies of carbon tetrachloride (CT), tetrachloroethylene (PCE) and 1,1,1-trichloroethane increased from 96%, 49% and 61% to 99.9%, 61% and 69% respectively. In a contaminated system, added substrates were provided reducing potential (electron activity) to drive dechlorination of chlorinated hydrocarbons Microscopic studies to determine shapes of bacterial populations were also carried out. Small and spherical shapes were observed in the cultures added along with acetate and humic acid, where as large size and long rod shape were found in the bacterial population supplemented by glucose. This study highlighted the role of supplemental substrates which provides high available free energy (reducing power) is conducive to the dechlorination of polychlorinated hydrocarbons in biotransformation and treatment of contaminated groundwater. Further work in the area of identification of the metabolic pathways for the chlorinated hydrocarbons and the electron transfer processes would elucidate the role of the auxiliary substrate towards reductive dechlorination as well as yield maximum results in the field clean-up works.

Doong et al. (1996) have studied the anaerobic biotransformation treatment of polychlorinated methane and ethene under various redox conditions. Two sets of batch-type experiments on the biotransformation of carbon tetrachloride (CT) and tetrachloroethylene (PCE) using different concentrations of acetate as the auxiliary substrate were conducted. Experiments were carried out under various redox conditions to evaluate the effects of the redox potential of the environment and substrate concentration. Results indicated that the disappearances of PCE (20 to 62%) and CT (22 to 99.9%) from their original concentrations with the redox potentials of the microcosms ranging from 188 to -263 mV. The threshold values of 60 and 70 mV were also identified for the obvious biotransformation (> 30%) of PCE and CT, respectively. Substrate concentration effect was emphasized only when the redox potential of the microcosms was below the threshold value. A higher substrate concentration produced more biomass than a lower substrate concentration. They concluded that redox potential of the environment was a more controlling environmental factor than the substrate concentration in determining the rate of biotransformation of chlorinated hydrocarbons.

Cheng et al. (1996) studied the effect of initial concentration of 2,4-dinitrotoluene on its biotransformation using ethanol-utilizing bacteria in a continuous flow laboratory fermentor with 2,4-DNT and ethanol as substrates. Under anaerobic conditions at 35°C, 2,4-DNT was converted into 2,4-diaminotoulene (DAT) along with two intermediate by-products 2-amino-4-nitrotoluene and 4-amino-2-nitotoluene. The concentration of 2,4-dinitrotoluene inhibits its own biotransformation as well as acetogenesis of ethanol. Propionate was formed during fermentation of ethanol without 2,4-DNT or low concentration of 2,4-DNT. This study explained the effect of substrate concentration and electron acceptors for effective biotransformation process.

You et al. (1996) studied the importance of surfactants and redox potential reducing agents on biotransformation of 1,1,1-trichloro-2,2-bis,(p-chlorophenyl) ethane [DDT]in aqueous and soil phase using mixed cultures. DDT transformed into 1,1-dichloro-2,2-bis,(p-chlorophenyl) ethane [DDD] upon removal of one aliphatic chlorine under unaltered conditions. The rate of transformation of DDT increased by the addition of non-ionic surfactants such as Triton X-114 or Brij 35. The addition of either surfactant or reducing agents did not extend the transformation of DDT, however addition of both surfactant and reducing agents extended DDT transformation by reducing accumulation of DDD and increasing accumulation of less chlorinated products.

Somsamak et al. (2001) demonstrated anaerobic biotransformation of methyl *tert*-butyl ether (MTBE), ethyl *tert*-butyl ether (ETBE) and methyl *tert*-amyl ether (TAME) (to a final concentration of 100 mg/l) under different anoxic electron accepting and sulfur reducing conditions, such as denitrification, sulfate reduction, Fe(III) reduction or methanogenesis. A stoichiometric amount of *tert*-butyl alcohol and *tert*-amyl alcohol accumulation indicated that *O*-demethylation was the initial step in MTBE and TAME biodegradation under sulfate-reducing conditions. No transformation of MTBE or TAME was observed under the other electron-accepting conditions over 3 years. Under all conditions tested, there was no biotic loss of ETBE observed. MTBE and TAME concentrations depleted upon refeeding the sulfate-reducing cultures.

Bhushan et al. (2003) studied enzyme catalyzed biotransformation of octahydro-1,3,5,7-tetranitro-1,3,5,7-tetrazocine (HMX), an explosive mediated by xanthine oxidase (XO). The rate of biotransformation was studied under both aerobic and anaerobic conditions were 1.6±0.2 and 10.5±0.9 nmolh^{-1} mg protein^{-1}, respectively. They found that anaerobic biotransformation rate was about 6-fold higher using NADH as an electron-donor compared to xanthine. The following products were obtained during biotransformation- nitrite (NO^-_2), methylenedinitramine (MDNA), 4-nitro-2,4-diazabutanal (NDAB), formaldehyde (HCHO), nitrous oxide (N_2O), formic acid (HCOOH), and ammonium (NH^-_4). A comparative study with native-, deflavo-, and desulfo-XO and the site-specific inhibition studies showed that HMX biotransformation occurred at the flavin adenine dinucleotide (FAD)-site of XO. Nitrite stoichiometry revealed that an initial single N-denitration step was sufficient for the spontaneous decomposition of HMX. This study showed effective degradation during anaerobic conditions and extended the fundamental knowledge to understand biotransformation process, which helps to design and optimize well controlled biochemical reactions for complete *in-situ* removal of explosives from soil and sediments.

8.7 Aerobic Biotransformation

Aerobic biodegradation of toxic pollutants is much faster than anaerobic biodegradation. The method is very well applied in microbial degradation of toxic contaminants like carbonaceous material and low molecular weight aliphatic hydrocarbons. In certain cases of aromatic and chlorinated compounds, this technique needs to be combined with other suitable methods as they produce toxic by products and must be treated and removed completely from the environment to maintain a safe and healthy ecosystem (Shen 1998).

The main factors which regulate rates of aerobic biotransformation are similar to other process requirements and depend upon the chemical nature of contaminant, availability of contaminant, degradation capability of microorganisms, amount of available oxygen, pH and temperature (Schultz et al. 1996).

8.7.1 Case Studies

Deeb and Cohen (2000) evaluated the impact of substrate interaction on aerobic biotransformation rates and mineralization potentials of gasoline monoaromatics and methyl tert-butyl ether (MTBE) compounds that commonly co-exist in groundwater contaminant plumes using mixed bacterial culture. Three toluene-grown cultures were shown to biotransform all six BTEX compounds (benzene, toluene, ethyl benzene, *O*-xylene, *m*-xylene, and *p*-xylene), both individually and in mixtures, over a broad range of concentrations (up to 80 mg/l) finally yielding CO_2. This study presented the effect of number of patterns, competitive inhibition and co-metabolism which occurred during aerobic biodegradation in ground water contaminated plumes.

8.8 Biological Fixation

Biological fixation is defined as a naturally occurring phenomenon that takes place during bioremediation process by converting inorganic and organic compounds into soil organic matter by utilizing living organisms and is mainly involved in nutrient cycles such as carbon, nitrogen, sulfur and phosphorous to maintain ecological balance (Frank et al. 2003). Various factors such as temperature, pH, substrate concentration, dissolved oxygen, and the presence of toxic and inhibitory substances are responsible for the biological fixation of different compounds (Dobereiner 1997).

Nitrogen Fixation

When a biological process fixes or incorporates nitrogen into microbial or plant biomass, it is known as nitrogen fixation. Nitrogen fixing bacteria fix about 60% of nitrogen gas into ammonia using prokaryotic organisms (diazotrophs). Nitrogenase is an enzyme used during N_2 fixation, which is inactivated in the presence of oxygen. Some cyanobacterial populations also perform N_2 fixation under aerobic conditions by developing special compartments within their cells known as heterocyst (Kessel and Hartley 2000). Cyanobacteria capable of creating heterocyst do not have photosystem II which produce oxygen during photosynthesis. In the absence of oxygen, non-heterocyst possessing cyanobacteria can also fix atmospheric N_2. *Azotobacter* species can also fix N_2 under microaerophilic conditions i.e. even in the presence of very low concentration of oxygen (Weidner et al. 2003).

Carbon Fixation

Micro algae like *Chlorella* and *Synechocystis* sp. play a major role in carbon fixation through Calvin cycle (Dubbs and Tabita 2004). Biological fixation of CO_2 reduces the level of pollution and maintains CO_2 and energy balance (Falkowski 1997).

Sulfur Fixation

Sulfur fixation is used to fix sulfur compounds necessary for building disulphide linkage in protein utilizing bacteria having a capability to reduce sulfate compounds in the absence of oxygen. Sulfur present in the environment is used in electron transport through iron/sulfur cluster. It is also used in the catalytic site of several enzymes and co-enzymes and is finally used as anti-carcinogenic and anti-septic compounds (Delgado and Follett 2002).

Phosphorus Fixation

Phosphate in the environment is found in the form of variety of compounds (sugar phosphates, lipids, nucleic acids, and free nucleotides) and is fixed in living organisms including plant cells (Smith et al 2000).

Humification

The biological processes by which organic matter decomposes into form humic substances (naturally occurring heterogeneous organic substances).

8.8.1 Case Studies

Chen et al. (2003) studied biological fixation for carbon and nitrogen using microorganisms under natural system. The action of microorganisms plays a major role in efficient nutrient cycling of soil and it converts complex organic compounds such as proteins, carbohydrates and cellulose into a usable form that plant can utilize for growth. Microbes can also help to stabilize the soil by physical binding of soil particles together by releasing gloxmalin which helps to bind the clay and organic material together. This process increases the soil productivity. This study revealed the importance of microorganisms in biological fixation by improving soil quality.

Lombi et al. (2002) carried out laboratory experiments to study biological fixation of metals under *in-situ* conditions using bauxite residues. Three soil amendments (red mud, beringite and lime) were applied to two soils polluted by heavy metals. Oil seed rape, wheat, pea and lettuce were grown successfully in pots on untreated and amended soils and metal uptake was determined. In both the soils, all three amendments reduced phytotoxicity of heavy metals, and increased plant growth and decreased metal concentration in plants. They found that red mud treatment also increased soil microbial biomass and was able to immobilize heavy metals into the soil.

8.9 Enzyme-Catalyzed Treatment

Bioremediation is a continually evolving process with progressively occurring improvements. One such improvement is the use of enzymes produced from microorganisms, plants, and animals for removal of toxic compounds from industrial waste water streams (Klibanov et al 1980; Maloney et al 1986; Ghioureliotis and Nicell 1999).

In most of the microbial bioremediation and phytoremediation treatment technologies, toxic compounds are converted into non-toxic end products via enzymatic reactions which take place within or outside the cell by means of secretion of intracellular or extracellular enzymes (Bewley 1996). There are several factors - pH, temperature and contaminant (substrate) concentrations which regulate enzymatic reactions (Nannipieri et al 2002).

8.9.1 Limitations of Enzyme Catalyzed Treatment

Enzymatic treatment has several limitations such as high cost of treatment and potential formation of toxic residual products which remain in the aqueous phase. Besides, enzymes are also susceptible to permanent inactivation by formation of various undesirable side products during the treatment process (Nicell et al 1993; Heck et al 1992; Aitken et al 1994).

8.9.2 Case Studies

Xu et al., (2005) studied oxidative polymerization of 1-naphthol using horseradish peroxidase (HRP) enzyme in pH 7. They found addition of peroxidase enzyme resulted in production of both soluble and insoluble naphthol polymerization products (NPP). Soluble NPP were more polar in nature than naphthol and comprised of trimers and tetramers. Insoluble NPP oligomers consist of dimers, trimers and tetramers were non-polar than naphthol. LC/MS characterization of insoluble NPP revealed that the nature of compounds such naphthoquinone, 2-hydroxy,1-4- naphthoquinone, dimeric 4-4'-bi-1-naphthol and trimeric naphthol. This study confirmed that the overall toxicity was reduced due to formation of NPP and followed by their precipitation.

Bhandari and Xu (2001) extensively studied the degradation and immobilization of phenolic contaminants by peroxidase enzyme treatment in surface soils. The sorption and desorption of phenol, o-cresol, 2,4-dichlorophenol (DCP) and 1-naphthol was studied in two surface soils. They added these compounds as single solutes or binary mixtures. Addition of HRP resulted in sorption and contaminant solubility was increased in the order of naphthol < DCP < cresol < phenol. They found that reduction in contaminant desorption by the addition of HRP. This study concluded that enzyme treatment resulted in production of hydrophobic polymers, due to their lower solubility in water, readily portioned on to the solid phase. HRP reduced the mobility of the phenolic contaminants in soil surface which leads to polymerization of phenols and subsequent adsorption on soil particles.

They also investigated the extractability of the parent solutes and their polymerization products using ^{14}C-labelling. They quantified the retained contaminant on soil particles as water-extractable, methanol extractable, humic/fulvic (HA/FA) acid-bound and soil/humin bound. They found that 2-20% of the solute retained on soil and remained bound to the HA/FA and soil/humin components in unamended soils after 7 day contact period. During peroxidase enzyme treatment the amount of solute bound was as high as 40-75%. The alkali-extractable HA/FA component contained the largest fraction of radioactivity in the peroxidase amended soils. They found that the soil organic matter content was the predominant factor which controlling the extent of sorption of the parent components. The polymerization products are retained on the clay content and their particle surface area also contributed to the retention of these compounds. High molecular weight oligomers produced during peroxidase-mediated polymerization of phenols associate strongly with soil components and were incorporated into the soil organic matter by oxidative coupling reactions. The enhancement of bound residue formation occurred with more soluble phenols, which indicated that peroxidase-mediated immobilization can be more effective for the containment of phenols that were highly mobile in soil and groundwater (Xu and Bhandari., 2003).

Novozymes and Mitsubishi Chemicals (2004) have developed an enzyme which degrades bioplastic polybutylene succinate (PBS) in an effective and ecofriendly manner. In Japan, enormous amount of PBS has been used by farmers for covering the cropland. The biodegradation of PBS takes several months preventing farmers to use land during this period. Spreading the new enzyme on bioplastic resulted in its degradation within 12 hours (Website: http://www.novozymes.com). This study confirmed the potential application of enzymes for faster and effective biodegradation of target compounds than conventional processes.

8.10 Biological Reactors

Bioreactors are enclosed vessels used in biodegradation of toxic chemicals and operated under controlled parameters with continuous monitoring system. Several types and design of bioreactors are available namely, batch bioreactor, continuous bioreactor, sequential batch biofilm reactor (SSBR), membrane bioreactor (MBR), fluidized bed reactor (FBR), biofilm reactor (BFR), and airlift bioreactor. Use of a particular type of bioreactor mainly depends upon nature of contaminants and site (Nakhla and Suidan 1995).

Bioreactors are widely used methods due to their cost-effectiveness and eco-friendly approach. Besides, they can easily degrade recalcitrant compounds and provide high-quality treatment by complete removal of toxic substances. They can be operated both under aerobic as well as anaerobic conditions, by using minimal manpower and generating minimum sludge (Vischetti et al 2004).

8.10.1 Factors Affecting Biodegradation in Reactors

In order to achieve effective and complete biodegradation of toxic compounds, certain essential requirements of bioreactors are to be fulfilled. Biodegradation of toxic compounds in a bioreactor depends on various parameters such as dissolved oxygen, pH, temperature, aeration, nutrient requirements, contaminant concentration and microorganisms (White and Schnabel 1998).

8.10.2 Limitations of Biological Reactors

Although bioreactors are used very commonly, they have certain limitations which makes the whole process lengthy. These limitations are:

- Prolonged period of acclimatization and time required for process optimization and standardization (Gander et al. 2000);
- Extensive design of bioreactor;
- Require large working area above ground level;
- Lack of control in volatile compounds;
- Poor understanding of microbial biokinetics and
- Factors influencing bacterial action.

8.10.3 Case Studies

Sajc and Novakovic (2000) studied bioconversion of anthraquinones using immobilized plant cells in airlift bioreactors. They used alginate immobilized *Frangula alnus* cells, a continuous aqueous phase (nutrient), dispersed solvent phase (n-hexadecane or silicone oil) and gas bubbles to achieve the bioconversion of contaminant.

8.11 Monitored Natural Attenuation

Recent years have seen an increased interest in developing novel strategies and processes of biodegradation for the removal of contaminants from contaminated sites (Suarez and Rifai 1999). Implementation of non-structural solutions such as monitored natural attenuation (MNA) can be used as a sole remedial alternative to soil and ground water contaminated sites (Lee and Batchelor 2003). Design of ground water monitoring in natural attenuation process were illustrated in Figure 8.3.

The popularity of this approach will increase due to complexities of available subsurface systems and costs associated with conventional engineered approaches. Selection of natural attenuation to treat contaminated sites depends mainly upon the ability to reach remediation goals with in reasonable timeframes (Small et al. 1998). Under favorable geochemical conditions, natural attenuation of ground water contaminants can result in significant reduction in cost over other remedial approaches. Besides it is a less intrusive and disruptive process (Nobre and Nobre 2004).

Natural attenuation is the term used for all natural processes which are responsible for remediation of pollutants from contaminated sites. Naturally occurring physical, chemical and biological processes can transform contaminants into harmless form or immobilize them in the subsurface thereby reducing the concentration of contaminants in water (Lorah and Olsen 1999).

The U.S. Environmental Protection Agency defines monitored natural attenuation as the "reliance on natural attenuation processes (within the context of a carefully controlled and monitored site cleanup approach) to achieve site-specific remediation objectives within a time frame that is reasonable compared to that offered by other more active methods. The 'natural attenuation processes' that are at work in such a remediation approach include a variety of physical, chemical, or biological processes that, under favorable conditions, act without human intervention to reduce the mass, toxicity, mobility, volume, or concentration of contaminants in soil or groundwater. These *in-situ* processes include biodegradation; dispersion; dilution; sorption; volatilization; radioactive decay; and chemical or biological stabilization, transformation, or destruction of contaminants." (EPA, OSWER Directive 9200.4-17P).

Non-destructive attenuation processes are dispersion, dilution, volatilization, precipitation and sorption to soil particles (EPA-600-R-98-128). Some of the indicators used to monitor natural attenuation processes are reduction of contaminant

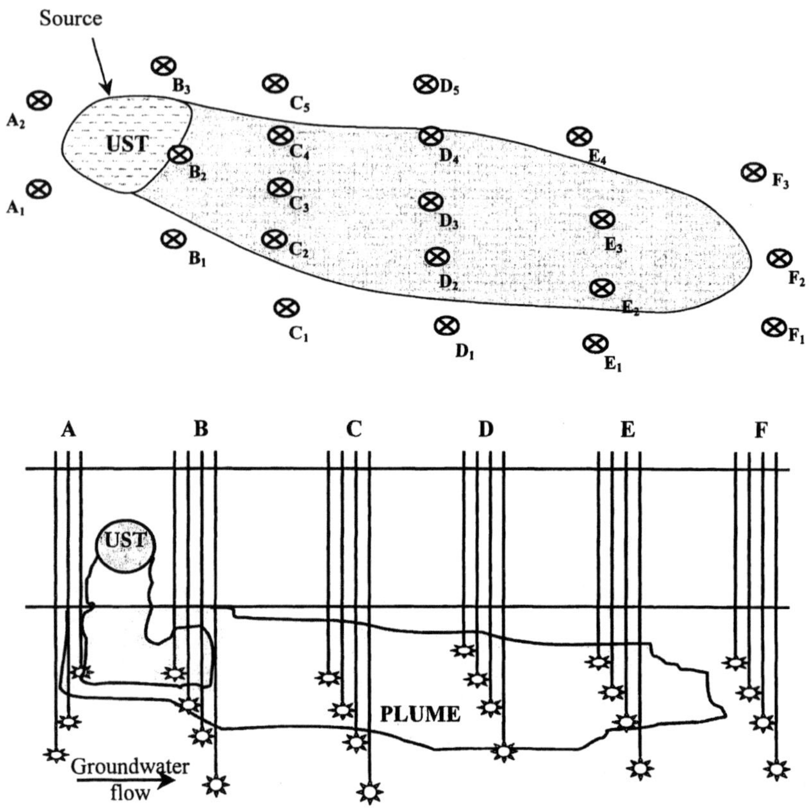

Figure 8.3 (Top) Plan view of optimal ground water monitoring. The monitoring points located as follows: up gradient A1 and A2; source zone B1 through B3; mid-plume (transverse to flow) either C1 through C5 or D1 through D5; plume toe: E1 through E4; plume centerline B2-C3-D3-E3 and underground storage tank (UST). **(Bottom)** Schematic diagram of longitudinal cross section of optimal ground water monitoring network process (note: figure not to scale).

concentrations, depleted oxygen levels, nitrate, and sulfate, elevated levels of methane in ground water and presence of daughter products and are important indications for successful progress of degradation of contaminant (Lorah and Olsen 1999). Despite the positive outcomes, large uncertainties may arise from (i) complex and unknown source histories at such sites (spatial scale, timing and magnitude of spills and infiltration) and (ii) poorly quantified rates of hydrodynamic dispersion at the plume fringe.

8.11.1 Factor Affecting Monitored Natural Attenuation

The factors which affect MNA processes are microbial physiology, concentration of contaminant, presence of electron acceptor, pH, aerobic, anaerobic and co-metabolic contaminant degradation. These factors are useful for process improvement as well as optimization (Brady et al. 1998).

8.11.2 Case Studies

Thronton et al. (2000) carried out natural attenuation of landfill leachate in laboratory column containing Triassic sandstone aquifer material from English midlands. The sorption and degradation of leachate constituents which consisted of dissolved organic matter (DOM) and 13 organic micro pollutants were studied under acetogenic and methanogenic conditions. The study indicated an eightfold decrease in micropollutant sorption in acetogenic leachate (up to 280 days) and six fold increase in methanogenic leachate (150-353 days than their half life >2400 days). This behavior reflects a combination of interactions between the micropollutants, leachate DOM and aquifer mineral fraction. Degradation of organic fractions occurred under Mn and SO_4^{2-} reducing conditions. Workers concluded that aquifer materials have large capacity for *in-situ* bioremediation of organic pollutants.

Berger et al. (2000) conducted studies to understand the controls on chemical evolution of drainage in a historic mining district, New Mexico. The drainage near waste rock pile was acidic in nature (pH ranges from 3.0-5.0) and carried high concentrations Zn (5.0 ppm), Al (0.050-0.20 ppm), Cu (1.3 ppm), Pb (0.015 ppm), Fe (0.3 ppm) and SO_4 (250 ppm). As drainage flowed towards Pecos River, pH increased to greater than 7 and heavy metal content decreased. A process model of natural attenuation in this drainage showed the main controls on pH and concurrent mixing with tributary streams. This study concluded that natural attenuation of contaminant takes place in two distinct pathways by precipitation of Al, Cu, Fe and Pb and dilution of Zn, Mg, Mn and SO_4.

Clement et al. (2002) studied degradation of dense non-aqueous phase liquids (DNAPL) using natural attenuation process. DNAPL compounds studied were mainly chlorinated ethene and ethane originated from chlorinated solvent manufacturing plants and from other refineries. In this study, workers followed various steps in implementing U.S.EPA guidelines for natural attenuation. The first stage of monitored natural attenuation (MNA) assessment was to collect field data useful to complete a biodegradation scrolling analysis as recommended in the protocol. The analysis results indicated that the site had potential for natural attenuation process. The second stage was a detailed conceptual model developed to identify various contaminants transport pathways and exposure points. It was useful to assess whether the contaminants were attenuating at a reasonable rate along these transport paths so that MNA can be considered as a feasible remedial option. The site data indicated that chlorinated ethane and ethane plumes were degrading and will attenuate within 1000 ft down gradient from the source. Therefore, they considered MNA as one of the feasible remedial processes for this site.

Erses and Onay (2003) applied natural attenuation to determine fate and behavior of heavy metals co-disposed with municipal waste under methanogenic conditions. Two landfill simulating reactors, with leachate recirculation and without leaching circulation were used. These reactors were operated under constant room temperature at 32° C and filled with typical composition of municipal solid waste. After the onset of methanogenic conditions, selected heavy metals including iron, copper, nickel, cadmium and zinc were added according to the quantity suggested for co-disposal. The study showed that 90% of all heavy metals were precipitated from reactors within 10 days due to establishment of highly reducing environment and formation of sulfides from sulfate which provided heavy metal precipitation. This confirmed the effectiveness of co-disposal of waste under methanogenic conditions.

Nobre and Nobre (2004) performed bioremediation of chlorinated organic pollutants using natural attenuation process. After a major accidental release of 1,2-dichloroethane (1,2-DCA) in the soil, a system was implemented for plume migration control which had 300 m long physical barrier (cement-bentonite diaphragm wall) and 12 extraction wells. Results confirmed natural degradation of 1,2-DCA into vinyl chloride as well as ethane under natural anaerobic conditions on-site. This study concluded that natural attenuation could be employed for degradation of chlorinated compounds and effective degradation can be achieved by combing with biosparging system.

Lee and Batchelor (2004) studied anaerobic reduction of chlorinated ethylenes [Concentrations expressed as electron equivalents relative to ethane (PCE, 10 equiv/mol; TCE, 8 equiv/mol; DCEs, 6 equiv/mol; VC, 4 equiv/mol; ethylene, 2 equiv/mol)] conditions in contaminated soil. This study was carried out to understand behavior of chlorinated ethylenes in natural systems, including systems modified to promote attenuation of contaminants. The reductive capacities of iron-bearing sulfide (pyrite), hydroxide (green rust; GRSO4), and oxide (magnetite) minerals for Cr(VI) and tetrachloroethylene (PCE) were 1-3 orders of magnitude greater than those of iron-bearing phyllosilicates (biotite, vermiculite, and montmorillonite). The reductive capacities of surface soil (plains of central Texas) were similar and slightly greater than those of iron-bearing phyllosilicates. The reductive capacity of iron-bearing soil minerals for Cr(VI) was roughly 3-16 times greater than that for PCE, implying that Cr(VI) is more susceptible to being reduced by soil minerals than is PCE. GR_{SO4} has the greatest reductive capacity for both Cr(VI) and PCE followed by magnetite, pyrite, biotite, montmorillonite, and vermiculite. They found this order was the same for both target compounds, which indicates that the relative reductive capacities of soil minerals are consistent. The reductive capacities of pyrite and GRSO4 for chlorinated ethylenes decreased in the order: trichloroethylene (TCE)>PCE>cis-dichloroethylene (c-DCE)>vinyl chloride (VC). Fe(II) content in soil minerals was directly proportional to the reductive capacity of soil minerals for Cr(VI) and PCE, suggesting that Fe(II) content is an important factor that significantly affects reductive transformations of target contaminants in natural systems. Rate of reductive dechlorination was induced using dithionate and Fe(II) in contaminated soil sites. The study results indicated higher reductive degradation of chlorinated ethylene. The main transformation products in reductive dechlorination were acetylene, ethylene and ethane. Chlorinated intermediates were not found above detection limit. Formation of chloride was observed in all reduced soil suspensions, but not in control. This study concluded that degradation of chlorinated compounds was also possible using natural attenuation under reductive degradation and sorption of chlorinated compounds in soil.

8.12 Conclusion

Bioremediation is effective and continuously evolving remediation technology to manage a large variety of pollutants. Bioremediation comprises various technologies like bioventing, biosparging, biostimulation, bioaugmentation, bioreactors, enzyme catalysis and natural attenuation systems. Each technology possesses inherent characteristics which result in its application at a particular site, for example, natural attenuation methods would be a positive approach to handle comparatively dilute waste streams vis-à-vis utilization of bioreactors for concentrated wastes. In the last 20 years, significant research efforts have developed a detailed process understanding of ecological, biochemical and genetic basis of microbial contaminant degradation, with a view to enhance microbial capabilities and design more effective bioremediation processes. Furthermore, in future, site assessment procedures must become more integrated and combine key elements of biology, chemistry, hydrogeology and process modeling to develop sustainable solutions to land contamination. In the time to come, bioremediation will continue to be a feasible and economical option to remediate contamination.

References

Aelion, C. M., Widdowson, M. A., Ray, R. P., Reeves, H. W. and Shaw, J. N. (1995) Biodegradation, vapor extraction, and air sparging in low-permeability soils. In:

Hinchee, R. E., Miller, R. N., Johnson, P. C. (Eds.), *In-situ Aeration: Air Sparging*, pp.127-134.

Aelion, C. M., Widdowson, M. A., Ray, R. P., Reeves, H. W. and Shaw, J. N. (1996) Simplified methods for monitoring petroleum-contaminated ground water and soil vapor. *J. Soil Contam.* 5(3): 225-241.

AFCEE., (1994) *Bioventing Performance and Cost Summary*, Draft. Brooks AFB, TX. http://www.afcee.brooks.af.mil/er/ert/bioventing.htm

Ahlet, R. C. and Peters. R. W. (2001) Treatment of PCB-contaminated Soils: I. Evaluation of *in situ* reductive dechlorination of PCBs. *Environmental Progress*. 20(2): 108-116.

Aitken, M. D., Massey, J. I., Chen, T., and Heck, P. E., (1994) Characterization of reaction products from the enzyme catalyzed oxidation of phenolic pollutants. *Water Res*. 28: 1879-1889.

Bachman, J., Kanan, S. M., and Patterson, H. H., (2001) Monitoring laboratory-scale bioventing using synchronous scan fluorescence spectroscopy: analysis of the vapor phase. *Environmental Pollution*. 113: 155-162.

Berger, A. C., Bethke, C, M., and Krumhansl, J. L., (2000) A process model of natural attenuation in drainage from a historic mining district. *Applied Geochemistry*. 15: 655-666.

Bewley, R. J. F., (1996) Field implementation of in situ bioremediation: key physiochemical and biological factors. In: Stotzky, G., Bollag, J, M., (Eds.), *Soil Biochemistry*. Marcel Dekker, New York. 9: 473-541.

Bhandari, A., and Xu, F., (2001) Impact of Peroxidase Addition on the Sorption-Desorption Behavior of Phenolic Contaminants in Surface Soils. *Environ. Sci. Technol.* 35 : 3163-3168

Bhushan, B., Paquet, L., Halasz, A., Spain, J. C., and Hawari, J., (2003) Mechanism of xanthine oxidase catalyzed biotransformation of HMX under anaerobic conditions. *Biochemical and biophysical research communications*. 306: 509-515.

Boersma, P. M., Piontek, K. R., Newman, P. A. B., (1995) Sparging effectiveness for ground water restoration. In: Hinchee, R. E., Miller, R. N., Johnson, P. C. (Eds.), *In-situ Aeration: Air Sparging*. pp. 39-46.

Brady, P. V., Brady, M. V., and Borns, D., (1998) Natural attenuation, *Lewis publishers, Boca Raton, FL.* pp-245.

Chapelle, F. H., (1999) Bioremediation of petroleum hydrocarbon contaminated ground water: The perspectives of History and hydrology. *Ground water*. 37(1): 122-132.

Chen, G., Zhu, H., and Zhang, Y., (2003) Soil microbial activities and carbon and nitrogen fixation. *Research in Microbiology*. 154: 393-398.

Cheng, J., Kanjo, Y., Suidan, M. T., and Venosa, A. D., (1996) Anaerobic biotransformation of 2,4-dinitrotoluene with ethanol as primary substrate: mutual effect of the substrates on their biotransformation. *Water Res.* 30 (2): 307-314.

Christopher, H., and Plowman, R., (1997) Bioaugmentation: Put microbes to work. *Chemical engineering magazine*. Pp 74-82.

Clayton, w. S., Brown, R. A., and Bass, D. H. (1995) Air sparging and bioremediation: the case for *in-situ* mixing. In: Hinchee, R. E. (Ed.), *In-situ aeration: Air sparging, Bioventing, and related remediation processes*. Battlelle press, Columbus, OH, pp. 75-86.

Clement, T. P., Truex, M. J., and Lee, P., (2002) A case study for demonstrating the application of U.S.EPA's monitored natural attenuation screening protocol at a hazardous waste site. *Journal of Contaminant Hydrology*. 59: 133-169.

Corinne Cluis. (2004) Junk-greedy Greens: phytoremediation as a new option for soil decontamination. *BioTeach Journal*. 2: 61-67.

Deeb, R. A., and Cohen, L., (2000). Aerobic Biotransformation of Gasoline Aromatics in MultiComponent Mixtures. *Bioremediation Journal 4*(1): 1-9.

Delgado, J. A., and Follett, R. F., (2002) carbon and nutrients cycles. *J. Soil Water Conserv.* 57: 455-464.

Dobereiner, J., (1997) Biological nitrogen fixation in the tropics: social and economic contributions. *Soil Biol. Biochem*. 26 (5-6): 771-774.

Doong, R., and Wu, S., (1995) Substrate effects on the enhanced biotransformation of polychlorinated hydrocarbons under anaerobic condition. *Chemosphere*. 30 (8): 1499-1511.

Doong, R., Wu, S., and Chen, T., (1996) Anaerobic biotransformation of polychlorinatedmethane and ethene under various redox conditions. *Chemosphere*. 32 (2): 377-390.

Dubbs, J. M., and Tabita, F. R., (2004) Regulations of nonsulfur purple phototrophic bacteria and the interactive control of CO_2 assimilation, nitrogen fixation, hydrogen metabolism and energy generation. *FEMS Microbiology Reviews*. 28: 353-376.

Eguchi M., Myoga, H., Sasaki S., and Miyake Y., (1998) Biotreatability studies for remediation of TCE-contaminated groundwater. *Bioremediation Phytoremediation*. Lewis Publishers, Monterey, Battelle Press, pp. 187–192.

Eguchi M., Myoga, H., Sasaki S., and Miyake Y., (2000) Influences of providing substances on trichloroethylene degradation of bioremediation through methane. *J. Soc. Water Environ.* 23(7): 421–426.

Eguchi, M., Kitagawa, M., Suzuki, Y., Nakamuara, M., Kawai, T., Okamura, K., Sasaki, S., and Miyake, Y., (2001) A field evaluation of *in-situ* biodegradation of trichloroethylene through methane injection. *Water Res.* 35 (9): 2145-2152.

Erses, A. S., and Onay, T. T., (2003) In situ heavy metal attenuation in landfills under methanogenic conditions. *Jounal of hazardous materials*. B99: 159-175.

Espinosaa, L. M., and Stephensona, T., (1996) Grease Biodegradation: is bioaugmentation more effective than natural populations for start up? *Water science and technology*. 34 (5-6): 303-308.

Evans, F. F., Rosado, a. S., Sebastian, G. V., Casella, R., Machado, P. L. O. A., Holmstrom, C., Kjelleberg, S., van Elsas, J. D., and Seldin, L., (2004) Impact of oil contamination and biostimulation on the diversity of indigenous bacterial communities in soil microcosms. *FEMS Microbiology Ecology*. 49: 295–305.

Evans, G. M. and Furlong. J. C. (2003) *Environmental Biotechnology: theory and applications*. West Sussex, UK, John Wiley and sons.

Falkowski, P., (1997) Evolution of the nitrogen cycle and its influence on biological sequestration of CO_2 in the oceans. *Nature*. 387: 272-273.

Ferguson, J. F., and Pietari, J. M. H., (2000) Anaerobic transformations and bioremediation of chlorinated solvents. *Environmental Pollution*. 107: 209-215.

Filler, D. M., (1997) Thermally enhanced bioventing of petroleum hydrocarbons in cold regions. *PhD Thesis*. Department of Civil and Environmental engineering, University of Alaska, Fairbanks.

Filler, D. M., Lindstrom, J. E., Braddock, J. F., Johnson, R. A., and Nickalaski, R., (2001) Integral biopile components for successful bioremediation in the Arctic. *Cold Regions Science and Technology*. 32: 143-156.

Flathman, P. E., and Jerger, D. E. (1994) *Bioremediation Field Experience*. Environmental Research Laboratory, Ada, OK: Lewis Publishers, CRC Press, Inc.

Frank, I. B., Lundgren, P., and Falkowski, P., (2003) Nitrogen fixation and photosynthetic oxygen evolution in cyanobacteria. *Research in Microbiology*.154: 157-164.

Gander, M., Jefferson, B., and Judd, S., (2000) Aerobic MBRs for domestic waste water treatment: a review with cost considerations. *Sep. Purif. Technol.* 18: 119-130.

Ghioureliotis, M., and Nicell, J. A., (1999) Assessment of soluble products of peroxidase-catalysed polymerization of aqueous phenol. *Enzyme and Microbial Technology*. 25: 185-193.

Hattan, G., Wilson B.and Wilson. J. T. (2003) Performance of conventional remedial technology for treatment of MTBE and benzene at UST sites in Kansas. *Bioremediation Winter*. 85-94.

Head, M. A., and Oleszkiewicz, J. A., (2004) Bioaugmentation for nitrification at cold temperatures. *Water Research* 38 (2004) 523–530.

Heck, P. E., Massey, J. I., and Aitken, M. D., (1992) Toxicity of reaction products from enzymatic oxidation of phenolic pollutants. *Water Science and Technology*. 26(9-11): 2369-2371.

Hwang, S. and Cutright. T. J. (2002) Biodegradability of aged pyrene and phenenthrene in a natural soil. *Chemosphere*. 47: 891-899.

Hwang, S., and Cutright, T. J., (2002) Biodegradability of aged pyrene and phenenthrene in a natural soil. *Chemosphere*. 47: 891-899.

Jianlong, W., Xiangchun, Q., Libo, W., Yi, Q., and Hegemann, W., (2002) Bioaugmentation as a tool to enhance the removal of refractory compound in coke plant wastewater. *Process Biochemistry* 38: 777-781.

Johnson, R. L., Johnson, P. C., McWhorter, B. B., Hinchee, R. E., Goodman, I., (1993) An overview of in situ air sparging. *Groundwater Monit. Rev.* 13(4): 127-135.

Kessel, C. V., and Hartley, C., (2000) Agricultural management of grain legumes: has it led to an increase in nitrogen fixation? *Field Crops Research*. 65: 165-181.

Kirtland, B. C., and Aelion, C. M., (2000) Petroleum mass removal from low permeability sediment using air sparging/soil vapor extraction: impact of continuous or pulsed operation. *Journal of Contaminant Hydrology*. 41: 367-383.

Klibanov, A. M., Alberti, B. N., Morris, E. D., and Felshin, L. M., (1980) Enzymatic removal of toxic phenols and anilines from waste water. *Journal of Applied Biochemistry*. 2: 414-421.

Leahy, J. G., and Colwell, R. R., (1990) Microbial degradation of hydrocarbons in the environment. *Microbiol. Reviews*. 54:305-315.

Lee, W., and Batchelor, B., (2003) Reductive capacity of natural reductants. *Environ. Sci. Technol.* 37: 535-541.

Lin, Q., and Mendelssohn, I. A., (1998) The combined effects of Phytoremediation and biostimulation in enhancing habitat restoration and oil degradation od petroleum contaminated wetlands. *Ecological Engineering*. 10: 263-274.

Loffler, F. E., Champine, J. E., Ritalathi, K. M., Sprague, S. J., and Tiedje, J. M., (1997) Complete reductive dechlorination of 1,2-Dichloropropane by anaerobic bacteria. *Applied and Environmental Microbiology*. 63(7): 2870-2875.

Lombi, E., Zhao, F., Wieshammer, G., Zhang, G., and McGrath, S. P., (2002) In situ fixation of metals in soils using bauxite residue: biological effects. *Environmental pollution*. 118: 445-452.

Lorah, M. M., and Olsen, L. D., (1999) Natural attenuation chlorinated volatile organic compounds in a fresh water tidal wetland: field evidence of anaerobic biodegradation. *Water Res.* 35: 3811-3827.

Maloney, S. W., Manem, J., Mallevialle, J., and Fiessenger, F., (1986) Transformation of trace organic compounds in drinking water by enzymatic oxidative coupling. *Environmental Science and Technology*. 20: 249-253.

Marschner, P., Yang, C. H., Lieberei, R., and Crowley, D. E., (2001) Soil and plant specific effects on bacteria community composition in the rhizosphere. *Soil Biology and Biochemistry*. 33: 1437-1445.

McCray, J. E., and Falta, R. W., (1996) Defining the air sparging radius of influence for ground water remediation. *Journal of Contaminant Hydrology*. 24: 25-52.

McCray, J. E., and Falta, R. W., (1997) Numerical simulation of air sparging for remediation of NAPL contamination. *Ground water*. 35(1): 99-110.

Meagher, R. B. (2000) Phytoremediation of toxic elemental and organic pollutants. *Current Opinion Plant Biology* 3: 153-162.

Mesania, F. A., and Jennings, A. A., (2000) Modeling soil pile bioremediation. *Environmental Modeling and software*. 15: 411-424.

Mihopoulos, P. G., Suidan, M. T., Sayles, G. D., (2000) Vapor phase treatment of PCE by lab-scale anaerobic bioventing. *Water Resource Res.* 34 (12): 3231-3237.

Mihopoulos, P. G., Suidan, M. T., Sayles, G. D., and Kaskassian, S., (2002) Numerical modeling of oxygen exclusion experiments of anaerobic bioventing. *Journal of Contaminant Hydrology*. 58: 209-220.

Mitsubishi and Novozymes Chemicals (2004). Novozymes expands the frontiers. Website: http://www.novozymes.com (cited 08 Feb. 2006).

Nakhla, G. F., and Suidan, M. T., (1995) Anaerobic toxic wastes treatment: dilution effects. *Journal of Hazardous Materials*. 42: 71-86.

Nakhla, G., and Niaz, M., (2002) Pilot scale *in-situ* Bioremediation of gasoline-contaminated ground water: Impact of process parameters. *Environmental Progress*. 21(1): 37-46.

Nannipieri, P., Kandeler, E., and Ruggiero, P., (2002) Enzyme activities and microbiological and biochemical processes in soil. In: Burns, R. G., Dick, R. P., (Eds.), Enzymes in the Environment. *Activity, ecology and Applications*. Marcel Dekker, New York, pp. 1-33.

Nicell, J. A., Bewtra, J. K., Biswas, N., and St. Pierre, C. C., (1993) Enzyme catalyzed polymerization and precipitation of aromatic hydrocarbons from aqueous solution. *Can J Civ Eng*. 20: 725-735.

Nobre, R. C. M., and Nobre, M. M. M., (2004) Natural attenuation of chlorinated organics in a shallow sand acquifer. *Journal of Hazardous Materials*. 110: 129-137.

Norris, R. D., Hinchee, R. E., Brown, R. A., McCarty, P. L., Semprini, L., Wilson, J. T., Kampbell, D. H., Reinhard, M., Bower, E. J., Borden, R. C., Vogel, T. M., Thomas, J.M., and C. H. Ward. (1994) *Hand book of Bioremediation*. Boca Raton, FL: CRC Press.

Norris, R.D., Hinchee, R.E., Brown, R.A., McCarty, P.L., Semprini, L., Wilson, J.T., Kampbell, D.H., Reinhard, M., Bower, E.J., Borden, R.C., Vogel, T.M., Thomas, J.M., and C.H. Ward. (1993) *In-situ Bioremediation of Ground Water and Geological Material: A Review of Technologies*. Ada, OK: U.S. Environmental Protection Agency, Office of Research and Development. EPA/5R-93/124.

Quan, X., Shi, H., Liu, H., Wang, J. and Qian, Y. (2004) Removal of 2,4-dichlorophenol in a conventional activated sludge system through bioaugmentation. *Process Biochemistry*. 39: 1701-1707.

Quan, X., Shi, H., Liu, H., Wang, J., and Qian, Y., (2004) Removal of 2,4-dichlorophenol in a conventional activated sludge system through bioaugmentation. *Process Biochemistry* 39: 1701–1707.

Riser-Roberts, E. (1992) *Bioremediation of Petroleum Contaminated Sites*. NCEL, Port Hueneme, CA: C. K. Smoley Publishers, CRC Press.

Sajc, L., and Novakovic, G. V., (2000) Extractive bioconversion in a fourphase external-loop airlift bioreactor. *AIChE Journal*. 46(7): 1368-1375.

Schultz, T. W., Bryant, S. E., and Kissel, T. S., (1996) Toxicological assessment in tetrahymena of intermediates in aerobic microbial transformation of toluene and p-xylene. *Bull. Environ. Contam. Toxicol*. 56: 129-134.

Shen, Y., (1998) *In vitro* cytotoxicity of BTEX metabolites in Hela cells. *Arch. Environ. Contam. Toxicol*. 34: 229-234.

Shen, Y., and West, C., (1998) Toxicity of aromatic aerobic biotransformation products of toluene to Hela cells. *Bull. Environ. Contam. Toxicol*. 60: 177-184.

Sleep, B., (1998) Modeling transient organic vapor transport in porous media with the dusty gas model. *Adv. Water Resources* 22 (3): 247-256.

Small, M. C., (1998) Risk based corrective action, natural attenuation, and changing regulatory paradigms. *Bioremediation J.* 2: 221-225.

Smith, C. K., Coyea, M. R., and Munson, A. D., (2000) Soil carbon, nitrogen and phosphorus stocks and dynamics under disturbed black spruce forests. *Ecol. Appl.* 10: 75-88.

Somsamak, P., Cowan, R. M., and Haggblom, M. M., (2001) Anaerobic biotransformation of fuel oxygenates under sulfate-reducing conditions. *FEMS Microbiology Ecology.* 37: 259-264.

Suarez, M. P., and Rifai, H. S., (1999) Biodegradation rates for fuel hydrocarbons and chlorinated solvents in ground water. *Biodegradation J.* 3: 337-362.

Suzuki Y. (1996) National project of bio-remediation technological development for restoration of in situ geopollution due to organochlorine compound. *Proceeding of the Sixth Symposium on Geo-Environments and Geo-Technics,* pp. 325–330.

Suzuki Y. (1998) National project of bio-remediation technological development (Part 2) for restoration of in situ geo-pollution due to organochlorine compound.*Proceeding of the eight Symposium on Geo-Environments and Geo-Technics*, pp. 43–48.

Thakur. I. S. (2004) Screening and identification of microbial strains for removal of colour and adsorbable organic halogens in pulp and paper mill effluent. *Process Biochemistry.* 39: 1693-1699.

Thronton, S. F., Bright, M. I., Lerner, D. N., and Tellam, J. H., (2000) Attenuation of landfill leachate by UK Triassic sandstone aquifer materials 2. Sorption and degradation of organic pollutants in laboratory columns. *Journal of Contaminant Hydrology.* 43: 355-383.

Trindade, P. V. O., Sobral, L. G., Rizzo, A. C. L., Leite S. G. F., and Soriano. A. U. (2005) Bioremediation of a weathered and recently oil-contaminated soils from Brazil: a comparison study. *Chemosphere.* 58: 515-522.

Trindade, P. V. O., Sobral, L. G., Rizzo, A. C. L., Leite, S. G. F., and Soriano, A. U., (2005) Bioremediation of a weathered and a recently oil-contaminated soils from Brazil: a comparison study. *Chemosphere.* 58: 515-522.

Truax, D. D., Britto, R., and Sherrard, J. H., (1995) Bench-scale studies of reactor based treatment of fuel contaminated soils. *Waste Management.* 15(5-6): 351-357.

U.S. Environmental Protection Agency (EPA). (1992) *A Technology Assessment of Soil Vapor Extraction and Air Sparging.* Cincinnati, OH: Office of Research and Development. EPA-600-R-92-173.

U.S. Environmental Protection Agency (EPA). (1998) Wiedemeier, T. H., Swanson, M. A., Moutoux, D. E., Gordon, E. K., Wilson, J. T., Wilson, B. H., Kampbell, D. H., Haas, P. E., Miller, R. N., Hansen, J. E., and Chapelle, F. H. *Technical protocol for evaluating natural attenuation of chlorinated solvents in ground water.* EPA-600-R-98-128.

U.S. Environmental Protection Agency (EPA). (1999) OSWER Directive 9200.4-17P, April 21 1999. Office of Solid Waste and Emergency Response, p41. *Use of monitored natural attenuation at Superfund, RCRA corrective action, and underground storage tank sites* EPA-9200.4-17P.

U.S. Environmental Protection Agency (EPA). (2000) *Abstracts of Remediation Case Studies, Volume 4,* June, 2000, EPA 542-R-00-006.

U.S. Environmental Protection Agency (EPA). (2000) *Engineered Approaches to In Situ Bioremediation of Chlorinated Solvents: Fundamentals and Field Applications,* July, EPA-542-R-00-008.

Vischetti, C., Capri, E., Trevisan, M., Casucci, C., and Perucci, P., (2004) Biomassbed: a biological system to reduce pesticide point contamination at farm level. *Chemosphere.* 55: 823-828.

Vogel, T. M., (1996) Bioaugmentation as a soil bioremediation approach. *Current Opinion in Biotechnology.* 7: 311-316.

Webb, W. S., and Phelan, J. M., (1997) Effect of soil layering on NAPL removal behavior in soil heated vapor extraction. *Journal of Contaminant Hydrology.* 27: 285-308.

Weidner, S., Puhler, A., and Kuster, H., (2003) Genomics insights into symbiotic nitrogen fixation. *Current Opinion in Biotechnology.* 14: 200-205.

Weston, Inc., Roy F. (1988) *Remedial Technologies for Leaking Underground Storage Tanks.* University of Massachusetts, Amherst, MA: Lewis Publishers.

White, D. M., and Schnabel, W., (1998) Treatment of cyanide waste in a sequencing batch biofilm reactor. *Water Research.* 32(1): 254.

Widdowson, M. A., Haney, O. R., Reeves, H. W., Aelion, C. M., Ray, R. P., (1997) Multilevel soil-vapor extraction test for heterogenous soil. *J. Environ. Eng.* 123: 163-168.

William J. D. (1997) *Groundwater Geochemistry: Fundamentals and Applications to Contamination*, Lewis Publishers, Boca Raton. pp.51–55.

Xu, F., and Bhandari, A., (2003) Retention and Extractability of Phenol, Cresol, and Dichlorophenol Exposed to Two Surface Soils in the Presence of Horseradish Peroxidase Enzyme. *J. Agric. Food Chem.* 51: 183-188.

Xu, F., Koch, D. E., Kong, I. C., Hunter, R. P., and Bhandari, A., (2005) Peroxidase-mediated oxidative coupling of 1-naphthol: Characterization of polymerization products. *Water Research.* 39: 2358–2368.

You, G., Sayles, G. D., Kupferle, M. J., Kim, I. S., and Bishop, P. L., (1996) Anaerobic DDT biotransformation: enhancement by application of surfactants and low oxidation reduction potential. *Chemosphere.* 32 (11): 2269-2284.

Acknowledgement

The authors are sincerely thankful to the Natural Sciences and Engineering Research Council of Canada (Grants A4984, STP235071, Canada Research Chair) for financial support. The views and opinions expressed in this article are those of authors.

CHAPTER 9

Bioremediation with Fungi

Mausam Verma, Satinder K. Brar, R. D. Tyagi, and R. Y. Surampalli

9.1 Introduction

Fungi are important bioremediation agents. They have been successfully used at laboratory, pilot and commercial levels with frequent modifications. Their immense potential as bioremediators has stimulated the scientific community to further categorize fungi on the basis of their remediation capabilities, e.g. saprotrophs (white rot and brown rot fungi) and mycorrhizal fungi (Meharg and Cairney, 2000). Potential contaminants like munition wastes, pesticides, organochlorines, polychlorinated biphenyls (PCBs), polycyclic aromatic hydrocarbons (PAHs), synthetic dyes, wood preservatives and synthetic polymers can be degraded or converted into less nuisance forms via fungal treatments (Pointing, 2001). Even though bacteria, algae, and plants are able to perform decontamination, their applications are limited by their low treatment efficacies. In most cases, xenobiotic chemicals are extremely resistant to biodegradation by native flora and fauna (Fernando and Aust, 1994). Several ligninolytic fungi, however, have been shown to decompose a large variety of recalcitrant compounds due to their non-specific enzyme systems (Novotny et al., 2004). Ligninolytic fungi appear to be equally effective in treating contaminants associated with soil or water. However, their proliferation in contaminated soil is highly susceptible to competition from native microbes, thereby warrants, bulking agents (as supplementary nutrient) like peat moss, bran flakes, and pine wood shavings (Meysami and Baheri, 2003).

Mycorrhizal symbioses (ectomycorrhizal, arbuscular, ericoid and orchid fungi) of fungi help in degradation of many xenobiotics by a consortium of large group of fungi, which are otherwise resistant to attack by a single or, a few organisms (Meharg and Cairney, 2000; Allen et al., 1995; Clapp et al., 1995; Perotto et al., 1996; Gardes and Bruns, 1996; Dahlberg et al., 1997; Liu et al., 1998). The need of mycorrhizal symbioses arose from some of the drawbacks of phytoremediation technology which has serious environmental implications, e.g., bioaccumulation of contaminants into plants and their subsequent entry into wild-life food chains (Anderson et al., 1993). In contrast to application of white rot fungi (WRF), the natural presence of mycorrhizal fungi in rhizosphere would be a definite advantage. However, slower growth, dependency on plant/tree root systems, and fungal biomass are some of the constraints of mycorrhizal fungi in bioremediation (Lappin et al., 1985; Donnelly and Fletcher, 1994; Meharg and Cairney, 2000).

Benevolent speciation of metals in environment by fungi is another widely studied and implemented mode of bioremediation (Lovley and Coates, 1997; Eccles, 1999). Furthermore, in addition to metals, speciation and mobility of other elements, including carbon, nitrogen, sulfur and phosphorus are fundamental to biogeochemical cycles implications for plant productivity and human health (Gadd, 1999, 2002, 2004; Verrecchia and Dumont, 1996). In general, metal speciation ability of microorganisms is due to their mobilization or immobilization processes that control transportation of metal species between soluble and insoluble phases (White et al., 1997, 1998; Sreekrishnan and Tyagi, 1994; Vachon et al., 1994).

Currently, there are some commercialized/proprietary processes, claiming simultaneous execution of bioremediation and growth factor enhancement of soil for agricultural purposes (Walker, 2003). These processes normally comprise of application of fertilizer and pesticide in combination with specific carbohydrate-based surfactant. It is proposed that the presence of adequate environmental conditions for native microbial flora would stimulate the remediation process and consequently plant growth promotion. There are several other types of commercial/proprietary formulations for soil remediation designed for different requirements, interestingly, a majority of them either contain inocula of fungi or, growth enhancers for bioremediator fungi (Gill, 1996, 1997; Bennett et al., 2001; Walker, 2003).

This chapter will broadly circumscribe important developments in saprotroph and mycorrhizal fungi associated soil bioremediation processes aimed at application and their known modes of action. Few case studies are presented for well accounted processes.

9.2 Saprotrophic Fungal Processes

White-rot fungi (WRF) are the most studied fungi among saprotrophs (WRF, brown-rot fungi, litter-decaying fungi) for soil bioremediation (Figure 9.1). It is a physiological classification rather than taxonomic, comprising those fungi that have a panoply of lignin degrading enzymes (Pointing, 2001). The name white-rot signifies the appearance of wood attacked by these fungi, in which lignin removal results in a white appearance of the substrate. The lignin degrading characteristic of WRF facilitate in soil bioremediation in two ways – (1) capability to degrade a wide range of highly recalcitrant organopollutants with structural similarities to lignin (Figure 9.2, Pointing, 2001); (2) WRF as well as other microbial flora nourish from bioavailable substrate obtained from lignin degradation. Mostly basidiomycetes, and a few ascomycete genera (Figure 9.1) are capable of white-rot decay (Eaton and Hale, 1993). Table 9.1 lists some WRF-mediated soil bioremediation processes, grouped on the basis of pollutants treated.

Various recalcitrant pollutants described in the following sections, have been categorized on the basis of their end-use and similarity in chemical class.

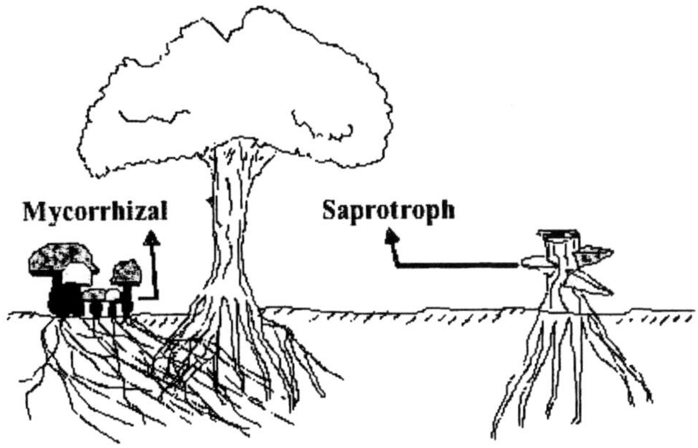

Figure 9.1 Schematic sketch of Saprotrophs, and mycorrhizal fungi.

9.2.1 Synthetic Dyes, Pesticides and Polychlorinated Biphenyls

These pollutants are introduced into environment by agricultural sanitization, textile dyeing, paints, refineries and electrical industries (Novotny et al., 2001). These pollutants are of great environmental concern because of their abundance and high toxic, carcinogenic and reproductive effects on animal and humans. About 632 million tons of PCBs have been produced for transformer oil (1929–1970, Rojas-Avelizapa et al., 1999; Duke et al., 1970; INE., 1995; Hutzinger et al., 1974). Inadequate use of these oils, lack of proper storage and disposal of spent oils, and accidental spillage has caused serious contamination problems (INE, 1995). Similarly, there are several other major sources of these pollutants (Meharg and Cairney, 2000).

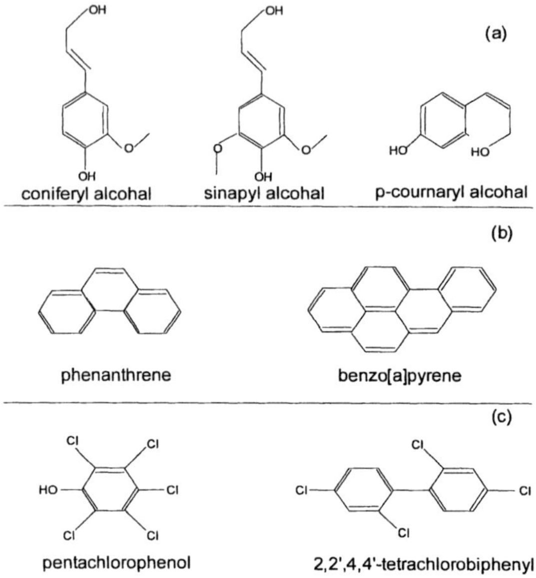

Figure 9.2 (a) Lignin monomers, (b) polycyclic aromatic hydrocarbons (PAH), and (c) halogenated compounds mineralized by the ligninolytic enzyme system of white-rot fungi (Pointing, 2001).

In particular, pesticides are persistent in the environment in the form of organochlorines, organophosphate and are linked to toxic effects and population declines at higher trophic levels (Alloway and Ayres 1993). Fortunately, bacteria and several genera of soil fungi (e.g. *Fusarium, Penicillium*) obtained from pesticide contaminated soils are now known to degrade pesticides with great efficacy (Twigg and Socha, 2001; Wong et al., 1992).

9.2.1.1 Mechanisms of Bioremediation

The oxidation of lignin takes place in the secondary metabolic process of WRF (Figure 9.3), in order to access wood polysaccharides bound in lignin-carbohydrate complexes (Jeffries, 1990). On the other hand, organophosphate insecticides are not generally persistent, and certain WRF e.g., *Phanerochaete chrysosporium* have been demonstrated to mineralize chloropyrifos, fonofos, and terbufos (Bumpus et al. 1993). The action of lignin-modifying enzymes was unclear, however, hydrolytic cleavage of

Table 9.1 White rot fungi based bioremediation of different pollutants.

Saprotroph Fungi (White Rot and Brown Rot)	Bulking Agent(s)/ Synthetic Media	Pollutant	Pollutant Concentration	Reference
Aspergillus fumigatus, A. sydowii, A. terreus, A. clavatus, A. niger, Fusarium oxysporum, F. decemcellulare, Penicillium oxalicum, P. restrictum, P. simplicissimum, P. verruculosum, P. montanense, P. pinophilum, P. janthinellum, Amorphoteca resinae, Neosartorya fischeri, Paecilomyces variotii, Talaromyces flavus, Graphium putredinis, Cunninghamella echinulata, Eupenicillium ochrosalmoneum,	KCl, 250 mg l^{-1}; NaH_2PO_4, 1 g l^{-1}; $MgSO_4$, 0.5 g l^{-1}; $NH_4\ NO_3$, 1 g l^{-1}, Chloramphenicol, 100 mg l^{-1}	Arabian light crude oil (BAL 150)	1 g l^{-1}	Chaillan et al., (2004)
Coriolopsis gallica, Bjerkandera adusta, Pleurotus pilmanarius, P. Ostreatus, Phanerochaete chrysos, Trametes versicolor	Pine wood shavings, peat moss and bran flakes	Wheathered crude oil	4000-14000 ppm	Meysami and Baheri, (2003)
Phanerochaete chrysosporium	1% (w/v) malt extract; 1% (w/v) glucose; 0.2% (w/v) peptone; 0.2% (w/v) yeast extract; 1.5% (w/v) agar	TNT (2,4,6-trinitrotoluene)	2-10 μg l^{-1}	Stahl and Aust, (1993a, 1993b); Zheng and Obbard, (2002)
Pleurotus osteratus	Spent mushroom compost wetted with fish oil as surfactant	Creosote (containing 16 different PAHs)	1900 mg PAH/kg creosote	Eggen, T. (1999)
Lentinula edodes	Spent mushroom compost	PAHs	--	Buswell, (1994)
Phytophthora sp., *Mucor* sp., *Cladosporium* sp., *Coniothyrium* sp., *Doratomyces* sp., *Fusarium* sp., *Phialophora* sp., *Scedosporium* sp., *Sphaeropsis* sp., *Stachybotrys* sp., *Trichoderma* sp.	BSM–glycerol medium: Glycerol, 10 g l^{-1}; KH_2PO_4, 2 g l^{-1}; $CaCl_2.2H_2O$, 0.14 g l^{-1}; $MgSO_4.7H_2O$, 0.07 g l^{-1}; thiamine hypochloride, 2.5 mg l^{-1}; Tween 80, 7.5 ml l^{-1}; D- diammonium	PAHs	800–1100 mg PAHs /kg soil	Potin et al., (2004a, 2004b)

Saprotroph Fungi (White Rot and Brown Rot)	Bulking Agent(s)/ Synthetic Media	Pollutant	Pollutant Concentration	Reference
	tartrate, 1.84 g l^{-1}; yeast extract, 1 g l^{-1}; $FeSO_4.7H_2O$, 70 mg l^{-1}; $ZnSO_4.7H_2O$, 46.2 mg l^{-1}; $MnSO_4.H_2O$, 35 mg l^{-1}; $CuSO_4.5H_2O$, 7 mg l^{-1}.			
Absidia cylindrospora, A. spinosa, Acremonium murorum, Alternaria alternata, Aspergillus flavipes, A. fumigatus, A. niger, A. oryzae, A. parasiticus, A. terreus, A. ustus, Botryotrichum piluliferum, Cladosporium herbarum, Cunninghamella bainieri, Cylindrocarpon lucidum, Doratomyces asperulus, D. stemonitis, Fusarium culmorum, F. lateritium, F. oxysporum, F. solani, Gliocladium roseum, G. virens, Humicola grisea, Minimedusa polyspora, Mucor hiemalis, Neosartorya fischeri, Paecilomyces lilacinus, Penicillium canescens, P. janczewskii, P. montanense, P. restrictum, P. simplicissimum, Phoma eupyrena, P. exigua, Pseudallescheria boydii, Seimatosporium sp., Stachybotrys bisbyi, S. chartarum, Talaromyces flavus, Trichoderma hamatum, T. harzianum, T. koningii. Trichurus spiralis, Verticillium tenerum, Westerdykella dispersa, Zygorhynchus heterogamus	GS* liquid medium supplemented with 5 g l^{-1} glucose	Fluorene	100 µg /g soil	Garon et al. (2004)
Kuehneromyces mutabilis and Agrocybe aegerita	Mycelial mat grown over malt agar	Pyrene	80 µg / g soil	Sack and Fritsche (1997)
Pleurotus ostreatus, Phanerochaete chrysosporium,	Freshly ground, dried birch sticks – steel	PAH (gasworks	5–30 µg / g soil	Andersson et al. (2000)

Saprotroph Fungi (White Rot and Brown Rot)	Bulking Agent(s)/ Synthetic Media	Pollutant	Pollutant Concentration	Reference
Hypholoma fasciculare	net-sieved (7 mm) – use as a fungal substrate	plant, Ystad, Sweden)		
Actinomyces sp. and *Paecilomyces* sp.	KNO_3, 1 g l^{-1}; FeCl3, 0.02 g l^{-1}; MgSO4, 0.2 g l^{-1}; NaCl, 0.1 g l^{-1}; CaCl2, 0.1 g l^{-1}; K2HPO4, 1 g l^{-1}; yeast extract, 0.05 g l^{-1}, transformer oil, 1% v/v (after sterilization)	Transformer oil (PCB content, 88% w/v)	7000 mg / kg soil	Rojas-Avelizapa et al. (1999)
Pleurotus osteratus	--	PCB commercial mixture Delor 103 (1.02 mg/ml in acetone)	10.21 g / kg soil	Kubatova et al. (2001)
Irpex lacteus and *Pleurotus ostreatus*	Growth medium – malt extract/glucose medium, malt extract broth medium, and low nitrogen mineral medium; Support medium – polyurethane foam/pinewood chips/coarsed-milled straw	Remazol Brilliant Blue R (Synthetic dye)	150 μg / g soil	Novotny et al. (2001)
Phanerochaete chrysosporium, *Trametes versicolor* and *Pleurotus ostreatus*	Growth medium – malt extract/glucose medium, yeast extract/peptone/glucose medium, and phenazine induction medium; Support medium – polyurethane foam/pinewood chips/coarsed-milled straw	PCB mixture (Delor 106) and a number of synthetic dyes	150 μg / g soil	Novotny et al. (2004)

Saprotroph Fungi (White Rot and Brown Rot)	Bulking Agent(s)/ Synthetic Media	Pollutant	Pollutant Concentration	Reference
Aspergillus flavus, A. fumigatus, Fusarium avenaceum, F. compactum, F. equiseti, F. oxysporum, F. proliferatum, F. semitectum, F. solani, Penicillium spinulosum	--	Commercial pesticide - Sodium monofluoroacetate (1080)	20 mM	Twigg and Socha (2001)

* Galzy and Slonimski, (1957)

the organophosphates, fenitrothion and fenitrooxon by non-ligninolytic fungus *Trichoderma viridae* (Baarschers and Heitland 1986) have been well known. Thus, it could be inferred that WRF have been able to degrade a wide range of pesticides simply due to their diversity in enzymes.

In vitro role of extracellular peroxidases and laccases in oxidizing recalcitrant compounds have been well documented (Figure 8.3) but the functional significance of individual enzyme levels *in vivo* is poorly understood (Novotny et al., 2001). In a majority of cases, the complexity of biodegradation processes where various interactions may determine the rate-limiting step renders enzymatic actions unclear. However, this has not been an obstacle in developing WRF based soil bioremediation of PCBs (Pointing, 2001; Van Acken et al., 1999). Unfortunately, there are many other pesticides which are known to contaminate soils, have not been studied well and need extensive research to develop effective bioremediation approaches.

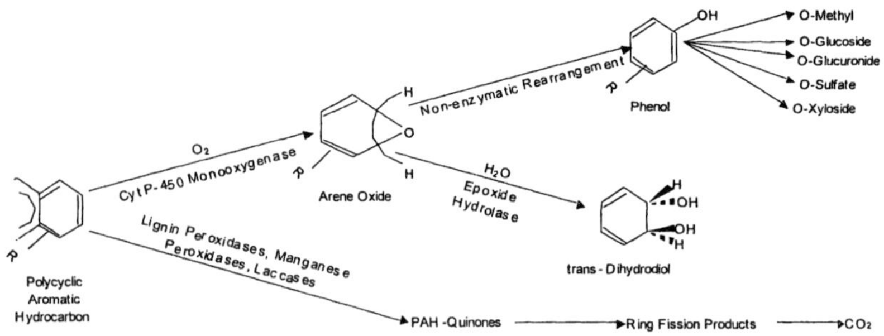

Figure 9.3 Fungal metabolism mechanisms of polycyclic aromatic hydrocarbons (Cerniglia, 1997).

9.2.1.2 Case Studies

While some soil microorganisms are ubiquitous and commonly occur in a variety of moist soils (Kelly, 1965; Bong et al., 1979), it is worthwhile to emphasize that fungal species have been more efficient in pesticide degradation even in arid and semi-arid soil conditions. Some case studies for fungal bioremediation of synthetic dyes, pesticides and polychlorinated biphenyls are presented below:

Many WRF have been efficient in mineralizing organochlorine pesticides, such as DDT, 2,4,5-trichlorophenoxyacetic (2,4,5-T) and 2,4-dichlorophenoxyacetic (2,4-D). Under ligninolytic growth conditions, many WRF strains, e.g., *Phanerochaete Chrysosporium*, *Pleurotus ostreatus*, *Phellinus weirii*, and *Polyporus versicolor* are able to mineralize 5.3–13.5% of added ^{14}C-radiolabeled DDT, dicofol, and methoxychlor over 30 days (Bumpus and Aust 1987).

Highly recalcitrant pesticides like the chlorinated triazine herbicide 2-chloro-4-ethylamine-6-isopropylamino-1,3,4-triazine (atrazine) have been transformed by white-rot fungi *P. chrysosporium* (Mougin et al. 1994) and *Pleurotus pulmonarius* (Masaphy et al. 1993), yielding hydroxylated and *N*-dealkylated metabolites. The presence of healthy consortia of microorganisms in soil assures that certain pesticides can be used safely in environments, albeit under recommended application dosage.

In a safety study regarding pest control programs in Australia and New Zealand carried out by Twigg and Socha, (2001), role of fungal species of *Fusarium, Penicillium* and *Aspergillus* was assessed.. These researchers performed bioremediation of sodium monofluoroacetate (a commercial invertebrate pesticide) contaminated soil at pH 5.6 and 6.8 to distinguish the role of fungi in pesticide degradation in comparison to bacterial consortia and found that *Fusarium oxysporum* had greatest pesticide degradation ability (approximately 45% degradation of pesticide within 12 d). Degradation of the pesticides appeared to cease in a 28 d time course trial.

Novotny et al., (2001) successfully demonstrated that *in vivo* degradation of a broad selection of recalcitrant compounds (dyes, polyaromatic hydrocarbons–PAHs, PCBs) under a variety of conditions is possible, except that the enzyme levels are sufficiently high. These researchers used growth media like malt extract/glucose medium, yeast extract/peptone/glucose medium, and phenazine induction medium to produce mycelial mass and subsequently immobilized them to a support media like polyurethane foam, pinewood chips or coarse-milled straw before application on contaminated soil (Novotny et al., 2004).

Previously, specially designed tube reactors were used to evaluate biodegradation experiments (Novotny et al., 1999). Each reactor was made of a tube (dia. 3.5 cm, length 24 cm) divided by a fine nylon net into two compartments. In one compartment, straw was inoculated with the WRF and fungal mycelium grew through the net to other compartment containing sterilized soil (10 g) that was contaminated with 100 μl of a Delor 103 solution in acetone. Subsequently, the system was aerated and moistened throughout the remediation period (2 month). Kubatova et al., (2001) could obtain a maximum removal efficiency of 40% of Delor 103 in two months using such a remediation system. However, such configurations are only applicable for micro-scale experiments and can not be feasible for field application.

Rojas-Avelizapa et al., (1999) used a mixed culture including fungi to degrade transformer oil. In liquid media amended with emulsifier Triton X-100 and supplemented with mineral salts, yeast extract (50 mg l^{-1}) and the transformer oil as sole carbon source, about 75% degradation occurred within 10 days. However, such studies could only provide partial solution of isolating or screening potential microbial strains for PCBs pollutants removal. Therefore, future studies should be more oriented towards field application of these potential fungi and their feasibility in soil bioremediation of PCBs.

9.2.2 Crude Oils and Polycyclic Aromatic Hydrocarbons

Crude oil and its derivatives like polyaromatic hydrocarbons (PAHs) are highly toxic environmental contaminants introduced into the environment in huge quantities by several point sources (Meharg et al., 1998). The incidences of soil-contamination by PAH have been mostly noted for accidental spillage of crude oil besides, their perennial sources of emissions like, coal gasification processes, refineries and polymer industries (Leyval and Binet, 1998; Nicolotti and Egli, 1998). Different strains of bacteria have the ability to degrade simple PAHs like naphthalene (Lisowska and Dlugonski, 1999; Milstein et al., 1988; Katayama and Matsumura, 1991; Hofrichter et al., 1993; Lamar et al., 1993; Sack and Giinther, 1993).

9.2.2.1 Mechanisms of Bioremediation

In addition to the complexity of higher PAHs, temporal bioavailability of PAHs renders them more illusive towards bioremediation and requires concerted efforts of microbial consortia, which could simultaneously make PAHs bioavailable as well as degrade them, a phenomenon which is much easier for fungal species owing to their lytic enzyme systems (Bollag et al., 1992; McFarland et al., 1992; Eggen, 1999).

WRF produce an array of enzymes depending on their genetic constitution and environmental conditions. Some key degradation enzymes like lignin peroxidase (LiP), manganese-dependent peroxidase (MnP), manganese-independent peroxidase (MIP), and laccase are well documented (Lamar, 1992; Vyas et al., 1994; Bogan and Lamar,

1996; Kotterman et al., 1996; Eggen, 1999). Extensive research has been focused on the role of these enzymes in bioremediation. Despite these serious efforts, the exact metabolic processes of biodegradation of PAHs are unavailable. For example, the PAHs degrading ability of *Pleurotus osteratus*, in the absence of LiP (Hatakka, 1990; Waldner et al., 1988) has been loosely correlated with laccase activity (Kerem et al., 1992; Thurston, 1994). In whole cultures (*Pleurotus ostreatus*, *Bjerkandera adusta*), attempts to link laccase activity with oxidation of phenanthrene and pyrene (Bezalel et al. 1996a) and the production of oxidative enzymes with PAH metabolism have been unsuccessful (Schutzendubel et al., 1999). It is likely that cytochrome P-450 monooxygenase produced by *Pleurotus osteratus* is responsible for the initial attack on PAHs, followed by subsequent degradation by laccase in similar manner to non-ligninolytic fungi (Bezalel et al., 1996b).

9.2.2.2 Case Studies

There are numerous examples where WRF have been incorporated in soil with or, without bulking/synthetic support media to treat PAHs like pyrene and fluorine (Sack and Fritsche, 1997; Garon et al., 2004). The use of PAH amended fungi (fungus pre-acclimatized to PAH or pollutant) culture may be more effective in bioremediation compared to non-acclimatized cultures (Garon et al., 2004). Certain fungi (*Gliocladium roseum, G. virens, Penicillium janczewskii, Stachybotrys chartarum, Trichoderma koningii* and *Zygorhynchus heterogamous*) show significant adaptability towards biodegradation of pollutants, while others (*Mucor hiemalis, Aspergillus terreus* and *A. ustus*) are not affected by acclimatization. Some case studies for fungal bioremediation of crude oils and polycyclic aromatic hydrocarbons are presented below:

Buswell, (1994) suggested use of spent mushroom culture (i.e. *Pleurotus ostreatus* – oyster mushroom, *Lentinula edodes* – shiitake mushroom; a by-product from commercial mushroom growers) as fungal inoculum for WRF in soil bioremediation. Further, Eggen, (1999) compared spent mushroom culture to colonized mushroom substrate (obtained from two commercial mushroom growers) before mushroom (fruiting body) production as an alternative for fungal inoculum. Eggen, (1999) investigated the modes of inoculum application, e.g., layering the fungi with soil and mixing fungi and soil. The PAH degradation potential of two commercial sources of fungi was dependent on the number of aromatic rings in the compounds. Spent mushroom compost was more effective than fungal substrate generated before mushroom production. Moreover, mixing of fungal inoculum and soil favored degradation of 4- and 5-ring compounds rather than layered incubation. After 7 weeks incubation at room temperature: 86% of total 16 PAHs, 89% of 3-ring PAHs, 87% of 4-ring PAHs and 48% of 5-ring PAHs were obtained.

A deuteromycete fungus, *Cladosporium sphaerospermum*, isolated from soil of an aged gas manufacturing plant was investigated by Potin et al., (2004a, b), to degrade PAHs. The average PAH (including high molecular weight PAHs) degradation capacity of this strain was 23%, after 4 weeks of incubation in aged PAH-contaminated soil. However, in liquid culture, it rapidly degraded benzo(α)pyrene during its early exponential phase of growth (18% after 4 days of incubation). In addition, only laccase activity was detected in liquid culture in the absence or in presence of benzo(α)pyrene. Therefore, bioremediation by *C. sphaerospermum* might be a potential and effective treatment approach for aged PAH-contaminated soils.

Laboratory scale experiments on axenic fungal biodegradation of aliphatic and aromatic hydrocarbons using a soil contaminated with 10% crude oil showed 65–74% reduction in 90 days (Colombo et al., 1996). Normal alkanes were almost completely degraded in the first 15 days, whereas slow dissociation of aromatic compounds (phenanthrene and methylphenanthrenes) was a rate–limiting step. *Aspergillus terreus* and *Fusarium solani*, isolated from oil-polluted areas, were more efficient in contaminant removal. Simultaneous multivariate analysis of parameters established a molecular weight dependant reactivity trend of oil components during biodegradation.

Meysami and Baheri, (2003) studied methods to support fungal growth and proliferation in soil contaminated with a weathered crude oil. They determined the ligninolytic enzyme activity and toxicity threshold of several white-rot fungi and attributed them to their hydrocarbon degradation ability. Approximately 100 g of artificially contaminated soil was inoculated with 7 days old mycelia of *Bjerkandera adusta* (approximately 60 ml buffer was added to 15 ml of mycelium) in jars at room temperature with loose caps until fungi mycelia had grown over the bulking agent surface completely. This was followed by complete mixing of contents and subsequent incubation for another 2 weeks. Pine wood chips, peat moss and Kellogg's bran flakes were examined for their properties as bulking agents and solid amendments. All strains developed severe toxicity at concentrations higher than 10000 ppm, a limitation in remediation process. The highest ligninolytic enzyme activities were shown by two strains of *Bjerkandera adusta* UAMH 7308 and 8258. A mixture of peat moss with bran flakes resulted in the best bulking agent composition for white-rot fungi, whereas, in the absence of any bulking agent, soil bioremediation was nominal. A maximum of 50% total petroleum hydrocarbon (TPH) reduction could be achieved by most of the fungal cultures within 4-5 weeks.

9.2.3 Munitions Waste

In-situ (Lamar et al., 1993) and *in-vitro* (Barr and Aust, 1994b; Field et al., 1993) use of WRF in biodegradation of persistent organic chemicals containing aryl rings (e.g., explosives) has been extensively investigated. The role of lignolytic enzymes induced by white rots under stress conditions of carbon and nitrogen limitations (Barr and Aust, 1994b; Field et al., 1993) has been quoted as a principal one in bioremediation. It has been demonstrated by several researchers that WRFs have the ability to degrade explosive compounds like, 2,4,6- trinitrotoluene (TNT). For instance, TNT contamination of old munitions manufacturing and storage sites have been subjected to bioremediation utilising the white rot fungus *Phanerochaete chrysosporium* (Meharg et al., 1997a). White rots are promising solutions for the *in-situ* degradation of TNT in contaminated soils (Fernando et al., 1990; Michels and Gottshalk 1994; Stahl and Aust, 1993a, 1993b; Spiker et al., 1992).

9.2.3.1 Mechanisms of Bioremediation

The initial degradation mechanisms of TNT are common in all microorganisms (Meharg et al., 1997a). According to this mechanism, one of the nitro groups on aryl ring is reduced to form hydroxylamino-dinitrotoluenes and subsequent reduction results in the formation of amino-dinitrotoluenes (Michels and Gottshalk 1994; Goruntzy et al., 1994; Higson, 1992). The initial degradation of TNT by *P. chrysosporium* is subtly distinct from its degradation of many other organic chemicals, with respect to the role of lignolytic enzymes (Field et al., 1993). Also, the initial attack of *P. chrysosporium* on TNT is via reduction of nitro groups on the aryl ring takes place (Michels and Gottshalk 1994; Stahl and Aust, 1993a). However, once initial reduction occurs, lignolytic enzymes of *P. chrysosporium* can further degrade amino-dinitrotoluenes through oxidative attack, rather than subsequent reduction (Michels and Gottshalk 1994; Stahl and Aust, 1993a). Hydroxylamino-dinitrotoluene, an early intermediate of TNT reduction, inhibits *P. chryosporium* lignases, preventing its further degradation (Michels and Gottshalk 1994), thereby, limiting the potential of the fungi as bioremediators.

Stahl and Aust (1993b) concluded that reduction of TNT by *P. chryosporium* occurs via a redox potential driven reduction at plasmalemma. They conducted extensive assays of various cytological components to verify possible reduction of TNT and concluded that only intact *P. chryosporium* mycelium could reduce TNT. Therefore, it was established that TNT reduction occurs via plasmalemma redox potential.

9.2.3.2 Field Application

In spite of extensive research on understanding biochemical mechanisms of explosives degradation using WRF, present feasibility studies are not in unison with their future application. Some case studies for fungal bioremediation of munitions waste are presented below:

Craig et al., (1995) reported that WRF based processes were the least efficient among; (1) composting, (2) anaerobic bioslurry and (3) aerobic bioslurry for TNT degradation (Figure 9.4). In addition, selection criteria for explosive-contaminated soil bioremediation is also very narrow for WRF based process (Craig et al., 1995).

9.2.4 Metals

Several methods for metal bioremediation have been investigated by many researchers.

9.2.4.1 Mechanisms of Bioremediation

Treatment of metal-contaminated soil could take place via processes like heterotrophic (chemoorganotrophic) leaching – a strain of *Penicillium simplicissimum* has been employed to leach Zn from insoluble ZnO laden industrial filter dust by induced production of citric acid (>100 mM) (Schinner and Burgstaller, 1989; Franz et al., 1991, 1993); siderophore-mediated metal solubilization, where, metal is adsorbed to the biomass and or precipitated, with biomass separated from a soil slurry by flocculation (Gadd, 2004), resulting in a complete decrease in bioavailability of Cd, Zn and Pb (Diels et al., 1999); biomethylation of toxic metal complexes – several fungal species can methylate arsenic compounds such as arsenate [As(V), AsO_4^{3-}], arsenite [As(III), AsO_2^-] and methylarsonic acid (CH_3H_2 AsO_3) to volatile dimethyl- [$(CH_3)_2HAs$] or trimethylarsine [$(CH_3)_3As$] (Tamaki and Frankenberger, 1992), thereby, rendering them less bioavailable to soil flora and fauna; redox transformations – most fungi can mobilize metals, metalloids and organometallic compounds by oxidation/reduction processes (Gadd, 1993; Gharieb et al., 1999; Lovley, 2000). Reduction of Hg(II) to Hg(0) by fungi results in diffusion of elemental Hg out of cells (Silver, 1996, 1998; Hobman et al., 2000) which can be utilized to mobilize Hg from contaminated soils.

Besides, biotransformation of metal and its complexes, fungi can also perform metal and metal-complex sorption to cellular surfaces, and even cationic species can be accumulated within cells via membrane transport systems of varying affinity and specificity. Inside cells, metal species may be incorporated within intracellular structures depending on the species concerned and the fungi (Gadd, 1996; White et al., 1997; Gadd and Sayer, 2000). In addition, there are numerous studies performed on free-living, pathogenic and plant symbiotic fungi associated with formation of calcium oxalate crystals from solubilised calcium (Gadd, 1999; Gharieb et al., 1998). This is important for biogeochemical processes in soils, acting as a buffering factor for calcium and phosphate availability. Other than calcium, fungi can also produce other metal oxalates and metal-bearing minerals, e.g., Cd, Co, Cu, Mn, Sr and Zn (White et al., 1997; Gadd, 1999; Sayer et al., 1999).

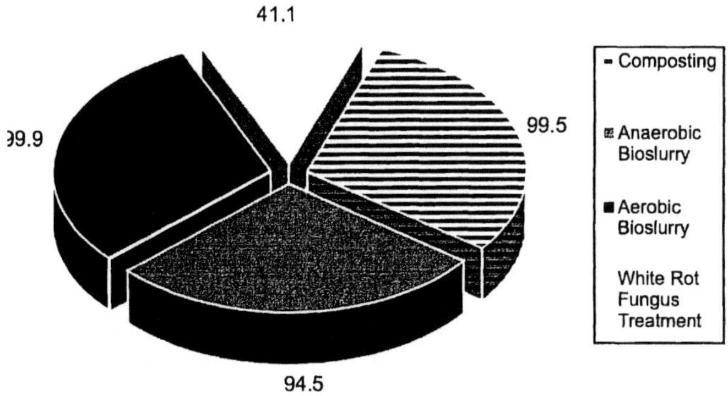

Figure 9.4 Comparison chart for bench scale treatability as % destruction and removal efficiency of various processes for TNT contaminated soil bioremediation (Craig et al., 1995).

9.2.4.2 Case Studies

Metals bioremediation by fungi has been researched to a great extent by various researchers with very few field applications. Some case studies for fungal bioremediation of metals are presented below:

Barclay et al., (1998) demonstrated biodegradation of metallocyanide complexes by mixed fungi cultures by exploring the potential of different consortia like *Fusarium solani*, *Trichoderma polysporum*, *F. oxysporum*, *Scytalidium thermophilum*, and *Penicillium miczynski*. Under acidic conditions (pH 4), the metallocyanide complex degradation was slower (28 d for 90–95% cyanide removal), whereas, at pH 7 similar degradation was obtained within 5 days.

From the preceding examples, it becomes apparent that processes based on these fungal strains have the potential to provide efficient management and or degradation of toxic pollutants and or its derivatives. However, in order to fabricate a practical and sustainable approach, proper measures should be taken before adapting WRF based bioremediation process in place of efficient physical and chemical methods or others.

9.3 Mycorrhizal Fungal Processes

The evolution of different guilds (fungal consortia) of mycorrhizal fungi (Figure 9.1) appears to have been driven by prevailing environmental conditions (Read, 1991, 1992). This has helped in the development and exploration of mycorrhizal processes, benevolent for soil remediation (Table 9.2). Although the naturally occurring symbiosis between higher plants and mycorrhizal fungi have been very well documented in many studies, the significance of functional diversity in mycorrhizal fungi (ectomycorrhizal, arbuscular, ericoid and orchid fungi) is still unclear (Cairney and Meharg, 1999; Cairney and Burke, 1994). It is believed that functional diversity of fungi is important for its symbiotic relationship with the host plant and to ecosystem level, which has the potential to offer an array of benefits to their plant hosts (Allen et al., 1995). Notably, in most of cases, the mycobionts present in rhizosphere are able to detoxify xenobiotics, thereby enhancing the quality of soil. Furthermore, close association of fungi to rhizosphere also includes nutritional needs of the fungi, thereby, resulting in a sustainable "eco-system", which could be of great interest in soil bioremediation. Mycorrhizal fungi dominate microbial ecology of heathlands, boreal and temperate forest biomes (Smith and Read, 1990), this fact is one of the many reasons that makes them suitable candidates in soil bioremediation. In the following sections, several pollutants have been discussed in relation to mycorrhizal fungi bioremediation.

9.3.1 Synthetic Dyes, Pesticides and Polychlorinated Biphenyls

Mycorrhizal fungi, in particular, ectomycorrhizal (ECM), and ericoid collected from rural locations have been efficient remediators of a wide range of PCBs (Green et al., 1999; Meharg et al., 1997b; Donnelly et al., 1993). They have an innate ability to catabolize PCBs and pesticides like atrazine, 2,4-dichlorophenoxyacetic acid and chlorophenols (Green et al., 1999).

9.3.1.1 Mechanisms of Bioremediation

Though mycorrhizal fungi may not degrade PCBs to yield energy, they may co-metabolize them as a consequence of consuming cyclic compounds, exuded by plants. For example, plant phenolics, such as catechin and coumarin co-metabolize degradation of PCBs by bacteria (Salt et al., 1998).

9.3.1.2 Case Studies

The recommended dose of pesticide in field might be deleterious to a newly reclaimed calcareous soil with low populations of ECM (Abd-Alla et al., 2000). Therefore, information about pesticide contamination could be very important in assessing the soil bioremediation potential of ECM. Some case studies for fungal

Table 9.2 Mycorrhizal soil fungi and respective pollutants.

Mycorrhizal Fungi	Host Plant and Medium	Pollutant	Concentration	Reference
Amanita pantherina, Amphinema bissoides, Armillaria mellea, Cenococcum geophilum, Hebeloma crustuliniforme, Laccaria amethystea, L. bicolor, L. laccata, Pisolithus tinctorius, Tricholoma vaccinum	Black poplar (*Populus nigra*), Norway spruce (*Picea abies*)	Crude oil	0.1–50 g kg^{-1}	Nicolotti and Egli (1998)
Trichoderma harzianum, Penicillium simplicissimum, P. janthinellum, P. funiculosum and P. terrestre	Basal salts medium	Pyrene	100 mg l^{-1}	Saraswathy and Hallberg, (2002)
Aspergillus niger	--	2,4-D and MCPA	--	Faulkner and Woodcock, (1964)
Arbuscular mycorrhizal fungi	--	Pesticides – Afugan, Brominal, Gramoxone, Selecron and Sumi Oil	0.13–3 mg/kg soil	Abd-Alla et al. (2000)
Arbuscular mycorrhizal fungi	--	PAHs	8–10 g l^{-1}	Cabello (1997); Leyval and Binet (1998)
Trichoderma sp.	--	Penta-chlorophenol, endosulfan and DDT	--	Katayama and Matsumura (1991)
Amanita muscaria, A. rubescens, A. Spissa, Bysporia terrestris, Gautieria crispa, G. othii, Hebeloma crustuliniforme, H. hiemale, H. sinapizans, Lactarius deliciosus, L. deterrimus, L.	--	Phenanthrene, anthracene, fluronthene, pyrene, perylene, 4-fluorobiphenyl, TNT, 2,4-dichlorophenol,	--	Gramss et al. (1999); Donnelly and Fletcher (1995); Green et al. (1999); Meharg et al. (1997a,b)

Mycorrhizal Fungi	Host Plant and Medium	Pollutant	Concentration	Reference
torminosus, Morchella conica, M. elata, M. esculenta, Paxillus involutus, Piloderma croceum, Radiigera atrogleba, Suillus granulatus, S. variegatus, Tricholoma lascivum, T. terreum		chlorpropham,		
Trichoderma koningii and *Fusarium culmorum*	Rye plant	$CdCl_2$	--	Kurek and Majewska (2004)
Suillus bovines	--	Cd and Zn	30–200 fungal/soil concentration ratio	Colpaert and Van Assche (1992)

bioremediation of synthetic dyes, pesticides and polychlorinated biphenyls are presented below.

Biodegradation of organochlorine herbicides 2,4,5-T and 2,4-D have been also found to be mediated by mycorrhizal fungi association. However, the role of ligninolytic enzymes of *Phanerochaete chrysosporium* in this process was not confirmed (Ryan and Bumpus 1989; Yadav and Reddy 1993). Meanwhile, catabolic degradation of 2,4-D and 2-methyl-4,6-dichlorophenoxyacetic acid (MCPA) by *Aspergillus niger* (not a white-rot fungus) has been already established in previous studies on these herbicides (Faulkner and Woodcock 1964). Katayama and Matsumura, (1991) had shown degradation potential of rhizosphere-competent fungus *Trichoderma* sp. against several synthetic dyes, pentachlorophenol, endosulfan, and DDT.

9.3.2 Crude Oils and Polycyclic Aromatic Hydrocarbons

Fungi in general have been successful in degradation of Polycyclic Aromatic Hydrocarbons (PAHs), and crude oil. There are increasing number of examples, where ectomycorrhizal (ECM) fungi have been shown to degrade major environmentally deleterious PAHs. The ratio of ECM fungi screened for being a potential PAHs remediator is high (e.g., in a study, 33 ECM species out of 42 screened were efficient remediators, Meharg and Cairney, 2000).

9.3.2.1 Mechanisms of Bioremediation

Lower (2–3) chlorinated PCBs were readily degraded by majority of ECM species screened, and a limited number of species were able to degrade 4–5 chlorinated biphenyls (Donnelly and Fletcher, 1995). Gramss et al., (1999) have extensively studied selective PAH degradation by many ECM species and found that 4–5 ring PAHs were preferentially degraded by some species. Also, pollutants like chlorpropham, dichlorophenol, trinitrotoluene and monofluorobiphenyl have also been shown to be degraded by many ECM fungi (Meharg and Cairney, 2000; Green et al., 1999). Nevertheless, their degradation rate for the pollutants in comparison to WRF were somewhat low (Gramms et al., 1999), owing to inactive growth phase of the mycelia (Meharg and Cairney, 2000).

The PAH degradation efficiencies for ECM fungi have been high, for example, 90% trinitrotoluene (Meharg et al., 1997a), 95% monofluorobiphenyl (Green et al., 1999) and 50% benzo-(a)-pyrene from solution culture (Braun-Lullemann et al., 1999). It would be imperative to suggest that, isolation of ECM fungi from unpolluted soils may express PAH-degrading activities in their natural habitats.

In addition to saprotrophs, there is evidence of moderate to high tolerant arbuscular mycorrhizal (AM) fungi towards soil contaminated with PAHs as high as 8–10 gl^{-1} (Cabello 1997; Leyval and Binet, 1998). Ganesan et al., (1991) found decreased diversity of AM fungal propagules in undisturbed soils in comparison to coal wastes, lignite spoils and calcite mine spoils, indicating negative effect of PAH on fungal population, whilst it indicated natural selection of fungal strains tolerant to PAH toxicity.

9.3.2.2 Case Studies

In a sagebrush-grassland ecosystem, alterations in AM fungal populations associated with application of waste water from oil shale processing units was observed by Stahl and Williams (1986). They observed good response from AM fungal taxa with respect to sensitivity towards crude oil pollutants – showing both decrease as well as increase in density of spores in the treated soils. Nevertheless, decreased competition with other AM fungal taxa might be accepted as one of the valid reasons behind this phenomenon.

In order to assess the damage caused to ecosystems in Trecate's affected area Nicolotti and Egli (1998) studied the effect of artificial crude oil (88% hydrocarbons, 8% PAHs) contamination on ECM infection on the plant *Populus nigra* and *Picea abies*. The ECM infection was reduced in case of *Populus nigra* seedlings, due to oil contamination, however, for soils exposed to oil contamination prior to planting seedlings, ECM infection was stimulated, instead. Oil generally had no significant effect on percentage infection of *Picea abies*, but there was a decrease in infection in the soils exposed to oil for longer period of time (> 5 weeks) before planting. In addition, for diverse ECM fungi species exposed to same crude oil in axenic culture, some exhibited dose-dependent toxicity to crude oil, while others showed no response or stimulated growth, suggesting differential sensitivity to crude oil by ECM fungi (Nicolotti and Egli, 1998).

Another interesting finding of Nicolotti and Egli, (1998) was that after contamination, seedling growth of two plants reduced with time. Under normal conditions, soil would have recovered with time due to decrease in phytotoxic compounds. They hypothesized that seedling growth should be affected by mycorrhizal colonization. As a matter of fact, after contamination, the ectomycorrhizal infection potential on spruce and poplar decreased over time, contributing to seedlings better growth during contamination period. This incidence suggested that a crude oil spill in mixed agriculture or, forest area does not cause long-term environmental damage in contrast to coastal ecosystems. Many of the mycorrhizal fungi are able to survive in contaminated soil (Barr and Aust, 1994a). They use crude oil as their principal nutrient (Nicolotti and Egli, 1998), thereby, contributing to soil bioremediation.

There are ample evidences of various other soil fungi contributing to PAHs degradation e.g., *Trichoderma* sp., *Penicillium* spp., *Gliocladium* sp., whilst affecting mycorrhizal fungi both positively and/or, negatively (Azcbn-Aguilar and Barea, 1997). Saraswathy and Hallberg, (2002) reported a maximum of 75% removal for pyrene (4-ring PAH) at 50 mg l^{-1} for axenic cultures of *Trichoderma* sp. and *Penicillium* spp. They also claimed that these soil fungi used pyrene as sole carbon source.

Therefore, it could be inferred that PAHs/crude oil contamination could be detoxified to a greater extent by naturally occurring mycobionts of mycorrhizal fungi and other soil fungi.

9.3.3 Munitions Waste

In contrast to WRF, ectomycorrhizal (ECM) basidiomycetes also have a considerable potential to facilitate degradation of explosive chemical compounds (Donnelly and Fletcher, 1995; Donnelly et al., 1993, 1994). Meharg et al., (1997a) demonstrated that ECM fungi have the capacity to biotransform TNT.

9.3.3.1 Mechanisms of Bioremediation

The munitions bioremediation potential of ECM was exhibited by both intact mycelial mass and extra-cellular enzymes under symbiotic conditions with the host plant. Meharg et al., (1997a) found that the biotransformation rate decreased with nitrogen limitation, contrary to basidiomycetes, whilst no decrease was observed under short term carbon starvation. This is crucial in the context to soil bioremediation since under symbiotic conditions, ECM fungi would have carbon and nitrogen sufficient environment.

Similar to *P. chryosporium* (Dass et al., 1995), ECM basidiomycetes may also release considerable amounts of non-specific extracellular proteases in liquid culture (Griffiths and Caldwell, 1992). Dass et al., (1995) concluded that *P. chryosporium* produces a number of proteases under different nutritional regimes, each having different abilities to degrade extracellular enzymes. Therefore, for *P. chryosporium* culture, in absence of protease inhibitors, there is greater possibility that other enzymes present might be degraded. Thus, TNT biodegradation capacity of culture filtrate must be viewed as a minimum as some residual proteolytic activity could have been present in filtrates. Further, the fact that extracellular enzymes are capable of degrading TNT could be regarded as additive phenomenon to plasmalemma redox mediated reduction or cytosolic biotransformation. In fact, intact cells were much more efficient at biotransformation than a cocktail of cell components. Furthermore, the apparent role of ECM basidiomycetes to biotransform TNT via extracellular enzymes has considerable significance in the potential use of ECM as bioremediators. It is expected that the efficiency of ECM fungi in biodegradation would augment in the presence of extracellular enzymes. Finally, extracellular enzymes may modify the soil environment which are rather inaccessible to fungal hyphae, hence, contributing to efficacy in soil bioremediation.

9.3.4 Metals

In soil, bioavailability of metals is dependant upon its various forms such as free metal ions, soluble metal complexes (sequestered to ligands), exchangeable metal ions, organically bound metals, precipitated or insoluble compounds such as oxides, carbonates and hydroxides, or they may form part of the structure of silicate minerals (Leyval et al., 1997). Furthermore, toxicity of metals in soil depends on their bioavailability, defined as their ability to be transferred from soil compartment to a living organism (Juste 1988).

Under axenic conditions, metal uptake and accumulation in the mycelium of ectomycorrhizal fungi have also been very efficient. For example, fungal/soil concentration ratios around 200 and 80 for Cd, and 40 and 30 for Zn of non-tolerant and metal-tolerant isolates of *Suillus bovinus,* respectively were reported by Colpaert and Van Assche (1992). ECM fungi have also been known to increase availability of metals in the rhizosphere by solubilizing minerals, including metal-containing rock phosphates, by production of organic acids or proton extrusion (Leyval et al. 1997).

Bradley et al. (1981, 1982) showed that ericoid mycorrhizal colonization led to a significant decrease in metal content of the shoot and an increase in the plant roots grown in sand amended with Cu or Zn. This explains mycorrhizal protection against excess heavy metal uptake by plants, in an ecosystem, where heavy metal availability may be higher due to soil acidity (Leyval et al. 1997). Consequently, excess translocation of metals within mycorrhizal fungi would exert more toxicity on fungal biomass, thereby, decline of fungal occurrence (propagule density) and infectivity in metal-polluted soils are possible. In fact, this has been used as bioindicator of soil contamination (Grodzinskaya et al. 1995). On the other hand, mycorrhizal colonization of plant roots after soil remediation can be an indication of metal detoxification/non-bioavailability.

These finding collectively suggest that mycorrhizal fungi based soil bioremediation of PCBs is possible, however, due to their higher susceptibility and slower metabolism, mycorrhizal fungi would be inefficient in contrast to white-rot fungi.

9.4 Conclusion

Fungi demonstrate an excellent ability to carry out soil bioremediation. White rot fungi need supplementary-nutrient source other than pollutants. Hence, once the growth supporting bulking agent(s)/substrate(s) (wood chips, peat moss, cereal flakes) deplete(s), the soil must be reinoculated with fresh white rot fungal inoculum, a major drawback for its large-scale sustainable use. A self-sustaining establishment of fungal-consortia in soil is possible with mycorrhizal fungi-plant symbiosis. In fact, all tree species are usually infected by mycorrhizal symbionts and ectomycorrhizal mycelia are extensive in forest soils. In some instances, almost half the soil biomass could be constituent of mycorrhizal symbionts. Both WRF and mycorrhizal fungi have their own advantages and disadvantages. For example, although WRF inoculum density can be regulated, their success is greatly dependent on treatment location, e.g., cellulose-rich residues would support WRF growth. Similarly, mycorrhizal fungi are comparatively slower and their growth mainly relies on rhizosphere of plant and other soil microbial consortia. Whilst, WRF are more extensively studied and applied in soil bioremediation, there are only a few mycorrhizal fungi, screened against a limited number of pollutants. Nevertheless, WRF showed greater efficiency in soil bioremediation in contrast to mycorrhizal fungi, limited bioavailability of pollutants would be better dealt with sustainable ecological niche of mycorrhizal fungi in soil. In other words, WRF are at their best but potential of mycorrhizal fungi should also be intelligently realized to facilitate sustainable remediation of soil.

References

Abd-Alla, M.H., Omar, S.A., Karanxha, S. 2000. The impact of pesticides on arbuscular mycorrhizal and nitrogen-fixing symbioses in legumes. *Applied Soil Ecology.* 14,

191–200.

Allen, E.B., Allen, M.F., Helm, D.J., Trappe, J.M., Molina, R., Rincon, E. 1995. Patterns and regulation of mycorrhizal plant and fungal communities. *Plant and Soil.* 170, 47–62.

Alloway, J.B., Ayres, D.C. 1993. Chemical principles of environmental pollution, Chapman and Hall, London.

Anderson, T.A., Guthrie, E.A., Walton, B.T. 1993. Bioremediation in the rhizosphere. *Environmental Science and Technology.* 27, 2630–2636.

Andersson, B.E., Welinder, L., Olsson, P.A., Olsson, S., Henrysson, T. 2000. Growth of inoculated white-rot fungi and their interactions with the bacterial community in soil contaminated with polycyclic aromatic hydrocarbons, as measured by phospholipid fatty acids. *Bioresource Technology.* 73, 29–36.

Azcbn-Aguilar, C., Barea, J.M. 1997. Applying mycorrhiza biotechnology to horticulture: significance and potentials. *Scientia Horticulturae.* 68, 1–24.

Barclay, M., Hart, A., Knowles, C.J., Meeussen, J.C.L., Tett, V.A. 1998. Biodegradation of metal cyanides mixed and pure cultures of fungi. *Enzyme and Microbial Technology.* 22, 223–231.

Barr, P.D., Aust, S.D., 1994a. Mechanisms white rot fungi use to degrade pollutants. *Environmental Science and Technology.* 28, 78–87.

Barr, P.D., Aust, S.D. 1994b. Pollutant degradation by white rot fungi. Rev. Env. *Contam. Toxicol.* 138, 49–72.

Baarschers, W.H., Heitland, H.S. 1986. Biodegradation of Fenitrothion and Fenitrooxon by the fungus *Trichoderma viridae. J. Agric. Food. Chem.* 34, 707–709.

Bogan, B.W., Lamar, R.T. 1996. Polycyclic Aromatic Hydrocarbon degrading capabilities of *Phanerochaete laevis* HHB-1625 and its extracellular ligninolytic enzymes. *Applied Environmental Microbiololgy.* 62, 1597–1603.

Bollag, J.-M., Myers, C.J., Minard, R.D. 1992. Biological and chemical interactions of pesticides with soil organic matter. *The Science of the Total Environment.* 123/124, 205–217.

Bezalel, L., Hadar, Y., Cerniglia, C.E. 1996a. Mineralization of Polycyclic Aromatic Hydrocarbons by the white-rot fungus *Pleurotus ostreatus. Applied Environmental Microbiololgy.* 62, 292–295.

Bezalel, L., Hadar, Y., Fu, P.P., Freeman, J.P., Cerniglia, C.E. 1996b. Initial oxidation products in the metabolism of pyrene, anthracene, fluorene, and dibenzothiophene by the white rot fungus *Pleurotus ostreatus. Applied Environmental Microbiololgy.* 62, 2554–2559.

Bennett, J.W., Childress, A.M., Wunch, K.G., Connick, Jr. W.J. 2001. Fungal compositions for bioremediation. United States Patent 6,204,049.

Bradley, R., Burt, A.J., Read, D.J. 1981. Mycorrhizal infection and resistance to heavy metal toxicity in *Calluna vulgaris. Nature.* 292, 335–337.

Bradley, R., Burt, A.J., Read, D.J. 1982. The biology of mycorrhiza in the Ericaceae. VIII. The role of mycorrhizal infection in heavy metal resistance. *New. Phytol.* 91, 197–209.

Braun-Lullemann, A., Huttermann, A., Majcherczyk, A. 1999. Screening of ectomycorrhizal fungi for degradation of polycyclic aromatic hydrocarbons. *Applied Microbiology and Biotechnology.* 53, 127–132.

Bumpus, J.A., Aust, S.D. 1987. Biodegradation of DDT [1,1,1-trichloro-2,2-bis(4-chlorophenyl)ethane] by the white-rot fungus *Phanerochaete chrysosporium. Appl. Environ. Microbiol.* 53, 2001–2008.

Bumpus, J.A., Kakar, S.N., Coleman, R.D. 1993. Fungal degradation of organophosphorous insecticides. *Appl. Biochem. Biotechnol.* 39, 715–726.

Chaillan, F., Flèche, A.L., Bury, E., Phantavong, Y., Grimont, P., Saliot, A., Oudot, J. 2004. Identification and biodegradation potential of tropical aerobic hydrocarbon-degrading microorganisms. *Research in Microbiology.* 155, 587–595.

Cabello, M.N. 1997. Hydrocarbon pollution: its effect on native arbuscular mycorrhizal fungi (AMF). *FEMS Microbiology Ecology.* 22, 233–236.

Cairney, J.W.G. and Burke, R.M. 1994. Fungal enzymes degrading plant cell walls: their possible significance in ectomycorrhizal symbiosis. *Mycol. Res.* 98, 1345–1356.

Cerniglia, C.E. 1997. Fungal metabolism of polycyclic aromatic hydrocarbons: past, present and future applications in bioremediation. *Journal of Industrial Microbiology & Biotech.* 19, 324–333.

Clapp, J.P., Young, J.P.W., Merryweather, J.W., Fitter, A.H. 1995. Diversity of fungal symbionts in arbuscular mycorrhizas from a natural community. *New Phytologist.* 130, 259–265.

Colombo, J., Cabello, M., Arambarri, A.M. 1996. Biodegradation of aliphatic and aromatic Hydrocarbons by natural soil microflora and Pure cultures of imperfect and lignolitic fungi. *Environmental Pollution.* 94, 355–362.

Colpaert, J.V., Van Assche, J.A. 1992. The effects of cadmium and the cadmium-zinc interaction on the axenic growth of ectomycorrhizal fungi. *Plant Soil.* 145, 237–243.

Craig, H.D., Sisk, W.E., Nelson, M.D., Dana, W.H. Bioremediation of Explosives-Contaminated Soils: A Status Review. Presented in: *Proceedings of the 10th Annual Conference on Hazardous Waste Research.* pp. 164–179. Kansas State University, Manhattan, Kansas, May 23-24, 1995.

Dahlberg, A., Jonsson, L., Nylund, J.-E. 1997. Species diversity and distribution of biomass above and below ground among ectomycorrhizal fungi in an old-growth Norway spruce forest in south Sweden. *Canadian Journal of Botany.* 75, 1323–1335.

Dass, S.B. Dosoretz, C.G., Reddy, C.A., Grethlein, H.E. 1995. Extracellular proteases produced by the wood-degrading fungus *Phanerochaeta chrysosporium* under lignolytic and non-lignolytic conditions. *Arch. Microbiol.* 163, 254–258.

Diels, L., DeSmet, M., Hooyberghs, L., Corbisier, P. 1999. Heavy metals bioremediation of soil. *Mol. Biotechnol.* 12, 149–158.

Dittmann, J., Heyser, W., Bucking, H. 2002. Biodegradation of aromatic compounds by white rot and ectomycorrhizal fungal species and the accumulation of chlorinated benzoic acid in ectomycorrhizal pine seedlings. *Chemosphere.* 49, 297–306.

Donnelly, P.K., Fletcher, J.S. Potential use of mycorrhizal fungi as bioremediation agents. In: *Bioremediation Through Rhizosphere Technology* (Edited by T. A Anderson & J. IL Coats), pp. 93-98. American Chemical Society, Washington DC (1994).

Donnelly, P.K., Fletcher, J.S. 1995. PCB metabolism by ectomycorrhizal fungi. Bull. *Environ. Contain. Toxicol.*, 54, 507–513.

Donnelly, P.K., Entry, J.A., Crawford, D.L. 1993. Degradation of atrazine and 2,4-dichlorophenoxyacetic acid by mycorrhizal fungi at three nitrogen concentrations *in vitro*. *Appl. Environ. Microbiol.*, 59, 2642–2647.

Duke, T.W., Lowe, J.I., Wilson, A.J. 1970. A polychlorinated biphenyl (Arochlor 1254) in the water, sediment and biota of Escambia Bay, Florida. *Bull. Environ. Contam. Toxicol.* 5, 3171–80.

Eccles, H. 1999. Treatment of metal-contaminated wastes: why select a biological process? *Trends Biotechnol.* 17, 462–465.

Eggen, T. 1999. Application of fungal substrate from commercial mushroom production – *Pleuorotus ostreatus* – for bioremediation of creosote contaminated soil. *International Biodeterioration & Biodegradation.* 44, 117–126.

Faulkner, J.K., Woodcock, D. 1964. Metabolism of 2,4-dichlorophenoxyacetic acid by *Aspergillus niger* Van Tiegh. Nature. 203, 865–866.

Fernando, T., Bumpus, J.A., Aust, S.D. 1990. Biodegradatiun of TNT (2,4,6-trinitrotoluene) by *Phanerochaete chrysosporium*. Appl. Environ. Microbiol. 56, 1666–1671.

Field, J.A., Jong, E., Feijoo-Costa, Bont, G.J.A.M. 1993. Screening for lignolytic fungi applicable to the biodegradation of xenobiotics. *Trend. Biotech.*, 11, 44–49.

Franz, A., Burgstaller, W., Schinner, F. 1991. Leaching with *Penicillium simplicissimum*: influence of metals and buffers on proton extrusion and citric acid production. *Appl. Environ. Microbiol.* 57, 769–774.

Franz, A., Burgstaller, W., Muller, B., Schinner, F. 1993. Influence of medium components and metabolic inhibitors on citric acid production by *Penicillium simplicissimum*. *J. Gen. Microbiol.* 139, 2101–2107.

Gadd, G.M. 1993. Microbial formation and transformation of organometallic and organometalloid compounds. *FEMS Microbiol. Rev.* 11, 297–316.

Gadd, G.M. 1996. Influence of microorganisms on the environmental fate of radionuclides. *Endeavour.* 20, 150–156.

Gadd, G.M., Sayer, J.A. 2000. Fungal transformations of metals and metalloids. In: Lovley, D.R. (Ed.), *Environmental Microbe –Metal Interactions.* Am. Soc. Microbiol, Washington, pp. 237– 256.

Gadd, G.M. 1999. Fungal production of citric and oxalic acid: importance in metal speciation, physiology and biogeochemical processes. *Adv. Microb. Physiol.* 41, 47–92.

Gadd, G. M. 2004. Microbial influence on metal mobility and application for bioremediation. *Geoderma.* 122, 109– 119.

Galzy, P., Slonimski, S. 1957. Variations physiologiques de la levure au cours de la croissance sur l'acide lactique comme seule source de carbone. *Cri. Rev. Acad. Sci. Paris.* 245 sér. D, 2423–2426.

Ganesan, V., Ragupathy, S., Parthipan, B., Rajini Rani, D.B., Mahadevan, A. 1991. Distribution of vesicular-arbuscular mycorrhizal fungi in coal, lignite and calcite mine spoils of India. *Biology and Fertility of Soils.* 12, 131–136.

Gardes, M., Bruns, T. 1996. Community structure of ectomycorrhizal fungi in a Pinus muricata forest: above- and below-ground views. *Canadian Journal of Botany.* 74, 1572–1583.

Garon, D., Sage, L., Wouessidjewe, D., Seigle-Murandi, F. 2004. Enhanced degradation of fluorene in soil slurry by *Absidia cylindrospora* and maltosyl-cyclodextrin. *Chemosphere.* 56, 159–166.

Gharieb, M.M, Kierans, M., Gadd, G.M. 1999. Transformation and tolerance of tellurite by filamentous fungi: accumulation, reduction and volatilization. *Mycol. Res.* 103, 299–305.

Gill, P.E. 1996. Process for bioremediation of soils. United States Patent 5,525,139.

Gill, P.E. 1997. Spill clean-up process. United States Patent 5,609,668.

Goruntzy, T., Drzyzga, O., Kahl, M.W., Bruns-Nagel, D., Breitung, J., von Loew, E., Blotevogel, K.H. 1994. Microbial degradation of explosives and related compounds. *Crit. Rev. Microbiol.* 20, 265–284.

Gramss, G., Kirsche, B., Voigt, K.D., Gunther, T., Fritsche, W. 1999. Conversion rates of five polycyclic aromatic hydrocarbons in liquid cultures in fifty-eight fungi and the concomitant production of oxidative enzymes. *Mycological Research.* 103, 1009–1018.

Green, N.A., Meharg, A.A., Till, C., Troke, J., Nicholson, J.K. 1999. Degradation of 4-fluorobiphenyl by mycorrhizal fungi as determined by 19F nuclear magnetic resonance spectroscopy and ^{14}C radiolabelling analysis. *Applied and Environmental Microbiology.* 65, 4021–4027.

Griffiths, R.P. and Caldwell, B.A. Mycorrhizal mat communities in forest soils In: *Mycorrhizas in Ecosystems.* (Edited by D. J. Read, D. H. Lewis, A. H. Fitter, and I. J. Alexander), pp 98-105. CAB International, Wallingford (1992).

Hatakka, A. 1990. Lignin-modifying enzymes from selected white rot fungi: Production and role in lignin degradation. *FEMS Microbiology Reviews.* 13, 125–135.

Higson, F.J. 1992. Microbial degradation of nitroaromatic compounds. *Adv.Appl. Microbiol.* 37, 1–19.

Hobman, J.L., Wilson, J.R., Brown, N.L. 2000. Microbial mercury reduction. In: Lovley, D.R. (Ed.), *Environmental Microbe – Metal Interactions*. Am. Soc. Microbiol, Washington, pp. 177– 197.

Hofrichter, M., Gunther, T., Fritsche, W. 1993. Metabolism of phenol, chloro- and nitrophenols by the *Penicillium* strain Bi 7/2 isolated from a contaminated soil. *Biodegradation*, 3, 415–421.

Hutzinger O, Safe S, Zitko V. *The chemistry of PCBs*. Cleveland, OH: CRC Press. 1974, 7–39.

INE. *Bifenilos policlorados*. Mexico, DF: Instituto Nacional de Ecologia, Secretary of the Environment and Natural Resources, 1995 in Spanish.

Juste, C. 1988. Appréciation de la mobilité et de la biodisponibilité des éléments en traces du sol. *Sci. Sol.* 26, 103–112.

Katayama, A. and Matsumura, F. 1991. Photochemically enhanced microbial degradation of environmental pollutants. *Environ. Sci. Technol.*, 25, 1329–1333.

Kerem, Z., Friesem, D., Hadar, Y. 1992. Lignocellulose degradation during soild-state fermentation: *Pleurotus ostreatus* versus *Phanerochaete chrysosporium*. *Applied Environmental Microbiology*. 58, 1121–1127.

Kotterman, M.J.J., Wasseveld, R.A., Field, J.A. 1996. Hydrogen peroxide production as a limiting factor in xenobiotic compound oxidation by nitrogen-sufficient cultures of *Bjerkandera* sp.. Strain BOS55 overproducing peroxidases. *Applied Environmental Microbiology*. 62, 880–885.

Kubatova, A., Erbanova, P., Homolka, L., Nerud, F., Sasek, V. 2001. PCB congener selective biodegradation by the white rot fungus *Pleurotus ostreatus* in contaminated soil. *Chemosphere*. 43, 207–215.

Kurek, E., Majewska, M. 2004. In vitro remobilization of Cd immobilized by fungal biomass. *Geoderma*. 122, 235–246.

Lamar, R.T., Evans, J.W., Glaser, J.A. 1993. Solid-phase treatment of a pentachlorophenol-contaminated soil using lignin-degrading fungi. *Environ. Sci. Technol.*, 27, 2566–2571.

Lamar, R.T. 1992. The role of fungal lignin-degrading enzymes in xenobiotic degradation. *Biotechnology*. 3, 261–266.

Lappin, H.M., Greaves, M.P., Slater, J.H. 1985. Degradation of the herbicide mecoprop (2-(2-methyl-4-chlorophenoxy) propionic acid) by a synergistic microbial community. *Applied and Environmental Microbiology*. 49, 429–433.

Leyval, C., Binet, P. 1998. Effect of polyaromatic hydrocarbons in soil on arbuscular mycorrhizal plants. *Journal of Environmental Quality*. 27, 402–407.

Leyval, C., Turnau, K., Haselwandter, K. 1997. Effect of heavy metal pollution on mycorrhizal colonization and function: physiological, ecological and applied aspects. *Mycorrhiza*. 7, 139–153.

Lisowska K, Dlugonski J. 1999. Removal of anthracene and phenanthrene by filamentous fungi capable of cortexolone 11-hydroxylation. *J. Basic Microbiol.* 39, 117–25.

Liu, G., Chambers, S.M., Cairney, J.W.G. 1998. Molecular diversity of ericoid mycorrhizal endophytes isolated from Woollsia pungens (Cav.) F. Muell. (Epacridaceae). *New Phytologist.* 140, 145–153.

Lovley, D.R. 2000. Fe(III) and Mn(IV) reduction. In: Lovley, D.R. (Ed.), *Environmental Microbe –Metal Interactions.* Am. Soc. Microbiol, Washington, pp. 3–30.

Lovley, D.R., Coates, J.D. 1997. Bioremediation of metal contamination. *Curr. Opin. Biotechnol.* 8, 285– 289.

Masaphy, S., Levanon, D., Vaya, J., Henis, Y. 1993. Isolation and characterization of a novel atrazine metabolite produced by the fungus *Pleurotus pulmonarius*, 2-chloro-4-ethylamino-6-(1-hydroxyisopropyl)amino-1,3,5-triazine. *Appl. Environ. Microbiol.* 59, 4342–4346.

McFarland, M.J., Qiu, X.J., Sims, J.L. 1992. Remediation of petroleum impacted soils in fungal compost bioreactors. *Water Science Technology.* 25; 197–206.

Meharg, A.A., Cairney, J.W.G. 2000. Ectomycorrhizas – extending the capabilities of rhizosphere remediation? *Soil Biology & Biochemistry.* 32, 1475–1484.

Meharg, A.A., Wright, J., Dyke, H., Osborn, D. 1998. Polycyclic aromatic hydrocarbon (PAH) dispersion and deposition to vegetation and soil following large scale chemical fire. *Environmental Pollution.* 99, 29–36.

Meharg, A.A., Dennis, G.R., Cairney, J.W.G. 1997a. Biotransformation of 2,4,6-trinitrotoluene (TNT) by ectomycorrhizal basidiomycetes. *Chemosphere.* 35, 513–521.

Meharg, A.A., Cairney, J.W.G., Maguire, N. 1997b. Mineralisation of 2,4-dichlorophenol by ectomycorrhizal fungi in axenic culture and in symbiosis with pine. *Chemosphere.* 34, 2495–2504.

Meysami, P., Baheri, H. 2003. Pre-screening of fungi and bulking agents for contaminated soil bioremediation. *Advances in Environmental Research.* 7, 881–887

Michels, J., Gottshalk, G. 1994. Inhibition of lignin peroxidase of *Phanerochaete chrysosporium* by hydroxylamino-dinitrotoluene, an early intermediate in the degradation of 2,4,6-tfinitrotoluene. *Appl. Environ. Microbiol.* 60, 187–194.

Milstein, O., Trojanowski, J., Hiittermann, A., Gressel, J. 1988. Catabolism of single ring aromatic acids by four *Aspergillus* species. *Microbios.* 55, 7–16.

Mougin, C., Laugero, C., Asther, M., Dubroca, J., Frasse, P. 1994. Biotransformation of the herbicide atrazine by the white-rot fungus *Phanerochaete chrysosporium. Appl. Environ. Microbiol.* 60, 705–708.

Nicolotti, G., Egli, S. 1998. Soil contamination by crude oil: impact on the mycorrhizosphere and on revegetation potential of forest tress. *Environmental Pollution.* 99, 37–43.

Novotny, C., Svobodova K., Erbanova P., Cajthaml, T., Kasinath A., Lang, E., Sasek, V. 2004. Ligninolytic fungi in bioremediation: extracellular enzyme production and degradation rate. *Soil Biology and Biochemistry.* 36, 1545–1551.

Novotny, C., Rawal, B., Bhatt, M., Patel, M., Sasek, V., Molitoris, H.P. 2001. Capacity of *Irpex lacteus* and *Pleurotus ostreatus* for decolorization of chemically different dyes. *Journal of Biotechnology.* 89, 113–122.

Novotny, C., Erbanova, P., Cajthaml, T., Rothschild N. Dosoretz, C., Sasek, V. 2000. *Irpex lacteus*, a white rot fungus applicable to water and soil bioremediation. *Appl. Microbiol. Biotechnol.* 54, 850–853.

Ousley, M.A., Lynch, J.M., Whipps, J.M. 1994. Potential of *Trichoderma* spp. as consistent plant-growth stimulators. *Biol Fertil Soils.* 17, 85–90.

Perotto, S., Actis-Perino, E., Perugini, J., Bonfante, P. 1996. Molecular diversity of fungi from ericoid mycorrhizal roots. *Molecular Ecology.* 5, 123–131.

Pointing, S.B. 2001. Feasibility of bioremediation by white-rot fungi. *Applied Microbiology and Biotechnology.* 57, 20–33.

Potin, O., Veignie, E., Rafin, C. 2004a. Biodegradation of polycyclic aromatic hydrocarbons (PAHs) by *Cladosporium sphaerospermum* isolated from an aged PAH contaminated soil. *FEMS Microbiology Ecology.* 51, 71–78.

Potin, O., Rafin, C., Veignie, E. 2004b. Bioremediation of an aged polycyclic aromatic hydrocarbons (PAHs)-contaminated soil by filamentous fungi isolated from the soil. *Int. Biodeterior. Biodegrad.* 54, 45–52.

Read, D.J. 1991. Mycorrhizas in ecosystems. *Experientia.* 47, 376–391.

Read, D.J. The mycorrhizal mycelium. In: *Mycorrhizal Functioning: an Integrated Plant-Fungal Process.* (Edited by M. J. Allen), pp 102-133. Chapman and Hall, New York 1992.

Rojas-Avelizapa, N.G., Rodriguez-Vazquez, R., Enriquez-Villanueva, F., Martinez-Cruz, J., Poggi-Varaldo, H.M. 1999. Transformer oil degradation by an indigenous microflora isolated from a contaminated soil. *Resources, Conservation and Recycling.* 27, 15–26.

Ryan, T.P., Bumpus, J.A. 1989. Biodegradation of 2,4,5-trichlorophenoxyacetic acid in liquid culture and in soil by the white rot fungus *Phanerochaete chrysosporium. Appl. Microbiol. Biotechnol.* 31, 302–307.

Sack, U., Gunther, T. 1993. Metabolism of PAH by fungi and correlation with extraceilular enzymatic activities. *J. Basic Microbiol.* 33, 269–277.

Sack, U., Fritsche, W. 1997. Enhancement of pyrene mineralization in soil by wood-decaying fungi. *FEMS Microbiology Ecology.* 22, 77–83.

Saraswathy, A., Hallberg, R. 2002. Degradation of pyrene by indigenous fungi from a former gasworks site. *FEMS Microbiology Letters.* 210, 227–232.

Schinner, F., Burgstaller, W. 1989. Extraction of zinc from an industrial waste by a *Penicillium sp. Appl. Environ. Microbiol.* 55, 1153–1156.

Schutzendubel, A., Majcherczyk, A., Johannes, C., Huttermann, A. 1999. Degradation of fluorene, anthracene, phenanthrene, fluor-anthene, and pyrene lacks connection to the production of extracellular enzymes by *Pleurotus ostreatus* and *Bjerkandera adusta. International Biodeterioration and Biodegradation.* 43, 93–100.

Silver, S. 1996. Bacterial resistances to toxic metal ions—a review. *Gene.* 179, 9–19.

Silver, S. 1998. Genes for all metals – a bacterial view of the periodic table. *J. Ind. Microbiol. Biotech.* 20, 1–12.

Smith, S.E., Read, D.J. 1990. Mycorrhizal symbiosis, 2nd ed. Academic Press, San Diego, Calif.

Spiker, J.K., Crawford, D.L., Crawford, R.L. 1992. Influence of 2,4,6-trinitrotoluene (TNT) concentration on the degradation of TNT in explosives-contaminated soils by the white rot fungus *Phanerochaete chrysosporium*. *Appl. Environ. Microbiol.* 58, 3199–3202.

Sreekrishnan, T.R., Tyagi, R.D. 1994. Heavy metal leaching from sewage sludges: a techno-economic evaluation of the process options. *Environ. Technol.* 15, 531–543.

Stahl, J.D., Aust, S.D. 1993a. Metabolism and detoxification of TNT by *Phanerochaete chrysosporium*. *Biochem. Biophys. Res. Comm.* 192, 477–482.

Stahl, J.D., Aust, S.D. 1993b. Plasma membrane dependent reduction of 2,4,6-trinitrotoluene by *Phanerochaete chrysosporium*. *Biochem. Biophys. Res. Comm.* 192, 471–476.

Tamaki, S., Frankenberger, W.T. 1992. Environmental biochemistry of arsenic. Rev. *Environ. Contam. Toxicol.* 124, 79–110.

Thurston, C.F. 1994. The structure and function of fungal laccases. *Microbiology.* 140, 19–26.

Twigg, L.E., Socha, L.V. 2001. Defluorination of sodium monofluoroacetate by soil microorganisms from central Australia. *Soil Biology & Biochemistry.* 33, 227–234.

Vachon, P.R., Tyagi, R. D., Auclair, J.C., Wilkinson, K.J. 1994. Chemical and biological leaching of aluminium from red mud. *Environ. Sci. Technol.* 28, 26–30.

VanAcken, B., Godefroid, L.M., Peres, C.M., Naveau, H., Agathos, S.N. 1999. Mineralization of 14C-U ring labeled 4-hydroxylamino-2,6-dinitrotoluene by manganese-dependent peroxidase of the white-rot basidiomycete *Phlebia radiata*. *Journal of Biotechnology.* 68, 159–169.

Verrecchia, E.P., Dumont, J.-L. 1996. A biogeochemical model for chalk alteration by fungi in semiarid environments. *Biogeochemical.* 35, 447– 470.

Vyas, B.R.M., Bakowski, S., Sasek, V., Matucha, M. 1994. Degradation of anthracene by selected white-rot fungi. *FEMS Microbiology Ecology.* 14, 65–70.

Waldner, R., Leisola, M.S.A., Fiechter, A. 1988. Comparison of ligninolytic activities of selected white-rot fungi. *Applied Microbiology and Biotechnology.* 29, 400–407.

Walker, R.T. 2003. Surfactant coated products and methods for their use in promoting plant growth and soil remediation. United States Patent 6,565,860.

White, C., Sayer, J.A., Gadd, G.M. 1997. Microbial solubilisation and immobilization of toxic metals: key biogeochemical processes for treatment of contamination. *FEMS Microbiol. Rev.* 20, 503– 516.

White, C., Sharman, A.K., Gadd, G.M. 1998. An integrated microbial process for the bioremediation of soil contaminated with toxic metals. *Nature Biotechnol.* 16, 572–575.

Yadav, J.S., Reddy, C.A. 1993. Mineralization of 2,4-dichlorophenoxyacetic acid (2,4-D) and mixtures of 2,4-D and 2,4,5-trichlorophenoxyacetic acid by *Phanerochaete chrysosporium*. *Appl. Environ. Microbiol.* 59, 2904–2908.

Zheng, Z., Obbard, J.P. 2002. Oxidation of polycyclic aromatic hydrocarbons (PAH) by the white rot fungus, *Phanerochaete chrysosporium*. *Enzyme and Microbial Technology*. 31, 3–9.

Acknowledgements

The authors are sincerely thankful to Natural Sciences and Engineering Research Council of Canada (Grants A4984, Canada Research Chair) for financial support. The views and opinions expressed in this article are those of authors.

CHAPTER 10

Phytoremediation

Pascale Champagne

10.1 Introduction

Conventional remediation approaches for contaminated soils, sediments and groundwater are based on technologies developed over the past two decades. These include a wide variety of physical, chemical and thermal treatment alternatives and combinations thereof, as well as engineering strategies to accelerate or reduce mass transport in the contaminated matrix. The success of these highly engineered technologies is attributable to their relative insensitivity to heterogeneity in the contaminant matrix, their ability to be effective over a wide range of oxygen concentrations, pH, pressure, temperature, and osmotic potentials, and their ability to produce relatively rapid contaminant mitigation rates (Cunningham *et al.*, 1997). However, several sites covering large areas remain contaminated with no remediation plan in sight simply because site remediation using conventional approaches would be too expensive. Phytoremediation provides a possible economic solution for several such sites. The fundamental principles of phytoremediation for exploitation as an environmental remediation technology were first proposed in the late 1970s. Since then, research, technology development, and application of phytoremediation as a method for remediating contaminated soils and groundwater has increased significantly due to its relatively low-cost and versatility (Carman and Crossman, 2001).

The term *phytoremediation* is derived from the Greek prefix "phyto" meaning plant, and the Latin suffix "remedium" meaning to cure. Phytoremediation is defined as the *in-situ* or *ex-situ* engineered use of plants to remove or control contaminants, or to foster contaminant breakdown by microorganisms in the plant. It is a passive technology that involves immobilizing, removing, sequestering or degrading organic and inorganic contaminants in soil and water via plant-influenced biological, chemical and physical processes. As a technology, it is considered to be sustainable, especially compared to conventional approaches to contaminated site remediation and/or management, because it is considered to be primarily solar powered.

10.2 Elements of Phytoremediation

The fate of contaminants in the plant-soil system is determined by a complex series of biochemical, and biophysical reactions and/or interactions. The potential fate of contaminants during phytoremediation is illustrated in Figure 10.1. The pathways and kinetics of microbial and plant-based transformations, degradation and mineralization for the majority of organic and inorganic contaminants found in the environment are not completely known to date. A description of the mechanisms illustrated in Figure 10.1 has been presented in Chapter 2.

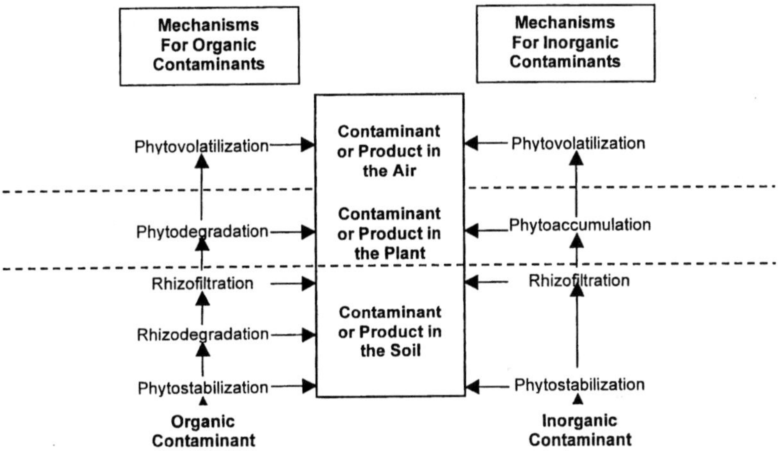

Figure 10.1 Potential fate of organic and inorganic contaminants during phytoremediation.

Contaminants that have the potential to be mitigated using phytoremediation include metals, metalloids, radionuclides, salts, nutrients, xenobiotic organic chemicals, as well as air-borne particles or odor-causing chemicals. While heterotrophic microorganisms mineralize organic contaminants for their energy, carbon and nutrient content, plants use sunlight as their energy source and carbon dioxide as a carbon source while transforming and storing metabolic products. Plants also require catabolic enzymes to break down lignin, cellulose, coumarins, flavonoids, and a variety of other complex phyto-compounds. Hence, for every biomolecule that plants, animals or microorganisms produce, a microbial or botanical enzyme process should exist which could potentially mineralize that compound. When xenobiotic compounds, which are not products of biosynthesis, have natural analogs with similar chemical structures and properties, these compounds can be mitigated via passive processes such as phytoremediation (McCutcheon and Schnoor, 2003).

10.2.1 The Plant-Soil System

Phytoremediation takes advantage of the selective uptake capabilities of plant root systems, along with the translocation, bioaccumulation, contaminant storage and degradation capabilities of entire plant systems (Suthersan, 2002). The portion of soil in close proximity to the roots of growing plants which is often characterized as a zone of increased microbial activity, is known as the rhizosphere. Microorganisms in the plant rhizosphere can enhance the availability of contaminants for adsorption or uptake by the plant root system and may contribute to the degradation of organic pollutants. An important characteristic is that plants have the ability to exert limited control over the rhizosphere, local biogeochemistry, availability of water and nutrients, and the local microclimate (McCutcheon and Schnoor, 2003).

Plants produce biomass and have tissues involving specialized enzymes and cells in specific areas that can transform, conjugate, and store toxins, as well as readily process and store water, energy, and nutrients. As plants are stationary, many have evolved elaborate defences against pests and developed tolerances to changes in their soil-water microenvironment for their survival (McCutcheon and Schnoor, 2003). Microorganisms lack this level of organization and can die off rapidly under periods of environmental stress. However, they can reproduce and increase in numbers rapidly when the water, substrate, energy source and nutrients required for growth become available once again.

Only plants possessing appropriate biochemical pathways can be effective in phytoremediation applications. To provide a viable phytoremediation option, selected plants must be able to grow and thrive in the presence of contaminants that may be deleterious to many other types of vegetation and must tolerate local conditions at the site. The ability to grow rapidly and being a perennial rather than an annual could also be advantageous. Another important consideration in selecting plant species for phytoremediation is the size, activity, and species composition of the rhizosphere community, as well as the volume occupied by the rhizosphere (Alexander, 1999).

10.2.1.1 Plants

Plants employ carbon dioxide to photosynthesize carbon biomass, produce energy and release oxygen to the surrounding environment. They typically grow by sending their roots into a soil environment, from which they take up water and dissolved inorganic constituents. The roots release exudates into the root zone and transpire water to the atmosphere. Plants require thirteen essential inorganic nutrients (N, P, K, Ca, Mg, S, Fe, Cl, Zn, Mn, Cu, B and Mo) for growth, development, and reproduction. These essential nutrients can be taken up actively by the plant via transport proteins associated with the root membrane or passively with the water absorbed by the plant. Other non-essential inorganic constituents can also be taken up by the plant via active or passive transport. Since these are not essential nutrients and may be toxic to plant functions at high concentrations, plants have developed various mechanisms to sequester and stabilize these and other potential contaminants. Considerable variations in the ability to take up and tolerate various contaminants exist between different plant species.

Atmospheric carbon dioxide is converted to organic matter, or photosynthates, using reductants generated during photosynthesis. These products of photosynthesis are translocated throughout the plant and are incorporated into the biomass, metabolized to produce energy during cell respiration or exuded into the root zone. Roots receive 30-60% of the net photosynthetic carbon, from which 10-20% is released by rhizodeposition. Exudates consist primarily of low molecular weight and high molecular weight organic acids, typically 10-20 mM total concentration in the roots in the form of lactate, acetate, oxalate, succinate, fumarate, malate, citrate, isocitrate and aconitate. The remainder of the dissolved organics in plant roots is composed of sugars (90 mM), amino acids (10-20 mM) and a number of secondary plant metabolites (Jones, 1998; Singer et al., 2003).

Another important contribution of plants in the transformation of contaminants is derived from degradative enzymes released by plant roots to their surrounding environment. The enzymes are usually wall-associated enzymes, which partially transform substances into products that can be more readily taken up by plant roots or rhizosphere microorganism (Gianfreda and Rao, 2003). The presence of these enzymes can significantly increase transformation of contaminants occurring in the rhizosphere, particularly xenobiotic compounds.

In some plants, oxygen is transported, by diffusion or convective air flow, to the roots and rhizomes (Guntenspergen et al., 1989). A portion of the oxygen leaks from the root system into the rhizosphere which creates an oxidizing environment. The translocation of oxygen in the plant root zone is illustrated in Figure 10.2. The redox state of the rhizosphere can have a significant effect on the intensity of oxygen released through the roots (Stottmeister et al., 2003).

Plants may directly influence the biodegradation of contaminants by promoting transformations related to plant growth. They can also enhance microbiological cometabolism through the rhizodeposition of cosubstrates as exudates, which then induce the necessary enzymes for the transformation of contaminants. Plants may also have an indirect effect on biodegradation processes by generating microenvironments that are conducive to increased genetic exchange of degradative plasmids. They may also physically alter the soil environment for xenobiotic degradation by altering pH and redox conditions in the rhizosphere or by influencing the spatial arrangement of microbial communities that develop in the root zone (Crowley et al., 1997).

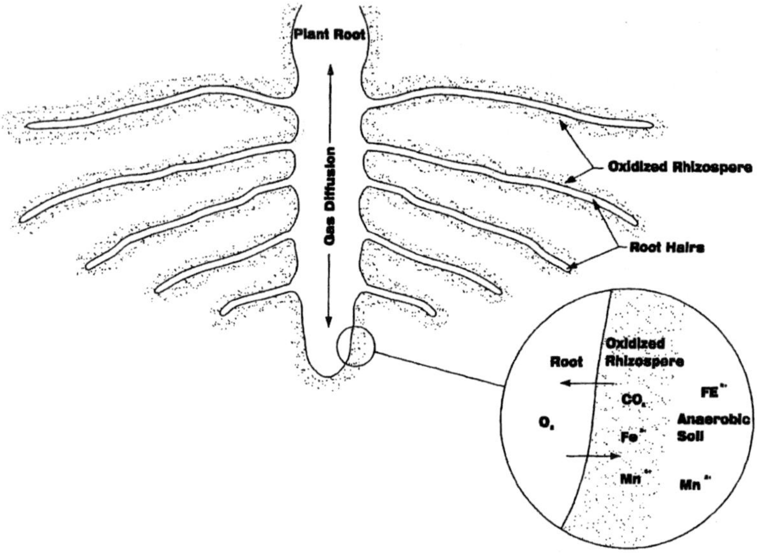

Figure 10.2 Oxygen transfer in the rhizosphere environment.

10.2.1.2 Rhizosphere

In natural plant ecosystems, indigenous soil microorganisms present in the rhizosphere are found to exist in mutually beneficial relationships with plants. The rhizosphere includes the root surfaces and the adjacent soil, which are extensively colonized by microorganisms, particularly bacteria and mycorrhizal fungi. However, the rhizosphere does not extend far from the root surface because the physical and chemical factors that make this zone unique are spatially limited.

The rhizosphere effect is often expressed as the ratio of the number of microorganisms in the rhizosphere soil to the number of microbes in the non-rhizosphere soil. Although the R/S ratio commonly ranges from 5 to 20, it can be greater than 100 in some cases. The microbial composition in the rhizosphere is complex, and is a function of the plant species, soil type, and growth period of the plant (Carman and Crossman, 2001). It is divided into the inner rhizosphere, at the root surface, and the outer rhizosphere, embracing the adjacent soil. The microbial population is typically largest and most active in the inner zone where the biochemical interactions are most pronounced and root exudates are most concentrated (Suthersan, 2002). Diverse species of heterotrophic microorganisms are brought together in large numbers, and these may either enhance the stepwise transformation of recalcitrant compounds by the consortia of microorganisms, or provide an environment that is conducive to genetic exchange and gene rearrangement leading to the remediation of contaminants. The principal microbial consortium generally includes bacteria, actinomycetes, and mycorrhizal fungi. Mycorrhizal fungi grow in symbiotic association with the plant and are particularly important in the rhizosphere because they have unique enzymatic pathways that help to degrade organics that could not otherwise be transformed by bacteria alone (Suthersan, 2002; Carman and Crossman, 2001). The rhizosphere microbial community composition is dependent on root type, plant age, plant species, environmental conditions, substrates, root exudates and soil type, as well as factors such as plant root exposure history to various contaminants. Generally, the primary root microbial population is determined by the habitat created by the plant and the secondary microbial population depends upon the activities of the initial population (Anderson and Coats, 1995).

Plant growth promoting microorganisms can positively influence plant growth and development either directly or indirectly. Microorganisms directly promote plant growth by transforming organic and inorganic constituents to a form that is more readily available to the plant or by facilitating the uptake of nutrients from the environment by the plant. Indirect promotion of plant growth or plant protection occurs when microorganisms prevent or decrease the incidence of deleterious effects by phytopathogenic organisms (Glick, 2003).

Plants, in combination with microbes, alter redox conditions, pH and organic matter content of sediments and can, therefore, impact the chemical speciation and mobility of some constituents, particularly metals. In some soil environments, the biogeochemistry of the rhizosphere can change on a seasonal basis (Jacob and Otte, 2003). As can be seen in Figure 10.3, constant changes, both physical and chemical, at the root-soil interface can lead to alterations in the soil matrix and microbial environment. Differences between the rhizosphere and the bulk soil in terms of substrate, carbon dioxide and oxygen concentrations, moisture content, osmotic and redox potentials, and pH substantially influence microbial community structure and activity, where constituents may be mobilized or immobilized depending on the actual combination of factors (Crowley et al., 1997; Anderson and Coats, 1995).

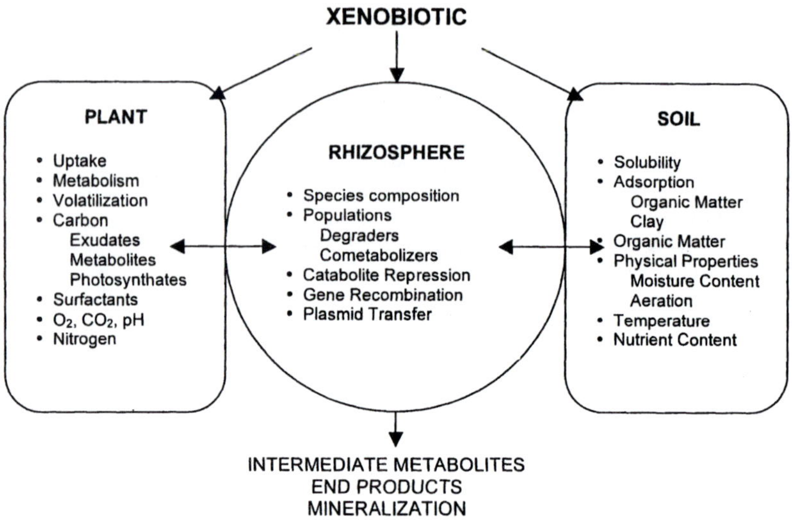

Figure 10.3 Potential plant-rhizosphere-soil interactions in the degradation of a xenobiotic contaminant.

It is primarily the continuous supply of oxygen and exudates released from the roots that result in the increase of rhizosphere microbial population densities well beyond those of the surrounding bulk soil, as well as the attraction of mobile bacteria and fungal hyphae that stimulate an array of positive, neutral or negative interactions with plants. As can be seen from Figure 10.4, the exudates serve as primary or secondary substrates for soil microorganisms. The decomposition of organic acids and many other compounds is often reported to be 2 to 3 times greater in the rhizosphere than in the bulk soil.

In some cases, the release of structural analogues of various xenobiotics contained in root exudates, cell wall components and lysates, as well as secondary plant metabolites may select for or stimulate microbial transformation reactions which metabolize or cometabolize xenobiotic compounds (Crowley et al., 1997; Singer et al., 2003). This phenomenon may be of importance when the microorganisms that generally employ the xenobiotic compound as a carbon source are absent or require a long acclimation period prior to reach a population density necessary for rapid degradation of the compound. Rates of cometabolism depend on the availability of substrates provided by rhizodeposition in different root zones, as well the induction of the enzymes necessary for the degradation of the xenobiotic compound. Spatial variations in nutrient availability and differences in the composition of plant derived exudates in different root zones can impact both growth-linked and cometabolic degradation of the contaminants (Crowley et al., 1997).

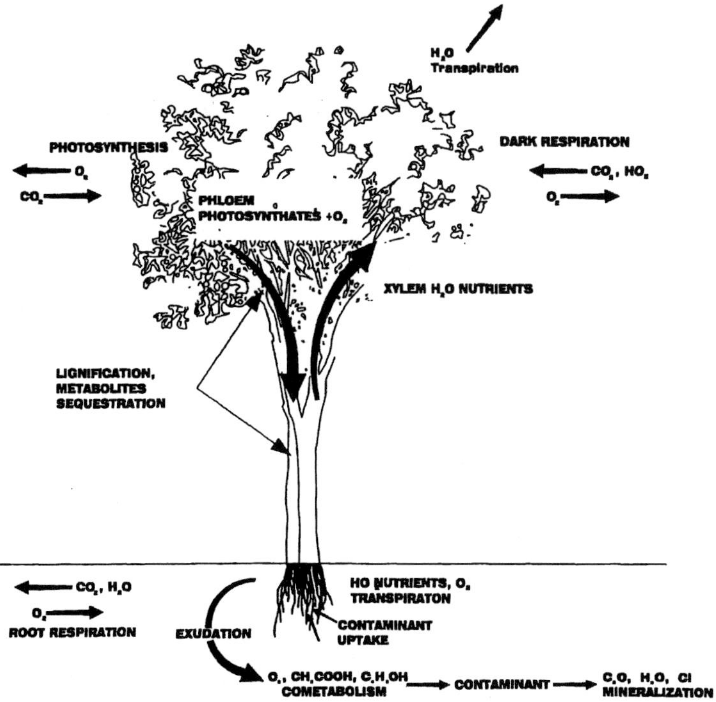

Figure 10.4 Mass flow through the plant-soil-air continuum.

10.2.2 Phytoremediation Processes

Phytoremediation applications can be classified based on the processes impacting the mitigation of contaminants in the plant-soil system: degradation, extraction, immobilization, containment or a combination of these processes (Table 10.1). Such plant-mediated processes can include extraction of contaminants from soil or groundwater, concentration of contaminants in plant tissues, biotic or abiotic degradation, volatilization or transpiration of volatile contaminants from plants to the air, contaminant immobilization in the root zone, hydraulic control of contaminated groundwater, and control of runoff, erosion and infiltration by vegetative covers.

Table 10.1 Summary of phytoremediation processes.

Mechanism	Process Goal	Media	Contaminants	Plant Type
Phytostabilization	Containment erosion control	Soils, sediments Sludges	Metals: As, Cd, Cr, Cu, Pb, Zn	Herbaceous species, grasses, trees, wetland species
Rhizodegradation	Tranformation mineralization	Soils, sediments sludges, groundwater	Organic compounds (TPH, PAHs, BTEX, chlorinated solvents, pesticides, PCBs)	Herbaceous species, grasses, trees, wetland species
Phytoaccumulation	Extraction accumulation	Soils, sediments sludges	Metals: Ag, Au, Cd, Co, Cr, Cu, Hg, Mn, Mo, Ni, Pb, Zn, Radionuclides: ^{90}Sr, ^{137}Cs, ^{239}Pu, $^{234,238}U$	Herbaceous species, grasses, trees, wetland species
Phytodegradation	Tranformation mineralization	Soils, sediments, sludges, groundwater surface water	Organic compounds, chlorinated solvents, phenols, pesticides, munitions	Algae, herbaceous species, trees, wetland species
Phytovolatilization	Extraction and release to air	Soils, sediments, sludges, groundwater	Chorinated solvents, MTBE, metals: (Se, Hg, As)	Herbaceous species, trees, wetland species
Evapotranpiration	Containment erosion control	Groundwater, surface water, stormwater	Water soluble organics and inorganics	Herbaceous species, grasses, trees, wetland species

10.2.2.1 Degradation

Contaminant mitigation in plant-soil systems can result from phyto-degradation and rhizodegradation mechanisms. Some plants have the ability to uptake toxic compounds and in the process of metabolizing the available nutrients, detoxify the pollutants (USEPA, 2000). Phytodegradation (Figure 2.12) illustrated the breakdown of contaminants taken up by plants through metabolic processes and transformation of contaminants outside the plants by enzymes produced and released by the plants. In many cases, contaminants are degraded, used as nutrients, and incorporated into plant tissues. In other cases, however, metabolic intermediates or end-products are re-released to the environment depending on the type of contaminant and plant species (Suthersan 2002).

Plant transformation pathways differ depending on plant species and tissue type. They are generally categorized as reduction, oxidation, conjugation or sequestration. The Green-Liver model has been proposed to describe the metabolic pathways of herbicides, pesticides, explosives and other nitoraromatic compounds (Burken, 2003). Plants produce a large number of enzymes as a result of primary and secondary metabolism and have been shown to uptake and metabolize organic contaminants to less toxic compounds. Plant enzyme systems can be constitutive or induced and can play a role in transformation pathways, and plant adaptation and/or tolerance to adverse growth conditions. Enzymes that have been demonstrated to be useful in phytodegradation include nitroreductase (ammunitions and pesticides), dehalogenases (chlorinated solvents and pesticides), phosphatase (pesticides), peroxidases (phenols), laccases (aromatic amines), cytochrome P-450 (pesticides and chlorinated solvents), and nitrilase (herbicides). In some instances, contaminant degradation by plant enzymes has been shown to take place in environments in which microorganisms could not survive due to elevated contaminant concentrations, suggesting that phytodegradation might be an effective remediation alternative even under conditions which do not support biodegradation (Suthersan, 2002).

Rhizodegradation (Figure 2.15) involves the breakdown of contaminants in the soil through microbial activity enhanced by the rhizosphere, where microorganisms consume and degrade or transform organic constituents as a source of nutrients. The increased microbial populations and diversity in the rhizosphere is due to the release of plant exudates, including oxygen, nutrients, sugars, alcohols, amino acids, and enzymes. The exudates are considered to be the primary source of energy for microbes in the rhizosphere. In the process of metabolizing these substances, xenobiotic contaminants can be either metabolized directly or co-metabolized by the microbes. Roots also release organic material in the form of decaying roots and mucigel, and gelatinous substance that is a lubricant for root penetration through soil during growth (Carman and Crossman, 2001).

10.2.2.2 Extraction

Aquatic plants and algae are known to accumulate metals and other contaminants from solution. Large differences in removal rates have been reported and these are generally a function of plant species and strain differences, cultivation methodology, and process control techniques. Naturally immobilized plants such as attached algae and rooted plants and those easily separated from suspension such as filamentous micro-algae, macro-algae, and floating plants have been noted to have high adsorption capacities (Suthersan, 2002). The removal of metals and other ions from solution typically involves a rapid initial phase followed by a second longer phase. The initial phase is generally attributed to physical and chemical processes including chelation, ion exchange and chemical precipitation, while the second phase, which is longer, is attributable to biological processes including intracellular uptake and translocation of the contaminants into plant shoots (Carman and Crossman, 2001). Whether or not organic contaminants are extracted by plants is generally dictated by Briggs's Law.

The processes by which contaminants can be extracted from the surrounding water, soil or sediments involve phytoaccumulation, rhizofiltration and, in some cases, phytovolatilization mechanisms. Phytoaccumulation (Figure 2.11) is a process that employs plant species that are known to uptake, translocate, and accumulate contaminants in their roots, shoots and leaves. Once they have reached a desired growth level, the plants are harvested, thereby removing the target contaminants from the site. This technique yields a mass of plant and contaminant that must subsequently be transported for further disposal or recycling (USEPA, 2000). Both organic and inorganic contaminants can be extracted by plants, however, the fate of these compounds once extracted by the plant are very different. Plants that grow in environments with high concentrations of inorganic contaminants such as metals can either adapt to accumulate, exclude or avoid the metals. The phytoaccumulation of metals relies on the tendency of some metals to relocate from the soil or water to the plant tissues, hence inorganic contaminants generally accumulate in the roots and shoots of plants. Hyperaccumulator plants avoid the toxic effects of metals such as clorosis, necrosis, disruption of chlorophyll synthesis, alteration in water balance, and stunted growth by binding the metals to cell walls, pumping metal ions into vacuoles, or complexing heaving metals with organic acids. Plant species that exclude inorganic contaminants do not prevent uptake of the constituent, but restrict translocation, and the detoxification of the metal occurs in the plant roots. Mechanisms of detoxification for excluder plant species include immobilization of heavy metals on cell walls, exudation of chelate ligands, or formation of a redox or pH barrier at the plasma membrane (Carman and Crossman, 2001).

When the roots from plants are primarily responsible for absorption, concentration and precipitation of metals, radionuclides and organics, the process is referred to as rhizofiltration (Figure 2.16). Rhizofiltration is similar to phytoaccumulation in that it also removes and/or concentrates rather than degrade contaminants. In contrast to phytoaccumulation, which uses plants grown in soil or sediments, rhizofiltration is generally implemented in an aquatic environment for the purpose of contaminant removal (Carman and Crossman, 2001).

Organic contaminants, once extracted by plants, are generally broken down further and metabolized or volatilized from the leaf tissue. Inorganic contaminants can also sometimes be methylated. Phytovolatilization (Figure 2.14) is the uptake and transpiration of contaminants by plants, with the release of the contaminant or a modified form thereof to the atmosphere. This process generally occurs as plants take up water, and organic and inorganic contaminants. Some of these contaminants can pass through the plants to the leaves and volatilize to the atmosphere.

10.2.2.3 Immobilization

Immobilization using plants either binds contaminants to the soil reducing their bioavailability or immobilizes them by removing the means of transport (Suthersan, 2002). Phytostabilization (Figure 2.13) refers to the use of certain plant species to immobilize contaminants in the soil and groundwater through absorption and accumulation by roots, adsorption onto roots or precipitation within the rhizosphere and the physical stabilization of soils. Phytostabilization strategies are often implemented to reduce the mobility of heavy metal and high molecular weight organics and prevent their migration to the groundwater or air.

A second type of phytostabilization refers to the physical process of establishing or re-establishing a vegetative cover on sites lacking natural vegetation. The physical containment of contaminants by plants is generally achieved through reduced soil erodibility, decreased wind blown dust potential and, at times, reduced contaminant solubility with the addition of soil amendments. This type of phytostabilization can take the form of contaminant binding within a humic molecule (humification), physical sequestration of metals or root accumulation in non-harvestable plants. Certain plants sequester large concentrations of metals in their roots, and while harvesting and removal is difficult or impractical, the contaminants present a reduced environmental risk while they are bound in the roots (USEPA, 2000).

10.2.2.4 Hydraulic Containment

Hydraulic containment is an application of phytoremediation to capture groundwater and soil pore water, leading to contaminant containment or immobilization. The basis of phytocontainment (Figure 2.17) is that consumption of water during evapotranspiration by plants can be used to control the migration of a contaminant plume in a groundwater body (USEPA, 2000). Water uptake rates of certain plants, particularly phreatophytes, can be significant enough to suppress the water table and in some instances, create zones to capture contaminated groundwater. Phytocontainment is restricted to the deepest rooting depth of vegetation, which is a function of the plant species, soil conditions, and water availability. Roots will generally extend only to a depth necessary to maintain plant viability and will not extend more than a foot or two into the water table. In some cases, hydraulic containment using vegetation can be more effective than conventional pump and treat systems, which require that individual pores be interconnected to allow removal by pumping (Carman and Crossman, 2001). Plant roots have the ability to penetrate microscopic-scale pores in the soil matrix for contaminant removal. Phytoremediation of groundwater plumes is ideal for contaminants that are water soluble, organic contaminants that are leachable, and when inorganic contaminants are present at concentrations that are not phytotoxic (Suthersan, 2002).

10.3 Phytoremediation Applications

For effective application of phytoremediation technologies, the mechanisms described in Chapter 2 and the previous section need to be exploited in specific design applications tailored for a given contaminated site the associated remediation objectives. In many cases, hydraulic containment is a management goal to ensure that contaminants do not migrate off-site or impact other receptors. Another objective for site remediation might be stabilization, accumulation, reduction, degradation, metabolism, or mineralization of specific contaminants to decrease the risks associated with exposure to these contaminants. Phytoremediation applications are considered appropriate for sites where the following conditions exist (ITRC, 2001):

- sufficient area exists for establishing an effective vegetative cover;
- treatment can be applied over long periods of time;
- concentrations of contaminants are nontoxic to plants;
- other methods of remediation are not cost effective or practical;
- a transition from a primary treatment to a longer term strategy may be desired;
- vegetation can be used as a final cap for closing or restoring the site.

10.3.1 Wetland Systems

In general terms, wetlands are lands where a depth of water covers the soil, or where water is present either at or near the surface of the soil or within the root zone, consistently or intermittently throughout the year, including during the growing season. The presence of water at or near the soil surface is at a frequency and duration sufficient to contribute to the formation of hydric soils that are characteristic of wetlands (NAS,

1995; Mitsch and Gosselink, 1993), and the establishment of plants adapted for life in saturated soils and animal communities living in such soils and on its surface. Wetlands are characterized by a range of properties which make them attractive for the management of contaminants in water. The treatment mechanisms of wetland systems involve physical, chemical and biological processes occurring in the soil-water matrix, in plants and in the rhizosphere. Wetlands are valuable because they can greatly influence the quality of water and its flow. They improve water quality by intercepting surface runoff and removing or retaining inorganic constituents, processing organic matter, and reducing suspended sediments.

10.3.1.1 Types of Wetlands

Natural wetlands are poorly drained, transitional areas between deeper open water and dry land. Often located in low-lying areas, wetlands receive runoff water and overflow from rivers, streams and tides, and these areas of land are cyclically, intermittently or permanently inundated or saturated with fresh, brackish or saline water. On the other hand, constructed wetlands are man-made systems designed using natural materials of soil, water, and biota to simulate and optimize the physical, chemical and biological processes and functions of a natural wetland to achieve a desired water quality or habitat objective. Constructed treatment wetlands are typically divided into two basic types: free water surface wetlands (FWS) and subsurface flow wetlands (SSF).

Free water surface wetlands (Figure 10.5) are shallow excavations or shallow earth banked lagoons, typically densely vegetated basins and underlain by a subsurface barrier, liner or compacted clay to prevent seepage. Soil or another suitable medium such as gravel or organic matter provides a growing medium to support roots of emergent vegetation, and water at a relatively shallow depth (3 to 24 inches) flowing through the unit (Suthersan, 2002; McCutcheon and Schnoor, 2003). A system of pipes or channels distributes the wastewater over the inlet end of the wetland, and a collection channel collects the treated effluent at the outlet. The wastewater flows along the surface, where it comes in contact with the bacterial populations on the surface of the media and plant stems.

In subsurface flow wetlands (SSF), the flow moves through a matrix (1 to 3 feet deep) of soil, gravel, sand, organic substrate or a mixture that supports vegetation roots and biofilms (Suthersan, 2002; McCutcheon and Schnoor, 2003). Water can flow in a horizontal or vertical flow configuration, where wastewater is mitigated by physical and chemical processes, as well as microbiological degradation.

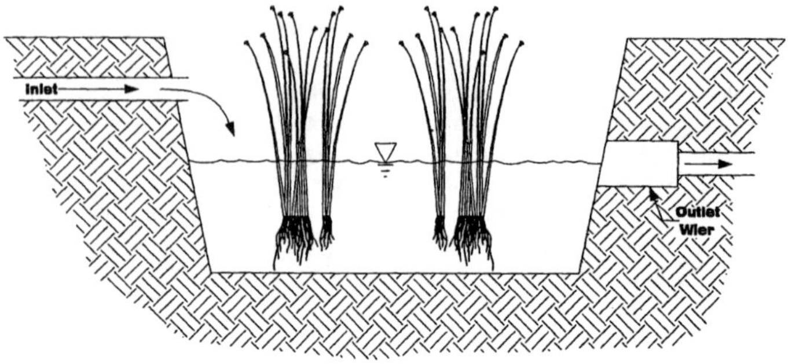

Figure 10.5 Free water surface wetland system.

Horizontal flow wetlands (Figure 10.6) are shallow excavations with a synthetic or clay liner. In a horizontal flow system the wastewater is fed through an inlet and passes the filter matrix under the surface of the bed in a relatively horizontal path until it reaches the outlet zone. The matrix is planted with marsh plants, such as the common reed, *Phragmites australis*. The primary removal mechanisms are filtration and biodegradation, where dense populations of bacteria growing on the roots of the marsh plants contribute to the treatment process.

In vertical flow systems (Figure 10.7), the wastewater is applied over the entire surface area through a distribution system and the flow passes through the matrix in a relatively vertical path. Marsh plants such as *P. australis*, are typically planted in the matrix. The wastewater is dosed on the bed in large batches, thereby flooding the surface to a depth of several centimeters. At the base of the excavation are drainage pipes which are usually turned up so they reach the surface at their ends, allowing air to move in and out of the wetland. The wastewater then slowly percolates downward through the substrate matrix, which acts as a filter. Between dosing times, oxygen can diffuse through the pores of the matrix. The treated water is typically collected in a bottom drainage system and flows into an outlet well.

10.3.1.2 Wetland Processes

Wetlands provide a diverse niches and micro-environments that play important roles in the mitigation of contaminants. Various processes occur within the vegetation, the water column, on the wetland substrate, in the soil, and in concentrated areas of microbial activity on plant stems and roots, in the water column and within the soil and substrate matrices. These processes may be biological, chemical or physical, although considerable overlap and interaction occurs between them. General wetland processes involved in the mitigation of organic and inorganic contaminants are illustrated in Figure 10.8.

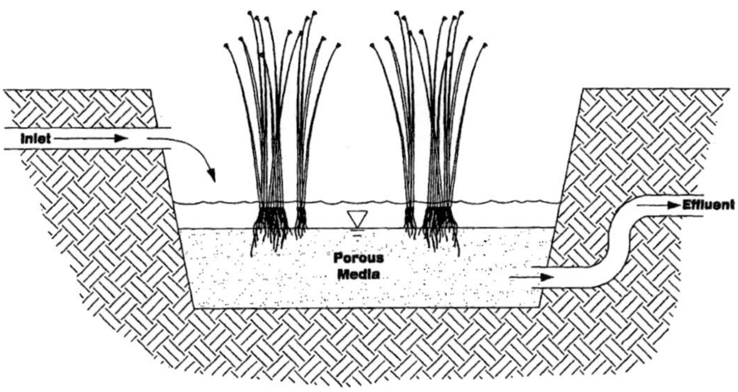

Figure 10.6 Subsurface horizontal flow wetland system.

In natural or constructed wetland treatment systems both biotic and abiotic mechanisms are involved. Biological mechanisms can be vegetative and microbiological in nature, with the remainder of the mitigation occurring via chemical and physical processes, principally at the interfaces of the water and soil, the soil and the root, or the vegetation and the water. The effectiveness of all of these processes varies with residence time. Longer residence times have beneficial effects by permitting more contaminants to be removed. However, very long residence times can also have detrimental effects as the redox potential of the sediments may change, allowing the release of nutrients and pollutants back into the water column (NSW DLWC, 1998).

Chemical and biological wetland processes occur at rates dependent upon environmental factors including: temperature, oxygen, pH and salinity. At lower temperatures, process rates may be slower, cease altogether or even reverse, releasing nutrients under certain environmental conditions. Wetland processes occur under the influence of interrelated and constantly changing environmental conditions, which ultimately impact the performance of a wetland treatment system. Two of the most important cyclic changes in ambient conditions which impact wetland treatment and function are diurnal and seasonal changes.

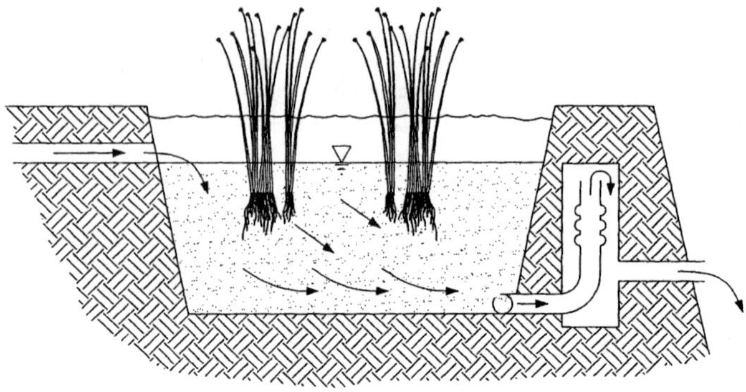

Figure 10.7 Subsurface vertical flow wetland system.

Diurnal changes usually result in variations in wetland temperature profiles and dissolved oxygen concentrations. Lower night temperatures slow various chemical reactions and microbial activity, in comparison to warmer daytime temperatures. Photosynthesis adds oxygen, which determines the direction of many wetland processes, to the water column by day, but this oxygen is reduced and possibly depleted overnight due to microbial respiration. Seasonal changes cause variations in daylight hours (photoperiod) and temperature. During the growing season, emergent and submerged vegetation from water and sediments uptake nutrients at high rates. By the time the vascular plants die, they have translocated a portion of nutrient material to the roots and rhizomes. Cold weather reduces plant and microbial activity in the water and sediments, affects community selection, and slows down biological and chemical processes as well as their resultant biogeochemical nutrient cycling kinetics (Werker et al., 2002; Newman et al, 2000). In the spring, excessive runoff combined with cool temperatures leads to high flow rates and reduced nutrient and metal retention, and hence diminished nutrient retention from the fall to early spring (Mitsch and Gosselink, 2000).

Fluctuating water levels, a characteristic of wetlands, effectively control oxygen concentrations in the wetland water column and sediments, and the oxidation-reduction (redox) chemistry of the wetland system. Oxygen diffuses slowly in water, and is often used by microbial activity faster than it can be replenished. This affects root respiration, and impacts nutrient availability. The redox conditions are, therefore, governed by hydroperiods and play a significant role in nutrient cycling and availability, pH, solubility or insolubility of nutrients and metals within the wetland system, vegetation composition, sediment and organic matter accumulation, degradation, and inorganic constituent availability (Connell et al., 1984). The salinity of water within the wetland can increase as water levels decrease, and contaminants may become concentrated. Changes in salinity may impact some of the removal and retention processes in wetlands. Elevated salt concentrations create increased competition between cations and metals for binding sites on soils and organic substrates. Often, inorganic contaminants such as metals are desorbed into the overlying water column (Connell et al., 1984).

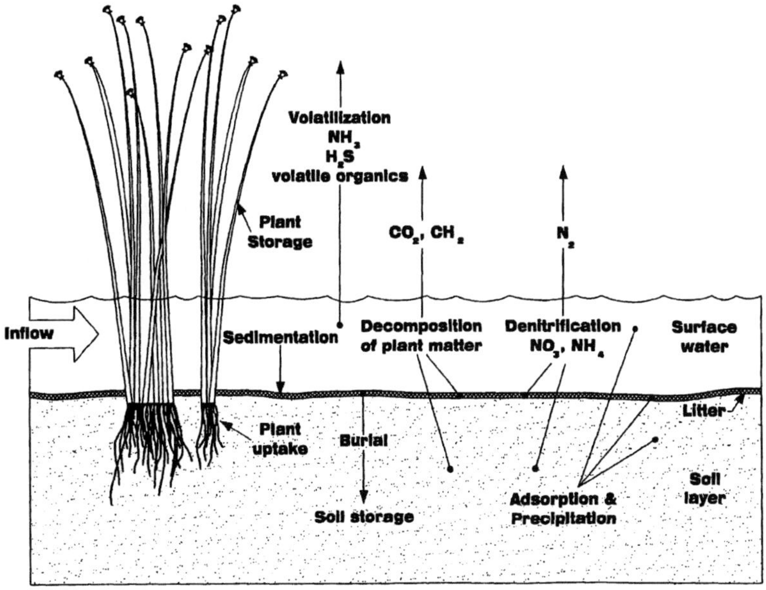

Figure 10.8 Wetland processes controlling the mitigation of contaminants.

The pH of water and soils in wetlands also exerts a strong influence on the direction of many biological and chemical reactions. Organic wetland soils tend to be acidic, while mineral wetland soils are more neutral or alkaline. The consequence of flooding previously drained wetland soils is usually to push the pH toward neutrality, whether it be formerly acidic or alkaline. Many processes, including biological transformations, cation exchange, and solubility of solids and gases, are pH dependent, and less effective when the pH is out of an optimal range, (NSW DLWC, 1998).

Physical and Chemical Processes

In wetland systems, the physical removal of contaminants associated with particulate matter in wastewater occurs efficiently primarily via sedimentation and filtration processes. Water typically moves very slowly through wetlands due to the resistance provided by rooted and floating plants in free water surface systems and due to the soil, sediment and substrate matrix in subsurface flow systems, and is consequently, settled, filtered, and retained in the wetland system.

A wide range of chemical processes are involved in the removal and mitigation of contaminants in wetland treatment systems. Exposure to light and atmospheric gases, can lead to the breakdown of organic compounds and kill pathogens. Chemical reactions between constituents leading to their transformation and subsequent precipitation from the water column as insoluble compounds are important chemical processes within wetland treatment systems. Another important precipitation reaction that occurs in wetland soils is the formation of metal sulfides, which are highly insoluble compounds and represent an effective means for immobilizing many metals in wetlands treatment systems.

One of the most important chemical removal processes in wetland soils, sediments, substrates and plant surfaces is sorption, which results in short-term retention or long-term immobilization of several types of contaminants. Contaminants may be removed by adsorption, ion exchange, precipitation or complexation. Wetland soils and vegetation generally have a high cation exchange capacity (CEC) for a variety of chemical constituents. Wetland organic soils contain humic substances, which are a product of the undecomposed vegetation in the wetland soils, contain a large number of phenolic (C_6H_4OH), carboxyl (-COOH) and hydroxyl (-OH) functional groups, which are hydrophilic and serve as pH dependent ion exchange sites.

Biological Processes

In addition to physical and chemical mitigation processes, biological processes involving wetland microorganisms and vegetation also contribute significantly to the mitigation effectiveness of treatment wetlands. The most widely recognized and utilized biological process for contaminant removal in wetlands is biological uptake. Contaminants that are present in the form of essential plant nutrients, such as nitrate, ammonium and phosphate, are readily taken up by wetland plants. However, many wetland plant species take up and transform a variety of organic and inorganic constituents. The rate of contaminant removal by plants varies, depending on plant growth rate and concentration of the constituent in plant tissues. Algae may also provide a significant amount of nutrient uptake, but the storage of nutrients and other organic and inorganic constituents may be short-term, due to the short life-cycle of algae. Microorganisms in the soil also provide uptake and short-term storage of nutrients and other contaminants.

As a result of vegetation and microbial death, detritus and litter accumulates at the soil surface, of wetland systems. Some of the nutrients, metals or other elements previously removed from the water by wetland biota are then recycled back to the water and soil by leaching and decomposition. Leaching of water-soluble constituents may occur rapidly upon the death of the plant or plant tissue, while a more gradual loss of constituents occurs during the decomposition of detritus by bacteria and other organisms. Recycled contaminants may be flushed from the wetland in the surface water, or may be removed again from the water by biological uptake or other phytoprocesses.

Although microorganisms may provide a measurable amount of contaminant uptake and storage, it is their metabolic processes that play the most significant role in the removal of organic compounds. Many of the microbiological processes can be facilitated by or occur in conjunction with plants and algae present in the system, and other phytoprocesses and the biological processes are often part of larger element cycles occurring within the wetland system. Microbial activity will depend on the concentration and nature of the substrate undergoing transformation, as well as the presence and availability of the suitable enzymes.

Microbial decomposers, primarily soil bacteria, utilize the carbon (C) in organic matter as a source of energy, converting it to carbon dioxide (CO_2) or methane (CH_4) gases. This provides an important biological mechanism for removal of a wide variety of organic compounds. The efficiency and rate of organic compound degradation by microorganisms is highly variable for different types of organic compounds. Respiration using O_2 as an electron acceptor dominates under aerobic conditions, while under the anaerobic conditions often found in wetlands, organic decomposition can occur in the presence of any number of other terminal electron acceptors, including NO_3^-, Mn^{2+}, Fe^{3+}, SO_4^{2-} and CO_2.

10.3.1.3 Role of Wetland Vegetation and Microbial Communities

Wetland systems offer a diversified range of biological activities. The primary function of algae and hydrophytic plants is the uptake of dissolved nutrients and contaminants from water, which they employ to produce new plant biomass. The interaction of wetland soil, sediment vegetation and aquatic constituents also provides a desirable environment for the growth and reproduction of microorganisms, which can play an active role in contaminant mitigation. In addition to playing an instrumental role in the reduction and cycling of nutrients and mitigation of metals and other contaminants, wetland vegetation influences wetland hydrology and hydraulics through the system. Macrophytes reduce flow velocities, which enables sedimentation, physical filtering of sediment particles and decreases erosion.

Algae

Algae are unicellular and multicellular photosynthetic bacteria and plants that are highly diverse and can adapt to a wide range of aquatic habitats, including wetlands. They are dependent on light, carbon, typically CO_2, and other nutrients for their energy, growth and reproduction. When these elements are not limiting, algae can generate large populations and contribute significantly to the overall food chain, the transformation, storage and cycling of wetland nutrients, the fixation of contaminants through sorption and settling as well as contaminant mitigation (Kadlec and Knight, 1996). In wetlands with few emergent macrophytes, filamentous algal mats can develop, which tend to control DO and CO_2 concentrations in the wetland water column. However, when they are shaded by macrophytes, algae tend to play a lesser role in the processes of the wetland system.

When algae growth is substantial, they can play a significant role in the short-term fixation and immobilization of nutrients and hydrocarbons in wetland systems, followed by a gradual release through algal death and decomposition. From a wetland functión perspective, this form of nutrient cycling can lead to the immobilization and transformation of some contaminants which can be used for growth by the algae, while the gradual release would result in an attenuation of the discharge concentrations of contaminants. Long-term algal fixation of contaminants into wetland sediments via sorption, settling and burial may also be significant, particularly in low-nutrient wetlands where the microbial decomposition rates of algal biomass would be low (Kadlec and Knight, 1996).

Higher Plants

The types of higher plants involved in wetland systems can be classified as: emergent, floating or submerged plants based on their predominant growth form. In emergent plant species, most of the above ground part of the plant emerges above the waterline and into the air. They are rooted in the sediment soil where they have an extensive root and rhizome structure. Emergent plants generally have structural components which allow them to be self-supporting and to transport oxygen to the roots and rhizomes. The presence of macrophytes in wetlands is also important because they provide surface area for microbial attachment and growth both in the water column and in the root zone or rhizosphere. Submerged species are also rooted in the wetland soil, but have buoyant stems and leaves that cover large surface areas filling a niche in the water column above the wetland sediment layer. Floating species have been adapted to float on the surface of the wetland due to buoyant leaves and stems. They are not generally rooted in the wetland soil, and roots typically dangle below the surface within the water column (Guntenspergen et al., 1989).

Macrophytes are the dominant vegetative structural component of most wetland systems. They include vascular plants with tissues which result from specialized cells. As oxygen is often limiting in flooded environments, macrophytes have adapted to saturated conditions of wetland systems where they thrive, despite the challenges of an anaerobic environment, and contribute to its treatment capacity. These wetland plants have adopted other strategies, such as long, oxygen transporting tubes (e.g., emergent reeds), the ability to float on shallow water (e.g., lilies), or buttresses trunks (e.g., Cypress trees) to obtain and transport oxygen to areas of the plant where it is required for various plant functions. Wetland plants are adapted to changing redox conditions. They often contain arenchymous tissue (spongy tissue with large pores) in their stems and roots that allows air to move quickly between the leaf surface and the roots. Oxygen released from wetland plant roots oxidizes the rhizosphere and allows processes requiring oxygen, such as organic compound breakdown, decomposition, and denitrification, to occur (Steinberg and Coonrod, 1994).

Microorganisms

Wetland microorganisms remove soluble organic matter, coagulate colloidal material, stabilize organic matter and convert soluble and colloidal organic matter. These microbial populations can have a significant influence on the soil and water chemistry of wetland systems. Different microoganisms have a variety of tolerances and requirements for dissolved oxygen, temperature, pH and nutrients, which can influence important transformations of nitrogen, iron, manganese, sulphur and carbon (Suthersan, 2002). Bacteria and fungi are generally the first organisms to colonize and begin the sequential decomposition of organic solids and typically have the first access to dissolved constituents (Kadlec and Knight, 1996).

In wetland systems bacteria are often found on solid plant surfaces, decaying organic matter and in the soil and sediment environment. Most bacteria are heterotrophic which implies that they obtain their food and energy requirements for growth from organic compounds, and a few are autotrophic, and they synthesize organic molecules from inorganic carbon (CO_2). Fungi are heterotrophic organisms. They are also abundant in wetland environments and they play an important role in wastewater treatment. Most aquatic fungi are molds and are associated with the detritus and sediment layers, where they can degrade the dead organic matter. Fungi are ecologically important in wetlands because they mediate a significant proportion of the recycling of carbon and other nutrients. If fungi were inhibited through the action of toxic metals and other chemicals in the wetland environment, nutrient cycling of scarce nutrients would be reduced greatly limiting the primary productivity of algae and higher plants (Kadlec and Knight, 1996).

10.3.2 Terrestrial Systems

Phytoremediation can be used in the remediation of terrestrial systems contaminated with a variety of pollutants including metals, pesticides, solvents and various organic chemicals. There are over 400 different species considered suitable for use in phytoremediation which employ a number of mechanisms in combination to remove contaminants. The species selected to provide soil, sediment or groundwater remediation must tolerate the contaminant concentrations expected at the affected site. The specific remediation mechanisms involved vary depending on the constituents to be treated, site conditions, other environmental factors, and the remedial goals. In addition, several of these mechanisms may operate in series or in conjunction with each other.

In typical phytoremediation tree stand systems, the contaminated soils, sediments and groundwater are generally exposed to the root systems of the plants and trees. Therefore, these systems are able to employ plant as well as rhizosphere remediation processes. The remediation of sites contaminated with metals typically takes advantages of the ability of certain plant species to remove or stabilize contaminants via bioaccumulation, phytoextaction, rhizofiltration or phytostabilization. Conversely, organic contaminants including pesticides, solvents and lubricants are generally removed via phytodegradation, rhizodegradation and phytovolatilization (Evans and Furlong, 2003). In addition to tree stand systems, other types of terrestrial phytoremediation applications have been employed to achieve specific remediation objectives. These include vegetative cover, hydraulic barriers and riparian buffer systems.

10.3.2.1 Vegetative Cover Systems

A vegetative cover is an engineered phytoremediation system that is constructed as an alternative to a traditional vegetated cap, particularly at landfill sites. The main objectives of these cover systems are to intercept rain, minimize the downward infiltration of rainfall and snowmelt into the waste, protect groundwater by reducing leachate production, and reduce erosion from the surface. These covers employ the natural evapotranspiration process of shallow and deep rooted plants to control the infiltration of water into waste materials and create a zone from which the plants can extract pore water within waste materials, hence, exhibiting a form of hydraulic control by plants. They can be designed using mixed communities of plants and trees that maximize rain interception (Table 10.2) and transpiration capacities (Table 10.3).

Table 10.2 Typical plant rain interception capacities.

Plant Name	Plant Type	Magnitude and Duration of Rain	Interception Capacity
Natural Pasture	Mixed grasses	389 mm/5 months	14-19%
Alfalfa	Agricultural crop	Unspecified	36%
Tall Panic Grass	Prairie species	12.7 mm/30 min.	57%
Little Blue Stem	Prairie species	12.7 mm/30 min.	50-60%
Birch	Tree species	350 mm/5 months	10%
Ash	Tree species	38 mm/(Unknown time)	24%
Spruce-Fir	Tree species	272 mm/5 months	30%

Table 10.3 Typical plant transpiration rates.

Plant Name	Plant Type	Transpiration Rate
Perennial Rye	Typical lawn grass	6.9 mm/d
Alfalfa	Agricultural crop	10.5 mm/d
Common Reed	Wetland species	11.2 mm/d
Great Bulrush	Wetland species	21.9 mm/d
Sedge	Wetland/prairie species	48.2 mm/d
Prairie Cordgrass	Praire species	12.1 mm/d
Cottonwood	2-year old tree	2.0-3.75 gpd per tree
Hybrid Poplar	5-year old tree	20-40 gpd per tree
Cottonwood	Full, mature tree	50-350 gpd per tree
Weeping Willow	Full, mature tree	200-800 gpd per tree

As can be seen from Figure 10.9, during a precipitation event, the initial precipitation is collected on the surfaces of plant leaves based on their respective interception capacity (Table 10.2). Once the capacity of the species to intercept precipitation has been exceeded, any additional precipitation will fall to the ground surface where it will infiltrate into the soil or runoff. Once the precipitation has ended, the water intercepted on the plant leaves and within the root zone will be removed via evapotranspiration (ITRC, 2001). The water holding capacity of the soil and the waste penetrated by the root system provides the storage capacity necessary for infiltration of water during precipitation and the dormancy period of the vegetation (Carman and Crossman, 2001).

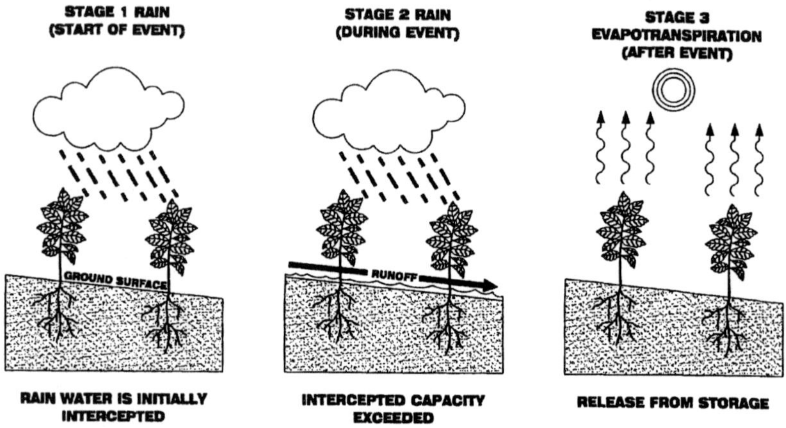

Figure 10.9 Schematic of an infiltration control from a vegetative cover.

In addition to the ability of plants to intercept precipitation and minimize infiltration, densely rooted groundcover plants and grasses have also been reported to enhance the biological breakdown of the underlying waste or contaminants through the enhanced biodegradation of organics in shallow surface soils. The primary mechanisms involved in this application are rhizodegradation and phytodegradation (Evans and Furlong, 2003). Halophytes and hyperaccumulator vegetation can be used to remediate shallow soils containing salts and metals, respectively. The primary mechanism involved in vegetative covers for inorganic contaminants is either phytostabilization or phytoaccumulation. Typical metals that have been treated with vegetative cover applications include Pb, Cd, Zn, Ni, Se, As, and Cu, as well as, Na, Mg and Ca chloride salts. The range of effectiveness for both the inorganic and organic remediation applications is generally 1 to 2 feet below the ground surface. However, depths down to 5 feet have been reported under some conditions (Olsen and Fletcher, 1999). In contrast, traditional caps do not enhance natural biodegradation processes and, in some instances, have been found to limit biological growth by altering biogeochemical conditions in the waste and decreasing natural attenuation processes (Carman and Crossman, 2001).

In general, vegetative covers are not appropriate for sites that produce landfill gas in chronic, large or uncontrolled amounts. However, they are porous and permeable to gases and can allow passive gas venting if necessary, while traditional landfill covers are essentially impermeable and may require elaborate gas venting systems (Carman and Crossman, 2001).

10.3.2.2 Hydraulic Barriers

In addition to using plants as a vegetative groundcover, wetland plants and deep-rooted tree species can be used to generate hydraulic barriers to control surface water and groundwater movements as well as physically stabilize the soil environment. For groundwater plume control, a relatively large number of deep-rooted, high-transpiring plants or trees are concentrated at the down-gradient edge of the plume. The basic premise behind this phytoremediation application is that the deep-rooted plants or trees access the groundwater and cause a local depression in the water table through uptake and transpiration This is illustrated conceptually in Figure 2.17. This depression is sufficient to prevent the migration of the contaminated groundwater plume beyond the boundary of the tree stand. The amount of groundwater that can be taken up by a tree stand is dependent upon the age of the trees, the depth of groundwater, the soil conditions, and the climatic region (ITRC, 2001). The hydraulic control exhibited by plants as a result of evapotranspiration, particularly by tree species, has also been applied to overcome localized water-logging, particularly on land used for agricultural or amenity purposes. The planting regime may involve the establishment of close tree stand groupings which then function as single elevated withdrawal points.

Other plant-based processes can also take place simultaneously to remediate the surrounding soil, sediments and groundwater. Organic contaminants are generally taken up by the plants at lower concentrations than they are found *in situ*, primarily due to membrane barriers at the plant root hairs. For those contaminants not taken up by the plant, the establishment of a hydraulic containment regime will increase the soil concentration of the contaminants as a result of evapotranspiration (Evans and Furlong, 2003).

10.3.2.3 Riparian Buffer

Riparian buffers are vegetated areas that protect adjacent water resources from non-point source pollution, provide bank stabilization, and habitats for aquatic and other wildlife. They are intended to minimize the migration of contaminants into water courses, and are typically used along the banks of rivers, or around the perimeter of affected sites. Riparian buffers can be established or maintained under a variety of land use patterns such as natural, agricultural, forested, suburban, and urban. A cross-sectional view of a typical riparian buffer is shown in Figure 10.10. Natural riparian buffers are composed of grasses, trees, or both types of vegetation. Poplar and willow species have been reported to be particularly effective in reducing the wash-out of nitrates and phosphates making them useful as control measures for surface water contamination from agricultural fertilizer residues along watercourses (Evans and Furlong, 2003).

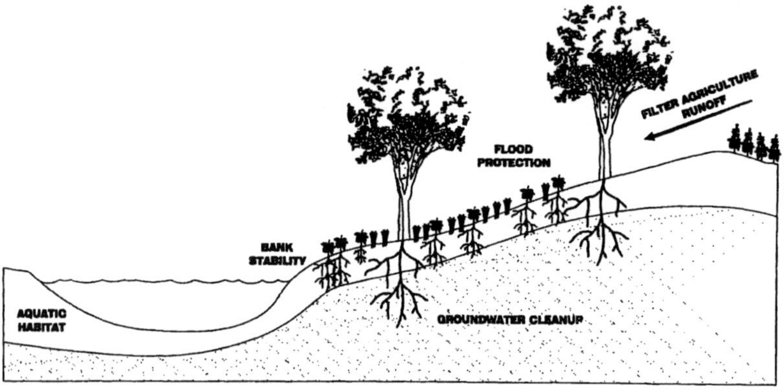

Figure 10.10 Schematic of a riparian buffer system.

Hydrology is the most important factor that determines the effectiveness of riparian buffers. Adequate width of the buffer is necessary to control the hydrology through the riparian buffer. While buffers that are too narrow may not be sustainable, and may not provide adequate pollution control or adequate stream bank protection, buffers that are wider than required may limit or impact land use. Phytostabilization, rhizodegradation, phytoaccumulation, phytodegradation, and phytovolatilization are the phytoremediation mechanisms responsible for contaminant mitigation in riparian buffers. The roots of the trees, shrubs and grasses in the riparian buffer provide an energy source for bacteria that promote the stabilization, accumulation, reduction, degradation, metabolism, or mineralization of the contaminants (ITRC, 2001).

10.3.3 Hydroponic Systems

Phytoremediation can also be applied as an *ex situ* technology. For instance, pump-and-treat systems could supply contaminated water as the influent into an area where plants are cultivated. As an alternative, the contaminated stream could be passed through artificial, planted systems known as hydroponic systems (Figure 10.11).

Hydroponic systems generally use an artificial soil medium, such as sand mixed with perlite or vermiculite. The plant species are planted into the artificial soil and the stream of water requiring treatment is passed through the system.

Plants can also be grown without an alternate soil medium, where the plants are raised with their roots directly in the flowing water stream and supported with a wire mesh or other physical structures. This phytoremediation method, also referred to as rhizofiltration, has only been applied to waters affected by inorganic contaminants. Depending on the concentration of contaminant, the flowrate of the contaminated water, and the area available for constructing these systems, the treatment of the contaminated stream may be conducted entirely above ground. The plants used for remediation are typically grown separately in greenhouses using a nutrient solution. Once the plants have developed large root systems, the contaminated water is diverted from one waste site and brought to the plants, or the plants are cultivated and transplanted to the contaminated site. When the roots or plants become saturated with contaminants, the plants are harvested and replaced with a new batch of plants. An advantage of these systems is that they can be operated throughout the year because the pants are grown in greenhouses.

10.4 Phytoremediation of Inorganic Contaminants

The fate of inorganic contaminants, such as metals and radionuclides, in phytoremediation applications is typically governed by the natural abilities of certain plant species to remove or stabilize the metals via phytoaccumulation, rhizofiltration, phytostabilization or, in some cases, phytovolatilization. Most inorganic contaminants have multiple chemical forms in soil, sediment and water matrices, and the chemistry of these contaminants and their mobility impacts their toxicity in the environment (Suthersan, 1997). The uptake of inorganic contaminants is dependent on their relative bioavailability in the contaminated matrix. Changes in soil chemistry such as the decrease in pH or the limiting of anion input into the system can increase the availability of many metals for root uptake. Changes in the microbial community of the rhizosphere can also have a significant impact on metal uptake by plant roots (Cunnignham and Lee, 1995).

10.4.1 Phytoaccumulation

Phytoaccumulation, also known as phytoextraction, refers to the mechanisms responsible for the uptake of metals and radionuclides from soil or sediments by plant roots followed by their translocation to the shoot and above-ground plant tissues. Many plants can accumulate metals essential for growth and development such as Fe, Mn, Zn, Cu, Mg, Mo and Ni, while others have the capacity to accumulate heavy metals with no known biological function such as Cd, Cr, Pb, Co, Ag, Se and Hg. Certain species termed hyperaccumulators, have the innate ability to accumulate large quantities (2-5%) of metals compared to most ordinary plants, at typically at concentrations that are 10 to 500 times higher than most crops. For example, the accumulation of metals such as Ni, Zn and Cu, observed in hyperaccumulators can reach 105% of their dry weight, an order of magnitude higher than concentrations found in non-accumulating plants (Garbisu and Alkorta, 2001). Hyperaccumulator plants usually have shoot to root metal concentration ratios of more than 1, while non hyperaccumulator plants generally have higher metal concentrations in their roots than in shoots (McGrath et al., 2001).

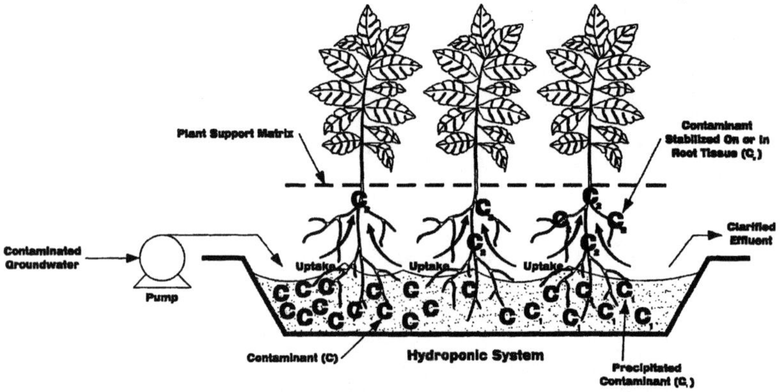

Figure 10.11 Schematic of a hydroponic rhizofiltration system.

Higher plants employ two basic strategies to tolerate heavy metals and radionuclides in their environment – exclusion, where uptake and/or root to shoot transport of metals are restricted, and accumulation where metals are accumulated and detoxified in the plant shoots. Metal exclusion is the most common strategy of metal tolerant species. Approximately 400 plant species have been identified as metal hyperaccumulators, which represent less than 0.2% of all angiosperms. Hyperaccumulator plants tend to be relatively small in size and have relatively slow growth rates. Approximately two-thirds of these are known Ni accumulators, while plant species able to hyperaccumulate Cd, Co, Cu, Pb and Zn account for a smaller fraction of these hyperaccumulator species (McGrath et al, 2001). Hyperaccumulators are by definition tolerant to the metals and radionuclides they accumulate in their shoots. These plant species can contain metals in their leaves and stalks at levels that are approximately 100 times those found in non accumulator plants growing in the same soil, with some metal species and metal complexes exceeding conventional plant concentrations by a factor of more than a thousand (Suthersan, 1997). For example, *Thlaspi caerulescens* has been reported to accumulate Zn up to 2,000 to 4,000 mg/kg. The Indian Mustard plant *Brassica juncea*, hem dogbane (*Apocynum sp.*) and common ragweed also have been noted to accumulate significant concentrations of Pb. *Aeollanthus subcaulis var. lineris* and *Paspalum notatum* have can accumulate Cu and Cs, respectively (Suthersan, 1997).

In their wild form, hyperaccumulator plants are often found in naturally metal-rich areas, where their unusual ability and adaptation in tolerating elevated metal concentrations, offers a selective advantage from an evolutionary perspective. Currently the best candidates for removal by phytoaccumulation are Cu, Ni and Zn, since these are the metals most readily taken up by the majority of hyperaccumulator plant species (Evans and Furlong, 2003). Phytoaccumulation can be applied to sites with contaminants down to a depth of 24 inches. If contamination is deeper, deep-rooted trees, such as poplars, should be used for the phytoaccumulation of the metal contaminants. These trees can accumulate metals by sequestration (Suthersan, 1997).

Heavy metals, generally have low solubilities in soil and sediment environments. There are several potential benefits to using chelates in chemically-assisted phytoaccumulation. Some of these benefits include increasing the solubility of metals in soils and sediments, overcoming the diffusional limitations of certain metals in the rhizosphere, and facilitating root to shoot translocation of the metal. The formation of metal-chelate complexes prevents precipitation and sorption of the metals thereby maintaining their availability for plant uptake. The addition of chelates to the soil or sediments can also bring metals into solution through desorption of sorbed species, dissolution of Fe and Mn oxides, and dissolution of precipitated compounds. Chemically-assisted phytoaccumulation is beneficial when the metal to be extracted, such as Pb, U or Au, is of very low bioavailability, thereby allowing the establishment of a large plant biomass prior chelate application. On the other hand, metals such as Cu, Zn and Cd, which are typically more bioavailable in soil and sediments, can cause phytotoxicity at concentrations requiring phytoremediation. The development of phytotoxic conditions prior to chelate application may significantly inhibit plant growth, and may thus limit the success of chemically-assisted phytoaccumulation. Another concern associated with chemically-assisted phytoaccumulation is the potential for the metal chelates to leach the underlying aquifer (McGrath et al., 2001).

Chelating agents that have been used as soil extractants include ethylenediaminetetraacetic acid (EDTA), hydroxyethylenediaminetriacetic acid (HEDTA), diethylenetriaminepentaacetate (DTPA), 1,2-cyclohexane diamino tetraacetate (CDTA), ethylene glycol tetraacetic acid (EGTA), Etilendiamin di(O-Hidroxypheyl acetic acid) (EDDHA), N-(2-Hydroxyethyl)iminodiacetic acid (HEIDA),

[S,S']-ethylenediaminedisuccinic acid (EDDS), nitrilotriacetic acid (NTA), N,N'-bis (2-hydroxybenzyl) ethylenediamine-N,N'-diacetic acid (HBED), citric acid, and malic acid (Alkorta et al., 2004). Plants also secrete to the rhizosphere natural metal-chelating molecules that can mobilize soil-bound metals. These include Fe-chelating phytosiderophores, and possibly metal chelating proteins related to metallothionins or phytochelatins. Some metals may be transported to the shoots complexed to organic acids such as citrate (Garbisu and Alkorta, 2001). Mycorrhizae have been shown to either reduce or enhance metal uptake by plants.

Once inside the plant, most metals become fairly insoluble as they form carbonate, sulfate or phosphate precipitates immobilizing them in extracellular and intracellular compartments. Unless the metal ion is transported as a non-cationic metal chelate, extracellular transport is further limited by the high cationic exchange capacity of cell walls (Raskin et al., 1997). After the plants have grown for an appropriate length of time, they are harvested and the accumulated metal is permanently removed from the contaminated site. It has been reported that harvesting done at different seasons in a year had pronounced differences in accumulation concentrations (Suthersan, 1997).

Case Study

A brownfield site in Trenton, NJ was the focus of a Superfund Innovative Technology Evaluation (SITE) demonstration project addressing an urban area contaminated from the manufacturing of lead acid batteries and other industrial activities between 1947 and 1987 (USEPA, 1999). The facility was abandoned since 1989. The soils at the site consisted primarily of gravelly sand and miscellaneous debris. Site investigations identified Pb concentrations in the upper 24 inches of soil exceeding a regulatory limit of 400 mg/kg. Lead contamination ranged from 200 to 1,800 mg/kg across the site. Forty percent of the lead contaminated soil exceeded 400 mg/kg and 7 % exceeded 1000 mg/kg. The soil pH varied significantly from 5.1 to 7.1, and 72% of the total soil lead was considered to be biologically available for plant uptake..

Phytoremediation was implemented on a 4,500 ft^2 area of land by Edenspace Systems Corporation. The phytoremediation demonstration project evaluated 3 crops grown in 30 x 57 ft^2 plots and compared to the results of those obtained from a 30 x 40 ft^2 control plot. Three crops of *Brassica juncea* (Indian Mustard) were grown during the 1996 season and each one was harvested after 6 weeks of growth. Harvested plant tissue samples were collected to evaluate the amount of Pb in each crop and the corresponding soil samples. The plants were dried and removed from the site. After the third crop of *Brassica juncea* had been harvested, areas that exceeded 400 mg/kg were reduced from 40% of the total area to 28%. None of the treated area exceeded 800 mg/kg and only 25 m^2 exceeded 600 mg/kg by the end of the growing season. The results of the study also demonstrated that the most significant impact of phytoremediation was in the top layer of soil (0-15 cm), in which the average lead concentration decreased from 429 to 373 mg/kg, resulting in a 13% reduction. Although most surface soil were generally below 400 mg/kg initially, in areas exceeding 600 mg/kg, phytoremediation had a substantial impact resulting in a decrease in lead concentration from 736 to 539 mg/kg (21%).

One section of the site was treated with 2 mol EDTA/m^2 in fall of 1996 to enhance the mobility and bioavailability of the Pb in soils. In the spring of 1997, soil samples were taken down to a depth of 120 cm to evaluate the effects of precipitation and snow melt from the previous winter, on the distribution of both lead and EDTA within the soil matrix. It was noted that the total soil Pb concentration decreased with depth, 878 mg/kg was found at the surface compared to 15 mg/kg at 120 cm depth. Soil samples also indicated that EDTA was primarily effective in the upper 30 cm of the soil.

Two crops of *Brassica juncea* plants were grown for a 6 week period and harvested over the spring and summer of 1997 and one sunflower plant crop was grown in the summer of 1998. The average lead concentrations accumulated in above-ground plant tissues samples obtained from the two *B. juncea* crops were 830 mg/kg and 2,300 mg/kg, respectively. The increase in concentrations of Pb in the above ground portion of the plants was primarily attributed to the chelating agent.

Case Study

In 2003, a field demonstration study for gold phytomining was carried out at the Fazenda Brasileiro mine in Bahia, Brazil (Anderson et al., 2004). The mine was active for 15 years, with underground and open pit operations producing 100,000–150,000 ounces of gold annually. The high market value of gold and relatively high concentrations found in spent heap leach pads and waste dumps make this metal an ideal candidate for phytomining. In order for gold phytomining to be a potentially viable process, the aim is to harvest a crop with a dry biomass of 10 t from 1 ha of land, and to extract a dry biomass gold concentration of 100 mg/kg, yielding approximately 1 kg of gold per hectare.

In the summer of 2003, a 15 m × 15 m × 0.5 m ore plot was constructed on top of a waste-rock pile at the Fazenda Brasileiro mine. The base of the plot was lined with a geo-textile and the plot filled with 150 tons of low-grade oxide ore, 50% of which was an unamended oxide ore, while the other 50 % was amended with 8 kg/ton of calcium hydroxide. Calcium hydroxide is added to the oxide ore during crushing to maintain the stability of the gold cyanide complex during heap leaching. Urea and a NPK fertilizer were applied across the plot areas. Five 0.5 m wide rows each of *B juncea* and *Zea mays* (corn) were seeded across both ore preparations and the plot was watered twice daily using a sprinkler irrigation system.

Species of *B. juncea* and *Z. mays* were tested for their ability to grow on an oxidized ore pile containing 0.6 g gold per ton of ore and to absorb mobilized gold after cyanide and thiocyanate treatment. A six-week growing cycle was used in the trial, but was found to be insufficient for the two species to attain maximum biomass. Under ideal conditions, *B. juncea* and *Z. mays* were expected to yield 10–15 tons/ha, and 30 tons/ha, respectively. The observed differences in recorded biomass between the unamended and $Ca(OH)_2$ amended plots indicated that phytotoxicity had occurred in the amended plot due to the high pH of the lime-amended ore.

Accumulated metal concentrations were significantly higher in *B. juncea* than in *Z. mays*. For *B. juncea*, NaCN had a greater impact on enhancing gold uptake than KCN, however, no such difference was apparent for the uptake of copper. The highest average gold assay achieved was 39 mg/kg for *B. juncea* with the unamended ore. Five individual *B. juncea* plants were harvested and it was noted that gold concentrations varied throughout the plant indicating that *B. juncea* translocates gold to all plant organs when using CN treatment. Higher gold concentrations were observed in the upper stem and leaf tissues of plant samples compared to basal biomass.

Variability in the plant-gold concentration was found to be a function of the viariability in the ore-gold concentration. An average Au concentration of 39 mg/kg was achieved in *B. juncea* after NaCN treatment of an ore containing 0.6 mg/kg gold. The highest individual Au concentration determined through the analysis of selected biomass was 63 mg/kg (NaCN treatment of *B. juncea*). Although a higher pH may lead to an increased stability of the Au-CN complex, increasing pH via lime amendment did not appear to enhance Au uptake and lead to reduced biomass production. Cyanide treatment showed limited toxicity to the plants and promoted the translocation of Au from plant roots to the uppermost stem and leaf organs of individual plants. Thiocyanate did not induce Au accumulation above 2 mg/kg.

Phytomining is currently not sufficiently competitive with conventional Au recovery processes such as heap-leaching from low-grade materials. However, for applications such as the reclamation of a spent heap-leach pile or for a small tonnage ore deposit, phytomining could be a useful approach to institute revegetation practices that would also recover residual gold in the pile. Should the method be found sustainable for continued application for 4 or 5 growth cycles, the generation of organic material in the pile will lead to better and more hardy vegetation on the surface.

10.4.2 Phytostabilization

The primary goal in phytostabilization is to reduce the mobility of contaminants, such as heavy metals and radionuclides, by minimizing soil erosion, decreasing the potential for wind-blown dust and decreasing contaminant solubility through the addition of soil amendments. The addition of vegetation at contaminated sites significantly reduces soil erodibility due to wind and water. The dense vegetative cover and extensive root system can effectively hold the soil in place and provide a stable cover against erosion (Suthersan, 1997). The ability of plants to transpire large volumes of water can also be applied to provide effective hydraulic control, thereby preventing the migration of contaminants or leachate offsite or toward the groundwater table or receiving waters.

Phytostabilization is similar to phytoextraction and rhizofiltration in that it makes use of processes of uptake and accumulation by adsorption onto, or precipitation around the roots of plants. One distinguishing characteristics of this phytoremediation application is that there is no harvesting of vegetation to permanently remove the contaminants from the site. In this sense, phytostabilization immobilizes contaminants, concentrating and containing them within a living system, rather than transforming them to less toxic compounds or removing them permanently from the site. The accumulation of heavy metals or radionuclides locks these contaminants within the plant biomass or within the rhizosphere, thereby reducing their bioavailability and preventing their migration off-site (Evans and Furlong, 2003).

The immobilization of heavy metals and radionuclides *in situ* can represent a favorable remediation alternative at sites with low contamination levels or vast contaminated areas where large scale conventional mitigation approaches or other *in situ* applications are not feasible. Phytostabilization is appropriate for radionuclide contaminated sites allowing the contaminants to be held in place, thereby reducing the risk associated with secondary contamination and exposure. This application can represent a particularly beneficial alternative if radionuclides concerned have relatively short half lives. Phytostabilization may be especially useful in controlling tailings from strips and open pit uranium mines (Dushenkov, 2003).

Another important factor in enhancing phytostabilization is the addition of amendments including a variety of alkalizing agents, phosphates, organic matter and biosolids (Table 10.4) to render the metals insoluble and unavailable for leaching (Berti and Cunningham, 2000). Amendments such as lime and gypsum are added to influence the pH of the soil. Binding conditions can be influenced by adding Fe, Mn, or Al compounds. To maintain or raise the organic matter content of the soil, materials such as peat, manure or mulch can be added. The amendments sequester the metals into the soil matrix, while the plants maintain the stabilized matrix in place (Suthersan, 1997).

Table 10.4 Soil amendments employed in phytostabilization.

Amendment Type	Amendments	Target Contaminants	Suggested Mode of Inactivation
Phosphate materials	H_3PO_4, apatite, calcium orthophosphates, Na_2HPO_4, KH_2PO_4, other phosphate fertilizers	Pb	Formation of insoluble metal phosphate minerals, such as Pb pyromorphites
Hydrous Fe oxides	Iron rich or other byproducts containing hydrous Fe oxides, isolated hydrous Fe oxides	As, Cd, Cu, Ni, Pb, Zn	Sorption of contaminants on oxide surface exchange sites, co-precipitation, or formation of contaminant-Fe compounds
Organic materials	Manures, composts, sludges, and other biosolids	As, Cd, Cu, Pb	Sorption of contaminants on exchange sites, or incorporation into the organic material
Inorganic clay minerals	Synthetic zeolites, natural aluminosilicates, or aluminosilicate byproducts from burning of coal refuse	As, Cd, Cu, Mn, Ni, Pb, Zn	Sorption of contaminants on mineral surface exchange sites, or incorportation into the mineral structure

Case Study

The Palmerton Zinc Superfund Site is a former zinc smelting operation located in Palmerton, PA (Li and Chaney, 1998). Zinc smelting began in 1898 and the plant operated continuously until 1980. During production, it is estimated that an average of 3,575 tons of Zn and 47 tons of Cd were emitted each year over an 80-year span. It is reported that these pollutants were carried as far as 30 miles from theoperation. In the area surrounding the smelting operation, soil concentrations of Zn and Cd range from 26,000 to 80,000 mg/kg and 900 to 1,500 mg/kg, respectively. The combination of high soil Zn and acidic soil pH created a relatively phytotoxic soil environment, leading to the absence of a vegetative cover, which facilitated wind and water erosion of metal-contaminated particles.

A four-year study was undertaken at the site to investigate effects of applying limestone and Fe-rich limed biosolids compost to the soil on plant growth and metal accumulation. A site in the vicinity of the Palmerton smelter was selected. Soil Zn and Cd concentrations of 12,650 mg/kg and 880 mg/kg were measured at a depth of 15 cm. A metal tolerant turfgrass species Merlin red fescue (*Festuca rubra L.*) was compared to three non-tolerant turfgrasses, Bonanza tall fescue (*Festuca arundinaceae Schreb.*),

Touchdown Kentucky bluegrass (*Pao pratnesis L.*) and Kentucky-31 tall fescue (*Festuca arundianceae Schreb.*). Eight tons/ha of limestone and 224 tons/ha of compost by dry weight were applied to the treatment plots six months prior to planting in the first year of the experiments. Control and limestone plots were fertilized with NPK fertilizer at a rate applicable for the establishment of a new lawn. Three months after planting, the three non-tolerant turfgrass species showed significant signs of enhanced growth in the limestone treatment compared to that in control plots. However, the three species gradually died and did not become reestablished after the first growing season. Throughout the study, the percent of organic matter, nitrogen level and cation exchange capacity in the compost treated plots were consistently higher than in the limestone and control treatments. The soil pH in the compost treated pots also became slightly alkaline. Dissolved Zn and Cd concentrations in the compost treatment were lower than in untreated soil indicating that the compost treatment was effective at decreasing Zn and Cd phytotoxicity.

Only the Merlin red fescue was found to survive on the high Zn soil in the Palmerton area. However, the leaf color was fairly light compared to the dark green leaf color observed in the compost treatment vegetation suggesting that growth of the metal tolerant species on untreated Palmerton soil was affected by the heavy metal concentrations. Merlin grown on treated and untreated soils demonstrated a remarkable ability to exclude Zn and Cd. Significant differences in shoot Zn and Cd concentrations were noted for Merlin red fescue depending on the soil treatments. Shoot tissues of Merlin exhibited significantly higher Zn and Cd concentrations in control soils than in compost treatments over a three-year period. The compost application was effective at decreasing Zn and Cd phytotoxicity, and all turfgrasses exhibited excellent growth in the compost-treated plots. The Bonanza and Kentucky-31 tall fescue species grew rapidly and covered almost 100% of the soil within three months in the compost treatment. On the other hand, Zn-tolerant Merlin plants had a slow seedling emergence and slow growth in both treated and untreated plots.

Case Study

An 18-mile stretch of Whitewood Creek was contaminated with heavy metals, primarily As and Cd as a result of 130 years of gold mining near Whitewood SD (Pierzynski et al., 1994). Average concentrations of 1250 mg/kg total As and 9.4 mg/kg fo total Cd were measured in the tailings from the gold mine, and the pH of the tailings ranged from 3.9 to 5.4. In 1991, an experimental plot of 3,100 hybrid poplar trees was established along Whitewood Creek, to vegetate the mine tailings to reduce wind-blown dust and minimize the potential vertical and lateral migration of As and Cd into nearby Whitewood Creek. A commercial fertilizer was employed at recommend rates to ensure a vigorous growth of cuttings. To determine the concentration of As and Cd in the leaf, stem and roots of the poplars, samples were collected at the end of the first growing season.

In a parallel study, cuttings were grown in a laboratory plant incubator, planted in various mixtures of tailings with 100%, 50% and 0% tailings, and amended with Hoagland R growth fertilizer. The 50% mixture was grown in tailings and vermiculite (50:50 w/w), while the 0% mixture was composed solely of peat. The poplar cuttings grew in all three mixtures, however cuttings rooted in 100% tailings grew more slowly than those planted in the 0% and 50% treatments.

The trees in the experimental field plots along Whitewood Creek exhibited considerable growth during the first season. Samples of leaves, stems and roots collected from the field and the laboratory studies were used to compare the translocation and uptake of As and Cd in the trees. Samples of poplar leaves indicated that the vegetation in the experimental plot did not accumulate significant amounts of As or Cd, and that the accumulation rates noted in the field were generally lower than the rates observed in the laboratory trees. Concentrations of As and Cd were also measured in native vegetation and were reported to be on the same order of magnitude as those measured in the poplar trees. However, the As concentration was higher in leaves of lambsquarter (14 mg/kg) and native cottonwood (1.6 mg/kg) than in trees in the experimental field plot. Field and laboratory investigations demonstrated that the tailings could be vegetated and stabilized with hybrid poplars.

10.4.3 Rhizofiltration

Rhizofiltration is a process by which inorganic contaminants present in the soil water are absorbed into, or adsorbed or precipitated onto plant roots. The main difference between this application and phytostabilization is that rhizofiltration is generally employed to mitigate contaminants in the groundwater rather than within the soil matrix. The plants used in this application are typically grown hydroponically and gradually acclimatized to the specific characteristics of the water requiring treatment. Once the growth of the plants has been established, they are transplanted on the contaminated site. Harvesting may take place once the plants have become saturated with contaminants.

Many interactions have been reported between contaminants and plant roots with the possible metal removal mechanisms including extracellular precipitation, cell wall precipitation and adsorption, and intracellular uptake followed by vacuolar deposition or cytoplasmic compartmentalization. Intracellular uptake of inorganic contaminants may employ the same mechanisms as those responsible for the uptake of essential ions including membrane transport systems which fail to discriminate among different ions of similar ionic radii and ionic charge such as aqueous pores, ion efflux pumps, ion selective channels, and proton-anion contraports. Phytochelatins are the primary metal binding polypeptides of plants that are constitutively expressed and play a central role in the accumulation and/or detoxification of excess metals in plant cells. This phenomenon can further assist rhizofiltration applications as it increases metal specificity to binding domains in the plant roots. Plants may contain several phytochelatins to increase binding capacity of several metal ions. A number of metals are known to induce phytochelatin synthesis, however, metal binding by phytochelatins has only been demonstrated for Cu, Zn, Pb, and Cd (Dushenkov and Kapulnik, 2000).

The biological filter formed by active plant roots that have a large surface area is very effective in sorbing inorganic contaminants from water. Plants selected for rhizofiltration should exhibit characteristics that provide the maximum metal removal from a contaminated stream in conjunction with producing minimal secondary waste requiring disposal. The plants used in rhizofiltration applications should produce a significant root biomass or surface area which can be grown hydroponically, and the plant roots should be able to accumulate significant amounts of the target contaminant and tolerate it in high concentrations.

Case Study

Uranium (U) contamination of groundwater poses a serious environmental problem in uranium mining areas and in the vicinity of nuclear processing facilities (Dushenkov et al., 1997). In laboratory experiments, four plant cultivares were tested for their ability to remove U from the Ashtabula site water with concentrations of 56 g/L. Sunflower (*Helianthus annuus* L.) cv. SF-187 and cv. Mammoth giant, beans (*Phaseolus coccineus*), and Indian mustard (*B. juncea*) seedlings were cultivated hydroponically in an aerated nutrient solution. Once established, the plants were exposed to 750 mL of Ashtabula site water (containing $UO_2(C_2H_3O_2)_2$, $SrCl_2{\cdot}6H_2O$, and $CsNO_3$) in plastic jars. Both sunflower cultivars were observed to remove more than 95% of U from the solution in a 24 hour period, while beans and *B. juncea* were less effective in U removal. After 48 h of treatment, however, the beans had also reduced U concentrations to the proposed drinking water limit of 20 mg/L. Analysis of U distribution in the sunflower (cv. SF-187) demonstrated that more than 99% of the metal was concentrated in the roots and the quantity of U transported to shoots was negligible.

Growth chamber experiments were conducted to determine metal removal from water by sunflower roots. Four-week-old sunflower plants were used to treat 750 mL of water that containing 200 mg/L Cs, 200 mg/L Sr, or 600 mg/L U. The concentrations of Cs, U and Sr were reduced considerably within a few hours. Each metal exhibited different removal kinetics, and the sunflower roots were most effective in removing U. The U concentration decreased 10-fold to 63 mg/L in 1 hour and after 48 hours, an equilibrium of 10 mg/L was reached. A significant reduction in Cs concentration was noted within 6 hours, while after 24 hours almost all of the Cs had been removed, yielding a concentration of less than 3 mg/L. Strontium concentration was reduced to 35 mg/L within 48 hours and continued to decline for the following 48 hours, reaching a level of 1 mg/L.

Miniature batch rhizofiltration systems were used to determine the impact of U concentration and solution pH on U concentrations in plant tissues. Sunflower plants were grown hydroponically for 3 weeks with the plant roots placed on floating Styrofoam platforms so that the roots were exposed to the U solution. The 3-week old sunflower plants were then transferred to miniature batch rhizofiltration systems containing U concentrations of 10, 30, 90, 810, or 2430 mg/L. The experiment was conducted simultaneously at pH levels of 5.0 and 7.0. The plants were grown for one week in the U solution. In most cases, uranium levels in the sunflower shoots remained below 2 mg/g dry weight. The U concentrations in roots were found to increase linearly with increasing solution concentration. Sunflower roots adsorbed more U at pH 5 than at pH 7. A bioaccumulation coefficient, expressed as the ratio of metal concentration in the plant tissue (mg/kg DW) to the metal concentration in the solution (mg/L), was computed. The bioaccumulation coefficient for sunflower shoots was low, while the sunflower roots concentrated U from the solution up to 10,000-fold. The lower pH resulted in a significant increase in U bioaccumulation, where in the concentration range of 10 to 2,430 mg/L, the average bioaccumulation coefficient for U in sunflower roots were 6,624 ± 870 and 3,379 ± 430 for pH 5 and pH 7, respectively.

This rhizofiltration application was tested using a field-scale system with U contaminated water in a greenhouse constructed at a former uranium processing facility in Ashtabula, OH. The U concentration in the water fluctuated from 21 to 874 mg/L, with ^{238}U as the major isotope and pH ranging from 6.8-7.9. Plants were grown hydroponically using 33.5 × 43.5 cm^2 flats with a 2 cm deep feeder layer of Pro-Mix for 6 weeks. Well-established plants were moved to the rhizofiltration system and were used for 2 weeks for water treatment. A surge tank was used for the pretreatment of the source water. The water temperature was maintained at 24 C, and the pH was automatically adjusted to 5.5 ± 0.3. The system consisted of two skids. Each skid contained five sequentially connected tanks of 150 L each. To evaluate the performance of the field-scale system, different initial flow rates were tested. The field-scale rhizofiltration system reduced the U concentration from several hundred micrograms per liter to below the regulatory discharge level of 20 mg/L. The highest flow rate yielding discharge U concentrations below 20 mg/L was found to be approximately 1 L/min. The reduction in soluble U concentrations was correlated with an increase in U concentration in the roots. Under continuous flow, roots were able to accumulate more than 1.0% of U on a dry weight basis, producing a bioaccumulation coefficient of 30,000.

Results of the field-scale rhizofiltration operation showed that, from an average influent U concentration of 207 mg/L, the effluent concentration was generally reduced below 20 mg/L. Several spikes with the influent U concentration over 500 mg/L did not significantly affect rhizofiltration system performance. The greenhouse experiments and performance of the field-scale rhizofiltration system clearly demonstrated that it is possible to use hydroponically grown terrestrial plants to remove uranium from contaminated water. Using the rhizofiltration system based on sunflower plants, a total of more than 200,000 L of water was treated at Ashtabula, producing U concentrations below the regulatory discharge level of 20 mg/L.

10.4.4 Phytovolatilization

Phytovolatilization applications make use of the ability of some inorganic contaminants such as Se, As, and Hg to be biologically methylated forming volatile compounds that can subsequently be lost to the atmosphere. Removal pathways for phytovolatilization depend on the nature of the contaminant. For instance, both microorganisms and plants have been shown to play an important role in the volatilization of Se from soils. Some plants can remove Se from the soil by accumulating nonvolatile methyl selenate derivatives in their foliage. Others remove Se in the form of methyl selenate via volatilization. Mercury is a liquid at room temperature and can readily be volatilized. Bacteria can catalyze the reduction of the mercuric ion to elemental mercury using mercury reductase (Raskin et al, 1997).

Phytovolatilization of radionuclides relies heavily on the ability of the plant to transpire large quantities of water and has been investigated for radionuclides such as tritium (^3H). Tritium is a radioactive isotope of hydrogen and shares physical and chemical properties of hydrogen. It has a half-life of approximately 12 years and decays to a stable helium. Tritium is typically found in the environment as tritiated water that is integrated in the biogeochemical water cycle. It produces minimal external radiation exposure because it is a weak beta emitter and is easily shielded by air and skin. However, its absorption into the body via water and organic compound incorporation can present a health hazard (Dushenkov, 2003).

Case Study

Tritium released from nuclear facilities such as the Savannah River Site, Brookhaven National Laboratory, and other facilities has led to an increased interest in alternative remediation applications for the decontamination, isolation or removal of tritiated surface and ground waters (Negri and Hinchman, 2000). Throughout the years, the experience acquired at the Savannah River Site indicated that a 40% reduction in tritium concentrations could be achieved by releasing tritiated water to the atmosphere as water vapor rather than allowing it to flow offsite in surface water streams. This suggests an alternate ^3H removal method that may be cost-effective in minimizing ^3H exposure risks to public health.

At the Savannah River Site, tritiated water was applied as irrigation water to the forest floor as part of an engineered system designed to reduce the discharge of ^3H from the Burial Ground Complex southwest plume into Four Mile Branch. The system was considered to be a combination of enhanced evapotranspiration and hydraulic control. Tritium contaminated water was collected, relocated upgradient from the discharge point, and employed to irrigate plants. In this design, the tritiated water entered the hydrological cycle as it reached the soil on the forest floor. The water then either evaporated from the soil surface, was absorbed into the plant transpiration stream by plant roots, or was drained through the soil to the groundwater table. A small portion of ^3H that was absorbed by plant roots remained in the plant tissues in the form of readily exchangeable hydroxyl ions or incorporated into organic molecules through photosynthesis.

Case Study

Selenium (Se) is a metalloid that exists in a variety of oxidation states including selenide (Se^{2-}), elemental Se (Se^0), selenite (Se^{4+}), selenate (Se^{6+}), and organic and volatile Se compounds. Selenite and selenate, the oxidized forms of Se, are soluble, biologically available and potentially toxic. The selenide and elemental Se are insoluble and less biologically available. There are three potential fates for soluble Se entering a wetland system: (1) chemical reduction and immobilization in the sediments, (2) absorption by plants and accumulation in root and shoot tissues, and (3) volatilization by plants or microbes to the atmosphere.

The use of wetlands for the mitigation of Se from wastewater was initially suggested in a study of a 36-ha surface flow constructed wetland situated near the mouth of the San Pablo Bay, one of the five basins of the adjacent San Francisco Bay estuary system in California (Hansen et al., 1998). Analysis of the wetland inlet and outlet waters showed that the constructed wetland was successful in removing at least 70% of the Se from the wastewater. A field study was then conducted to determine Se partitioning among the sediments, water, and plant material. The results from the study accounted for 70% of the Se removed by the wetland, while the fate of the remaining 30% Se could not be identified. One explanation was that the Se may have been converted to volatile forms and released into the atmosphere. Both plants and microbes have been reported to take up Se in the form of selenate or selenite and metabolize it to volatile forms such as dimethyl selenium in a process referred to as biological volatilization. This form of Se removal presented an attractive alternative as it minimized the entry of Se into the food chain. Volatilization mainly takes place in the root zone and most of the volatilized Se is released directly into the atmosphere, minimizing plant-bound Se. In addition, most of the volatile forms of Se are relatively nontoxic.

In 1995, a subsequent 16-week investigation was undertaken to assess the potential importance of biological Se volatilization as a means of Se removal by constructed wetlands and to determine how effectively the 36-ha San Francisco Bay constructed wetland removed Se from an oil refinery effluent. The investigation focused on the first 12-ha section of the wetland, where oil refinery wastewater was supplied to the wetland. The oil refinery wastewater flowing into the constructed wetland contained 20-30 mg of Se L^{-1}; the water exiting the wetland contained < 5 mg/L of Se. The data indicated that nearly all of the Se in the inlet and outlet waters was in the dissolved form, primarily selenite. The amount of Se removed daily over the course of the study by the wetland was 156 ± 13 g per day, which translated to an 89.4 ± 0.6% removal expressed as a percentage of the mass per day of Se delivered in the inflow. Over the first half of the study period, most of the Se removed by the wetland was mitigated in the first 12-ha section of the wetland, which became progressively less effective from the ninth to the 16th week. The decrease in the effectiveness of Se removal was attributed to a seasonal decline in plant uptake and microbial reduction of Se associated with cooler temperatures in the fall.

Selenium concentrations were measured in roots and shoots, sediments, and sediment pore water. In sediments extracted from saltgrass and rabbitfoot grass sites approximately Se concentrations of 2 mg/kg were typically observed, while values of approximately 6 mg/kg were noted in the sediments of bulrush, cattail, or brass buttons sites. In most instances, Se concentrations were substantially higher in roots than shoots of all the species tested. Se concentrations in sediments and plant tissues were generally found to be orders of magnitude higher than those of the sediment pore water. The concentrations of Se in shoot tissues did not appear to differ significantly among species, typically ranging from 5.2 mg/kg in bulrush to 7.9 mg/kg in rabbitfoot grass.

The study demonstrated that biologically mediated volatilization may have accounted for as much as 10-30% of the Se removed. The highest mean rates of Se volatilization for vegetated sites were 190, 180, and 150 mg/m^2-day for rabbitfoot grass, cattail and saltmarsh bulrush, respectively. Rates for the dominant species, saltmarsh bulrush, fluctuated throughout the year with mean rates of volatilization of 150, 70, and 25 mg/m^2-day in February, June, and October, respectively.

One concern regarding the use of constructed wetlands for Se mitigation is that Se may build up in the wetland and reach toxic concentrations. One wetland management approach which can potentially address this issue is the periodic harvesting and removal of Se-containing biomass. Shoot removal may have the additional benefit of enhancing Se volatilization. Throughout the study, volatilization rates were measured over the saltmarsh bulrush-populated sites from which bulrush shoots were removed. Volatilization from bulrush plants harvested in September and October was measured continuously for 7 days immediately following harvesting, and it was noted that the harvested sites exhibited two to four-fold increases in volatilization.

10.5 Phytoremediation of Organic Contaminants

Gasoline and diesel oil are among the most widespread organic contaminants in soils and groundwater. Plants can absorb contaminants from the soil and detoxify xenobiotic compounds either via metabolic processes converting these into nontoxic byproducts or by incorporating them into stable cell components. However, despite their ability to metabolize some xenobiotics compounds, plants are generally constructive and not destructive organisms. Therefore, when compared to microbial metabolism, plants tend to act on a narrower range of substrates, perform simpler degradation steps and generally do not mineralize xenobiotic compounds (Cunningham and Lee, 1995). The interaction between the plants and their roots produces a soil environment rich in microbial activity that can change the availability or enhance the degradation of some organic contaminants. The primary mechanisms of organic contaminant phytoremediation are considered to be phytodegradation, rhizodegradation, and phytovolatilization (Evans and Furlong, 2003).

The fate of organic contaminants in the rhizosphere and the tendency for these contaminants to be taken up by plants can be predicted using Brigg's Law and information about the octanol-water partitioning coefficient (K_{ow}) of the contaminant. Moderately hydrophobic organic constituents ($0.5 < \log K_{ow} < 3$) such as volatile organic carbons, chlorinated solvents, and aliphatics are efficiently removed by direct plant uptake. Organic contaminants with $\log K_{ow} < 1$ are very water soluble and plants do not generally accumulate these compounds at a rate beyond passive influx in the transpiration stream. Compounds with high water solubility are generally considered mobile in both the plant xylem and phloem (Cunningham and Lee, 1995). Constituents with $\log K_{ow} < 0.5$ such as methyl tertiary butyl ether and 1,4-dioxane are considered to be too water soluble to be taken up by plant roots (Carman and Crossman, 2001). Less water soluble compounds with $\log K_{ow}$ between 1 and 4 can be taken up by plant roots and are generally considered mobile in the xylem, but immobile in the phloem. These types of compounds possess the physical properties which permit translocation in the soil zone, root uptake, and the potential to be metabolized by plants, making them targets for phytoremediation (Cunningham and Lee, 1995). Lipophilic constituents are not typically translocated to the shoot to a significant extent. Hence constituents with $\log K_{ow} > 3$, such as polycyclic aromatic hydrocarbons are generally considered to be too large, relatively immobile, and not suitable for plant uptake (Carman and Crossman, 2001).

10.5.1 Phytodegradation

In phytodegradation, organic contaminants are readily taken up from the soil or water, metabolized into plant tissues, and transformed to less toxic or non-toxic metabolites and by-products either internally via various metabolic processes or externally using enzymes secreted by the plants. The existence of extracellular enzyme pathways has allowed this application to be successfully employed for the remediation of chemicals such as chlorinated solvents, explosives and herbicides (Evans and Furlong, 2003). The overall metabolic process involved in phytodegradation is analogous to human metabolism of xenobiotic compounds, thus, a geen liver conceptual model is often employed to describe phytodegradation (Sandermann, 1994).

In plant tissues, contaminants are transformed via three biotransformation and compartmentalization phases involving a number of chemical and biochemical metabolic reactions. Phase I and II are considered to be biotransformation reactions, where Phase I involves the oxidation of the xenobiotic constituent and Phase II deals with the conjugation of Phase I products. The Phase II conjugation system is considered to be a xenobiotic constituent detoxification process, which involves metabolic activity and the accumulation of metabolites in the cells which could potentially lead to a reduction in Phase II detoxification activity. In Phase III, active processes catalyzed by membrane-bound ATP-driven pumps sequester toxic metabolites and contaminants in plant cells into vacuoles (Komive and Gullner, 2000).

Case Study

The Iowa Army Ammunition Plant (IAAP) located in Middletown, Iowa is a 20,000 acre site (31.25 mi^2) of which 8,000 acres are used for both munitions production and testing (Best et al., 1997a; 1997b; 1998). In 1948, a concrete impoundment was constructed near Line 1 to contain explosives production wastewater in a lagoon having an area of about 3.6 acres. The five-acre Line 800 Pink Water Lagoon Area was surrounded by an earthen berm and at one time was also used for sludge disposal. The lagoon basins and underlying groundwater had been contaminated with 2,4,6-trinitrotoluene (TNT) and hexahydro-1,3,5-trinitro-1,3,5-triazine (RDX) as well as many other compounds and heavy metals. The action levels were determined by the explosives contamination. Soil action levels were set at 47 mg/kg for TNT and 1.3 mg/kg for RDX. Action levels for aqueous concentrations were set at existing EPA health advisory lifetime levels of 2 mg/L for both TNT and RDX.

Preliminary studies investigating the potential of phytoremediation and active remediation at the site were conducted as a coordinated effort by the U.S. Army Corps of Engineers, Waterways Experiments Station, and the University of Iowa. Excavation was used initially to remediate the most highly contaminated areas. Approximately

100,000 cubic yards of soil were excavated from the Line 1 (24,000 yd^3) and Line 800 (76,000 yd^3) areas in early 1997. Following excavation, the potential impact on the local groundwater supplies was expected to have been greatly diminished and the U.S. Army Corps of Engineers anticipated a level of residual contamination equal to or less than the action levels. As a precaution, residual contamination requires further treatment. Studies were then conducted to assess the potential use of phytoremediation with poplars as a final polishing step for the soil and use of wetland systems to treat residual contamination in surface water and groundwater. The process of selecting the wetland treatment approach and obtaining construction and operational approvals included many steps and investigations. A full-scale wetlands treatment system is currently in place at the IAAP site.

Research was conducted on aquatic systems at Waterways Experiments Station in Vicksburg, Mississippi (Best et al., 1997; Larson, 1997; Best et al., 1998). These screening studies also investigated the potential of phytoremediation, from the perspective of aquatic-based systems. The goals of the research were as follows:

- quantify the ability of aquatic and wetland plants to treat TNT and RDX containing surface water or groundwater;
- investigate the transport and fate of the explosives in plants;
- determine the degradation by-products and their distribution in sediments; and
- select potential plant species for phytoremediation of the site.

From the initial wetland research involving the screening of plants, a number of species were determined to have the capability of removing TNT and RDX from aqueous systems. Plants that exhibited the greatest potential for use in wetland systems were the submerged plants cottontail and American pondweed, as well as the emergent plants common arrowhead, reed canary grass, and fox sedge. TNT removal occurred at a higher rate than RDX removal. In laboratory studies, TNT was not detected in plant tissues and TNT degradation products were below detection limits. RDX was found to bioaccumulate in a number of plants, reaching concentrations up to 1000 mg/kg on a dry weight basis (Best et al., 1997). RDX had also been detected in plant tissues from the IAAP site prior to the initial excavation in arrowhead roots (7 mg/kg) and in reed canary grass shoots (10 mg/kg) (Schneider et al., 1995). The detection of RDX in plant tissues was a concern going into implementation at the IAAP.

Following excavation of the Line 1 and Line 800 sites at the IAAP, the excavation pits were converted to constructed wetlands. The wetlands were considered to be a means to remediate the collected groundwater, provide ecological enhancement to the area, and avoid the costs associated with backfilling of the area with clean material. Sediment/seed material from nearby Stump Lake was used in each wetland to introduce indigenous species. Following the addition of the sediment, the wetlands were allowed to fill with the infiltrating groundwater.

Desorption of RDX from the contaminated soils tainted the groundwater that filled the wetland resulting in a rebound effect in the RDX concentration. In January 1998, the Line 1 wetland RDX concentration reached 760-830 ppb. RDX concentration declined in the wetland over the following months, reaching 1.6 ppb in August 1998. The following December, RDX reached 22 ppb in the Line 1 wetland. However, the concentrations were < 0.25 ppb the following summer. Water from the wetland was only released when RDX concentrations were < 2 ppb action level. Seasonal fluctuations in RDX concentrations were also observed in the Line 800 impoundment wetland. RDX levels in the Line 800 wetland reached 190-760 ppb in January 1998. RDX concentrations were below detection limits during the following summer months, and rebounded to 1.9 ppb in November 1998.

Possible reasons for the observed RDX increases in concentration during the winter month include decreased plant activity and biomass, decreased rates of enzymatic reactions, decreased physical-chemical reaction rates, decreased photolytic reactions as incident sunlight decreases, and/or increased concentration in the infiltrating groundwater. Residual soil and sediment concentrations of RDX were greater than 1.3 mg/kg soil, suggesting that this residual concentration could contribute to groundwater and surface water concentrations greater than 2 mg/L.

Bioaccumulation of RDX in plant tissues was a concern at the site. Thus, a sampling program of the plant material was undertaken in June 1999. Duplicate testing was conducted by the environmental engineering laboratories at the University of Iowa in Iowa City (Just, 1999) and the testing laboratories at the U.S. Army Corps of Engineers Waterways Experiments Station, Vicksburg Mississippi (Larson, 1999). The testing revealed that RDX and TNT were both below detection limits (< 1 mg/kg) in the tissues of the plant species tested, which included reed canary grass, arrowhead water plantain, and coon tail. The compound 1,3,5-trinitrobenzene, a photodegradation product, was detected in some plant tissues at < 30 mg/kg, which suggested that reactions other than those involved in plant-mediated processes contributed to the degradation of the explosives.

Case Study

It is believed that the dehalogenase enzyme, which oxidizes alkanes, alkenes, and methanes and their halogenated analogues, is involved in the oxidation of TCE to various metabolites. Dehalogenase ultimately mineralized TCE to carbon monoxide via an oxidative pathway. Some species or hybrids or poplar can produce more of this enzyme than others, and trees that generate large quantities of the enzyme generally have the ability to degrade TCE, while plants that produce this enzyme in smaller quantities tend to volatilize TCE.

The Aberdeen Proving Grounds in Maryland began serving as a U.S. Army weapons testing facility in 1918 (Chappell, 1997). Military weapons testing and past disposal activities resulted in the contamination of the soil and groundwater of the Proving Grounds. The J-Fields Toxic Pits Site had been used for many years as an open pit burning facility for munitions and chemical agents, which discharged large volumes of various chlorinated solvents. A plume of chlorinated solvents formed in the aquifer below the burning pits as a result of these activities. Concentrations of total VOCs in the groundwater range from < 20,000 μg/L in some areas to > 220,000 μg/L in others.

Several mitigative technologies were considered for cleaning the soil and groundwater at the site and phytoremediation was selected as the most applicable and cost effective remediation strategy. Phytoremediation application efforts at the J Fields site were initiated in September of 1995 and included a phytotoxicity assessment of the contaminants found on-site and a determination of potential nutrient deficiencies that would hinder tree growth. One hundred and eighty-three hybrid poplars (*Populus trichocarpa x Populus deltoids),* approximately one acre of trees, were planted over the areas of highest contaminant concentrations surrounding the leading edge of the plume in March and April of 1996, in an effort to provide treatment to the contaminated groundwater and hydraulic containment. The trees cost about $80/tree to install for a total of approximately $15,000 for 183 trees.

A plastic pipe was fitted around the upper roots of each tree to promote growth down to the saturated zone. To provide oxygen to the deeper root zone, a long piece of rubber tubing was inserted from the surface to the deeper roots and a drainage system was designed to redirect rainwater, thus, encouraging the trees to seek the contaminated groundwater as their primary source of water. Extensive monitoring of contaminant concentrations in the soil, water and plant tissues, as well as sap flow and transpiration rates, was undertaken to determine the fate of the contaminants. On a daily basis, the sap flow rate data demonstrated that maximum flow occurred throughout the morning and tended to increase with increasing solar radiation. In May of 1997, groundwater monitoring data indicated that the trees were effective at pumping large quantities of groundwater, generating a water table depression of approximately 2 feet beneath the trees compared to the data collected in April of 1996. Tree tissue samples indicated the presence of trichloroacetic acid (TCAA), a breakdown product of TCE. Evapotranspiration of chlorinated solvents, TCE and 1,1,2,2-tetrachoroethane was also reported at the site.

10.5.2 Rhizodegradation

The phytoremediation process by which contaminants are transformed by microbes in the rhizosphere of the plant-soil system is referred to as rhizodegradation. It is recognized as a much slower process than phytodegradation (Evans and Furlong, 2003). The oxygen concentrations, soil redox conditions, moisture content, root exudates and other soil properties in the rhizosphere can be manipulated by plant root activity, and consequently support a high microbial biomass capable of degrading a variety of compounds. The rhizosphere microbial community consists of microorganisms with different types of metabolisms and adaptive responses to variations in environmental conditions. Mycorrhizal fungi are typically associated with the plant roots and assist soil microorganisms in the biodegradation of organic contaminants due to their unique enzymatic pathways.

Rhizodegradation is responsible for the enhanced removal of organic contaminants with log $K_{ow} > 3$ such as TPH, PAHs and diesel oil as these are often considered to be too large and relatively immobile for plant uptake. The fate of many of these organic contaminants in the environment is associated with both abiotic and biotic processes, including chemical oxidation, bioaccumulation and microbial transformation (Singh and Jain, 2003).

Case Study

The Craney Island Fuel Terminal is the Navy's largest fuel storage facility in the United States, consisting of over 1,100 acres of underground and aboveground fuel storage tanks (Bank et al., 2000). A 15-acre bioremediation treatment cell was constructed to initiate remediation of contaminated sediments in 1995. The cell was underlain by layers of compacted clay, a synthetic geogrid, sand layers and a polyethylene liner. The petroleum contaminated soil was treated by placing and mixing 18 inches of contaminated soil in the cell. The soil was then aerated, irrigated and fertilized to encourage biodegradation of the hydrocarbon contaminants. Within the bioremediation treatment cell, a 0.5-acre area was delineated for phytoremediation and the area was filled with 2 ft of contaminated soil. Initial sampling at the site revealed TPH concentrations ranging from 3,036 to 7,923 mg/kg with a mean concentration of 4,551 mg/kg. White clover (*Trifolium repens L.*), tall fescue (*Festuca arundinacea*) and Bermuda/rye grass (*Cynod Dactylon L.*) were investigated for their ability to promote the degradation of TPH. Nitrogen and phosphorous amendments were provided at a 100:20:10 ratio of C:N:P.

All three plant species developed a good vegetative cover in the contaminated soil. The white clover was slower to produce a cover than other grasses. It grew relatively well during the first season, but did not thrive during the subsequent growing season. Bermuda grass was sodded and, as such, provided the most rapid cover and rooting. Perennial rye grass seeding was added to the Bermuda grass to enhance winter growth and treatment during Bermuda grass dormancy. By the 7th month, the Bermuda/rye grass treatment exhibited the most root production of all the treatments. Tall fescue also established a good cover and after 12 months; it exhibited the highest root mass and density of all the treatments.

The field study demonstrated that vegetated treatments enhanced TPH degradation compared to unvegetated treatments. TPH degradation was observed in the unvegetated plots, however, a noted increase in degradation of 9 to 19% was observed in the vegetated plots. After 24 months, white clover provided the highest rate of TPH degradation, followed by tall fescue and Bermuda/rye grass. Conversely, white clover exhibited the lowest root and aboveground biomass production. Hence, it was concluded that greater plant growth did not necessarily result in higher degradation rates. The white clover grew well in the first year, but significant root decomposition was observed in the second year of the study, when it did not thrive. These results supported the importance of root turnover rates, root exudation patterns, and the influence of the vegetation on the physical and chemical soil properties, as well as the establishment of a healthy vegetation cover, on TPH remediation.

Petroleum degraders were enumerated using the MPN technique at the end of the second growing season and indicated that the number of petroleum degraders were significantly higher in the vegetated plots than in the unvegetated controls. The highest number of petroleum degraders was found in the white clover plots, which also exhibited the lowest TPH concentrations. Data demonstrating carbon source utilization on the basis of chemical structure indicated that the microbial communities found in the white clover plots generally utilized the highest percentage of aromatics, carboxylic and amino acids. Similar patterns were observed in the Bermuda/rye grass and tall fescue plots, while microbial communities from the unvegetated plots did not appear to utilize these compounds efficiently.

At the end of the two-year phytoremediation study, it was concluded that a notable increase in the mitigation of TPH could be observed in vegetated treatments compared to unvegetated treatments. Physical changes in the soil profile were observed in vegetated plots and were attributed to the presence of the root biomass below the ground surface, which established a transport network for the distribution of nutrients, water and air in the soil. Important site management practices were recommended for the optimization of organic contaminant degradation using phytoremediation applications. These included fertilization, irrigation, as well as biomass removal and tillage.

Case Study

Phytoremediation was implemented at a Wisconsin industrial site to provide *in-situ* bioremediation of fuel oil contaminated soil (Carman, 1997; Carman and Crossman, 2001). Operations began at this site in the early 1900s. During the late 1970s, a below-ground pipe section transferring No. 2 fuel oil from an aboveground storage tank to a smaller day use tank failed resulting in a subsurface release of fuel oil. Approximately 15,000 gallons of fuel oil were recovered from shallow trenches constructed for contaminant recovery at the site, however, further site investigations revealed that highly contaminated soil with concentrations above 1000 ppm of diesel-range organics (DRO) remained in four generalized hotspots. The site was underlain by 3 to 15 feet of heterogeneous fill material, comprised of wood timbers, sawdust, construction debris, clay sand and gravel. The depth of the water table ranged from 3 to 9 feet, with the water table sloping west, toward the adjacent river. Three of the hotspots were situated below a hard-packed gravel equipment storage area and another was located below a vegetated area along the river.

Concentrations of DRO were highly variable across the site and within each of the four hotspots. However, the highest concentrations were found within the capillary fringe. The main objective of the phytoremediation application was to remediate soil and fill materials within four hotspots at the facility to below 1,000 mg/kg DRO. Excavation and treatment of the soil materials had been ruled out as a remediation option due to the potential costs associated with segregating construction debris from soil and risks associated with excavating at a facility with a long operational history. Results from microbiological enumerations indicated that a viable population of microbes capable of degrading fuel oil were indigenous to soil at the facility, which supported the potential use of *in situ* bioremediation and phytoremediation strategies for site remediation.

In an agronomical assessment, both hybrid poplars and willows showed good aerial growth during root development, but the willows demonstrated a stronger tendency to establish rooting within the DRO contaminated soil. Three hundred hybrid willow trees were planted at 8 feet intervals in the four hotspots in May 1996 and the trees exhibited fair to excellent growth during the first growing season. A fertilization program was implemented in the fall of 1996 and continued through 1999. This program included surface application of high nitrogen fertilizer in the rows between trees. Insecticide was also applied as needed (twice per year) to reduce insect damage. The willow trees planted as part of the phytoremediation program grew an average of 5 to 6 feet during the first 4 growing seasons.

In October 1999, soil samples were collected to determine the effectiveness of the phytoremediation program in reducing the DRO concentrations in the hotspots. Concentrations of DRO were found to have decreased by as much as 66 to 68 percent in two of the four hotspots. Soil samples were also collected outside the hotspots as controls and it was noted that the DRO concentrations did not decrease as significantly at these locations those inside the hotspots, suggesting that biological activities associated with the phytoremediation program were responsible for accelerating the decrease in DRO. In terms of technology implementation costs, it was found that for a treated volume of 4,500 yd^3, the total costs for the phytoremediation program at the Wisconsin site including implementation ($80,000) and annual operation and maintenance ($7,000) was approximately $21/$yd^3$.

10.5.3 Phytovolatilization

Phytovolatilization involves the uptake of the contaminants by plants and their release, or that of a metabolic transformation product, to the atmosphere through plant transpiration. This phytoremediation application generally relies on the transpiration pull of fast growing trees, which accelerates the uptake of the pollutants from groundwater. The transfer of contaminants from the soil or groundwater to the atmosphere is, perhaps, not as desirable a remediation alternative as *in situ* degradation, however, it may present a preferable option to prolonged exposure in the soil environment and the risk of groundwater contamination. In any case, very few contaminants are sufficiently water soluble, non-toxic to plants and volatile enough to reach atmospheric concentrations of concern. Phytovolatilization has been observed in the successful mitigation of contaminants such as TCE, PCBs and TPH. TCE, an organic compound used in engineering and other industries for degreasing, is a particularly mobile pollutant typically forming plumes which move beneath the soil's surface. In a number of studies, poplars have been shown to phytovolatilize approximately 90% of the TCE they take up (Evans and Furlong, 2003).

Case Study

Phytoremediation was selected as a remedial strategy for groundwater clean up at the Aberdeen Proving Ground J-Field Superfund site in Maryland (Hirsh et al., 2003). Significant levels of VOCs were detected in the groundwater in the vicinity of the Toxic Burning Pits area. The contaminants of concern at the site were the chlorinated volatile organic compounds 1,1,2,2-tetrachlorethane (1,1,2,2-TeCA) and trichloroethene (TCE). Several conventional groundwater remedial technologies were evaluated for remediation of the J-Field site including soil washing, soil-vapor extraction, groundwater pump-and-treat, and groundwater circulation wells. *In situ* biotic and abiotic remediation studies demonstrated that some VOC degradation occurred in a freshwater marsh located downgradient of the site. The J-Field was considered to possess unique conditions which supported the application of phytoremediation as a remediation strategy at the site. Theses included the remoteness of the site (restricted access and no groundwater users), low groundwater velocity, VOC plume under steady-state conditions, and a freshwater marsh with anoxic sediments to adsorb and biodegrade VOCs from the plume prior to surface water discharge.

Approximately 180 hybrid poplar trees (*Populus deltoides x Populus trichocarpa*) were planted at the J-Field site in the Spring of 1996 at the on-set of a pilot-scale phytoremediation study. The aim of the study was to evaluate the ability of the hybrid poplars to reduce dissolved phase VOC mass loading to the marsh via hydraulic containment of the VOC plume. A second objective was also to assess the ability of poplar trees to promote dissolved phase VOC mass reduction through VOC transpiration and biotranformation, and the enhancement of *in situ* biodegradation in the rhizosphere.

Five years of phytoremediation performance monitoring at J-Field supported the use of hybrid poplar trees to the remediate the dissolved-phase VOC plume. The data compiled during monitoring have demonstrated that the hybrid poplar trees (*Populus deltoides x Populus trichocarpa*) contribute significantly to the remediation of the VOC plume, transpire sufficient groundwater on a seasonal basis to influence groundwater flow direction, provide hydraulic containment of the VOC plume, and transpire VOC-contaminated groundwater. Detectable concentrations of VOCs in tree tissue and transpiration gas samples suggested that the VOCs were transformed within the trees and transpired to the atmosphere. Atmospheric monitoring indicated that the VOCs were photodegraded and diluted to very low concentrations once released to the atmosphere by the plants. Hence, the atmospheric pathway was considered to be of relatively low risk to human health and the environment.

Hydraulic containment of the VOC plume was supported by sap-flow data which, in 1999, indicated that 171 hybrid poplar trees transpired 4,130 L/day. These trees were expected to transpire 7,600 L/d after 30 years of growth. The groundwater data reported the formation of a discernible cone of depression beneath the hybrid poplar trees during the summer, when transpiration rates exceeded the rates of groundwater release from the aquifer. Tree-tissue sampling revealed that the trees contained measurable concentrations of TCE and 1,1,2,2-TeCA. The TCE concentrations decreased with increasing sample height along the tree length. Leaf tissue samples indicated the presence of 1,1,2,2-TeCA and dichloroacetic acid, and the transpiration gas samples showed that the hybrid poplar trees effectively volatilized 1,1,2,2-TeCA and TCE. Tree tissue and transpiration gas data indicated that the reduction in TCE was attributable to multiple processes including adsorption to roots and tree tissue, rhizosphere degradation, biotransformation via plant metabolism, and volatilization through trunk and leaves.

10.6 Design Considerations

The selection of a phytoremediation system for a contaminated site depends on the applicability of the approach at that particular site. Design of the system varies according to the contaminants, the conditions at the site, the level of clean up required and the plants employed. A thorough site characterization is required to provide the necessary data to design any type of phytoremediation system. The design considerations include contaminant concentrations, plant selection and root depth, contaminant treatability, area requirements, irrigation, agronomic inputs and maintenance, groundwater capture zone and transpiration rate, contaminant uptake rates, and the remediation time requirements (ITRC, 1999).

10.6.1 Contaminant Concentrations

The primary consideration in terms of contaminant concentration is that the contaminant concentrations cannot be phytotoxic or result in undesirable impacts on plant health or yield. Higher concentrations of contaminants may be tolerated more readily by plants than by soil microorganisms (Suthersan, 2002). Plants may not be able to mitigate all contaminants present at a given site. The composition of organic compounds (structure, log K_{ow}, degree of weathering, boiling point) and degree of adsorption are important factors in phytoremediation applications. High levels of contamination may result in the elimination of phytoremediation as a treatment alternative. Alternatively, the high contaminant concentrations and the resulting phytotoxicity could lead to a tiered remediation approach in which higher concentration contaminants could be handled with more expensive *ex situ* technologies that quickly reduce acute risk, while *in situ* phytoremediation could be used over a longer period of time to clean higher volumes of low contaminant concentrations (Tucker and Shaw, 2000).

10.6.2 Plant Selection and Root Depth

Plants are selected according to the application and the contaminants of concern. Major criteria for plant selection include the requirements for the method to be employed and the nature of the contaminants involved. For example, organic phytotransformation would require plant species which are hardy and fast growing, easy to maintain, have a high transpiration rate and have the ability to transform the contaminants to non-toxic or less toxic by-products. On some sites, the planting of grass species in conjunction with trees, often in between rows of trees to stabilize and protect the soil, may be the best route. Grasses tend to generate a tremendous amount of fine roots near to the surface and are helpful in controlling wind-blown dust, wash-off and erosion (Evans and Furlong, 2003).

Native, crop, forage, and other types of plants that can grow under regional and climatic conditions of a particular site are generally preferred. In temperate climates, phreatophytes such as hybrid poplar, willow, cottonwood, aspen are native to many regions, and selected due to their fast growth, ability to form deep roots down to the level of groundwater, and high transpiration rates. Plants used in phytoextraction include sunflowers and Indian mustard for lead, *Thlaspi spp.* (pennycress) for zinc, cadmium, and nickel; and sunflower and aquatic plants for radionuclides. In constructed wetland applications, emergent vegetation transpires water and is easier to harvest while submerged species do not transpire water but provide more biomass for the uptake and sorption of contaminants (ITRC 1999).

The establishment of an effective root zone is crucial in phytoremediation applications. Typical lawn grasses generally produce roots down to 1 foot below ground surface, while prairie grass species are known for their deep roots and can yield systems that reach 10 to 15 feet below ground surface. Alfalfa has been shown to produce roots down to 33 feet below the surface. For tree species, typical rooting depths range from 10 to 12 feet below ground surface. However, root systems down to 33 feet bgs, and even deeper have been reported as well.

For phytoremediation applications, it is typically recommended that the rooting depth of groundcover plants be in the 1 to 2 feet range. For trees, root depths downs to 12 feet are generally recommended, however, under circumstances where hydrogeological conditions are suitable, deeper rooting depths beyond these typical values may be reached. Engineered methods may be required to induce the development of deeper root systems, such as: sub-irrigation systems, soil mounding, the use of impermeable barriers to restrict rainwater infiltration, and specialized planting methods (ITRC, 2001).

10.6.3 Contaminant Treatability

Screening studies to determine contaminant treatability using particular plants are recommended to provide the necessary information to evaluate whether a particular phytoremediation system will achieve the treatment objectives. Treatability studies can provide toxicity and transformation data when different concentrations of contaminants are tested with proposed plant species and the fate of the contaminants in the plant system is assessed. (ITRC, 1999).

10.6.4 Capture Zone and Transpiration Rates

For applications involving groundwater remediation or hydraulic containment, a capture zone computation can be employed to estimate whether the trees can be effective at entraining the contaminant plume or providing a hydraulic barrier. In general, phreatophytes which exhibit high transpiration rates (Table 10.3) are employed in this application. With respect to groundwater mitigation, the goal is to create a water table depression where contaminants will flow to the vegetation root zone for uptake and treatment (ITRC, 1999). In terms of hydraulic containment, the objective is to prevent contaminant plume migration or to create a water table depression to minimize water logging. Typical evapotranspiration estimates can be used in conjunction with groundwater models to predict withdrawal necessary to achieve hydraulic control. The amount of vegetation required to achieve the same level of control can be computed by comparing the estimated groundwater withdrawal rate with plant water uptake and transpiration rates.

10.6.5 Area Requirements

If the size of the land area is limited, *in situ* phytoremediation technologies may be difficult or not applicable. To determine the necessary land requirements for a phytoremediation system, knowledge of the water uptake and transpiration rates throughout the life of the plant are necessary. For constructed wetland systems, the area required to conduct remediation is determined by the expected inflow of contaminated water into the system and the detention time required for the contaminants to settle out or stabilize within the sediments. These two parameters will allow the volumetric capacity of the wetland to be calculated. Depending on whether the system is a surface or subsurface design, the area required to contain the necessary volume can be determined (ITRC, 2001).

10.6.6 Irrigation, Agronomic Inputs and Maintenance

Irrigation will ensure a vigorous start to the system even under drought conditions and it may be necessary if a particular area does not receive sufficient rain to ensure a healthy vegetative growth throughout the growing season. Hydrologic modelling may be required to estimate the rate of percolation to groundwater during irrigation conditions to minimize contaminant migration. Agronomic inputs include the nutrients necessary for the growth of the vegetation and rhizosphere microbes. The surrounding soil should be characterized and amendments such as nitrogen, potassium, phosphorous, aged manure, sewage sludge, compost, straw and/or mulch added as required to ensure successful plant growth. Maintenance of phytoremediation systems may include the addition of fertilizers and complexing or chelating agents to bind metals to the soil (ITRC, 1999).

10.7 Potential and Limits of Phytoremediation Technologies

Phytoremediation technologies offer many advantages when compared to conventional approaches. Some of these advantages are presented in Table 10.5 (Glass, 2000). In terms of economics, phytoremedation applications are estimated to be at least 40% less costly than other *in situ* remediation strategies. For *ex situ* technologies, phytoremediation technologies are estimated to be 90% less costly than alternatives (Glass, 2000). The ability to reduce, stabilize, sequester, degrade, metabolize, and mineralize contaminants using phytoremediation applications can present a more rapid alternative than natural attenuation alone. Other advantages include a low maintenance, passive, *in situ*, solar driven system, potentially applicable in remote locations, decreased air and water emissions and secondary waste by-products, control of soil erosion, surface water runoff, infiltration, and dust emissions, potential to mitigate multiple or mixed contaminants, habitat creation or restoration can provide land reclamation upon completion, favorable public perception, increased aesthetics, reduced noise, and carbon dioxide and greenhouse gas sequestration.

Phytoremediation is a promising remediation alternative for sites where contamination is near the surface, in the top 3 to 6 feet, and in some instances, up to 20 feet. Plants with phytoremediation potential can be selected from wild species growing on contaminated sites, or alternatively, crop plants may be chosen due to their specific characteristics and affinity for a particular environment and contaminant.

Phytoremediation generally requires substantially lower capital and operating costs than competing technologies. Excluding site monitoring, the major costs of phytoremediation include tilling and preparation of the soil, planting the seeds, watering fertilization, weed and pest control, and harvesting and disposal of the biomass. Although there are few documented estimates of phytoremedation costs, the consensus estimate for the cost of treating liquid streams is in the range of $0.60 to $6.00 per 1,000 gallons treated. The lower end of this range is generally associated with the treatment of organic contaminants and the higher end is for the mitigation of metals and radionuclides. Average costs associated with various site remediation approaches are shown in Table 10.6 (Glass, 2000). The higher costs in the case of heavy metal remediation, generally include additional operating costs for harvesting, disposal of contaminated plant mass, and repeating the plant growth cycle (Suthersan, 2002).

Phytoremediation technologies also have some limitations (Table 10.5) and are not applicable to all sites. The primary limitations relate to the growth habits of the planted system, root penetration of the selected plants and amount of land available for planting. Phytoremediation systems designed to treat soil, sediment, surface water or groundwater generally require more land area than conventional methods. It is a time-consuming remediation alternative in comparison to more conventional remediation approaches and it can take several growing seasons to clean a site. In addition to relying on the life cycles of plants, planting and harvesting needs to be coordinated with growing seasons and local climatic conditions. Many naturally occurring hyperaccumulator plants are extremely slow growers and can take 13 to 16 years to clean a typical site.

Table 10.5 Advantages and disadvantages of phytoremediation.

Advantages	Disadvantages
Cost	*Time*
Low capital and operating costs	Slower than some alternatives; seasonally dependent
Metal recycling provides further economic advantages	Many natural hyperaccumulaors are slow growers
Performance	*Performance*
Permanent treatment solution	Biological methods are not capable of 100% reduction
In-situ application avoids excavation	May not be applicable to all mixed wastes
Capable of remediating bioavailable fraction	High Metal concentrations or other contaminants may be toxic
Capable of mineralizing organics	Soil phytoremediation applicable only to surface soils
Applicable to a variety of contaminants, including some recalcitrants	
Trees capable of high hydraulic pumping pressures	
	Space
	Groundwater, wastewater applications may require large available surface area
Other	*Other*
Public acceptance; aesthetically pleasing	Need to displace existing facilities
Compatible with risk-based remediation, brownfields	Regulators may be unfamiliar with the technology and its capabilities
Can be used during site investigation or after closure	Lack of recognized economic performance data

Phytoremediation technologies are limited by the length of the growing season, which is defined as the average first to average last frost dates for a particular region. Plants are typically dormant during winter periods, unless they are in a temperate climate where freezing temperatures are infrequently experienced. For vegetative covers designed to minimize infiltration, a reduced amount of infiltration occurs during the winter season because the precipitation is snow rather than rain. However during the spring thaw, a large influx of infiltration may result and must be considered in the design. The treatment efficiency of constructed wetlands is greatly reduced during winter, but remediation continues as long as the system is not completely frozen.

Table 10.6 Cost comparison of phytoremediation with other remediation technologies commonly used for the treatment of soil and groundwater contaminated by metals and radionuclides.

Soils	Cost ($US/ton)	Water	Cost ($US/1000 gal)
Chemical treatment	$100-$500	Ion exchange	$0.13-$1.75
Soil washing	$75-$200	Electrochemical reduction	$0.60
Soil flushing (in-situ)	$40-$190	Adsorption	$1.00-$20.00
Vitrification (reagent)	$75-$90	Membrane separation-filtration	$1.20-$5.80
Vitrification (thermal)	$250-$425	Chemical precipitation	$0.80-$1.20
Thermal desorption	$150-$500	Reverse osmosis	$3.00-$6.20
Thermal treatment	$170-$300	Activated carbon	$119-$210
Electrokinetics	$20-$200	Biosorption	$9-$3400
Incineration	$200-$1500	Phytoremediation	$0.60-$6.00
Landfilling	$100-500		
Phytoremediation	$25-$100		

For phytoremediation applications to be effective, the contaminated media must be in contact with the plant roots system, generally within approximately 20 feet of the land surface. The plants must either be able to extend their roots to the contaminant or the contaminated area must be moved to within range of the plant roots via contaminant mass transfer to the treatment or root zone. Phytoremediation applications will not be beneficial if plant phytotoxicity prevents the plants from rooting extensively, the contaminants are not readily bioavailable or leach quickly out of the rooting zone, or if the site under consideration is oxygen deficient. If vegetation is used for the purpose of extracting groundwater, the contaminants must then be located within a few feet of the water table surface.

Like most biological treatment methods, the use of plants does not produce 100% removal or reduction of contaminants, because reaction rates decrease as concentrations of the contaminant decrease. Vegetation that absorbs toxic heavy metals requires harvesting must be managed as a hazardous waste. Soil and sediment phytoremediation can be limited by contaminant solubility or availability (Glass, 2000). Some phytoremediation applications depend on solubility manipulation of the material to make it bioavailable to the plant. Such manipulation must be applied carefully to avoid migration of contaminants, which is another concern due to the potential risk associated with the mobilization of contaminants from the soil to the groundwater, or from the soil and groundwater into the air. Mobilization from soil to groundwater could occur as a result of excessive irrigation, addition of chelating agents or surfactants, chemical transformation of contaminant of concern, or pH manipulation (ITRC, 2001).

10.8 Conclusion

Soil and groundwater contamination by organic and inorganic constituents has become a matter of rising concern. Reclamation of these contaminated sites is typically expensive, energy consuming and laborious. Phytoremediation is considered to be a relatively new technology, but it has the potential to become a promising low-cost alternative when compared to other conventional treatment approaches. The various plant-based technologies that comprise phytoremediation have some similarities, differences, and a number of different possible applications. Contaminant remediation has been shown to take place through plant-based mechanisms as well as microbial processes within the rhizosphere, an enhanced area of biological activity resulting from plant-soil-microorganism interactions. Due to the nature and complexity of contaminants, their interactions with each other and with the environment, and their mitigation via physical, chemical and biological processes, each application will be site specific, and must be evaluated on a case-by-case basis. To date, only a few full-scale phytoremediation facilities have been implemented. However, there are number of laboratory-scale and pilot-scale projects being evaluated. In order to be successful as a remediation technology, continued research activities efforts are required to (i) better understand the removal mechanisms; (ii) improve the efficiency of phytoremediation treatment systems; (iii) understand and quantify the long-term performance and benefits of phytoremediation applications; and (iv) convince decision makers in business and administration of the large beneficial potential inherent to phytoremediation activities.

References

Alexander, M. (1999) *Biodegradation and Bioremediation.* Academic Press. New York.

Alkorta, I., Hernandez-Allica, J., Becerril, J. M., Amezaga, I., Albizu, I., Onaindia, M., Garbisu, C. (2004) Chelate-Enhanced Phytoremediation of Soils Polluted with Heavy Metals. *Reviews in Environmental Science and Bio/Technology*. 3:55-70.

Anderson, C., Moreno, F., Meech, J. (2004) A Field Demonstration of Gold Phytoextraction Technology. *Minerals Engineering*. 18:385-392.

Anderson, T. A., Coats, J. R. (1995) An Overview of Microbial Degradation in the Rhizosphere and its Implications for Bioremediation. Chapter 8 in *Bioremediation:*

Science and Applications. H. D. Skipper and R. F. Turco Eds. SSSA Special Publication No. 43. Soil Science Society of America, Inc.; American Society of Agronomy, Inc.; Crop Science Society of America, Inc.; Madison, Wisconsin.

Banks, M.K., Fiorenza, S., Oubre, C. L., Ward, C. H. (2000) *Phytoremediation of Hydrocarbon-Contaminated Soils*. Lewis Publishers. Boca Raton, FL.

Berti, W. R., Cunningham, S. D. (2000). Phytostabilization of Metals. Chapter 6 *in Phytoremediation of Toxic Metals: Using Plants to Clean up the Environment.* I. Raskin and B. Ensley Eds. Wiley-Interscience Publication Inc. New York.

Burken, J. G. (2003) Uptake and Metabolism of Organic Compounds: Green-Liver Model. Chapter 2 in *Phytoremediation: Transformation and Control of Contaminants.* S McCutcheon and J. Schnoor Ed. Wiley-Interscience. New York.

Carman, E. (1997) Phytoremediation of Fuel Oil Contaminated Soil in *In Situ and On-Site Bioremediation: Volume 3.* 4th International In Situ and On-Site Bioremediation Symposium. B. C. Alleman and A. Leeson, Eds. New Orleans, April 28-May 1.

Carman, E., Crossman T. (2001). Phytoremediation. Chapter 9 in *In Situ Treatment Technology.* 2nd Ed. Arcadis: Environmental Science and Engineering Series. Lewis Publishers. New York.

Chappell, J. (1997) *Phytoremediation of TCE Using Populus*. Report prepared for U.S. EPA Technology Innovation Office. Washington, D.C., 38 pp.

Connell, D. W. and G. J. Miller. 1984. *Chemistry and Ecotoxicology of Pollution.* John Wiley & Sons, NY.

Crowley, D. E., Alvey, S., Gilbert, E. S. (1997) Rhizosphere Ecology of Xenobiotic-Degrading Microorganisms. Chapter 2 in *Phytoremediation of Soil and Water Contaminants*. E. Kruger, T. Anderson and J. Coats Eds. ACS Symposium Series 664. ACS. Washington.

Cunningham, S., Shann, J., Crowley, D., Anderson, T. (1997) Phytoremediation of Contaminated Water and Soil. Chapter 1 in *Phytoremediation of Soil and Water Contaminants*. E. Kruger, T. Anderson and J. Coats Eds. ACS Symposium Series 664. ACS. Washington.

Cunningham, S. D., Lee, C. R. (1995) Phytoremediation: Plant-Based Remediation of Contaminated Soils and Sediments. Chapter 9 in *Bioremediation: Science and Applications*. H. D. Skipper and R. F. Turco Eds. SSSA Special Publication No. 43. Soil Science Society of America, Inc.; American Society of Agronomy, Inc.; Crop Science Society of America, Inc.; Madison, Wisconsin.

Dushenkov, S. (2003) Trends in Phytoremediation of Radionuclides. *Plant Soil.* 249:167-175.

Dushenkov, S., Kapulnick, Y. (2000) Phytofiltration of metals. Chapter 7 in *Phytoremediation of toxic metals - using plants to clean-up the environment*. I Raskin and B. D. Ensley Eds. New York, John Wiley & Sons, Inc.

Dushenkov, S., Vasudev, D., Kapulnik, Y., Gleba, D., Fleisher, D., Ting, K. C., Ensley, B. (1997) Removal of Uranium from Water Using Terrestrial Plants. *Environmental Science and Technology*. 31:3468-3474.

Dushenkov, V. Kumar, P. B. A. N., Motto, H., Raskin, I. (1995) Rhizofiltration: The Use of Plants to Remove Heavy Metals from Aqueous Streams. *Environmental Science and Technology*. 29:1239-1245.

Ensley, B. (2000). Rationale for Use of Phytoremediation. Chapter 1 in *Phytoremediation of Toxic Metals: Using Plants to Clean up the Environment.* I. Raskin and B. Ensley Eds. Wiley-Interscience Publication Inc. New York.

Evans, G. M., Furlong, J. C. (2003). *Environmental Biotechnology: Theory and Application*. John Wiley & Sons, Inc.

Garbisu, C., Alkorta, I. (2001) Phytoextraction: A Cost-Effective Plant-Based Technology for the Removal of Metals from the Environment. *Bioresource Technology*. 77:229-236.

Gianfreda, L., Rao, M. A. (2004) Potential of Extra Cellular Enzymes in Remediation of Polluted Soils: A Review. *Enzyme and Microbial Technlogy*. 35:339-354.

Glass, D. (2000). Economic Potential of Phytoremediation. Chapter 2 in *Phytoremediation of Toxic Metals: Using Plants to Clean up the Environment.* I. Raskin and B. Ensley Eds. Wiley-Interscience Publication Inc. New York.

Glick, B. R. (2003) Phytoremediation: Synergistic Use of Plants and Bacteria to Clean Up the Environment. *Biotechnology Advances*. 21:383-393.

Guntenspergen, G. R., Stearns, F., Kadlec, J. A. (1989) Wetland Vegetation. Chapter 5 in *Constructed Wetlands for Wastewater Treatment: Municipal, Industrial and Agricultural*. Lewis Publishers. Michigan.

Hansen, D., Duda, P. J., Zayed, A., Terry, N. (1998) Selenium Removal by Constructed Wetlands: Role of Biological Volatilization. *Environmental Science and Technology*. 32:591-597.

Hirsh, S. R., Compton, H. R., Matey, D. H., Wrobel, J. G., Schneider, W. H. (2003) Five-year Pilot Study: Aberdeen Proving Ground, Maryland. Chapter 20 in *Phytoremediation: Transformation and Control of Contaminants*. S McCutcheon and J. Schnoor Ed. Wiley-Interscience. New York.

Interstate Technology Regulatory Cooperation (ITRC) (2001) *Technical and Regulatory Guidance Document*. ITRC Work Group. Phytotechnologies Work Team.

Interstate Technology Regulatory Cooperation (ITRC) (2001) *Phytoremediation Decision Tree Document*. ITRC Work Group. Phytoremediation Work Team.

Jacob, D. L., Otte, M. L. (2003) Conflicting Processes in the Wetland Plant Rhizosphere: Metal Retention or Mobilization? *Water, Air, and Soil Pollution* 3:91-104.

Jones, D. (1998) Organic acids in the Rhizosphere – A Critical Review. *Plant Soil*. 205: 24-44.

Kadlec, R. H. and Knight R. L. (1996) *Treatment Wetlands*. Lewis Publishers, Washington, D.C.

Kömives, T., Gullner, G. (2000) Phytoremediation. Chapter 15 in *Plant-Environment Interactions*. 2nd Edition. R. Wilkinson Ed. Marcel Dekker Inc. New York.

Li, Y. M., Chaney, R. L. (1998) Case Studies in the Field – Industrial Sites : Phytostabilization of Zinc Smelter Contaminated Sites : The Palmerton Case. Chapter 11 in *Metal Contaminated Soils: In Situ Inactivation and Phytorestoration.* J. Vangronsveld and S. D. Cunningham, Eds. Springer. Landes Bioscience, Austin, TX.

McCutcheon, S., Schnoor, J. (2003) Overview of Phytoremediation and Control of Wastes. Chapter 1 in *Phytoremediation: Transformation and Control of Contaminants.* S McCutcheon and J. Schnoor Ed. Wiley-Interscience. New York.

McGrath, S. P., Zhao, F. J., Lombi, E. (2001) Plant and Rhizosphere Processes involved in Phytoremediation of Metal-Contaminated Soils. *Plant and Soil.* 232:207-214.

Mitsch, W. J., Gosselink, J. G. (2000) *Wetlands.* 3rd Edition. John Wiley & Sons Inc.

National Academy of Sciences (NAS) (1995) *WETLANDS: Characteristics and Boundaries.* National Academy of Sciences Press: Washington, D.C.

Negri, C. M., Hichman, R. R. (2000) The Use of Plants for the Treatment of Radionuclides. Chapter 8 in *Phytoremediation of Toxic Metals: Using Plants to Clean up the Environment.* I. Raskin and B. Ensley Eds. Wiley-Interscience Publication Inc. New York.

Newman, J. N., Clasen, J. C., Neafsey, J. A. (2000) Seasonal Performance of a Wetland Constructed to Process Dairy Milkhouse Wastewater in Connecticut. *Ecological Engineering* 14:181-198.

NSW Department of Land and Water Conservation (DLWC) (1998) *The Constructed Wetlands Manual.* Volume 1 & 2. Land and Water Conservation. Australia.

Olsen, P. E., Fletcher, J. S. (1999) Field Evaluation of Mulberry Root Structure with Regard to Phytoremediation. *Bioremediation Journal.* 3(1):27-33.

Piersynski, G. M., Schnoor, J. L., Banks, M. K., Tracey, J. C., Licht, L. A., Erickson, L. E. (1994) Vegetative Remediation at Superfund Sites. *Environmental Science and Technology – Mining and its Environmental Impact.* 1:49-69.

Rock, S. A., Sayre, P. (1999) Phytoremediation of Harzardous Wastes: Potential Regulatory Acceptability. *Environmental Regulations and Permitting.* Volume 8(3):33-42. John Wiley & Sons. New York, NY.

Sandermann, H. (1994) Higher Plant Metabolism of Xenobiotics: The "Green Liver" Concept. *Pharmacogenetics.* 4: 225-241.

Singer, A. C., Crowley, D. E., Thompson, E. P. (2003) Secondary Plant Metabolites in Phytoremediation and Biotranformation. *TRENDS in Biotechnology.* 21:123-130.

Singh, O. V., Jain, R. K. (2003) Phytoremediation of Toxic Aromatic Pollutants from Soil. *Applied Microbiology and Biotechnology.* 63:128-135.

Steinberg, S. L. Coonrod, H. S. (1994) Oxidation of the root zone by aquatic plants growing in gravel-nutrient solution culture. *Journal of Environmental Quality* 23: 907-913.

Stottmeister, W., Wiebner, A., Kuschk, P., Kappelmeyer, U., Kastener, M., Bederski, O., Müller, R. A., Moormann, H. (2003) Effects of Plants and Microorganisms in

Constructed Wetlands for Wastewater Treatment. *Biotechnology Advances* 22:93-117.

Suthersan, S. (2002) *Natural and Enhanced Remediation Systems.* Lewis Publishers, New York.

Suthersan, S. S. (1997) *Remediation Engineering: Design Concepts*. Lewis Publishers, Inc. Washington, D.C.

Tucker, R., Shaw, J. (2000) Phytoremediation and Public Acceptance. Chapter 3 *in Phytoremediation of Toxic Metals: Using Plants to Clean up the Environment.* I. Raskin and B. Ensley Eds. Wiley-Interscience Publication Inc. New York.

US EPA. (2001) *Brownfields Technology Primer: Selecting and Using Phytoremediation for Site Cleanup.* Solid Waste and Emergency Response Technology Innovation Office. Washington, D.C. EPA 542/R/01/006.

US EPA (2000) *Introduction to Phytoremediation*. National Risk Management Research Laboratory. Cincinnati, Ohio. EPA 600/R/99/107.

US EPA (1999) *Tech Trends*. USEPA Document EPA 542-N-99-005. August Issue No. 34.

Werker, A. G., Dougherty, J. M., McHenry, J. L., Van Loon, W. A. (2002). Treatment Variability for Wetland Wastewater Treatment Design in Cold Climate. *Ecological Engineering* 19:1-11.

CHAPTER 11

Metal Removal

J.F. Blais, S.K. Brar, N. Meunier, G. Mercier, R.D. Tyagi and R.Y. Surampalli

11.1 Introduction

Environmental contamination with heavy metals is a consequence of technological and industrial advances (Nriagu 1996; Nriagu and Pacyna 1988). The principal problem associated with this anthropogenic contamination is toxicity against all living organisms, in particular, humans (Baath 1989; Chapman et al. 2003). The risks associated with the presence of potentially toxic metals in the soil are rather well known and documented (Ferguson 1991; Salomons et al. 1995; Lippmann 2000; Allen 2002). It is, therefore, essential to remove or reduce the presence of these inorganic contaminants in order to diminish the possibility of uptake by plants, animals and humans and eventual accumulation in the food chain and also to prevent them from contaminating surface and groundwater by dissolution or dispersion (Adriano 1992; McLaughlin et al. 2000).

There is immense market potential for decontamination of soils and several sources have mentioned, estimated and measured tonnages of several sites worldwide (NATO/CCMS 1998; USEPA 1997). In United States, nearly half a million of potentially contaminated sites have been counted since the 80's and close to 217,000 of these required an extended intervention to clean sites (USEPA 1997). In fact, the government projected an investment of US $ 187 billion (1996 dollars) to completely restore these sites in an estimated time frame of 10 to 30 years. More than 31,000 of these industrial sites are contaminated solely by heavy metals (Moore and Luoma 1990).

In Europe, about 600,000 industrial contaminated sites were identified in several countries: Germany (240,000), Belgium (13,000), Denmark (15,000), France (200,000), Hungary (10,000), Sweden (10,000), Switzerland (3,000), Norway (3,350), Netherlands (110,000) and Czechoslovakia (3,000) (Tuin and Tels 1990; NATO/CCMS 1998). In England and Wales, more than 100,000 sites would be affected by contamination in comparison to 6,000 in Scotland (Davidson et al. 1998). Furthermore, there are several thousands of classified European sites, which were being used by military during first and second World War (NATO/CCMS 1998). Ensconce and Robertiello (1993) estimated restoration costs of these European sites to be more than US$ 130 billions (1993 dollars).

Prior to 1987, remediation methods for polluted soils and other matrices implied removal of contaminated environments and landfilling in cells for safety and or confinement of contaminants by physical barriers (USEPA 1997; Papassiopi et al. 1999; Mulligan et al. 2001a). In United States, 75% of remediation technologies used between 1987 and 1993 aimed at significant and permanent reduction of volume, toxicity and or mobility of contaminants on affected sites (Steele and Pichtel 1998). On the other hand, USEPA in 1993 mentioned that good functioning of Superfund program necessitated urgent elaboration of effective technologies for removal of inorganic contaminants in soils (Rampley and Ogden 1998). Since then, a world-wide thrust has been laid on research and development of technologies that favor high performance for removal of heavy metals (USEPA 1997) as they serve a final solution to soil contamination (NATO/CCMS 1998). Consequently, biological processes, electrokinetics or physicochemical technologies that allow removal or reduction of metal pollution in soils in a significant manner were exploited (Sims 1990; Steele and Pichtel 1998).

The present chapter portrays principal metal contaminants associated with the problems of soil and groundwater contamination. The actual state of the global problem of metal contaminated sites is then discussed, followed by a survey of the regulatory norms in this field. Thereafter, various technological options for remediation of metal contaminated sites will be presented. Finally, the methods of groundwater treatment contaminated with metals will be discussed.

11.2 Metal Contaminants in Soil and Groundwater

The contaminated soils consist of organic, inorganic components or a mixture of two. They normally comprise aliphatic hydrocarbons, aromatics and polychlorinated hydrocarbons, pesticides, cyanide and heavy metals.

In United States, metals range from 45 to 70% of total pollutants in contaminated sites (USEPA 1997). In effect, arsenic (As), cadmium (Cd), chromium (Cr), nickel (Ni), lead (Pb), and zinc (Zn), are on the list of common ten contaminants most often traced on the sites indexed by Superfund program and Department of Defense (Evanko and Dzombak 1997; USEPA 1997).

Mining operations and related technologies, metal treatment, automobile emissions (Harrison et al. 1981; Howard and Sova 1993), Pb based paints (Gooch 1993), industrial emissions and wastes (Oyler 1990) and recycling of batteries (Elliot and

Brown 1989; Papassiopi et al. 1999) constitute important sources of metal pollution. Several sites contaminated by heavy metals are located at military bases, shooting fields, armament factories and electroplating units (Marino et al. 1997). In fact, heavy metals represent five of the six contaminants most often recovered on military sites and principal three being Pb, Cd and Cr.

The following sections briefly describe some significant characteristics of the principal metal contaminants (As, Cd, Cr, Hg, Ni, Pb, Se, and Zn) found in the soils.

11.2.1 Arsenic

Arsenic (As), a metalloid, exists in the soil environment as arsenite (As(III)) or as arsenate (As(V)). Both forms of this element are toxic. Arsenate is less water soluble than arsenite and is relatively immobile in soil. The largest traditional use of arsenic has been in agriculture. Sodium arsenate has been the most commonly used arsenic containing veterinary product as an insecticidal product. Copper chrome arsenate (CCA) has also been largely used as a wood preservative. Moreover, arsenic is also used in the electronic industry. Symptoms associated with As toxicity notably include hyperkeratosis, liver dysfunction, myocardial ischemia and several cancers (Lippmann 2000). In the environnement, arsenic is subject to microbial oxidation, reduction, and methylation. Arsenite present in water can be biologically-oxidized in arsenate, which is precipitated out more readily by treatment with ferric ions (Williams and Silver 1984). Arsenite compounds can also be formed in anaerobic conditions by reduction of arsenate. For that reason, As-contaminated soils might be freed of some of the arsenic by promoting the reduction and thus increasing the leachibility of this pollutant (Dowdle et al. 1996).

11.2.2 Cadmium

Cadmium (Cd) is almost always found in the Cd(II) valence state. Cadmium originates from foundry fumes of Cu, Zn and Pb, electroplating factories, industries manufacturing Cd based alloys (Cd-Ni, Cd Ag-Cu, Cd-Au), nuclear power plants, manufacture and recycling sites of Ni-Pb batteries and various other industries (Khalid et al. 1998). The use of Cd-containing fertilizers in agriculture, along with metal-bearing pesticides, potentially contributes to further environmental pollution. Cadmium contamination of the water and air is also caused by mine runoff, and waste release from metal smelteries and other industries that utilize Cd in alkaline batteries, plastics and paints (Lippmann 2000). Cadmium is a very toxic element for human and animals. It is cumulative and is excreted with a half-life of 20 to 30 years. Symptoms of acute toxicity following Cd exposure include pulmonary edema, and pneumonitis. Oral ingestion results in renal necrosis and dysfunction (Viessman and Hammer 1993). Prostate and lung cancers have been observed in humans following long-term exposure to Cd compounds (Lippmann 2000). World-wide population was alerted by the case of paddy fields irrigated by mining drainage (Zn mines) where significant quantities of Cd were traced in rice. This rice consumption by farmers and local populations caused adverse health impacts (disease named as "itaï-itaï") (Kobayashi 1978).

11.2.3 Chromium

Chromium mostly exists as Cr (VI) or Cr (III) with chromite (FeCr2O4) as the most important Cr-containing ore. The principal contamination sources of chromium are electroplating procedures and disposal of various wastes (Smith et al. 1995). Leather tanning and wood preservation are other significant uses of chromium. The hexavalent form is toxic and highly soluble in water, whereas trivalent form is considerably less toxic, less water soluble, and hence less mobile in soil. Hexavalent chromium is readily absorbed by the skin where it acts as an irritant and immune-system sensitizer. Oral absorption results in acute renal failure. Chronic exposure to hexavalent chromium can result in lung cancer. Cr (VI) present in soils and the groundwater can be reduced biologically to Cr (III) following addition of an organic substrate (e.g. manure, benzoate, molasses) (Losi et al. 1994; Shen et al. 1996; Chirwa and Wang 1997). The reduction of hexavalent chromium could equally be done by generation of H_2S in anaerobic medium in the presence of sulphates (deFilippi 1994; Sulzbacher et al. 1997).

11.2.4 Mercury

Mercury (Hg) exists in three oxidation states: Hg(0), Hg(I), and Hg(II). Anthropogenic sources of mercury are numerous and occur worldwide (Lippmann 2000). Mercury is obtained by the mining and smelting of cinnabar ore. It is used in chloralkali plants, in paints as pigments or preservatives, in electrical equipment and batteries, in measuring equipment, in mercury quartz and luminescent lamps, in dental restoration materials, and in agriculture as fungicides. Mercury and its derivatives constitute toxins that accumulate slowly and are dangerous to life forms (Humenick and Schnoor 1974). In humans, ingestion of inorganic Hg compounds may cause gastrointestinal irritation, and kidney dysfunction. Chronic intoxication can also cause neurotoxic effects. Many microorganisms possess the ability to reduce divalent mercury to metallic Hg. The elemental mercury is volatile and thus moves into the overlying air. If the microbial activity is stimulated, they might form the basis for bioremediation. Divalent mercury can also be methylated. However, the resulting mono- and dimethyl mercury have high acute toxicity.

11.2.5 Nickel

The Ni (II) form is most commonly found in aquatic and soil environments (Watts 1998). The most abundant ores are the oxides and sulfides of Ni. Nickel is used extensively in alloys. Nickel is normally not very toxic to humans, but can cause headache and shortness of breath as acute effects. Chronic effects are observed on immune system resulting in allergic dermatitis. Nickel is phytotoxic and can adversely affect some fish species, particularly in fresh waters. Moreover, as with other metals, uptake of Ni from soil can cause accumulation of the metal in leaves, roots, and seeds of many plants.

11.2.6 Lead

Generally speaking, lead (Pb) is the principal inorganic contaminant of polluted soils originating from agricultural activities, urban and industrial activities like extraction and foundries (USEPA 1997; Sauvé et al. 1997; Steele and Pichtel 1998; Mercier et al. 2002a). In the United States, different epidemiological studies showed that 5.9% of American children had more than 10 μg Pb/L in their blood, which is above the threshold at which neurotoxicity occurs (USEPA 1994). Direct oral ingestion of contaminated soils and dust by children is the principal cause of lead absorption (Healy et al. 1982; Xintaras 1992). Lead is one of the heavy metals with relatively lower mobility and a tendency to adsorb to organic matter, clays and oxide/hydroxide complexes by forming precipitates (Khan and Frankland 1983) such as Pb carbonates ($PbCO_3$, $2PbCO_3Pb(OH)_2$, $4PbCO_3.2Pb(OH)_2$) (Mercier et al. 2002a; Zimdahl and Skogerboe 1977), orthophosphates of Pb, primary pyromorphites and plumbogummites (last three formed in anaerobic environment) (Nriagu 1974). The Pb contamination primarily originates from recycling sites and battery dumpyards municipal incinerator ashes, shooting range where it is present in different forms as $PbCO_3$, $PbSO_4$, PbO and PbO_2 (Hessling et al. 1990; Nedwed and Clifford 1997). In fact, prior to establishment of adequate pollution control regulations, recycling of batteries was promoted to extract Pb leaving residual acids and empty cases on sites. Pb was eventually piled up in heaps before being remelted on the same site or elsewhere. These methods led to Pb contamination of the order of thousands to several hundreds of thousands of mg Pb/kg soil (Nedwed and Clifford 1997). Chipping paint containing lead is also a significant source of contamination in soils near buildings and is a problem in old housing sites (USEPA 1994; Xintaras 1992).

11.2.7 Selenium

Inorganic selenium (Se) in soils, sediments, and waters is frequently present as Se (VI) in the form of selenate and Se (IV) in the form of selenite. The biological reduction of the selenate and selenite can be used for formation of elementary Se which is less toxic and insoluble. This strategy was proposed for the treatement of Se contaminated sites (Cantafio et al. 1996; Mattison 1992). The biougmentation of methylation of the selenium (CH_3SeCH_3), which occured naturally in the soils, was also proposed to decontaminate sites polluted by the toxic element. The dimethyl selenide generated was then eliminated by volatilization (Frankenberger and Karlson 1994).

11.2.8 Zinc

Zinc (Zn) is almost always found in the Zn (II) valence state. Zinc is used as an anti-corrosive agent for iron alloys and steel. It is a highly mobile metal present in groundwater in soluble forms at neutral and acidic pH (Smith et al. 1995). Major inputs to the aquatic environment are likely from the mining operations, manufacture of alloys, galvanizing plants and paper production. Zinc does not generally pose a serious risk to human health. However, it is a phytotoxic agent like copper and nickel.

11.3 Regulations Concerning Metal Contamination of Soils and Groundwater

In order to determine the degree of contamination of soils, most of the industrialized countries endowed themselves or are in the process of doing so, to establish laws and norms aiming at generic criteria on metal contamination in soils. Unfortunately, these criteria are not uniform for countries, a state/a province, and special attention must be paid to comprehend different terminologies before understanding environmental regulations.

In United States, a minimum criterion has been established by USEPA, leaving room for different states to apply stricter restrictions, if desired (USEPA 1992). This regulation such as Toxicity Characteristics Leaching Procedure (TCLP) stipulated in the RCRA land disposal restrictions developed by USEPA or based on risk analysis. The TCLP is a dilute acetic acid extraction which simulates generation of acid in a sanitary landfill site during decomposition of wastes. It does not therefore necessarily predict absorption of metals by human digestive system. Indeed, in contrast to simple total digestion of soils, TCLP allows evaluation of the mobility of contaminants in the soil before or after decontamination. Despite limitations of not representing effective mobility of contaminants, this method is highly recommended and used worldwide. Besides, in May 1998, USEPA promulgated obligatory treatment of contaminated soils (derived from American documents 63 FR 28555 and 40 CFR 268.49) that were previously confined for several years due to toxic wastes. For contaminated soils, USEPA regulates a treatment that would reduce the concentration by 90% or allow a respectable maximum contamination level defined as (10 times the universal treatment standard). Nevertheless, this new regulation remains less constraining for environmental intervention and hence new regulation for characterization of soils is based on leaching (TCLP) and not total concentration (mg/kg) employed in several countries.

However, different countries amended their regulations according to projected usage of soils. In this retrospective, Netherlands carried out such changes to fit decontamination markets. In fact, such measures will reduce restoration cost of 110,000 contaminated sites in this country from US $ 50 billion to 15-25 billion (dollars of 1996) (NATO/CCMS 1998). In United Kingdom (Table 11.1), concentrations of metal contaminants are compared with generic criteria of interdepartmental committee on redevelopement of contaminated sites (ICRCL) (Anon 2002). Nevertheless, this approach encompasses numerous limitations as it often refers to total contamination and not to mobility and bioavailability of contaminants.

Table 11.1 Generic criteria followed in United Kingdom according to ICRCL (Anon 2002).

Generic Criteria	Contaminant (mg/kg)							
	As	Cd	Cr*	Cu	Hg	Ni	Pb	Zn
Residential	10	3	600 (25)		1		450	
Parks, playgrounds and open spaces	40	15	1,000 (25)		20		750	
Green space				130		70		300
Green space, determined by extraction with 0.05 mol/L of EDTA				50		20		130

*Values in parentheses refer to Cr(VI) determined by extraction with 0.1 mol/L HCl at 37.5°C.

Table 11.2 compares the generic criteria used in some countries, provinces or states (Nedwed and Clifford 1997; NATO/CCMS 1998; Pronk 2000; Anon 2002). This table shows important variations in regulations of contaminated soils. Thus, Zn (for a similar criterion) varies between 300 and 1,500 mg/kg, Pb between 530 and 1,000 mg/kg and Hg between 1 and 20 mg/kg. Hence, soils judged as contaminated by certain countries may not be so for others. Similary, some metals are judged problematic (Ba and Sn) by a few countries and are not at all important for others.

11.4 Treatability of Soils and Contaminated with Heavy Metals

In order to develop adequate site restoration technologies for soils contaminated with heavy metals, knowledge of geochemical distribution in these matrices becomes a necessity. Thus, speciation of heavy metals in soils was the theme of several studies (Khan and Frankland 1983; Brown et al. 1986; Gibson and Farmer 1986; Calvet et al. 1990; Van Beschoten et al. 1997; Galvez-Cloutier and Dubé 1998; Papassiopi et al. 1999). The speciation of metals has significant effects on their water solubility, transport, toxicity, and treatment (Calmano et al. 2005).

Table 11.2 Comparisons of generic criteria established for some countries, provinces or states.

Country, Province or State	Contaminant (mg/kg)													
	Ag	**As**	**Ba**	**Cd**	**Cr**	**Co**	**Cu**	**Hg**	**Mo**	**Ni**	**Pb**	**Se**	**Sn**	**Zn**
Canada - Québec (1988)	40	50	2,000	20	800	-	500	10	40	500	1,000	10	300	1,500
Canada - Ontario (1989)	50	50		8	1,000	100	300	2	40	200	1,000	10		800
United States (1997)											1,000			
California (1997)		500		100	500			20			1,000			
Minnesota (1997)											500			
Florida (1997)		4		600	430						1,000			
Pennsylvania (1997)		3		20	300	20					500			
United Kingdom (2002)		40		15	1,000			20			750	6		300
Canada – B.C. (1989)	40	50	2,000	20	800	300	500	10	40	500	1,000	10		1,500
Netherlands (2000)		55	625	12	380	240	190	10	200	210	530			720
Italy (2002)				10	800		600			500	1,000			1,500
New Jersey (1990)		20	400	3	100		170	1	1	100	1,000	4		350

According to these studies, the chief mechanisms responsible for geochemical distribution of heavy metals in soils and sediments can be classified into four groups as, 1) adsorption to surface of clay minerals, oxides/hydroxides (Fe, Mn, Al) and organic matter, 2) precipitation -coprecipitation with secondary minerals (carbonates, oxides/hydroxides, sulfides), 3) complexation with organic matter and sometimes; 4) penetration of heavy metals into crystalline structure of primary minerals (Galvez-Cloutier and Dubé 1998; Papassiopi et al. 1999; Van Beschoten et al. 1997). These different modes of retention of heavy metals by soils play an important role at the level of potential transfer between solid and liquid phases and therefore in the selection of adequate restoration methods (Grasso et al. 1997; Papassiopi et al. 1999).

There are also a series of biotests (e.g. MicroTox, BIOMET), which were developed in order to predict the bioavailability of heavy metals in soils and other environment (OECD 1984; USEPA 1989; van der Lelie et al. 2005).

The selection of treatment method also depends on nature of contaminants, their concentration, decontamination level required, as well as physico-chemical characteristics of the matrix to be treated (USEPA 1997; Duchesne and Mercier 2003). It is therefore imperative to examine these different parameters before proposing a decontamination scheme for a specific soil. In this respect, Canadian researchers at Institut National de la Recherche Scientifique-Eau, Terre et Environnnement (INRS-ETE) proposed a five-step protocol to evaluate in the treatability of metal contaminated soils, sediments and sludges (Environment Canada 1998). Table 11.3 briefly summarizes the protocol of evaluation adopted.

Although the protocol considers only physico-chemical separation procedures for treatment of soils or sediments, yet it constitutes a logical and coherent basic approach, which could be extrapolated to evaluate chemical and/or biological separation procedures. In fact, a complete chain of decontamination of metal contaminated soils includes in several cases, integrated physico-chemical separation procedures to reduce volume load of contaminants as well as chemical and or biological methods of extraction (Ganguly et al. 1998).

Table 11.3 Protocol for evaluation of treatability of soils, sediments and sludges by mining/metallurgical technologies (Environment Canada 1998).

Step	Procedure
Sampling	Historical study of contamination Sampling and preservation
Characterization by mineralogical approach	Granulometric analyses Chemical analyses Mineralogical study
Confirmation of application of mining/metallurgical technologies	Study of characterization data by mineralogical approach
Elaboration of preliminary treatment scheme	Conciliation of characterization data with the conditions of use and performance of various mining/metallurgical equipments
Validation and optimisation of treatment scheme	Trials at laboratory and semi-industrial scale

11.5 Restoration of Soils Contaminated with Heavy Metals

The decontamination of soils polluted by toxic metals constitutes a complex challenge. Several technological approaches were thus developed and proposed to solve the problem of presence of toxic metals in polluted sites. These suggested methods of restoration included in situ as well as ex situ treatment processes. The Table 11.4 illustrates principal technologies of restoration of soils contaminated by toxic metals.

11.5.1 Solidification and Stabilization

Solidification implies a physical process of transformation of liquid substances into solid substances by addition of solidification agents such as cement, pozzolanic material, silicates, lime and other reactive minerals that are generally utilized and are cost-effective. Besides, Portland cement is commonly employed solidifying agent as it permits complete immobilization of several metals. On the contrary, stabilization technologies involve chemical reactions where chemical agents (such as polymers) cause precipitation of metals making them insoluble. Organic materials normally utilized as chemical agents comprise asphalt, polyethylene, paraffins, waxes and other polyolefins.

These processes can be performed on-site in closer proximity of the contaminated soils with the aid of containers or directly in soil (in situ). When the technology is applied on the specific site, it requires removal of gravel and sand and washing followed by mixing of soils with selected chemical agents. This mixture is then deposited in solidification moulds. Subsequently, elimination of products is performed in controlled discharge or in containment cells. When in situ technology is utilized, it is necessary to develop a method for efficient mixing of soils on-site and a system of storage and distribution of reagents. For example, pastes can be injected at higher speeds into soils at drill points. The radius thus treated is often several meters around the well.

Table 11.4 Restoration technologies used for treatment of soils polluted with heavy metals.

Type of Soil Treatment	Example of Technology
Ex-situ treatment	
Solidification/Stabilization	bituminization, asphalt, cement, polyethylene intrusion, vitrification
Isolation and confinement	on site isolation with membrane
Physical extraction	gravimetric separation, flotation, magnetic separation, pyrometallurgical separation
Chemical separation	acid extraction, leaching with chelating agents, biosurfactants or oxidants
Bacterial leaching	bioreactor leaching, heap biological leaching
In-situ treatment	
Electrokinetic separation	electrorestoration, electro-acoustic treatment
Phytoremediation	phytoextraction, phytostabilization
Excavation and off-site disposal	

The solidification and immobilisation technologies are effective as the soils treated by these technologies generally respect environmental norms for quantity of metals measured in soil leachates in comparison with concentrations obtained in leachates of non-treated soils. Thus, concentrations can often be reduced from several tens or hundreds of mg/L to less than 0.1 mg/L with an overall reduction of 95% (Freeman 1997).

However, solidification reactions are exothermic and it is necessary to treat gaseous emissions in case of volatile compounds. Further, they induce alkaline pH, limiting their usage in presence of amphoteric metals such as Pb and Al. These methods also bear a drawback of augmenting volume of material on-site. It is also to be noted that their effectiveness is not absolute as salting out of certain contaminants is possible even after long time treatment. Often, this treatment type is reserved for cases where other veritable method of decontamination is not applicable (USEPA 1993b).

Finally, vitrification (molten glass) processes are solidification methods that employ heat up to 1200°C to melt and convert waste soil into crystalline products. Heavy metals are incorporated into the glass structure which is, usually, a relatively strong, durable material that is resistant to leaching. Despite greater efficacy of vitrification, it encompasses several drawbacks. Thus, it cannot be used in the presence of soils saturated with water as heat would be used up in evaporation. Additionally, depth of soils to be treated is limited as one cannot exceed 6 m. Further, this technology does not allow revegetation of treated zones. The energy consumption is also very high ranging from 800 to 1,300 kWh per tonne of treated soil making it an expensive approach (USEPA 1993b).

11.5.2 Isolation and Confinement

Although isolation and confinement methods do not draw much practical application, yet they are employed as alternatives for soil decontamination owing to economical reasons. They consist of isolation and containment of contaminants in a controlled zone where migration of pollutants is inhibited or reduced to an acceptable minimum level. In these technologies, physical barriers of steel, cement, bentonite or other solid and inert material are erected as horizontal as well as vertical walls to contain contaminated soils. Establishment of an extraction system for groundwater flow is often required to avoid passage of contaminants under these barriers (Mulligan et al. 2001a).

11.5.3 Mechanical Separation

Mineral processing technologies can be efficiently used to separate inorganic contaminants from soils or sediments. These physical methods notably include particle size separation (screening and hydrocyclone), magnetism, gravimetric methods (Wilfley table, spiral and jig), and flotation. The applicability of these methods has been discussed by Mercier et al. (2001). The principal advantages associated with the use of these mining technologies are their relatively low costs and their capacity to treat large volumes of soil.

Table 11.5 summarizes four principal processes of physicochemical separation used to restore a site with relevant principles and conditions of use. Often, this type of treatment is followed by reprocessing of the most contaminated fractions by hydrometallurgical (chemical or biological technologies) and pyrometallurgical technologies (Mercier 2000; Mercier et al. 2002b).

Table 11.5 Physicochemical separation processes for decontamination of soils (Environment Canada 1998).

Technology	Principle	Condition of Use
Gravimetric concentration (dense medium separator, spiral, multiple gravity separator, etc.)	Difference in density between minerals	Particle sizes higher than 1 μm Difference between densities of materials to be separated higher than 1 g cm-3 Higher degree of release of inorganic contaminants
Flotation (flotation cell and column)	Hydrophobic and absorbent properties of surfaces of particles	Particle sizes higher than 5 μm Surfaces of particles to be separated showing natural or stimulated hydrophobic properties (addition of chemical reagents) Average degree of release of inorganic contaminants (unagglomerated forms, melted or chemically associated with matrix)
Magnetic separation (conveyor with magnetic head, wet separation drum)	Magnetic properties of minerals	Particle sizes higher than 75 μm Metal contaminants showing ferromagnetic properties (iron, magnetite) Average degree of release of inorganic contaminants
Electroacoustic separation (electrostatic fractionator or electrodynamics)	Difference in electrical conductivity between minerals	Particle sizes between 60 and 500 μm The materials to be separated must be perfectly dry Higher degree of release of inorganic contaminants

11.5.4 Pyrometallurgical Separation

The physical treatments of contaminated soils also include the pyrometallurgical processes using the furnaces (rotary kilns, arc furnaces, rotary hearth furnaces) at high temperatures (200 to 700 °C) to evaporate the contaminants (Vanbroekhoven et al. 2005). Once volatilized, metals must be recovered or immobilized. This approach was mainly applied for the recovery of mercury (Smith et al. 1995). The use of pre-treatment techniques by mechanical separation is often necessary in order to decrease the volume of soils to be treated thermally.

11.5.5 Chemical Separation

The removal of heavy metals from groundwater can be achieved by use of various chemicals, such as inorganic acids (HCl, H_2SO_4 or HNO_3), organic acids (acetic or citric acids), chelating agents (EDTA, DTPA, NTA) and oxidizing agents (NaClO, H_2O_2). These technologies of soil washing involve use of significant quantity of water in which the metals are solubilized by various chemicals. Subsequently, this process water must be treated in order to recover metals and, thus, to be able to recycle water and the solubilizing products during several soil washing cycles. Figure 11.1 shows a general outline of treatment of soil by chemical solubilization. Following sections describe various hydrometallurgical processes used, or proposed, to restore contaminated sites.

11.5.5.1 Acid Extraction

Extraction of metals from soils with inorganic acids such as sulfuric, nitric or hydrochloric acid was the objective of several research studies. For example, Tuin and Tels (1990) have listed 16 studies performed throughout the world on the use of these chemical reagents, with or without combination with heat treatment for decontamination of soils.

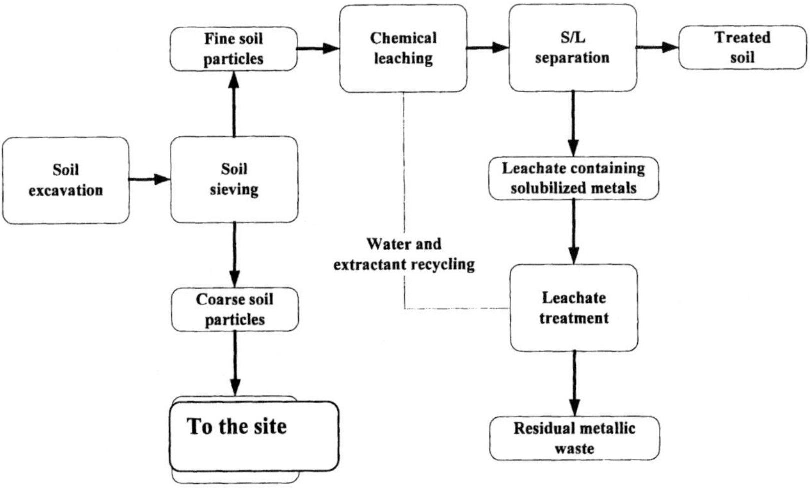

Figure 11.1 General scheme of soil treatment by chemical separation.

Prior to acid extraction, soils are usually first screened to remove coarse solids. Dilute acid (pH < 3) is then added into the soil in the extraction unit (batch- or continuous-stirred tank reactors). The residence time in the reactors varies depending on the contaminants, contaminant concentrations, soil type, but generally is short (less than 1 h). After reaction, the soil and extractant are separated using standard solid/liquid (S/L) separation equipments (e.g. hydrocyclones, settling tank, and filter). When the separation is complete, the solids are rinsed to remove acid and metals present between the soil particles. Finally, the soils are mixed with lime to neutralize any residual acid.

In general, these studies illustrated that the use of concentrated acids favour solubilization of metals, and in many cases, solubilization yields of metals remained modest. Moreover, use of concentrated acid solution considerably increased chemical costs, involved production of significant quantities of metal residues during neutralization of leachates and caused severe damage to soil structure (excessive solubilization of soil solids).

The extraction solution and washings streams are usually treated using precipitants, such as lime or sodium hydroxide. The precipitated metals during the liquid waste treatment are generally concentrated by dewatering and must then be conveyed to a hazardous waste storage site. In certain cases, it can be possible to recycle the metal residue generated during the chemical decontamination of soil. The treated effluents are normally recycled to the starting point of the treatment process to be used for other soil washing cycles. Other techniques of treatment can also be used for the treatment of acidic effluents charged with metals. For example, metals can separated by membrane processes (e.g. nanofiltration) (Ortega et al. 2005), adsorption/ ion exchange (Meunier et al. 2002, 2004a), and also by electrochemical processes (electrodeposition/ electrocoagulation) (Meunier et al. 2004b).

11.5.5.2 Leaching with Chelating Agents

Formation of soluble metal complexes by reaction with certain chelating agents constitutes another approach that has been seriously considered for removal of heavy metals from contaminated soils. Chelating agents are compounds having the property to strongly bind and in a more or less specific manner to a bivalent and trivalent positive ion (particularly metal) with which they form a compound (or chelate) in which central atom is linked to neighbouring atoms by at least two bonds by forming an annular structure. Ethylenediamine tetraacetic acid (EDTA), nitrilotriacetic acid (NTA) and diethylenetriaminepentaacetic acid (DTPA) are principal chelating agents that have been tested for removal of heavy metals (Fisher et al. 1998; Legiec 1997; Rampley and Ogden 1998; Samani et al. 1998; Zeng et al. 2005). Several other chelating agents have been also tested for metal removal from soils. For example, Chen et al. (1995) studied the effectiveness of more than 190 ligands like agents of target metal removal and recovery (Cd, Cu, Pb, Hg, Ni and Zn) from contaminated soils.

The driving factors favouring use of chelating agents are higher leaching capacity of these agents, higher thermodynamic stability and good solubility of metal complexes so formed, a weak adsorption of sequestering agents and metal compounds by soils and a greater stability of metal compounds which inhibit eventual chemical and biological transformations (Fisher et al. 1998).

Assink and Rulkens (1987) evaluated removal of metals from nine different soils on treatment with Na_3NTA at concentrations varying between 20 and 100 g/kg of soil and at pH adjusted between 3.0 and 3.5 with HCl. The yields of extracted metals were poor in comparison to those obtained by leaching in three stages with hydrochloric acid at pH 1.0. The removal yields of different metals with NTA were 87% As, 96% Cd, 74% Cu, 9% Cr, 14% Hg, 88% Pb and 43% Zn.

In their study, Tuin and Tels (1990) compared extraction of metals of clayey soils by hydrochloric acid washing and EDTA treatment (0.005 to 0.1 M). Out of six metals investigated (Cd, Cr, Cu, Ni, Pb and Zn), only Cd and Pb were extracted very well by treatment with chelating agent vis-à-vis treatment with acid. On the other hand, solubilization of metals increased with increase in EDTA concentrations in soil and also by partial acidification ($pH < 4$).

Further, recycling of sequestering agents or reduction in dosing concentrations was seriously explored because of high costs incurred. In this context, Brown and Elliot (1992) tested combined use of various electrolytes with EDTA. In the absence of added electrolytes, these researchers measured an extraction of approximately 65% of Pb present in soil samples treated for 5 h with 0.04 M EDTA at pH ranging between 5 and 9. Under these conditions, addition of 0.5 M lithium or sodium, ammonium perchlorate salts aided in augmentation of extraction yields to be around 80%. The use of magnesium or calcium perchlorate salts also aids in solubilisation of Pb, but in slightly acidic conditions (pH 4 to 6). However, economical repercussion of this approach was not discussed by these researchers.

Lastly, researchers at Office of Mines (USA) reported several problems associated with use of EDTA on contaminated sites at a large scale like: high costs of reagent, difficulties in filtering sands and clays and recycling of reagent. In fact, they contemplated that problems of filtration were caused by presence of colloids created by bonds formed between EDTA, metals and soil particles. Thus, two tests conducted (Firm Lee, Woodville, Wisconsin and ILCO Site, Leeds, AL) by Office of Mines confronted major problems during operation of pilot units, particularly at the stage of filtration (Nedwed and Clifford 1997).

Moreover, use of oxidizing agents for stabilization of contaminated soils is prone to controversy for in situ applications. Indeed, EDTA and its derived compounds are degraded very slowly in various soils influencing speciation and bioavailability of essential or toxic elements. The risk of pollution of groundwater with metal compounds and chelates must also be taken into account. Further, toxicological health risks to humans arising from the addition of possibly mutagen agents such as NTA in the environment strongly discouraged their use (Fisher et al. 1998).

11.5.5.3 Leaching with Biosurfactants

The surfactants of biological origin (biosurfactants) and diverse composition can also be employed for the removal of contaminants from soils and sediments. Thus, Mulligan et al. (2001b) reviewed various studies relating to the effectiveness of biosurfactants comprising trehalose lipids, rhamnolipids, sophorolipids, diglycosyl diglycerides, lipopolysaccharides, lipopeptides, ornithine, lysine peptides, phospholipids, sulfonyllipids and fatty acids. With regard to the removal of metals, the use of rhamnolipids and sophorolipids proves promising (Mulligan et al. 1999). The partial or total biodeterioration of these compounds of biological origin constitutes an advantage compared to the use of the chemical sequestering agents. Their production cost is generally higher than those of the chemicals representing a constraint with the use of these bioproducts.

11.5.5.4 Leaching with Other Chemical Compounds

In addition, certain metals such as Cr cannot be removed efficiently by leaching with inorganic acids. Even the use of sequestering agents such as EDTA does not allow good removal yields of Cr from clayey soils as observed by Tuin and Tels (1990). Several other chemical reagents (oxalic acid, sodium dithionite, hydroxyammonium hydrochloride, phosphate solutions) were also tested without any success (Tuin and Tels 1990). In fact, if soils are contaminated originally by oxidized form of Cr (Cr(VI)), this metal is usually present in soil in its reduced form (Cr(III)), following reaction with soil organic matter. Consequently, Cr(III) forms Cr hydroxides and later on very insoluble Cr oxides (Cr_2O_3). The formation of polymeric complexes of Cr hydroxides also reduces solubility of this contaminant.

When Cr in soil is mainly in Cr(III) form, use of an oxidizing agent could be a good alternative, because oxidation of Cr to its Cr(VI) form enhances the mobility of this element. Tuin and Tels (1990) proposed use of sodium hypochlorite (NaClO) and hydrogen peroxide (H_2O_2). NaClO was preferred to H_2O_2 as effectiveness of the latter is lowered by its natural decomposition. In acidic medium, oxidation of Cr by hypochlorite solution occurs according to the reaction presented below:

$$2\ Cr^{3+} + 3\ HClO + 4\ H_2O \rightarrow Cr_2O_7^{2-} + 3\ Cl^- + 11\ H^+ \qquad \text{(Eq. 11.1)}$$

In alkaline medium, oxidation of Cr occurs according to:

$$2\ Cr(OH)_3 + 3\ ClO^- + 4\ OH^- \rightarrow 2\ CrO_4^{2-} + 3\ Cl^- + 5\ H_2O \qquad \text{(Eq 11.2)}$$

With clayey soils, Tuin and Tels (1990) reported extraction yields of 43% for Cr after 0.5 h of treatment with sodium hypochlorite (0.46 M) at pH 8.5. This yield rises to 70% with a thermal treatment (80 °C).

The oxidizing agents can also be used as complements with inorganic acids in order to raise solubilization of various metals. Assink and Rulkens (1987) carried out tests on leaching of soils with hydrochloric acid preceded by an oxidation step carried out with hydrogen peroxide, sodium hypochlorite, potassium permanganate or by treatment with air or thermal treatment.

In addition, decontamination of soils polluted with heavy metals was also considered by leaching with organic acids. However, research undertaken by Adams and Alloway (1988) who tested use of acetic acid (0.05 M), reported lower yields (< 20%) for solubilization of metals (Cu, Mn, Ni, Pb and Zn) and associated costs were prohibitive.

The removal of metals from soils was also explored by using concentrated salt solutions allowing ion exchange. In this context, Adams and Alloway (1988) used ammonium nitrate (1 M) for five soil samples. However, yields of solubilization of metals (Cu, Mn, Ni, Pb and Zn) were far from encouraging (< 14%).

Finally, Nedwed and Clifford (2000) and Djedidi et al. (2005) have demonstrated the possibility of removing Pb from soils using a concentrated sodium chloride solution. Indeed, chloride ions are able to react with Pb and form soluble salts complexes, such as $PbCl_4^{2-}$, $PbCl^{3-}$ or $PbCl_2(aq)$. Likewise, Pb removal has been evaluated simultaneously using sodium chloride solutions and strong oxidizing agents (Lin et al. 2001).

11.5.6 Bacterial Leaching

The metabolic capacity of certain microorganisms allows passage of metals in solution from soils constituting a curious biological phenomenon studied by environmental microbiologists. This phenomenon catalyzed by bacterial activity is called bioleaching, and requires the presence of microorganisms able to proliferate in extreme ecosystems (strongly acidic pH, highly oxidizing conditions, and high concentration of metal ions in solution) capable of drawing their energy from oxidation of mineral sulfides.

Processes of extraction of metals by biological solubilization have been well studied and used for several years in the field of biohydrometallurgy (Guay et al. 1976; Torma 1986). Bioleaching of metals can be accomplished directly by microbial metabolism or indirectly by products of their metabolism (Lundgren and Silver 1980). A simple schematic of metal-microbial interactions, comprising bacterial leaching among others is presented in Figure 11.2. In the direct mechanism, leaching bacteria directly oxidize insoluble metal sulfides (MS) to soluble sulfate, according to following equation:

$$MS + 2\ O_2 \rightarrow M^{2+} + SO_4^{2-} \qquad \text{(Eq. 11.3)}$$

This direct oxidation of metal sulfides was shown for several metals: Cd, Ni, Zn, Co, Pb, Cu, Fe, Ga, Mn and Sb (Rossi 1990; Torma 1987).

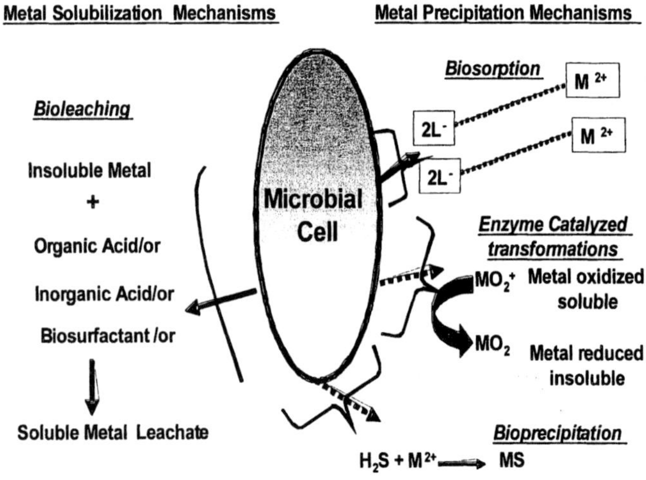

Figure 11.2 Schematic of metal-microbial interactions during different bioremediation processes.

In the indirect mode, metal sulfides are oxidized in a purely chemical way by Fe^{2+} ions that produce SO and metals in ionic form. The Thiobacillus then oxidizes this SO to H_2SO_4 and *Acidithiobacillus ferrooxidans* oxidizes Fe(II) to Fe(III). The cycle restarts with Fe(III) as summarized in equations below:

$$2\ Fe^{2+} + \frac{1}{2}\ O_2 + 2\ H^+ \rightarrow H_2O + 2\ Fe^{3+} \quad \text{(Eq. 11.4)}$$

The metal sulfides (MS) are oxidized by ferric ions according to the reaction:

$$MS + 2\ Fe^{3+} \rightarrow SO + M^{2+} + 2\ Fe^{2+} \quad \text{(Eq. 11.5)}$$

Subsequently, elemental sulphur is reoxidised to sulphuric acid by *Acidithiobacillus ferrooxidans* or other species of Thiobacillus or sulphur-oxidizing bacteria:

$$SO + 1.5\ O_2 + H_2O \rightarrow H_2SO_4 \quad \text{(Eq. 11.6)}$$

The pH of the medium decreases and the redox potential increases contributing to the solubilization of metallic oxides (MO) and carbonates (MCO_3):

$$MO + H_2SO_4 \rightarrow MSO_4 + H_2O \quad \text{(Eq. 11.7)}$$

$$MCO_3 + H_2SO_4 \rightarrow MSO_4 + H_2O + CO_2 \quad \text{(Eq. 11.8)}$$

The bioleaching of metals can be carried out by mesophilic microorganisms, such as, *Acidithiobacillus ferrooxidans* and *Acidithiobacillus thiooxidans* (Kelly and Wood 2000), or, by thermophilic microorganisms (temperature of growth higher than 45°C), such as *Acidianus*, *Sulfobacillus* and *Sulfolobus* bacteria (Zillig et al. 1980; Dufresne et al. 1993).

The bioleaching of metals present in contaminated soils can be carried out by using various systems of extraction (bioreactors of the agitated type, heap leaching, and percolation columns) (Tichi et al. 1998; Zagury et al. 2001). This technological approach however is seldom used since it is generally not very effective for the solubilization of certain metals, for example, lead, which is not very soluble in the presence of high sulfate concentrations (Blais et al. 2001).

Another leaching method that has potential for soil decontamination is the use of organic acids (citric and gluconic acids) to solubilize metals. These acids can be produced directly by microorganisms like the fungus, *Aspergillus niger* (Mulligan et al. 2001a).

11.5.7 Electrokinetic Separation

Processes of electrorestoration constitute another interesting technology to remove heavy metals from contaminated soil and groundwater (Penn and Savvidou

1996; West and Stewart 1996). The electrorestoration technology consists of the introduction of a pair of electrodes into soil at a given distance and to introduce electric current that will favour movement of contaminants towards one of the electrodes where they are recovered (Chung and Kang 1999). The electrokinetic processes involve different electrochemical mechanisms as illustrated in Figure 11.3.

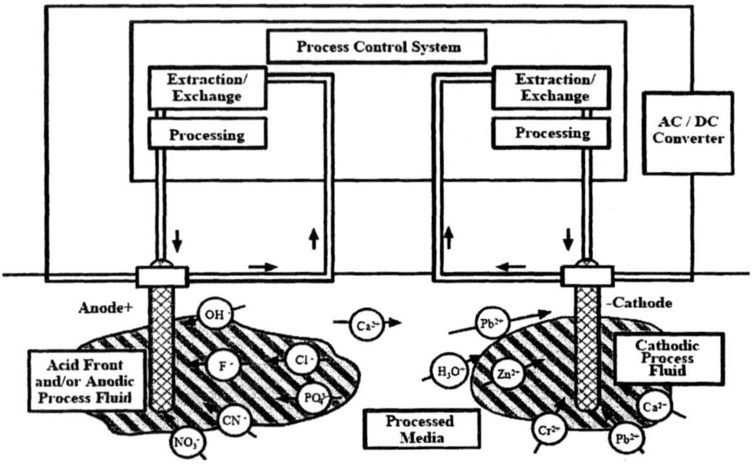

Figure 11.3 Schematic of a typical electro-kinetic treatment system (EPA-542-R-02-004, US EPA, 2002).

Electro-osmosis, which allows transportation of water, and electromigration, assures mobility of heavy metals, constitute the two principal forces responsible for extraction of contaminants by this technology (Acar et al. 1990; Ribeiro and Mexia 1997; Yeung 1994). Several electrochemical reactions between contaminants and soils such as electrolysis, adsorption and desorption of contaminants on clay particles, acidification of soil by transportation of H^+ ions and precipitation of inorganic species occur simultaneously (Acar and Alshawabkeh 1993; Rosand and Acar 1996; Yeung et al. 1997).

When electric current is induced in a saturated soil, water is electrolysed and gaseous oxygen and hydrogen ions are produced at the anode while hydrogen and hydroxide ions are produced at the cathode according to following reactions:

Anode : $2\ H_2O + 4\ e^- \rightarrow O_2 \uparrow + 4\ H^+$ (Eq. 11.9)

Cathode : $4\ H_2O + 4\ e^- \rightarrow 2\ H_2 \uparrow + 4\ OH^-$ (Eq. 11.10)

These reactions induce a decrease (acidification) in pH around the anode and increase (alkalinity) at the cathode. Thereafter, H^+ ions migrate towards the cathode by electrosmosis and electromigration while OH- ions direct themselves to anode. As the mobility of H^+ ions is greater than twice the mobility of OH^- ions, it produces gradual acidification of soil with the application of electrokinetics process. This acidification is sufficient to mobilize contained metals in soil as H+ ions tend to exchange with metallic ions adsorbed on clay particles. Further, lower pH allows dissolution of metallic precipitates (Chung and Kang 1999). Consequently, metals move towards the cathode by electrosmosis where they are pumped towards the surface with the help of two electrodes placed in recovery wells. The principal advantage associated with this technology is non-dispersion of the contaminant outside the treatment zone.

Electrokinetic processes were in particular used for the treatment of soils polluted by As, Cd, Co, Cr, Cu, Hg, Ni, Pb, Sr and Zn (Pamukcu and Wittle 1992; Lageman 1993; Ho et al. 1999; Suèr and Lifvergren 2003). In fact, a detailed review of the electroremediation of grounds polluted by heavy metals was presented recently by Ottosen and Jensen (2005).

11.5.8 Phytoremediation

Phytoremediation is the use of plants and their associated microorganisms, soil amendments and agricultural techniques for cleaning soils and water contaminated with toxic metals or organic pollutants (Schwitzguébel et al. 2005). Both decontamination (phytoextraction) and immobilization techniques (phytostabilization) are used for treatment of metal-polluted soils as illustrated in Figure 11.4. Phytoextraction consists of growing plants on contaminated sites so that polluting components infiltrate via radicle system of plant and accumulate in various parts (roots, stems, leaves). All plants (trees, herbs, grasses and crops) have the capacity to accumulate heavy metals from soil or water which are essential for their growth and development. These metals notably include Fe, Mn, Zn, Cu, Mg, Mo and Ni. Certain plants also have the ability to accumulate heavy metals which do not have known biological functions such as Cd, Cr, Pb, Ag, As and Hg.

Hyperaccumulating plants, such as *Alyssim, Arabidopsis, Urtica, Chenopodium, Polygonum* and *Thlaspi* are able to uptake and tolerate extraordinary levels of metals (Baker 1991; Vassilev et al. 2004). The criteria used for hyperaccumulation range from 100 mg/kg dry mass for Cd, to 1,000 mg/kg for Co, Cr, Cu, Ni, and Pb, to 10,000 mg/kg for Zn and Mn (Schwitzguébel et al. 2005).

Some tree species, mainly poplars (Populus) and willows (Salix), which are not hyperaccumulator plants, have also been seriously considered in phytoremediation because of their fast and easy growth, and their deep root system (Robinson et al. 2000;

Pulford and Watson 2003). Sunflower (*Helianthus annuus L.*) has a very good potential to accumulate moderate concentrations of heavy metals from contaminated soil and water (Lin et al. 2003; Madejon et al. 2003).

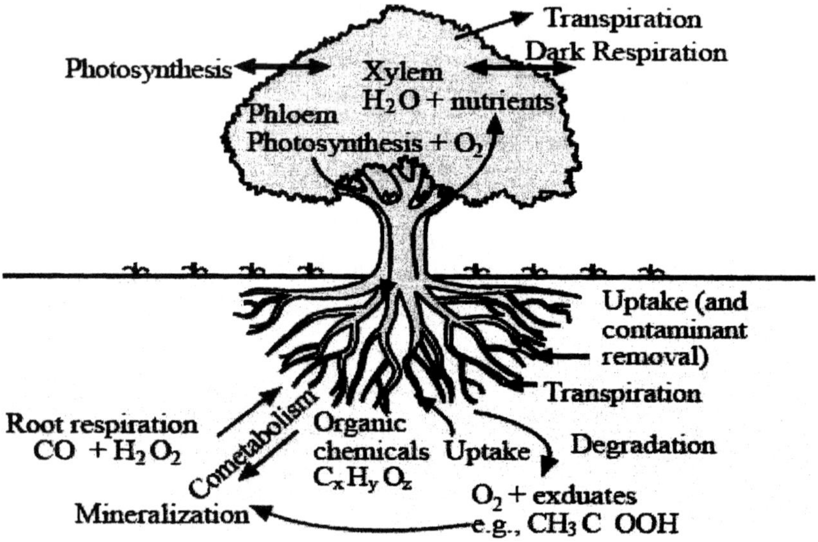

Figure 11.4 Phytoremediation of metal contaminated soils and groundwater (EPA-542-R-02-004, US EPA, 2002).

The technology of phytoremediation is very recent and made a debut in the beginning of 1990. It has been applied in situ and has been the subject of many research studies (Schwitzguébel et al. 2005). The application of this technology is however limited as contamination of soil should not exceed a certain depth so that roots of plants are in constant contact with metal pollutants. Moreover, climatic conditions and bioavailability of metals are pertinent factors to be evaluated. Similarly, it often takes a longer time to decontaminate a site due to growth limitations of the selected plant. It may sometimes be necessary to restore a site completely to proceed to several cycles of culture and harvest. Lastly, once contaminated, plants must be disposed of in an appropriate manner (Mulligan et al. 2001b).

11.5.9 Excavation and Off-Site Disposal

Finally, another way of management of the contaminated soils consists of excavation of material polluted by metals and/or other types of contaminants and transporting them to suitable disposal sites possessing containment infrastructures to avoid the migration of the contaminants in the receiving medium. This method represented the principal way of stabilization of contaminated soils before 1984 (FRTR 2005). With the recent development of various techniques of soil stabilization, the excavation and the disposal off site sites became less popular than in the past.

11.6 Treatment of Groundwater Polluted with Heavy Metals

In a way similar to decontamination of the soils, the treatment of groundwater polluted by metals can be done in situ, such as for example in the cases of the phytoremediation or the electrokinetic treatment. However, majority of the groundwater treatment options involve initial pumping of water and their subsequently treatment ex situ. Table 11.6 illustrates the principal techniques used and proposed for the removal of metals from groundwater.

Table 11.6 Ex-situ technologies for treatment of groundwater polluted with heavy metals.

Type of Groundwater Treatment	Example of Technology
Chemical precipitation	Precipitation as metal hydroxides, carbonates or sulfides, oxidation/reduction reactions
Bioprecipitation	Constructed wetlands, bioreactors (sulfate-reducing bacteria), for example, enzyme catalyzed transformations
Ion exchange	Synthetic resins, natural materials (zeolites, clays)
Adsorption	GAC, activated alumina, lignin, clays
Biosorption	Algae, bacteria, fungi, yeasts, sewage sludge, agro-industrial or forestry biomasses
Physical separation	Microfiltration, ultrafiltration, nanofiltration, reverse osmosis, membrane pervaporation, freeze crystallization, distillation
Electrochemical separation	Electrodeposition, electrocoagulation

11.6.1 Chemical Precipitation

The combination of precipitation/flocculation and sedimentation is a well-established technology for the removal of metals from groundwater, surface water and industrial effluents. In fact, according to the FRTR (2005) precipitation is becoming the most widely accepted methods for heavy metals removal from groundwater in pump and treat operations.

Metals can be removed by precipitation as metal hydroxides, sulfides or carbonates. Hydroxide precipitation with lime or sodium hydroxide is the most common method. Flocculating agents such as alum, iron salts and lime can be added to favorize the metal precipitates separation which is usually carried out in a settling tank. The resulting metallic sludge can be dewatered by filtration.

This technology encompasses various drawbacks, in addition to impossibility of recycling metals, precipitates of hydroxides tend to break when pH of solution varies. Moreover, presence of several kinds of metals in the same solution makes total precipitation more difficult due to minimal solubility of metals at different pH values (Couillard and Mercier 1992; Blais et al. 1999). Additionally, Cr(VI) is not recoverable by this technology, even cyanides cause interference and sequestering agents inhibit elimination of metals (Peters and Ku 1984).

Moreover, presence of highly acidic effluents containing sulphates involves formation of large quantities of secondary precipitates ($CaSO_4.2H_2O$) when alkaline agents such as calcium hydroxide (hydrated lime) are used. This increases quantity of metallic sludges considerably and raises costs of handling and disposal (Baltpurvins et al. 1997). They can also destabilize primary metal precipitate by inducing formation of colloids and hence solid/liquid separation stage becomes more difficult (Blais et al. 1999). It is thus preferable to use alkaline agents such as Na_2CO_3, NaOH, $(NH_4)2CO_3$ and NH_4OH, which form soluble by-products, when recovery of metals is necessary. Nevertheless, precipitation with hydroxides causes efficient removal of many metals like Cu, Zn, Fe, Mn, Ni and Co whereas it is often incomplete for Cd, Pb and Hg (Blais et al. 1999).

Precipitation with sulfides is carried out using reagents like Na_2S, NaHS, H_2S or FeS (Godd and Sund-Hagelberg 1985; McAnally et al. 1984; Robinson and Sum 1980). Metal sulfides exhibit lower solubility than their hydroxide counterparts, achieve more complete precipitation, and provide stability over a broad pH range. However, sulfide precipitation is generally more expensive than hydroxide precipitation. Moreover, there are also safety concerns associated with the possibility of H_2S formation in acidic conditions.

Reductive and oxidative techniques may also be used to detoxify or decrease the mobility of metal pollutants in groundwater. This appoach is commonly used in pump and treat operations, but can also be performed in situ by injection of the chemical into the groundwater (Evanko and Dzombak 1997). Potassium permanganate, hydrogen peroxide, hypochlorite or chlorine gas can be used for oxidization reactions, whereas reduction reactions are induced by the addition of alkali metals such as sodium, sulfur dioxide, sulfite salts and ferrous sulfate (Vanbroekhoven et al. 2005).

11.6.2 Bioprecipitation

The effluents containing metals can also be treated in biological systems in which metals are precipitated by the action of microorganisms. These systems also include passive systems of constructed wetlands (Lorion 2001), or, of the biological reactors comprising specifically sulphate reducing microorganisms (White et al. 1998; Alexander 1999).

When the groundwater or contaminated surface water are conveyed to a wetland ecosystem, metals are eliminated by different processes, viz., ion exchange, adsorption, absorption, and precipitation followed by different oxidation and reduction reactions of geochemical and biological origin. Principally, in wetland systems, the ion exchange reactions are attributed to contact of metals with humic acids and the organic substances. In addition, oxidation and reduction reactions catalyzed by bacteria present in the aerobic and anaerobic zones of the wetlands play a major role in the precipitation of metals in the forms of hydroxides and sulfides as presented in Figure 11.2. Precipitated and adsorbed metals elute in the stagnant water basins or are filtered by passage through the plants. The wetlands technology was very much used during the last several years in the treatment of the acid mining drainages (acid mine drainage: AMD) (USEPA 1993a).

Several types of reactors have been devised for the removal of metals by means of the sulfide generated by sulfate-reducing bacteria (Wildeman et al. 1995). The responsible microorganisms are notably species of *Desulfovibrio*. The sulfide formed by these anaerobes has also been used to treat metal-contaminated groundwater pumped through a biological treatment system (Barnes et al. 1994).

11.6.3 Ion Exchange

Various other techniques of metal recovery in solution, which are usually used in the treatment of industrial effluents, can also be employed in the removal of metals present in groundwater, during pump and treat operation. The use of ion exchange resins constitutes a good example of this fact. Ion exchangers are insoluble substances, composed of acidic or basic groups at molecular level, capable of modifying positive and negative ions at these groups, without alteration in their physical structure (Blais et al. 1999; Watts 1998) as shown in Figure 11.5.

In the beginning, ion exchangers used were natural substances containing alumino-silicates (zeolites, clays) (Eyde 1993; Kesraoui-Ouki et al. 1994). Nowadays most of the ion exchangers are of organic nature and termed as resins. Ion exchange resins are classified by their structure and properties. More specifically, exchange resins are classified as strong acid, weak acid, strong base, or weak base resins. Strong acid resins contain sulfonic exchange sites, and weak acid resins are characterized by carboxylic groups. Sulfonic group ($-SO_3^- H^+$) of a polystyrene resin, or, a chelating resin with iminodiacetic group are commonly used for extraction of metal cations from solutions (Vater et al. 1990; Duyvesteyn 1998). Commonly, treatment using diluted acid is employed for regeneration of resin (Brooks 1986).

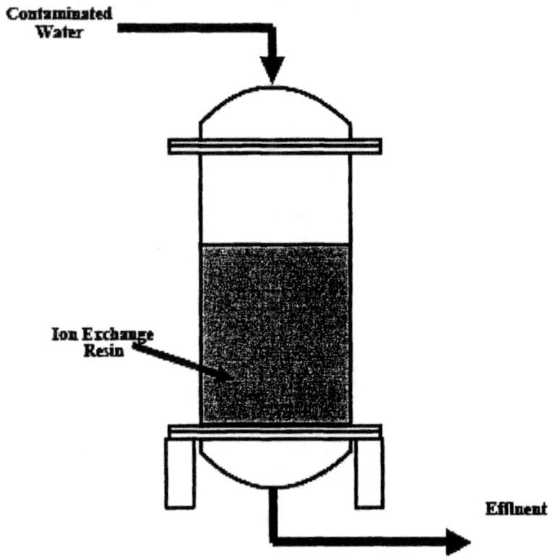

Figure 11.5 An ion-exchange system for treatment of a metal contaminated solution (US EPA, 2002).

Ion exchange is generally limited to effluents lower in insoluble colloidal particles, with total metal concentration <1000 mg/L and involves weak competition between cations (Brooks 1986). Therefore, this technique is used for advanced treatment and more rarely for detoxification of effluents rich in metals (Patterson 1988). The competition between cations is likely to be stronger and can prove to be an obstacle to allow complete treatment and attain regulatory standards for disposal. Finally, ion exchange is generally not very effective in the presence of highly acidic pH due to strong competition between H^+ ions and metal cations.

Developments in the field of metal recovery using ion exchange process also involve new chelate exchange resins having functionalities such as polyethylene imine dithiocarbamate, pyridyl imidazole, iminodiacetic-polystyrene and bispicolylamine (Brooks 1991). There is a whole range of ion exchange resins in market (Amberlite, Duolite, Dowex) bearing distinct characteristics with possibility for treatment of various types of effluents contaminated with metals.

11.6.4 Adsorption

Another example of technology which can be used for the elimination of metals following the pumping of groundwater is adsorption on natural or synthetic materials. Adsorption mechanisms are generally categorized as either physical adsorption, chemisorption, or electrostatic adsorption.

The most common adsorbent is granulated activated carbon (GAC). GAC is prepared from a number of carbon sources including wood, bituminous coal materials, lignite, and coconut shells (Watts 1998). Other natural and synthetic adsorbents notably include activated alumina, lignin and sorption clays. For last several years, adsorption has been the main objective of various research studies to evaluate its potential as a technology for removal of metals from solution. Bailey et al. (1999) and Babel and Kurniawan (2003) reviewed literature on capacity of low cost adsorbents like barks, lignin, chitosan, alginates, zeolite, clay, coal, fly-ash, peat moss, leaves, wool, cotton, natural oxide, and some others for metal removal from various types of effluents.

Adsorption phenomena are dependent on various experimental conditions, namely, pH, metal concentrations, ligand concentrations, competition between various ions and particle size (Bailey et al. 1999). The adsorption of metals by adsorption can be done in different systems, notably pulp units of agitated tanks, or, systems of percolation (columns or beds of percolation).

11.6.5 Biosorption

Many types of biomasses (yeasts, algae, fungi, bacteria and some other aquatic species) can also be used as adsorbent material for heavy metals recovery from effluents (Aksu and Açikel 1999; Al-Asheh and Duvnjak 1999; Alexander 1999; Atkinson et al. 1998). It implies use of live or dead biomass and/or their derivatives to adsorb metal ions with ligands or functional groups located on external surface of microbial cells (Volesky 1990; Wase and Forster 1997).

Examination of scientific literature reveals that use of biomass was largely studied for recovery of several metals like Ag, As, Cd, Co, Cr, Cu, Fe, Hg, Mn, Ni, Pb, Pd, Pt, Sn, U, Th and Zn (Brierley 1990; Prasetyo 1992; Kapoor and Viraraghavan 1995; Volesky and Holan 1995). Gardea-Torresdey et al. (2004) presented a review carried out on the use of the phytofiltration (adsorption on materials derived from plants) for the recovery of metals (Au, Cd, Co, Cr, Cu, As, Fe, Ni, Pb, Pt, Zn) from solution.

Mechanisms involved in the biosorption process include chemisorption, complexation, surface and pore adsorption-complexation, ion exchange, microprecipitation, hydroxide condensation onto the biosurface, and surface adsorption (Gardea-Torresdey et al. 2004). A simple mechanism of surface and pore adsorption-complexation is also presented in Figure 11.2. Spectroscopic studies and chemical modification have shown that cellular components including carboxyl, hydroxyl, sulfate, sulfhydryl, phosphate, amino, amide, imine, and imidazol moieties have metal binding properties and are therefore the functional groups in the biomass (Gardea-Torresdey et al. 2004).

11.6.6 Membrane Separation

Several types of membranes can be used for metal separation which includes ultrafiltration, nanofiltration, and Donnan membrane (Brooks 1991). Schematic of a simple membrane separation system is presented in Figure 11.6. Generally, membrane separation technologies are adequate for dilute solutions like washings. Literature reveals that these processes were applied for removal and/or recovery of various metals like Ag, Al, Cd, Co, Cr, Cu, Hg, Mn, Ni and Zn. On the other hand, membranes prove to be mechanically fragile and vulnerable to degradation by corrosion and oxidation. Finally, presence of insoluble particles or in suspension requires a preliminary stage of elimination as they clog membranes (Brooks 1991).

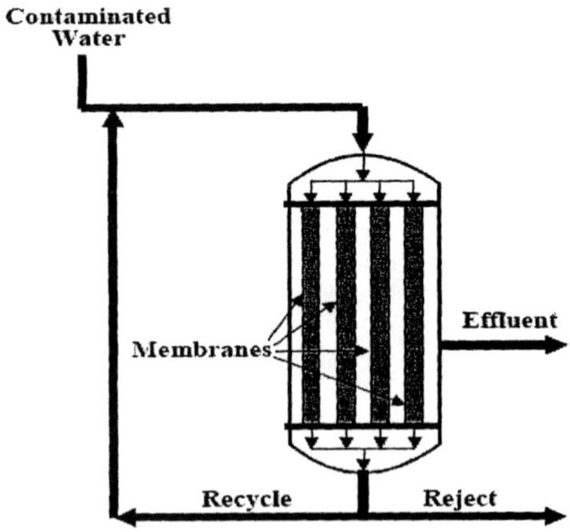

Figure 11.6 Schematic of a typical membrane separation system

Reverse osmosis and electrodialysis using semi-permeable membranes constitute two other membrane processes applicable for recovery of metal ions. As an example, use of reverse osmosis based on acetate membrane makes it possible to reach a removal rate between 95 and 98% (Brooks 1986). Electrodialysis process uses selective cation membranes which fit between electrodes in electrolytic cells. A continuous electrical current causes migration of ions and allows recovery of metals. Sulfonic polystyrene is often used as a selective membrane for cations (Blais et al. 1999).

11.6.7 Electrochemical Separation

It is also possible to recover several metals (Ag, Au, Cd, Co, Cr, Cu, Ni, Pb, Sn and Zn) by electrodeposition with insoluble anodes. In fact, in mining and metallurgical industries (leaching in heap, mining drainage) as well as processes of transformation of metals (plating and metal finishing wastes) and those from electronics industry, this technology is well established and largely used (Brooks 1991; Blais et al. 1999).

Thus, electrodeposition is usually used with monometallic solutions (only one metal) of minimal concentration of at least 1% (w/w) (Brooks 1986, 1991). On the other hand, installation of electrodes on larger surface, rotary electrodes, fluidized bed electrodes and other improvements in geometry of electrodes to support kinetics of deposition is necessary when the effluent to be treated contains less than 1% (w/w) of contaminant so as to improve recovery efficiency without increasing excessive energy consumption (Brooks 1991).

It is also possible to use an electrical current to recover several metals (Al, Ba, Ca, Cd, Cr, Cu, Fe, Mg, Mn, Ni, Pb, Ra, Zn) in solution using electrocoagulation (Persin and Rumeau 1989; Renk 1989; Blais et al. 1999). In fact, these treatment systems called electrocoagulation apply mainly to effluents slightly charged with metals, usually concentrations lower than 200 mg/L contrary to processes of electrodeposition. This technology delivers the coagulant in situ by anodic dissolution and produces subsequently, iron (or aluminium) hydroxides having considerable sorption capacity, while the simultaneous cathodic reaction allows for pollutant removal either by deposition on cathode electrode or by flotation (evolution of hydrogen at the cathode).

Electrolytic treatment is characterized by simple equipment, brief retention time and easy operation, which would also contribute to reducing the operating cost in large scale application (Rumeau 1989; Rajeshwar and Ibanez 1997; Wendt and Kreysa 2001). Exact specifications of system are established according to contaminants and hydraulic load desired. Treatment of various types of effluents (electroplating effluent, acid mining drainage, leaching process effluents, sewage sludge leachate) by this technology gave interesting results (Dalrymple 1994; Lee 1989; Meunier et al. 2004; Renk 1989; Rojo 1979).

Recently, hybrid process (chemical and electrochemical processes combined) called successive saline leaching and electrochemical Pb recovery (SSLELR) process has been proposed for soil decontamination (Djedidi et al. 2005). The process consists of treating soil by means of a saline leaching procedure using high concentration of chloride ions at pH 3.0, followed by Pb recovery via an electrolytic cell using mild steel electrodes. Leaching and electrochemical treatment have been successively carried out in a closed loop, with a small amount of residual sludge.

The electrolytic recovery of heavy metals solubilized from soils using EDTA (Juang and Wang 2000) or citrate (Arévalo et al. 2002) has been studied using a two-chamber cell separated by a cation-selective membrane. The results showed that the cell was effective to remove heavy metals from solution containing strong complexing agent.

11.7 Conclusion

Metals removal from soils and groundwater can be carried out by various physico-chemical and biological methods. Stringent regulations on metal contaminants in different environmental compartments have shifted the focus from simple containment to abatement procedures. Physico-chemical techniques have been used frequently for metals removal and recovery from soils and groundwater. More recently, efficient biological systems have been developed, and because of cost–effectiveness, these will be applied in the future. In the future, high metal concentrations would demand integrated treatment systems comprising physico-chemical (mining processes, chemical leaching, electrochemical processes) and/or biological methods (bioleaching, biosorption, phytoremediation).

Currently there are several processes for restoration of soils and groundwater contaminated by heavy metals in the market. These technologies integrate several physical, chemical or biological methods in their chain of treatment. In fact, it is possible to have information on several technologies via internet from principal governmental sites as well as specialized environmental sites. For example, United States Environmental Protection Agency (EPA) has a program entitled Superfund Innovative Technology Evaluation (SITE) established by EPA's Office of Solid Waste and Emergency Response and the Office of Research and Development which supports development of technologies likely to be used on sites identified in the data bank of Superfund. The SITE program is administered by National Risk Management Research Laboratory in the Land Remediation and Pollution Control Division. The documents relating to this program (SITE) can be consulted from "http://www.epa.gov/ORD/SITE/".

The Hazardous Waste Clean-up Information (CLU-IN) Web Site, which is located at "http://www.clu-in.org", provides information about innovative treatment and site characterization technologies to the hazardous waste remediation community. This site developed by EPA and launched in 1996 describes programs, publications, organizations, and different tools for federal and state personnel, consulting engineers, technology developers, remediation contractors, researchers and citizens. The Federal Remediation Technologies Roundtable (FRTR) established in 1990 aims at establishing partnerships between various levels of government, industries and public in order to develop, test and evaluate innovative remediation technologies. Several documents and computer tools of this organization are available at "http://www.frtr.gov". Finally, EPA REACH IT (REmediation And CHaracterization Innovative Technologies) sponsored by EPA's Office of Superfund Remediation Technology Innovation (OSRTI), is a new system to seek, see, download and print information on innovative remediation technologies, accessible at "http://www.epareachit.org/index3.html". A research carried out on this Web site in January 2005 made it possible to find 79 remediation technologies for soils contaminated with heavy metals and 57 groundwater treatment technologies polluted with toxic metals.

References

Acar, Y.B., and Alshawabkeh, A.N. (1993) "Principles of electrokinetic remediation". Environ. Sci. Technol., 27, 2638-2647.

Acar, Y.B., Gale, R.J., Putnam, G.A., Hamed, J., and Wong, R.L. (1990) "Electrochemical processing of soils : theory of pH gradient development by diffusion migration and linear convection". J. Environ. Sci. Health, A25, 687-714.

Adams, S.J., and Alloway, B.J. (1988) "The effects of variations in temperature on the extraction of trace metals from soils by selected reagents". Environ. Technol. Lett., 9, 695-702.

Adriano, D.C. (1992) "Biogeochemistry of trace metals". Lewis Publishers, Boca Raton, Florida.

Aksu, Z., and Açikel, Ü. (1999) "A single-stage bioseparation process for simultaneous removal of copper (II) and chromium (VI) by using C. vulgaris". Proc. Biochem., 34, 589-599.

Al-Asheh, S., and Duvnjak, Z. (1999) "Sorption of heavy metals from synthetic metal solutions and industrial wastewater using plant materials". Water Qual. Res. J. Can. 34(3), 481-503.

Alexander, M. (1999) "Biodegradation and bioremediation". Academic Press, San Diego, California, 453 p.

Allen, H.E. (2002) "Bioavailability of metals in terrestrial ecosystems : Importance of partitioning for bioavailability to invertebrates, microbes, and plants". Society of Environmental Toxicology and Chemistry (SETAC), Pensacola, Florida, 176 p.

Anon (2002) "Soil guideline values for lead contamination". Department for Environment, Food and Rural Affairs, London, United Kingdom.

Arévalo, E.F., Stichnothe, H., Thoming, J., and Calmano, W. (2002). "Evaluation of a leaching process coupled with regeneratioin/recycling of the extractant for treatment of heavy metal contaminated solids". Environ. Technol., 23, 571-581.

Assink, J.W., and Rulkens, W.H. (1987) Cleaning soils contaminated with heavy metals. In: "Environmental Technology". Martinus Nijhoff (Ed), Proc. Second Eur. Conf. Environ. Technol., Amsterdam, Netherlands, June 22-26, 1987, Dordrecht, Netherlands, pp. 502-512.

Atkinson, B.W., Bux, F., and Kasan, H.C. (1998) "Considerations for application of biosorption technology to remediate metal-contaminated industrial effluents". Water SA, 24, 129-135.

Baath, E. (1989) "Effects of heavy metals in soil on microbial processes and populations (a review)". Water Air Soil Pollut., 47, 335-379.

Babel, S., and Kurniawan, T.A. (2003) "Low-cost adsorbents for heavy metals uptake from contaminated water: a review". J. Hazard. Mater., B97, 219-243.

Bailey, S.E., Olin, T.J., Bricka, R.M., and Arian, D. (1999) "A review of potentially low-cost sorbents for heavy metals". Water Res., 33(11), 2469-2479.

Baker, A.J.M., Reeves, R.D., and McGrath, S.P. (1991) In situ decontamination of heavy metal polluted soils using crops of metal-accumulating plants: a feasibility study. In: "In situ bioreclamation". Hinchey, R.E., and Offenbach, R.F. (Eds), Butterworth-Heinemann, Boston, Massachussets, pp. 601-605.

Baltpurvins, K.A., Burns, R.C., Lawrance, G.A., and Stuart, A.D. (1997) "Effect of electrolyte composition on zinc hydroxide precipitation by lime". Water Res., 31, 973-980.

Barnes, L.J., Scheeren, P.J.M., and Buisman, C.J.N. (1994) "Emerging technology for bioremediation of metals". Means, J.L., and Hinchee, R.E. (Eds), Lewis Publishers, Boca Raton, Florida, pp. 38-49.

Blais, J.F., Dufresne, S., and Mercier, G. (1999) "État du développement technologique en matière d'enlèvement des métaux des effluents industriels". Rev. Sci. Eau, 12 (4), 687-711 (in French).

Blais, J.F., Mercier, G., and Chartier, M. (2001) "Décontamination à l'échelle pilote de sédiments pollués par les métaux toxiques par lixiviation chimique and biologique". Can. J. Chem. Eng., 79, 931-940 (in French).

Brierley, C.L. (1990) "Bioremediation of metal-contaminated surface and groundwaters". Geomicrobiol. J., 8, 201-223.

Brooks, C.S. (1986) "Metal recovery from industrial wastes". J. Metals, 38, 50-57.

Brooks, C.S. (1991) "Metal recovery from industrial wastes". Lewis Publishers Inc., Chelsea, Michigan.

Brown, G.A., and Elliot, H.A. (1992) "Influence of electrolytes on EDTA extraction of Pb from polluted soil". Water Air Soil Pollut., 62, 157-165.

Brown, P.A., Gill, S.A., and Allen, S.J. (2000) "Metal removal from wastewater using peat". Water Res., 34(16), 3907-3916.

Brown, R.M., Pickford, C.J., and Davison, W.L. (1986) "Speciation of metals in soils". Int. J. Environ. Anal. Chem., 18, 135-141.

Calmano, W., von der Kammer, F., and Schwartz, R. (2005) Characterization of redox conditions in soils and sediments: heavy metals. In: "Soil and sediment remediation. Mechanisms, technologies and applications". Lens, P., Grotenhuis T., Malina G., and Tabak H. (Eds), IWA Publishing, London, England, Chap. 6, pp. 102-120.

Calvet, R., Bourgeois, S., and Msaki, J.J. (1990) "Some experiments on extraction of heavy metals present in soil". Int. J. Environ. Anal. Chem., 39, 31-45.

Cantafio, A.W., Hagen, K.D., Lewis, G.E., Bledsoe, T.L., Nunan, K.M., and Macy, J.M. (1996) "Pilot-scale selenium bioremediation of San Joaquin drainage water with Thauera selenatis". Appl. Environ. Microbiol., 62, 3298-3303.

Chapman, P.M., Wang, F., Janssen, C.R., Goulet, R.R., and Kamunde, C.N. (2003) "Conducting ecological risk assessments of inorganic metals and metalloids: current status". Hum. Ecol. Assess, 9, 641-697.

Chen, T.C., Macauley, E., and Hong, A. (1995) "Selection and test of effective chelators for removal of heavy metals from contaminated soils". Can. J. Civ. Eng., 22, 1185-1197.

Chung, H.I., and Kang, B.H. (1999) "Lead removal from contaminated marine clay by electrokinetic soil decontamination". Eng. Geol., 53, 139-150.

Chirwa, E.M.N., and Wang, Y.T. (1997) "Hexavalent chromium reduction by Bacillus sp. in a packed-bed bioreactor". Environ. Sci. Technol., 31, 1446-1451.

Couillard, D., and Mercier, G. (1992) "Précipitations sélectives des métaux solubilisés biologiquement de boues aérobies d'épuration". Can. J. Chem. Eng., 70, 1021-1029 (in French).

Dalrymple, C.W. (1994) Electrocoagulation of plating wastewaters. In: "15th Conf. Pollut. Prevention Control", American Electroplaters and Surface Finishers Society, United States Environmental Protection Agency, Kissimme, Florida.

Davidson, C.M., Duncan, A.L., Littlejohn, D., Ure, A.M., and Garden, L.M. (1998) "A critical evaluation of the three-stage BCR sequential extraction procedure to assess the potential mobility and toxicity of heavy metals in industrially-contaminated land". Anal. Chim. Acta, 363, 45-55.

DeFilippi, L.J. (1994) "Remediation of Hazardous Waste Contaminated Soils". Wise, D.L., and Trantolo, D.J. (Eds), Dekker, New York, NY, pp. 437-457.

Djedidi, Z., Drogui, P., BenCheikh, R., Mercier, G., and Blais, J.F. (2005) "Laboratory study of successive soil saline leaching and electrochemical lead recovery". J. Environ. Eng. Div. ASCE, 131(2), 305-314.

Dowdle, P.R., Laverman, A.M., and Oremland, R.S. (1996) "Bacterial dissimilatory reduction of arsenic(V) to arsenic(III) in anoxic sediments". Appl. Environ. Microbiol., 62, 1664-1669.

Duchesne, J., and Mercier, G. (2003) "Établissement d'une méthode de caractérisation minéralogique décrivant les sols contaminés par le plomb". Can. Geotech. J., 40, 731-741.

Dufresne, S., Blais, J. F., and Guay, R. (1993) Municipal wastewater treatment plants sludges: Source of organic carbon-tolerant, sulfur-oxidizing Thiobacillus and Sulfobacillus strains. In: "Biohydrometallurgical Technologies". Torma, A.E., Apel, M.L., and Brierley, C.L. (Eds), Int. Biohydrometallurgical Symp., Jackson Hole, Wyoming, pp. 267-276.

Duyvesteyn, S. (1998) "Some recent developments in ion-exchange resins". JOM, 50(10), 62-63, 87.

Elliot, H.A., and Brown, G.A. (1989) "Comparative evaluation of NTA and EDTA for extractive decontamination of Pb-polluted soils". Water Air Soil Pollut., 45, 361-369.

Environnement Canada (1998) "Protocole d'évaluation de la traitabilité des sédiments, des sols and des boues à l'aide des technologies minéralurgiques". Fiche d'information Technologies Saint-Laurent, Environnement Canada, INRS-Géoressources and Centre de recherches minérales, Sainte-Foy, Québec, Canada, 4 p. (in French).

Evenko, C.R., and Dzomback, D.A. (1997) "Remediation of metals-contamined soils and groundwater". Technology Evaluation Report TE-97-01, Ground-Water Remediation Technologies Analysis Center, Pittsburgh, Pennsylvania, 53 p.

Eyde, T.H. (1993) Using zeolites in the recovery of heavy metals from mining effluents. In: "Proc. EPD Congress 1993". TMS-AIME Annual Meeting, Denver, Colorado, pp. 383-392.

Ferguson, J.E. (1991) "The heavy elements : chemistry, enviromental impact and health effets". Pergamon press, Oxford, United Kingdom.

Fisher, K., Bipp, H.P., Riemschneider, P., Leidmann, P., Bieniek, D., and Kettrup, A. (1998) "Utilization of biomass residues for the remediation of metal-polluted soils". Environ. Sci. Technol., 32, 2154-2161.

Frankenberger, W.T.Jr., and Karlson, U. (1994). "Selenium in the environment". Frankenberger, W.T.Jr., and Benson, S. (Eds), Dekker, New York, NY, pp. 369-387.

Freeman, H.M. (1997) "Standard handbook of hazardous waste treatment and disposal". McGraw-Hill, New York, NY.

FRTR (2005) Federal Remediation Technologies Roundtable. Website "http://www.frtr.gov", december 2005.

Galvez-Cloutier, R., and Dubé, J.S. (1998) "Approche méthodologique pour la transformation progressive d'un centre de transfert en centre de traitement de substances dangereuses. Partie 2, étude de faisabilité du lavage chimique des sols contaminés". Vecteur Environnement, 31, 30-37 (in French).

Ganguly, C., Matsumoto, M.R., Rabideau, A.J., and Van Benshoten, J.E. (1998) "Metal ion leaching from contaminated soils : model calibration and application". J. Environ. Eng. Div. ASCE, 124(12), 1150-1158.

Gardea-Torresdey, J.L., Tang, L., and Salvador, J.M. (1996) "Copper adsorption by esterified and unesterified fractions of Sphagnum peat moss and its different humic substances". J. Hazard. Mater., 48, 191-206.

Gardea-Torresdey, J.L., de la Rosa, G., and Peralta-Videa, J.R. (2004) "Use of phytofiltration technologies in the removal of heavy metals: a review". Pure Appl. Chem., 76(4), 801-813.

Gibson, M.J., and Farmer, J.G. (1986) "Multi-step sequential chemical extraction of heavy metals from urban soils". Environ. Pollut., 11, 117-135.

Godd, R.K., and Sund-Hagelberg, A.C. (1985) "Method of purifying waste aqueous solutions from heavy metals". U.S. patent, No. 4,503,017.

Gooch, J.W. (1993) "Lead-based paint handbook". Plenum Press, New York, NY.

Grasso, D., Butkus, M.A., O'Sullivan, D., and Nikolaidis, N.P. (1997) "Soil-washing design methodology for a lead-contaminated sandy-soil". Water Res., 31, 3045-3056.

Guay, R., Silver, M., and Torma, A.E. (1976) "Microbiological leaching of low grade uranium ore by Thiobacillus ferrooxidans". Eur. J. Microbiol., 3, 157-167.

Harrison, R.M., Laxen, D.P.H., and Wilson, S.J. (1981) "Chemical associations of lead, cadmium, copper, and zinc in street dusts and roadside soils". Environ. Sci. Technol., 15, 1378-1383.

Healy, M.A., Harrison P.G., Aslam, M., Davis, S.S., and Wilson, C.G. (1982) "Lead sulphide and traditional preparations: routes for ingestion, solubility and reactions in gastric fluid". J. Clinical Hospital Pharmacy, 7, 169-173.

Hessling, J.L., Esposito, M.P., Traver, R.P., and Snow, R.H. (1990) Results of bench-scale research efforts to wash contaminated soils at battery-recycling facilities. In: "Metals speciation, separation and recovery". Patterson, J.W., and Passino, R. (Eds), Lewis Publishers inc., Chelsea, New-York, NY, pp. 497-511.

Ho, S.V., Athmer, C., and Sheridan, P.W. (1995) "The Lasagna technology for in situ soil remediation". Environ. Sci. Technol., 33, 1086-1091.

Howard, J.L., and Sova, J.E. (1993) "Sequential extraction analysis of lead in Michigan roadside soils : mobilization in the vadose zone by deicing salts?". J. Soil. Cont., 2, 361-367.

Humenick, M.J., and Schnoor, J.L. (1974) "Improving mercury (II) removal by activated carbon". J. Environ. Eng. Div. ASCE, 100, 1249-1262.

Juang, R.S., and Wang, S.W. (2000). "Electrolytic recovery of binary metals and EDTA from strong complexed solutions". Water Res., 34, 3179-3185.

Kapoor, A., and Viraraghavan, T. (1995) "Fungal biosorption - an alternative treatment option for heavy metal bearing wastewaters : a review". Bioresource Technol., 53, 195-206.

Kelly, D.P., and Wood, A.P. (2000) "Reclassification of some species of Thiobacillus to the new designated genera Acidithiobacillus gen. nov., Halothiobacillus gen. nov. and Thermithiobacillus gen. nov.". Int. J. Syst. Evol. Microbiol., 50, 511-516.

Kesraoui-Ouki, S., Cheeseman, C.R., and Perry, R. (1994) "Natural zeolite utilisation in pollution control : A review of applications to metals' effluents". Tech. Biotechnol., 59, 121-126.

Khalid, N., Rahman, A., and Ahmad, S. (1998) "Adsorption of cadmium from aqueous solutions on rice husk". Radiochim. Acta, 83, 157-162.

Khan, D.H., and Frankland, B. (1983) "Chemical forms of Cd and lead in some contaminated soils". Environ. Pollut. Ser. B, 6, 15-31.

Kobayashi, J. (1978) Pollution by cadmium and the itai-itai disease in Japan. In: "Toxicity of heavy metals in the environment", Oehme, F.W. (Ed), Marcel Dekker, New York, NY, pp. 199-260.

Lageman, R. (1993) "Electroreclamation. Applications in the Netherlands". Environ. Sci. Technol., 27(13), 2648-2650.

Lee, S.K. (1989) "Electrochemical contaminant removal from aqueous media". U.S. patent, No. 3,926,754.

Legiec, I.A. (1997) "Pb mobility and extractant optimization for contaminated soils". Environ. Progress, 16, 88-92.

Lin, H.K., Man, X.D., and Walsh, D.E. (2001) "Lead removal via soil washing and leaching". J. Mineral Metals Material Soc., 53, 22-25.

Lin, J., Jiang, W., and Liu, D. (2003) "Accumulation of copper by roots, hypocotyls, cotyledons and leaves of sunflower (Helianthus annuus L.)". Bioresour. Technol., 86, 151-155.

Lippmann, M. (2000) "Environmental toxicants. Human exposures and their health effects". John Wiley & Sons, Inc., New York, NY, 987 p.

Lorion, R. (2001) "Constructed wetlands: passive systems for wastewater treatment". United States Environmental Protection Agency Technology Innovation Office, 24 p.

Losi, M.E., Amrhein, C., and Frankenberger, W.T.Jr. (1994) "Bioremediation of chromate-containated groundwater by reduction and precipitation in srface soils". J. Environ. Qual., 23, 1141-1150.

Lundgren, D.G., and Silver, M. (1980) "Ore leaching by bacteria". Ann. Rev. Microbiol., 34, 263-283.

Madejon, P., Murillo, J.M., Maranon, T., Cabrera, F., and Soriano, M.A. (2003) "Trace element and nutrient accumulation in sunflower plants two years after the Aznalcollar mine spill". Sci. Total Env., 307, 239-257.

Marino, M.A., Brica, R.M., and Neale, C.N. (1997) "Heavy metal soil remediation : the effects on attrition scrubbing on a wet gravity concentration process". Environ. Prog., 16, 208-214.

Mattison, P.L. (1992) "Bioremediation of metals: putting it to work". Cognis, Santa Rosa, California.

McAnally, S., Benefield, L., and Reed, R.B. (1984) "Nickel removal from a synthetic nickel-plating waste water using sulfide and carbonate for precipitation and coprecipitation". Sep. Sci. Technol., 19, 191-217.

McLaughlin, M.J., Zarcinas, B.A., Stevens, D.P., and Cook, N. (2000) "Soil testing for heavy metals". Commun. Soil. Sci. Plant Anal., 31, 1661-1700.

Mercier, G. (2000) "Disponibilité des métaux dans les sols and prévision du rendement d'enlèvement par des techniques minéralurgiques". Thèse de doctorat, Département de géologie et de génie géologique, Université Laval, Sainte-Foy, Québec, Canada and Thèse de doctorat, INSA-Toulouse, Toulouse, France, 277 p. (in French).

Mercier, G., Duchesne, J., and Carles-Gibergues, A. (2002a) "A simple and fast screening test to detect soils polluted by lead". Environ. Pollut., 118, 285-296.

Mercier, G., Duchesne, J., and Blackburn, D. (2001) "Prediction of metal removal efficiency from contaminated soils by physical methods". J. Environ. Eng. Div. ASCE, 127, 348-358.

Mercier, G., Duchesne, J., and Blackburn, D. (2002b) "Removal of metals from contaminated soils by mineral processing techniques followed by chemical leaching". Water Air Soil Pollut., 135, 105-130.

Meunier, N., Blais, J.F., and Tyagi, R.D. (2002) "Selection of a natural sorbent to remove toxic metals from acidic leachate produced during soil decontamination". Hydrometallurgy, 67, 19-30.

Meunier, N., Blais, J.F., and Tyagi, R.D. (2004a) "Removal of heavy metals from acid soil leachate using cocoa shells in a counter-current sorption process". Hydrometallurgy, 73(3/4), 225-235.

Meunier, N., Drogui, P., Gourvenec, C., Mercier, G., Hausler, R., and Blais, J.F. (2004b) Removal of metals in leachate from sewage sludge using electrochemical technology. Environ. Technol. 25, 235-245.

Moore, J.N., and Luoma, S.N. (1990) "Hazardous wastes from large-scale metal extraction". Environ. Sci. Technol., 24, 1278-1285.

Mulligan, C.N., Yong, R.N., and Gibbs, B.F. (2001a) "Remediation technologies for metal-contaminated soils and groundwater : an evaluation". Eng. Geol. 60, 193-207.

Mulligan, C.N., Yong, R.N., and Gibbs, B.F. (2001b) "Surfactant-enhanced remediation of contaminated soil: a review". Eng. Geol. 60, 371-380.

Mulligan, C.N., Yong, R.N., Gibbs, B.F., James, S., and Bennett, H.P.J. (1999) "Metal removal from contaminated soil and sediments by the biosurfactant surfactin". Environ. Sci. Technol. 33, 3812-3820.

NATO/CCMS (1998) "NATO/CCMS pilot study. Evaluation of demonstrated and emerging technologies for the treatment and clean up of contaminated land and grounwater : phase II, final report". United States Environmental Protection Agency, EPA/542/R/98/001a, Report number 219, North Atlantic Treaty Organization, Washington, DC.

Nedwed, T., and Clifford, D.A. (1997) "A survey of lead battery recycling sites and soil remediation processes". Waste Management, 17, 257-269.

Nedwed, T., and Clifford, D.A. (2000). "Feasibility of extracting lead from lead battery recycling site soil using high-concentration chloride solutions". Environ. Prog., 19, 197-206.

Nriagu, J. (1974) "Lead orthophosphates – IV, Formation and stability in the environnement". Geochim. Cosmochim. Acta, 38, 887-898.
Nriagu, J. (1996) "A history of global metal pollution". Science, 272, 22-24.
Nriagu, J.O., and Pacyna J.M. (1988) "Quantitative assessment of world-wide contamination of air, water, and soils by trace metals". Nature, 333, 134-139.
OECD (1984) "Guideline of the OECD for testing chemical products - Terrestrial plants, growth test - method # 208". Organisation for Economic Cooperation and Development, OECD Publications Service, Paris, France.
Ortega, L.M., Lebrun, R., Blais, J.F., and Hausler, R. (2005) Treatment of a soil acidic leachate containing metal ions by nanofiltration membranes. Sep. Purif. Technol., (submitted paper).
Ottosen, L.M. and Jensen, P.E. (2005) Electro-remediation of heavy-metal contaminated soil. In: "Soil and sediment remediation. Mechanisms, technologies and applications". Lens, P., Grotenhuis T., Malina G., and Tabak H. (Eds), IWA Publishing, London, England, Chap. 14, pp. 264-288.
Oyler, J.A. (1990) Remediation of metals-contaminated site near a smelter using sludge-fly ash amendments. In: "Proc. 44th Purdue Ind. Waste Conf. ". Lewis Publishers Inc., Boca Raton, Florida, pp. 75-82.
Pamukcu, S., and Wittle, J.K. (1992) "Electrokientic removal of selected heavy metals from soils". Environ. Prog., 11(3), 241-250.
Papassiopi, N., Tambouris, S., and Kontopoulos, A. (1999) "Removal of heavy mtals from calcareous contaminated soils by EDTA leaching". Water Air Soil Pollut., 109, 1-15.
Patterson, J.W. (1988) "Effect of metal treatment and recovery". Lewis Publishers Inc., Boca Raton, Florida, pp. 333-334.
Patterson, J.W., Allen, H.E., and Scala, J.J. (1977) "Carbonate precipitation for heavy metals pollutants". J. Water Pollut. Control Fed., 49, 2397-2410.
Penn, M., and Savvidou, C. (1996) Centrifuge modeling of the removal of heavy metals pollutants using electrokinetics. In: "Environ. Geotechnics". Vol. 2, Balkema, Rotterdam, Netherlands, pp. 1055-1060.
Persin, F., and Rumeau, M. (1989) "Le traitement électrochimique des eaux et des effluents". Tribune de l'Eau, 42, 45-56 (in French).
Peters, R.W., and Ku, Y. (1984) Removal of heavy metals from industrial plating wastewaters by sulfide precipitation. In: "Proc. 57th Industrial Waste Symposium". Water Pollution Control Federation Annual Conference, pp. 553-568.
Prasetyo, I. (1992) "Removal of toxic metals from aqueous solutions by biosorption". Ph.D. thesis, McGill University, Montréal, Québec, Canada, 117 p.
Pronk, J.P. (2000) "Circular on target values and intervention values for soil remediations". Ministry of Housing, Spatial Planning and Environment, DBO/1999226863, Netherlands, 51 p.
Pulford, I.D., and Watson, C. (2003) "Phytoremediation of heavy metal-contaminated land by trees – a review". Environ. Int., 29, 529-540.

Rajeshwar, K., and Ibanez, J. (1997) "Environmental electrochemistry – fundamentals and applications in pollution abatement". Academic Press, San Diego, California, 776 p.

Rampley, C.G., and Ogden, K.L. (1998) "Preliminary studies for removal of lead from surrogate and real soils using a water soluble chelator : adsorption and batch extraction". Environ. Sci. Technol., 32, 987-993.

Renk, R.R. (1989) Treatment of hazardous wastewaters by electrocoagulation. In: "3rd Annual Conference Proceedings". Colorado Hazardous Waste Management Society.

Ribeiro, A.B., and Mexia, J.T. (1997) "A dynamic model for the electrokinetic removal of copper from a polluted soil". J. Hazard. Mater., 56, 257-271.

Robinson, A.K., and Sum, J.C. (1980) "Sulfide precipitation of heavy metals". United States Environmental Protection Agency, EPA/600/2-80-139, Washington, DC.

Robinson, B.H., Mills, T.M., Petit, D., Fung, L., Green, S., and Clothier, B. (2000) "Natural and induced cadmium-accumulation in poplar and willow: implications for phytoremediation". Plant Soil, 227, 301-306.

Rojo, S.A. (1979) "Apparatus for removing impurities from wastewater". U.S. patent, No. 4,149,953.

Rosand, T., and Acar, Y.B. (1996) "Electrokinetic extraction of lead from spiked Norwegian marine clay". Geoenvironment, 2000, 1518-1533.

Rossi, G. (1990) "Biohydrometallurgy". Rossi, G. (Ed), McGraw-Hill Book Company, New York, NY, 607 p.

Rumeau, M. (1989) "Méthodes électrolytiques de traitement des effluents – première partie. Électro-coagulation, flottation, détoxication, chloration". L'eau et l'industrie, pp. 47-51 (in French).

Salomons, W., Förstner, U., and Mader, P. (1995) "Heavy metals". Springer-Verlag, Berlin, Germany.

Samani, Z., Hu, S., Hanson, A.T., and Heil, D.M. (1998) "Remediation of lead contaminated soil by column extraction with EDTA : II. modeling". Water Air Soil Pollut., 102, 221-238.

Sauvé, S., McBride, M.B., and Hendershot, W.H. (1997). "Speciation of lead in contaminated soils". Environ. Pollut., 98(2), 149-155.

Schwitzguébel, J.P., Nehnevajova, E., Federer, G., and Herzig, R. (2005) Metal accumulation and metabolism in higher plants: potential for phytoremediation. In: "Soil and sediment remediation. Mechanisms, technologies and applications". Lens, P., Grotenhuis T., Malina G., and Tabak H. (Eds), IWA Publishing, London, England, Chap. 15, pp. 289-309.

Shen, H., Pritchard, P.H., and Sewell, G.W. (1996) "Microbial reduction of Cr(VI) during anaerobic degradation of benzoate". Environ. Sci. Technol., 30, 1667-1674.

Sims, R.C. (1990) "Soil remediation techniques at uncontrolled hazardous waste sites : a critical review". J. Air Waste Manag. Assoc., 40, 740-730.

Smith, L.A., Means, J.L., Chen, A., Alleman, B., Chapman, C.C., Tixier, J.S., Brauning, S.E., Gavaskar, A.R., and Royer, M.D. (1995) "Remedial options for metals-contaminated sites. Lewis Publishers, Boca Raton, Florida.

Steele, M.C., and Pichtel, J. (1998) "Ex-situ remediation of a metal-contaminated superfund soil using selective extractants". J. Environ. Eng. Div. ASCE, 124(7), 639-645.

Suèr, P., and Lifvergren, T. (2003) "Mercury-contaminated soil remediation by iodide and electroreclamation". J. Environ. Eng. Div. ASCE, 129(5), 441-446.

Sulzbacher, K., Ecke, H., Lagerkvist, A., and Calmano, W. (1997) "Anaerobic reduction of hexavalent chromium in filter sludge of an electrochemical process". Environ. Technol., 18, 301-307.

Tichi, R., Rulkens, W.H., Grotenhuis, J.T.C., and Fajtl, J. (1998) "Bioleaching of metals from soils and sediments". Water Sci. Technol., 37(8), 119-127.

Torma, A.E. (1987) "Impact of biotechnology on metal extractions". Min. Proc. Extrac. Metall. Rev., 2, 289-330.

Torma, A.E. (1986) "Biohydrometallurgy as an emerging technology". Biotechnol. Bioeng. Symp., 16, 49-63.

Tuin, B.J.W., and Tels, M. (1990) "Distribution of six heavy metals in contaminated clay soils before and after extractive cleaning". Environ. Technol., 11, 935-948.

USEPA (1989) "Protocols for short term toxicity screening of hazardous waste sites". United States Environmental Protection Agency, Corvallis, Oregon.

USEPA (1992) "Considerations in groundwater remediation at superfund sites and RCRA facilities – update. Directive 9283.1-06". United States Environmental Protection Agency, Office of Solid Waste and Emergency Response, Washington, DC.

USEPA (1993a) "Constructed Wetlands-Based Treatment". United States Environmental Protection Agency, EPA/540/R-93/523, Washington, DC.

USEPA (1993b) "Standards for the use and disposal of sewage sludge, 40 CFR Parts 257, 403 and 503". Final rule, United States Environmental Protection Agency, Cincinnati, Ohio.

USEPA (1994) "Guidance manual for the integrated exposure uptake biokinetic model for lead in children (IEUBK)". United States Environmental Protection Agency, EPA/540/R/93/081, Triangle Park, North Carolina.

USEPA (1997) "Clean up the nation's waste sites : markets and technology trends". United States Environmental Protection Agency, EPA-542-R-02-004, Washington, DC.

USEPA (2002) "Arsenc treatment technologies for soil, waste,and water". United States Environmental Protection Agency, EPA/542/R/96/005, Washington, DC.

Van Beschoten, J.E., Matsumoto, M.R., and Young, W.H. (1997) "Evaluation and analysis of soil washing for seven lead-contaminated soils". J. Environ. Eng. Div. ASCE, 123(3), 217-224.

Van der Lelie, D., Taghavi, S., Diels, L., Geets, J., and Vangronsveld, J. (2005) Integrated approach to quantify bioavailable concentrations of heavy metals. In: "Soil and sediment remediation. Mechanisms, technologies and applications". Lens, P., Grotenhuis T., Malina G., and Tabak H. (Eds), IWA Publishing, London, England, Chap. 8, pp. 133-145.
Vanbroekhoven, K., Dejonghe, W., and Diels, L. (2005) Treatment of contaminated soil and groundwater: in situ immobilization of metals. In: "Soil and sediment remediation. Mechanisms, technologies and applications". Lens, P., Grotenhuis T., Malina G., and Tabak H. (Eds), IWA Publishing, London, England, Chap. 13, pp. 248-263.
Vassilev, A., Schwitzguébel, J.P., Thewys, T., van der Lelie, D., and Vangronsveld, J. (2004) "The use of plants for remediation of metal-contaminated soils". Scientif. World J., 4, 9-34.
Vater, C., Enders, R., Feudel, M., Bolz, R., and Jekel, M. (1990) "Purification of effluents from municipal waste incinerators by chelating ion exchange resins". Vom Wasser, 75, 47-58.
Viessman, W., and Hammer, M.J. (1993) "Water Supply and Pollution Control". 5th Edition, Harper Collins College Publishers, New-York, NY.
Volesky, B. (1990) "Biosorption of heavy metals". CRC Press, Boca Raton, Florida.
Volesky, B., and Holan, Z.R. (1995) "Biosorption of heavy metals". Biotechnol. Prog., 11, 235-250.
Wase, J., and Forster, C. (1997) "Biosorbents for metal ions". Taylor & Francis Ltd, London, England, 238 p.
Watts, R.J. (1998) "Hazardous Wastes. Sources, Pathways, Receptors". John Wiley & Sons, Inc., New York, NY, 764 p.
Wendt, H., and Kreysa, G. (2001) "Génie électrochimique – principes et procédés". Dunod, Paris, France, 386 p. (in French).
West, L.J., and Stewart, D.I. (1996) "Effect of zeta potential on soil electrokinetics". Geoenvironment, 2000, 1535-1549.
White, C., Sharman, A.K., and Gadd, G.M. (1998) "An integrated microbial process for the bioremediation of soil contaminated with toxic metals". Nature Biotechnol., 16, 572-575.
Wildeman, T., Gusek, J., Cevaal, J., Whiting, K., and Scheuering, J. (1995) "Bioremediation of Inorganics". Hinchee, R.E., Means, J.L., and Burris, D.R. (Eds), Battelle Press, Columbus, Ohio, pp. 141-148.
Williams, J.W., and Silver, S. (1984) "Bacterial resistance and detoxification of heavy metals". Enz. Microb. Technol., 6, 530-537.
Xintaras, C. (1992) "Analysis paper: impact of lead contaminated soil on public health". United States Department of Health and Human Services Agency for Toxic Substances and Disease Registry, Atlanta, Georgia.

Yeung, A.T. (1994) Electrokinetic flow processes in porous media and their applications. In: "Advances in porous media". Vol. 2, Corapcioglu, M.Y. (Ed), Elsevier, Amsterdam, Netherlands, pp. 309-395.

Yeung, Y.B., Scott, T.B., Gopinath, S., Menon, R.M., and Hsu, C. (1997) "Design, fabrication and assembly of an apparatus for electrokinetic remediation studies". Geotechnical Testing J., 20, 199-210.

Zagury, G.J., Narasiah, K.S., and Tyagi, R.D. (2001) "Bioleaching of metal-contaminated soil in semicontinuous reactor". J. Environ. Eng. Div. ASCE, 127(9), 812-817.

Zeng, Q.R., Sauvé, S., H.E. Allen and Hendershot, W.H. (2005) "Recycling EDTA solutions used to remediate metal-polluted soils". Environ. Pollut., 133, 225-231.

Zillig, W., Stetter, K.O., Wunderl, S., Schulz, W., Priess, H., and Scholz, I. (1980) "The Sulfolobus-Caldariella group: Taxonomy on the basis of the structure of DNA-dependent RNA polymerases". Arch. Microbiol., 125, 259-269.

Zimdahl, R.L., and Skogerboe, R.K. (1977) "Behavior of lead in soils". Environ. Sci. Technol., 11, 1202-1207.

CHAPTER 12

Performance Monitoring

Keith Lai, Rao Surampalli, R.D. Tyagi, Shankha Banerji, and S. Yan

12.1 Introduction

Performance monitoring is a major component of soil and groundwater remediation projects. Monitoring data can be used to evaluate whether a remediation technology will meet site-specific remediation objectives. Monitoring results are crucial in deciding the termination date of a remediation project and the implementation of contingency plans for further treatment. Performance monitoring also helps site managers examine changes in contaminated zones during remediation activities. Thermal treatment and surfactant/cosovlent flushing technologies can mobilize dense nonaqueous phase liquids (DNAPLs) beyond the treatment zone. Chemicals injected during *in situ* chemical oxidation (ISCO) can suppress native microorganisms in subsurface and support other microbial communities. Gas evolved from ISCO using hydrogen peroxide and biofilm formed during bioremediation may lead to plugging of soil matrix. Newer technologies, such as monitored natural attenuation (MNA), phytoremediation, and permeable reactive barriers (PRBs) often require more extensive performance monitoring than the more accepted remediation approaches (Gavaskar et al. 2000; USEPA 2000; USEPA 2004a).

Performance monitoring is necessary for evaluating the technical success of remediation technologies and can significantly improve the overall cost-effectiveness of soil and groundwater clean-up operations. In assessing the performance of the remediation technologies, there are two primary criteria: remediation effectiveness and efficiency. Effectiveness refers to the capability of the remediation technologies to achieve remediation objectives at contaminated sites (ITRC 2004). For example, if the remediation objective is a reduction in hexavalent chromium [Cr(VI)] mobility, the effectiveness is defined by whether the remediation technology can halt the continuous migration of Cr(VI). Normally, assessment of the remediation effectiveness involves quantifying reductions in contaminant mass, concentration, mobility, and/or toxicity (NRC 1997).

Efficiency of remediation technologies refers to the optimization of time, energy and costs expended towards the achievement of remediation effectiveness. It is typically assessed by comparing system operating parameters to the relevant design specifications. For example, ISCO systems are usually effective in reducing contaminant concentrations at the beginning stage but the rate of reduction in terms of time, energy and money expended becomes lower and less efficient with each successive injection. Thus optimization of the operating parameters to maintain good remediation efficiency is necessary (ITRC 2004).

To assess the remediation effectiveness and efficiency, performance monitoring is conducted specifically to collect data from contaminated sites. By definition, performance monitoring involves periodic measurement of physical and/or chemical site parameters to evaluate whether the remediation technologies perform as expected. Contaminant concentrations in groundwater, soil and soil vapor are the usual parameters measured during performance monitoring. However, each remediation technology has its own specific measuring parameters. For instance, monitoring of the nutrient concentrations in the subsurface is necessary to assess the performance of *in situ* bioremediation. The selection of monitoring parameters depends on remediation objectives, performance metrics and types of remediation technologies applied. Performance metrics are the environmental conditions and monitoring parameters measured to evaluate remediation progress or to confirm attainment of remediation objectives (ITRC 2004). Table 12.1 shows common performance metrics, methods and parameters required for monitoring the performance of remediation technologies for DNAPL source zone treatment.

This chapter discusses the remediation objectives, performance metrics and monitoring parameters generally involved during the performance assessment of common remediation technologies. Sampling systems commonly used described. Key monitoring parameters of remediation technologies such as pump-and-treat (P&T) systems, soil vapor extraction (SVE), air sparging (AS), surfactant/cosolvent flushing, *in situ* thermal remediation, ISCO, *in situ* bioremediation, MNA, PRBs and phytoremediation, are specifically mentioned (Bedient et al. 1999; USEPA 2004b; Lo et al. 2006a).

12.2 Remediation Objectives

Measuring the effectiveness of remediation technologies is a difficult undertaking if clear remediation objectives are not specified. Remediation goals can be quantitative or qualitative in nature but should at least be measurable so that remediation progress towards the goals can be monitored objectively. In most contaminated sites, remediation objectives are based on the overriding goals of most state and federal

Table 12.1 Performance metrics, monitoring methods and monitoring parameters for the performance monitoring of the remediation technologies for DNAPL source zone treatment.

Performance Metrics	Monitoring Methods	Monitoring Parameters	Derived Parameters
Adsorbed-phase reduction	Soil coring	Soil contaminant concentrations	Sorbed concentration
Dissolved-phase reduction	Groundwater sampling	Aqueous contaminant concentrations	Plume extent Change in NAPL composition
Source mass extracted	Effluent sampling	Volumes and contaminant concentrations of extracted fluids	Contaminant mass removed
Source mass destroyed	Groundwater sampling	Concentration of breakdown products (e.g., chloride)	Mass of DNAPL destroyed *in situ*
Source mass remaining	Soil coring	Soil contaminant concentrations and visual observation	DNAPL distribution Sorbed mass
	Tracer tests	Concentration of tracers	NAPL volume Saturation
	Groundwater sampling	Aqueous contaminant concentrations	Aqueous concentration
Mobility reduction	Soil coring	Soil contaminant concentrations and visual observation	NAPL presence and saturation
	Product gauging	NAPL thickness	NAPL thickness
Toxicity reduction	Soil coring	Adsorbed contaminant concentrations	DNAPL composition
	Groundwater sampling	Aqueous contaminant concentrations	Concentrations of toxic parameters
Mass flux/mass discharge reduction	Well transect sampling	Aqueous contaminant concentrations Hydraulic conductivities Hydraulic gradients	Darcy flux Mass flux Mass discharge
	Flux meter transects	Mass of resident tracers retained on sorbents Mass of contaminants retained on sorbents	Darcy flux Mass flux Mass discharge
	Integral pumping tests	Concentration-time series sampling from extraction well	Mass discharge Average plume concentration

regulatory programs, namely protection of human health and natural environment. Remediation objectives generally fall into three categories: short-term; intermediate-term; and long-term remediation objectives.

Short-term remediation objectives typically involve the alleviation of immediate risks to humans or natural resources through the prevention of further expansion of contaminant source zones. Therefore, short-term objectives customarily take account of the control of contaminant mobility and mitigation of contaminant migration. Intermediate-term remediation objectives are established to guide the remediation activity when complete removal of a contaminant source in one aggressive remedial effort is not feasible and the contaminants remaining in contaminated sites still exceed the regulatory standards (NRC 1994). Intermediate-term objectives may include: 1) depletion of a contaminant source adequately to allow for natural attenuation; 2) reduction of dissolved phase contaminant concentrations outside a source zone; 3) decrease in mass discharge rate or flux from a contaminant source; 4) reduction of the mass or volume of a contaminant source; and 5) prevention of migration of remediation fluids beyond a treatment zone. According to United States Environmental Protection Agency (USEPA), application of the information obtained from interim remediation actions for long-term risk management can significantly accelerate risk reduction from contaminant sites (USEPA 1997). It could take a year or several years for contaminated sites to meet the intermediate-term objectives.

Long-term monitoring is required to ensure that the interim treatment levels achieved are sustainable and are not subject to a rebound in contaminant concentrations in groundwater when post-treatment equilibrium is established in aquifers. Long-term remediation objectives focus on compliance with the regulatory treatment standards applicable to all contaminated media (i.e., groundwater, soil and soil vapor) at the sites. Achievement of the regulatory standards leads to the termination of remediation activities but in analogy to intermediate-term objectives, long-term monitoring is also required to ensure that compliance is sustainable. If long-term remediation objectives are not met or are not achievable, contingency plans are implemented requiring the consideration of secondary treatment.

12.3 Performance Metrics and Monitoring Parameters

Each remediation objective has its own set of performance metrics for assessing and monitoring the performance of the remediation technologies applied towards this objective. These metrics are neither equivalent to nor interchangeable with one another. They range from qualitative indicators of remediation progress to quantitative measures of specific factors following remediation (Table 12.1). There are three major categories of performance metrics: 1) qualitative estimation of the treatment progress of a contaminant source; 2) quantitative estimation of contaminant source mass reduction; and 3) quantitative evaluation of the effect of contaminant source treatment on factors, such as contaminant toxicity, mobility and plume strength.

12.3.1 Qualitative Estimation of Treatment Progress of Contaminant Source

One of the common remediation objectives is the reduction of site contaminant concentrations below recommended or required levels. The remediation activity is preliminary viewed as a success if there is a decrease in the average contaminant concentrations in soil, groundwater and/or soil vapor. However, these performance metrics of reduction in contaminant concentrations in the subsurface media can only qualitatively illustrate that progress is being made towards the contaminant source treatment. They do not provide quantitative data about the progress of the remediation operations.

12.3.1.1 Decrease in Contaminant Concentrations in Soil

Decrease in soil contaminant concentrations can be monitored through soil sample collection followed by chemical analyses of contaminant concentrations. Fraction of organic carbon (f_{oc}) of the soil samples is also a parameter monitored at sites contaminated by organic contaminants (Bedient et al. 1999). Pre- and post-treatment soil data are required to examine the changes in soil concentrations. To obtain reliable soil concentration data, large number of soil samples are collected to address site heterogeneities, and temporal and spatial variability of contaminant concentrations. Soil sampling methods may sometimes underestimate actual soil concentrations because the contaminated soils may mix with neighboring clean or less contaminated soil during sampling and analyzed as a whole sample. Due to the destructive nature of soil sampling, replicate samples cannot be collected from the exactly same location resulting in the introduction of variability in contaminant concentration estimates.

12.3.1.2 Decrease in Contaminant Concentrations in Groundwater

Monitoring the decrease in contaminant concentrations in groundwater requires sampling and chemical analyses of contaminated groundwater. Groundwater samples are collected via a network of monitoring wells or multilevel samplers installed in the middle and the periphery of the contaminant plume as well as upgradient and downgradient locations. Before collecting groundwater samples, the monitoring wells or multilevel samplers are purged to remove stagnant water and ensure that samples collected are representative of the groundwater condition in the aquifer. However, excessive purging of wells and samplers can alter the measured contaminant concentration at the sampling points by introducing water from adjacent locations. The representativeness of groundwater data are prone to be affected by the temporal and spatial variabilities associated with groundwater contaminant concentrations. The aqueous phase concentrations are also subject to slow release or dissolution of contaminants from the aquifer solids, which can lead to a rebound of groundwater concentrations soon after sampling.

12.3.1.3 Decrease in Contaminant Concentrations in Soil Vapor

Monitoring the reduction in contaminant concentrations in soil vapor is applicable when treating volatile contaminants in the vadose zone. Contaminant concentrations in extracted vapor steam and the vapor remaining in subsurface are monitored. During remediation, the soil vapor concentrations in the subsurface must be monitored over the entire area of a soil gas plume to check whether remediation is addressing the entire contaminant plume. Site heterogeneities, spatial or temporal variabilities of contaminant concentrations and the potential for rebound of soil vapor concentrations can impact monitoring.

12.3.2 Quantitative Estimation of Mass Reduction of Contaminant Source

Three major types of performance metrics including contaminant mass extracted from the subsurface, contaminant mass destroyed, and contaminant mass remaining in the subsurface after treatment are monitored to determine remediation effectiveness. This information is compared to baseline conditions to compute the percentage mass removal or fraction of contaminant mass remaining.

12.3.2.1 Contaminant Mass Extracted from Subsurface

This performance metric is applied at contaminated sites that involve extraction of contaminated groundwater and/or soil vapor for *ex situ* treatment. Such technologies include P&T systems, SVE, AS, *in situ* thermal remediation and surfactant/cosolvent flushing. The contaminant mass extracted is estimated by integrating a plot of the contaminant concentrations in the extracted groundwater and/or soil vapor against the cumulative volume of groundwater or soil vapor extracted. The area under the curve represents the mass of contaminant removed from the contaminant plume during extraction of contaminated groundwater or soil vapor.

12.3.2.2 Contaminant Mass Destroyed in Subsurface

This performance metric focuses on the amount of contaminant mass destroyed *in situ* by the remediation technologies, such as MNA, ISCO, bioremediation, and PRB by processes such as oxidation, reduction or biotransformation. Estimation of the contaminant mass destroyed is generally more complicated than the estimation of contaminant mass extracted since the former is usually determined indirectly by monitoring specific indicator parameters. For example, when a chemical oxidant such as potassium permanganate, $KMnO_4$ is injected into an aquifer containing dissolved trichloroethylene (TCE), the contaminant destruction can be indirectly estimated by monitoring chloride concentrations in the groundwater. Chloride ions are released when TCE is oxidized by $KMnO_4$ according to the reaction stoichiometry described in Eq. 12.1. Each 1 mg increase in chloride is ascribed to the degradation of 1.23 mg of TCE in the aquifer.

$$2KMnO_4 + C_2HCl_3 \rightarrow 2CO_2 + 2MnO_2 + 3Cl^- + H^+ + 2K^+ \qquad \text{(Eq. 12.1)}$$

Carbon isotope ratios may also be monitored to examine the mass of chlorinated solvents being destroyed (Song et al. 2002). Molecules of chlorinated solvents and other organic contaminants are composed of both light (^{12}C) and heavy carbon (^{13}C) atoms. However, there is a slight difference in the oxidation, reduction and biodegradation rates between ^{12}C and ^{13}C in which the former is transformed more quickly, thereby resulting in enrichment of ^{13}C in residual reactants and ^{12}C in products. By measuring the $^{12}C/^{13}C$ ratio of contaminants before and after the treatment, the amount of contaminant destroyed can be calculated using the Rayleigh model (Dayan et al. 1999; VanStone et al. 2004). In a similar manner, the ratio of chlorine isotopes (^{37}Cl and ^{35}Cl) can also be used as an indicating parameter for the destruction of chlorinated contaminants (Sturchio et al. 1998; Heraty et al. 1999).

12.3.2.3 Contaminant Mass Remaining in Subsurface

Estimation of the contaminant mass remaining in the subsurface after treatment is of overriding importance if the remediation objective is specifically linked to the fraction of contaminant mass remaining. The contaminant mass remaining in the soil matrix can be calculated by multiplying the contaminant concentrations in soil samples by the bulk density of the soil and the volume of the contaminant zone. The mass of contaminants remaining in groundwater can be estimated from the contaminant concentrations in the collected groundwater samples. Hydrologic and hydrogeologic data such as porosity and groundwater velocity can be used to compute the contaminant mass in groundwater using numerical models (Garabedient et al. 1991; Lai et al. 2006a). To estimate the mass of DNAPL remaining in subsurface, a partitioning interwell tracer test (PITT) can be applied which involves the injection of conservative and nonconservative tracers into one or more wells, and the subsequent recovery of the tracers from nearby extraction wells. The conservative tracer can pass through the DNAPL source zone freely without retardation; whereas the transport of nonconservative tracer is noticeably retarded by interaction with the DNAPL. Tracer responses observed at the monitoring wells or extraction wells can then be used to estimate the average DNAPL saturation and the total volume of DNAPL remaining in the subsurface (Jin 1995; Jin et al. 1995; Dwarakanath 1997).

12.3.3 Quantitative Evaluation of Remediation Effects on Toxicity, Mobility and Mass Flux of Contaminant Source

12.3.3.1 Decrease in Toxicity of Contaminant Source

For sites that contain several types of contaminants, monitoring of the overall reduction in the total contaminant mass may not be a suitable performance metric. This is because the reduction in overall contaminant mass may be due to the removal of less toxic contaminants, while highly toxic contaminants still remain in the site. In such cases, remediation should target the highly toxic contaminants in order to reduce the overall toxicity of the contaminant source. For these sites the remediation objective may be met if the concentration of the highly toxic contaminants is substantially reduced even though the site still contains high concentrations of less toxic contaminants.

12.3.3.2 Decrease in Mobility of Contaminant Source

Short-term remediation objectives customarily focus on reducing the further spread of the contaminant zone. To accomplish this goal, remediation technologies applied should be able to deplete the contaminant source sufficiently to reduce the concentration level to a point at which the contaminant source is relatively stable. For example, DNAPL saturation, which can be determined by analyzing the total DNAPL concentration (adsorbed, dissolved, vapor and NAPL phase) in the soil matrix or by PITT, is a good indicating parameter for the mobility of a DNAPL source. When DNAPL saturation drops to a residual saturation level, DNAPL source mobility reduces considerably.

12.3.3.3 Decrease in Mass Flux of Contaminant Source

From a risk perspective, the most important characteristic of a contaminant source is the contaminant mass capable of contributing to a larger dissolved phase plume. This characteristic, called the plume load, is defined as the discharge rate of the contaminant mass from a source to the surrounding groundwater. The total mass discharge rate can be obtained by integrating the localized mass flux, which is the flow rate of contaminant per unit area of the plane perpendicular to the direction of groundwater flow, across the entire source zone. A significant reduction in contaminant mass may not correspondingly lead to a substantial decrease in the plume load since it is also a function of the contaminant distribution and hydrodynamic structure of the subsurface. In some cases, the remediation activity can be terminated if the total mass discharge rate from the source zone is smaller than the natural attenuation capacity of the aquifer (Rao et al. 2001).

Four approaches are commonly used to estimate the total contaminant mass discharge rate. The first method is related to a continuous pumping from extraction wells. This method involves capturing the entire contaminant plume by one or more extraction wells pumping at a continuous rate. By knowing the pumping rate and measuring the steady contaminant concentration in the extracted groundwater, the total discharge rate of the contaminant mass can be calculated directly. This approach is cost-effective when there is a hydraulic containment system surrounding the contaminant plume.

Another estimation method, called the integrated pump test, is based on short-term and active pumping of wells located in a transect across the contaminant plume (Figure 12.1a). This technique was originally developed for calculating the natural attenuation rate constant at former gasworks site (Bockelmann et al. 2001; Teutsch et al. 2001). Well positions, pumping rates and pumping times are optimized to allow well capture zones to cover the entire groundwater flow downstream of the contaminated site. The wells are pumped and contaminant concentrations in the extracted water are measured as a function of time until the entire mass discharge at the transect location is determined or assumed to be removed.

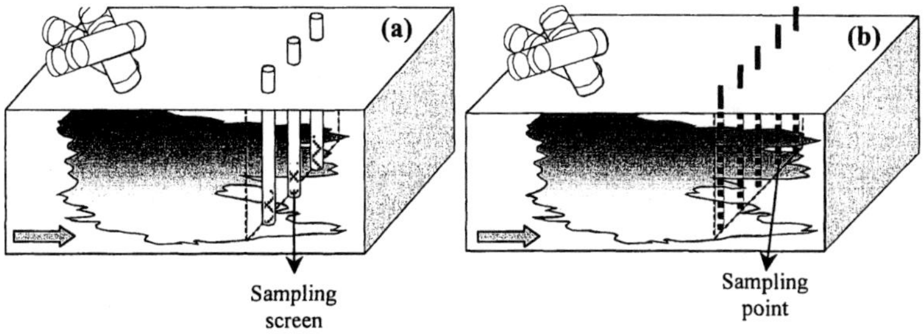

Figure 12.1. (a) Extraction wells and (b) multilevel samplers located in a transect across the contaminant plume

The most common method used to estimate the mass flux is through the groundwater sampling from multilevel samplers. Unlike extraction wells, multilevel samplers allow collection of contaminated groundwater from numerous, closely spaced sampling points along a transect of wells intersecting the plume and aligned perpendicular to the groundwater flow direction (Figure 12.1b). Measurement of contaminant concentrations in groundwater samples collected from the transect of multilevel samplers gives the geostatistical average concentration. The contaminant mass flux can be calculated using the velocity of groundwater flowing through the transect of multilevel samplers.

Passive borehole flux meter is another innovative technology for direct *in situ* measurement of both cumulative groundwater and contaminant fluxes. During the contaminant mass flux estimation, the meter is inserted into a well or borehole to intercept the groundwater flow. Hydrophobic and hydrophilic sorbents placed inside a permeable unit of the meter retain dissolved organic and inorganic contaminants in groundwater intercepted by the meter. The permeable unit also contains a known amount of soluble tracer that is leached from the meter at a rate proportional to the groundwater flux. After exposing the flux meter to the contaminated groundwater for a period of time, the meter is removed and the mass of contaminants sorbed by the sorbents and the tracer remaining are quantified. The contaminant mass sorbed is used to calculate the time-average contaminant mass flux and the mass of the residual tracer is used to compute cumulative groundwater flux (ESTCP 2003).

12.4 Groundwater Sampling for Performance Monitoring

A reliable monitoring network installed for groundwater sampling at contaminated sites is capable of providing data to demonstrate attainment of all remediation objectives. Specifications of the monitoring network design are based on information available about the processes and factors expected to control contaminant distribution. For instance, original contaminant source distribution, site geology and hydrology can cause spatial and temporal variability of plume shapes, which, in turn, can affect the selection of monitoring locations and frequencies and necessitate continual re-evaluation of the monitoring network. In addition, the density of sampling points in the monitoring network may depend on the spatial scale of the horizontal and vertical contaminant distribution in the subsurface. The distance between well transects is controlled by the changes in contaminant concentration along the plume, and the location of the contaminant source and distal portions of the plume. Subsurface stratigraphy, the vertical component of hydraulic gradients and vertical contaminant distribution can affect the elevations of sampling points. Ideally, the elevation interval of sampling points should allow groundwater sampling from different stratigraphic layers, i.e., from the core of the contaminant plume, as well as above and below the plume (USEPA 2004a). Transect-based monitoring networks are widely applied for performance monitoring because such networks can provide a better delineation of contaminant distribution, and its spatial and temporal variability (Bockelmann et al. 2001; Kao and Wang; Lai et al. 2006b). The transect approach also helps to locate groundwater flow lines and contaminant migration paths. Figure 12.2 illustrates the plan, side and front views of a series of well transects installed over a contaminant plume.

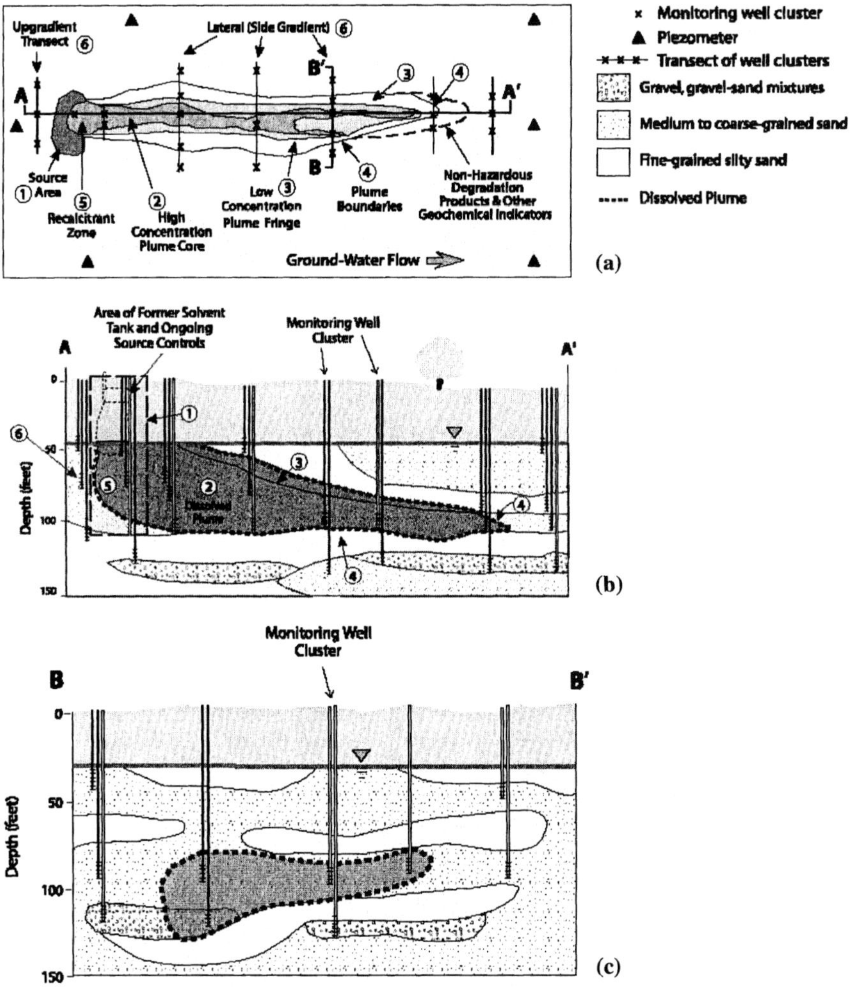

Figure 12.2. (a) Plan, (b) side and (c) front views of a monitoring network installed over a contaminant plume for monitoring performance of natural attenuation (USEPA 2004a).

12.4.1 Determination of Groundwater Sampling Locations

Groundwater sampling locations to be considered during performance monitoring of remediation technologies include: 1) contaminant source area; 2) transmissive zone with the highest contaminant concentration or hydraulic conductivity; 3) distal or fringe portions of the plume; 4) plume boundaries; and 5) recalcitrant zones (see Fig. 12.2). Transects of monitoring wells installed within and immediately downgradient of the source area can provide data on contaminant release from the source zone and enable estimation of reductions in contaminant concentrations over time. When containment technologies have already been applied for source control, the well transects can provide data to monitor the effectiveness containment technologies such as slurry walls, grout curtains or sheet piling. Such wells can also warn if a rise in water table transfers additional contaminant from the vadose zone into the aquifer.

Groundwater sampling in transmissive zones with the highest contaminant concentrations or hydraulic conductivity is of paramount importance because any change in conditions in these zones, such as an increase in contaminant release from source area or a change in groundwater velocity, may result in a rapid impact on downgradient receptors. The distal or fringe portions of the contaminant plume is the plume area where reduction of contaminant concentrations to a level required by remediation objectives may be attained most rapidly. A continuous increase in contaminant concentration in these zones can indicate a possible plume expansion. Sampling points installed around plume boundaries include points placed at the sidegradient and downgradient locations of the plume, around vertical plume boundaries (i.e., above and below the plume), and at locations between the plume boundary and possible receptors. Monitoring data obtained from these sampling points can show any unacceptable plume expansion and change in the direction of groundwater flow

Monitoring wells should also be installed in recalcitrant zones where contaminant reduction rates appear to be lower than the rate required to meet remediation objectives. Attainment of remediation concentration levels in these areas within accepted time frames may be impeded by site conditions (e.g., the presence of previously undetected contaminant source). Monitoring data from these areas is can help determine whether additional remedial action is required.

12.4.2 Determination of Screen Lengths of Monitoring Wells

The length of the well screen should be sized to sample the interval of interest determined by subsurface stratigraphy and contaminant concentrations. Well screens can match the stratigraphic intervals if these intervals are relatively small and contaminant concentrations are consistent throughout the vertical extent of the interval intersected by the screen. For example, a relatively homogeneous sand layer with 1 m of thickness and uniform contaminant concentration can be sampled by a 1-m long screen to intersect the whole interval. Intervals with significantly different hydraulic conductivities can be sampled by screens with different screen lengths. Well screens should also be sized to match with contaminant loadings in the subsurface. They should be sized to sample the most contaminated part of the plume. If the plume is highly heterogeneous (i.e., high spatial variation in contaminant concentration), application of long screens for groundwater sampling should be avoided. Such screens may produce erroneously low contaminant concentrations due to mixing with groundwater possessing dilute contaminant concentrations.

12.4.3 Determination of Groundwater Sampling Frequencies

A sampling frequency design for performance monitoring should

- give timely warning of impact to receptors,
- detect contaminant releases to groundwater to warn of possible plume expansion,
- detect changes in plume size,
- illustrate temporal variability of monitoring data, and
- provide sufficient data to reliably evaluate remediation progress toward the remediation objectives.

Generally, more frequent sampling is needed at the early stage of a remediation activity while less frequent sampling is appropriate as the remediation system stabilizes at close-to-optimum conditions (ITRC 2004). Factors customarily taken into account for designing the sampling frequency include: 1) anticipated contaminant travel time to receptors; 2) cyclic changes in contaminant concentrations and plume boundaries; 3) relevance of monitoring parameters, and 4) stability of the monitoring data. Short groundwater travel time from downgradient plume boundaries to receptors requires more frequent sampling. Groundwater sampling should also be more frequent for wells located near and downgradient of plume boundaries as compared to those located upgradient and at the core of the plume.

Climatic changes in recharge rates and groundwater flow characteristics (during wet and dry seasons) may lead to noticeably cyclic trends in contaminant concentrations and plume boundaries due to seasonal variations in the input of contaminants from the vadose zone to the aquifer. Under these circumstances, more frequent groundwater sampling may be required based on the historical variability in groundwater level at the sites and the recorded climatic variability (i.e., drought frequencies or periods of above average rainfall). Generally, monitoring data gathered over several years are needed to evaluate the cyclic changes and determine the suitable sampling frequency for capturing the changes.

The frequency at which monitoring parameters are measured depends upon the significance of the parameters in relation to the assessment of the remediation technology. If the parameters are not expected to significantly influence the performance evaluation of a technology, the relevant monitoring frequency can be substantially reduced. However, the entire suite of contaminants, site geochemical parameters and hydrogeological parameters should be measured at all sampling points during the remedial action if there is no specific reason for excluding these parameters. Observation of stable data trends over a period of several years allows reduction in the monitoring frequency of the relevant parameters. Furthermore, if two or more wells sampling the same zone are located closely together and consistently produce similar trends of data, reduction in the sampling frequency of these wells can be considered. However, if sudden changes in these parameters are observed, an increase in the monitoring frequency may be required to obtain enough information for understanding the changes and provide early warning of further changes.

12.5 Key Monitoring Parameters for Remediation Technologies

Successful evaluation of the effectiveness and efficiency of remediation technologies applied for soil and groundwater treatment depends highly on the parameters being monitored. Monitoring of appropriate parameters not only indicates the progress towards the attainment of remediation objectives, but can also point out the reasons for failure of a remediation technology. Criteria for selection of the monitoring parameters are based on the remediation objectives, specific site conditions, performance metrics to be evaluated and the type of remediation technology. In this section, key monitoring parameters of each soil and groundwater remediation technologies will be generally introduced.

12.5.1 Pump-and-Treat Systems

Pump-and-treat (P&T) has been the most common remediation technology used to contain groundwater contaminants, and/or restore polluted aquifers. In the United States, nearly 72% of Superfund site Records of Decisions have selected P&T systems as the prime groundwater remediation approach (USEPA 1992; USEPA 1994). The operating principles of P&T systems involve capturing and pumping contaminated groundwater by extraction wells for *ex situ* treatment (Figure 12.3a). Treated water is either infiltrated into the aquifer or pumped into surface water bodies. The inward hydraulic gradient created by extraction wells within the plume (Figure 12.3b), helps control the downgradient movement of contaminated groundwater, thereby preventing continued expansion of the plume (Eldho 2003). The performance of P&T systems can be further enhanced by installing physical containment systems and/or fluid injection systems to flush the contaminated groundwater towards the capture zone.

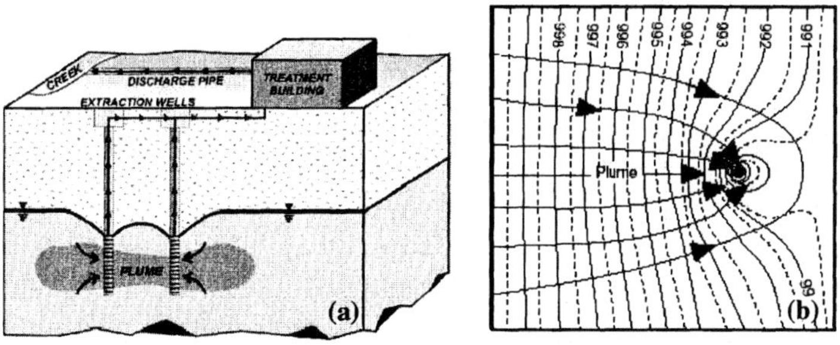

Figure 12.3. Schematic diagrams showing (a) the extraction wells and *ex situ* treatment unit of a P&T system (USEPA 2001a), and (b) hydraulic containment of contaminant plume by the extraction well (USEPA 1996)

12.5.1.1 Key Monitoring Parameters

Monitoring Hydraulic Containment. Evaluating the hydraulic containment of a contaminant plume by a P&T system involves monitoring the hydraulic gradient, measuring extraction and/or fluid injection rates, contaminant concentrations, and tracer movement. Table 12.2 summarizes these monitoring parameters, and the corresponding monitoring locations and purposes. An inward hydraulic gradient across the plume boundary would indicate that contaminated groundwater flow is inward, thereby ensuring the successful capture or containment of the plume. The inward hydraulic gradient can be estimated by comparing hydraulic head in piezometers located near the containment perimeter and downgradient of the extraction wells. It can also be evaluated by interpreting potentiometric surface maps determined from the hydraulic head data measured in wells within and outside the containment area. The hydraulic gradient of more permeable portions of the aquifer should be determined to monitor preferential flow of contaminated groundwater and migration of contaminants across the containment boundary. At the base of the contaminant plume or containment volume, the inward hydraulic gradient towards the extraction wells may be specified as upward hydraulic gradient. Such upward hydraulic gradients can prevent the possible downward migration of contaminants. The upward gradient can be determined by comparing the differences in hydraulic head at different depths or comparing potentiometric surfaces obtained at different elevations and stratigraphic layers.

Pumping and fluid injection rates should be monitored to ensure inward and upward hydraulic gradients in the containment zone. Monitoring of contaminant concentrations should be performed throughout the entire P&T operation to check

Table 12.2. Monitoring parameters of pump-and-treat (P&T) systems for the hydraulic containment of contaminant plume and aquifer restoration.

Locations	Parameters	Purposes
Hydraulic Containment of Contaminant Plume		
Containment perimeter primarily at downgradient locations of extraction wells Within and outside the containment area Permeable portion of the aquifer	Inward hydraulic gradient	Remedial performance and capture zone analysis
Base of the contaminant plume or containment volume	Upward hydraulic gradient	Prevention of downward contaminant migration
Extraction wells	Extraction rate	Creation of hydraulic gradient towards the extraction wells and capture zone analyses
Injection wells	Injection rate	Flushing of contaminated groundwater towards the extraction wells
Monitoring points along or near the potential downgradient containment boundary	Contaminant concentration in groundwater	Remedial performance
Monitoring points beyond the containment perimeter or all monitoring points	Tracer movement	Location of containment failure or delineation of capture zones
Aquifer Restoration		
As listed above	Hydraulic containment	Prevention of further spread of contaminant plume during restoration efforts
Extraction wells	Pumping rate	Pore volume of flushing
All monitoring points	Hydraulic gradient Tracer movement Groundwater flowrate	Determination and control stagnation zones
All monitoring wells in contaminated zone From soil (or rock) through borings in contaminated zone Influent and effluent of treatment unit Upgradient and sidegradient of contaminant plume if contaminants have migrated beyond the containment zone	Groundwater contaminants and chemicals of concern	Remedial performance/progress
Representative locations in the contaminated zone	Soil contaminant concentrations and organic carbon content	Remedial performance/progress

whether the temporal and spatial variations in contaminant distribution along or near the downgradient of the containment boundary are consistent with design assumptions. Failure of hydraulic containment is suggested if: 1) the estimated total contaminant mass in groundwater beyond the containment perimeter increases with time; 2) temporal changes of contaminant concentrations in perimeter or downgradient monitoring wells is inconsistent with the design under effective containment; and/or 3) comparatively retarded contaminants previously restricted to the containment area are detected in perimeter monitoring wells. In some contaminated sites, tracers may be periodically released into the containment zone where hydraulic control is considered least effective. Tracer detection in groundwater beyond the containment perimeter would indicate containment failure and identify the possible locations of failure. A P&T capture zone can also be delineated by releasing tracers in areas of uncertain capture followed by monitoring of the tracers present in the extracted groundwater.

Monitoring Aquifer Restoration. Aquifer restoration by P&T systems is more difficult to attain than plume containment due to tailing and rebound of contaminant concentrations during the P&T operation (USEPA 1994). This phenomenon is caused by the limited aqueous solubility of most contaminants, slow desorption from aquifer materials, and slow dissolution of contaminant precipitates. Therefore, P&T systems deployed for aquifer restoration have to be monitored extensively to identify problem areas and optimize system operations.Generally, the hydraulic containment of the contaminant plume is a prerequisite for the aquifer restoration. As shown in Table 12.2, all monitoring parameters for the hydraulic containment are also included for the monitoring of aquifer restoration. Therefore, further spread of the contaminant plume can be prevented during the restoration efforts. In addition to inward hydraulic gradient monitoring and capture zone analyses, monitoring of the pumping rate during aquifer restoration helps in determining the number of pore volumes (NPV) of groundwater flushed through the contaminant zone by a P&T system. Effective aquifer restoration by a P&T system requires sufficient NPV of groundwater flushing to remove both existing dissolved contaminants, and those that may desorb from aquifer materials or dissolve from precipitates or NAPLs. Measurement of the hydraulic gradient, tracer movement, and groundwater flowrate by a downhole flowmeter can identify stagnant zones induced by low hydraulic gradient.

Once the stagnant zone is identified, its size and duration can be minimized by changing extraction and/or fluid injection rates, and well locations. Groundwater and soil concentrations in the contaminant zone, and contaminant concentrations in the influent and effluent of the treatment unit should be analyzed periodically to monitor the remedial performance and progress towards attainment of remediation objectives. Concentrations of other chemicals in groundwater that can potentially affect the performance of the treatment unit should also be measured. For example, iron concentration in groundwater should be monitored if groundwater is aerated during the treatment process since iron may precipitate and clog the treatment unit.

12.5.2 Soil Vapor Extraction and Air Sparging Systems

12.5.2.1 Soil Vapor Extraction Systems

In SVE systems, vapor extraction wells are drilled near the contaminant source area in the vadose zone (Figure 12.4a). A vacuum is applied through the extraction wells to create a negative pressure gradient causing the movement of vapors towards the extraction wells. The extracted vapor is treated prior to discharge to the atmosphere (USEPA 2001b; ITRC 2004). SVE is only applicable for the contaminants with sufficiently high vapor pressures and Henry's Law constants. Low air permeability and heterogeneous distribution of air permeability in soils significantly limit the effectiveness of SVE for soil remediation.

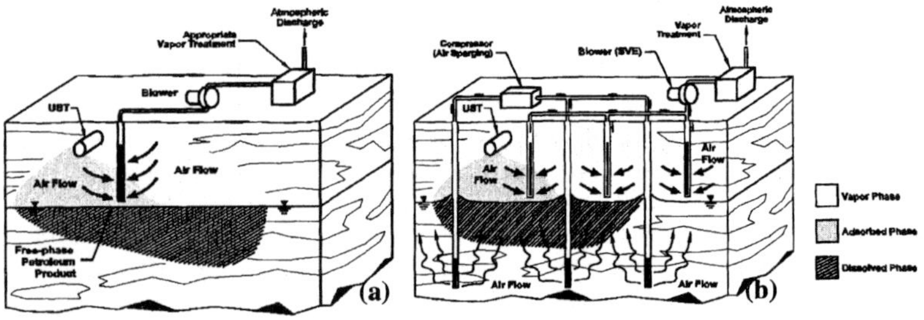

Figure 12.4. Typical (a) soil vapor extraction (SVE) systems, and (b) combined air sparging (AS) and SVE systems (USEPA 2004b).

12.5.2.2 Key Monitoring Parameters

Monitoring of extraction flow rates, vacuum pressures, and vapor-phase contaminant concentrations in extraction wells effluents, treatment unit manifolds and effluent stacks are required for optimization of contaminant mass removal during the operation of SVE systems (Table 12.3). Based on these monitoring data, the vacuum pressures can be adjusted to focus on extraction wells that produce vapors with high contaminant concentrations. In addition, periodic monitoring of extraction flow rates and vapor-phase contaminant concentrations from extraction wells and effluent stacks allow determination of the contaminant mass extracted by the SVE systems for examination of remedial progress. If the SVE systems exhibit asymptotic behavior with respect to both vapor contaminant concentration and cumulative mass removal, some modification of the SVE system may be needed. Modifications may include increasing flow to the extraction wells with high vapor-phase contaminant concentrations, ceasing vapor extraction from the wells with low vapor contaminant concentrations, and installing additional extraction wells (USEPA 2004b).

Table 12.3. Key monitoring parameters of soil vapor extraction (SVE) systems and combined air sparging (AS) and SVE systems.

Locations	Parameters	Purpose
Soil Vapor Extraction Systems		
Extraction wells, manifolds to vapor treatment unit and effluent stacks	Extraction flowrate, vacuum pressure, and vapor contaminant concentration	System optimization and remedial performance /progress
Combined Air Sparging and Soil Vapor Extraction Systems		
Air sparging wells (AS), groundwater monitoring wells (AS), extraction wells (SVE), manifolds to vapor treatment unit (SVE) and effluent stacks (SVE)	Sparging pressure (AS), injection flowrate (AS), hydraulic gradient (AS), extraction flowrate (SVE), vacuum pressure (SVE), vapor contaminant concentrations (SVE)	System optimization and remedial performance /progress
Groundwater monitoring wells	Groundwater contaminant concentrations	Remedial performance/progress

12.5.2.3 Air Sparging Systems

Air sparging (AS) is an *in situ* remediation technology that operates by stripping volatile organic contaminants and creating oxygenated conditions favorable for aerobic biodegradation. Air stripping requires injection of contaminant-free air into the aquifer through air sparging wells. When air stripping is the primary remediation mechanism, SVE systems are often used in conjunction with AS systems to remove the vapor vented to the vadose zone from the subsurface (Figure 12.4b). If biodegradation is the main remediation mechanism, SVE systems are usually not included. Air injection flow rates are controlled to prevent excessive generation of organic vapors in the soil gas or at adjoining locations (ITRC 2004; USEPA 2004b).

12.5.2.4 Key Monitoring Parameters

When combined AS-SVE systems are deployed, the SVE system is started up and optimized in advance to obtain monitoring data. This is done to prevent accumulation of organic vapors in soil gas or migration of organic vapor to nearby receptors during the operation of the AS system. After starting the AS system, the air injection flow rate, extraction flow rate, vacuum pressure, hydraulic gradient and vapor contaminant concentration are monitored to balance the injection flow rate and optimize the rate of contaminant mass removal (Table 12.3). To determine the mass of contaminant extracted and that remaining in the aquifer, contaminant levels in groundwater and the vapor phase in the groundwater monitoring wells, vapor extraction wells and effluent stacks of the treatment unit are measured periodically. The asymptotic behavior of the AS systems with respect to both dissolved-phase and vapor-phase concentrations is also identified (USEPA 2004b).

12.5.3 Surfactant/Cosolvent Flushing Systems

Surfactant/cosolvent flushing involves injection of chemicals into the vadose zone or aquifer to sweep the NAPL zone and extract the elutrate (i.e., a mixture of the injected fluids and swept contaminants) for treatment or recovery (Figure 12.5). Surfactants and cosolvents such as ethanol or isopropanol reduce the interfacial tension between the NAPL and the aqueous phase, and increase the solubility of NAPL components as they are flushed through the contaminated zone (ITRC 2004). The criteria for selection of surfactants and cosolvents for flushing depend on contaminant characteristics, such as hydrophobicity, density, viscosity, and interfacial tension. Site conditions including soil heterogeneity, hydraulic conductivity of aquifer, groundwater geochemistry and soil mineralogy also influence the selection of surfactant or cosolvent.

12.5.3.1 Key Monitoring Parameters

Despite proven abilities of surfactants and cosolvents to lower interfacial tension and increase contaminant solubility, small changes in electrolyte concentrations in groundwater can exert a significant influence on the performance of this technology (ITRC 2003). Thus periodic monitoring of surfactant/cosolvent concentrations, contaminant concentrations and groundwater characteristics (e.g., pH and specific conductance) is a key to ensure effective and efficient removal of the subsurface contaminant by this technology. Table 12.4 summarizes the key monitoring parameters for surfactant/cosolvent flushing technologies. Monitoring of the contaminant concentration in the extracted groundwater allows for the determination of contaminant mass extracted. If the surfactants or cosolvents are reused, their amounts being recovered can be calculated by monitoring the surfactant/cosolvent concentrations in the extracted groundwater and recovered fluids. Monitoring of surfactant/cosolvent concentrations in the injection solution is necessary to ensure that the injectionsolution meets design specifications. This step becomes more important if the surfactant/cosolvent is reused.

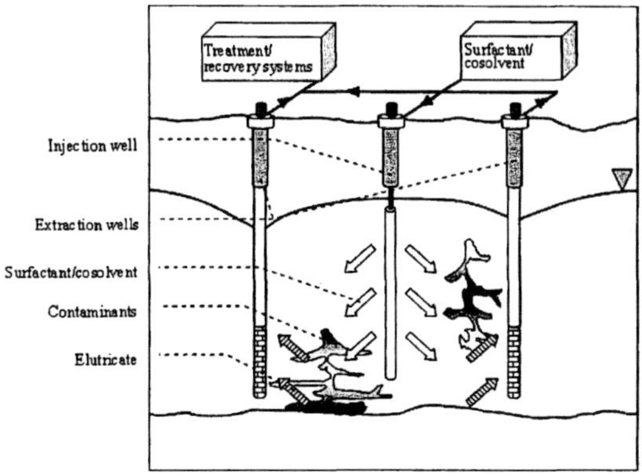

Figure 12.5 Schematic diagram showing the operation of surfactant/cosolvent flushing systems for groundwater remediation

Table 12.4. Key monitoring parameters for surfactant/cosolvent flushing technologies.

Locations	Parameters	Purpose
Injection wells, extraction wells, treatment and recovery units, and monitoring wells within and outside of the remediation zone	Dissolved contaminant concentration, surfactant/cosolvent levels, pH, temperature, specific conductance and electrolytic concentration	Remedial performance/progress, and fluid chemistry
Injection and extraction wells	Injection and extraction rates	Fluid flow properties
Injection and extraction wells, monitoring points	Groundwater levels	Aquifer properties
Injection, extraction, and wastewater processing fluid lines	Pressure	Fluid flow properties
All monitoring points	Free-phase NAPL levels	Aquifer properties

Monitoring of the contaminant concentration in the influent and effluent of the treatment unit becomes imperative when treated groundwater is re-introduced into the aquifer. Contaminant and surfactant/cosolvent concentrations should also be measured periodically from the monitoring points located below, above and around the remediation zone to ensure hydraulic capture of the flushed fluids and swept contaminants, and assess the remedial progress. Pre- and post-flushing monitoring of contaminant concentrations in the injection, extraction and monitoring wells is necessary to ascertain reductions in dissolved contaminant concentrations in the subsurface. Post-flushing monitoring should be implemented once the groundwater system has re-equlibrated. Long-term monitoring is also needed to assess any rebound in contaminant concentrations. Groundwater sampling from monitoring points outside the remediation zone should be performed to verify that contaminants and surfactants or cosolvents have not migrate out of the treatment zone.

Injection and extraction flow rates are crucial in ensuring proper performance of surfactant/cosolvent flushing systems. Unbalanced injection and extraction rates can lead to poor sweep efficiency and mounding of the groundwater table. Serious fluctuation of the groundwater table may affect the hydraulic control system and result in poor hydraulic capture of the flushed fluids. Accurate measurement of injection and extraction rates is necessary for precise determination of the contaminant mass extracted and the amount of surfactant or cosolvent recovered.

12.5.4 In Situ Thermal Remediation Systems

Thermally enhanced remediation systems are *in situ* technologies for removal of light NAPLs (LNAPL) or DNAPLs in soil and/or groundwater. Heat energy is delivered to the subsurface to vaporize and/or mobilize NAPL contaminants. A network of extraction wells in the vadose zone and aquifers is used to remove the contaminant vapors, mobilized NAPL, and contaminated groundwater for above ground treatment. During the remediation, continuous boiling of groundwater is desirable to create steam stripping and thereby offer sustained treatment of contaminated groundwater. The three common methods used to deliver thermal energy to vadose zones or aquifers include: 1) steam injection; 2) electrical resistance heating; and 3) thermal conduction heating (ITRC 2004).

12.5.4.1 Steam-Enhanced Extraction

Steam-enhanced extraction involves injection of steam around a pool of NAPL contaminants in the subsurface (Figure 12.6). Steam can be injected into the vadose zone, the saturated zone or both. When steam is injected into well bores, it heats the well bores and the geological formations around the steam injection zone. The injected steam condenses and as more steam is injected, the condensed hot water moves into the formation. Steam also enters the porous media and pushes the condensed hot water and cold water in front of it towards the contaminant zone. Contaminants are flushed and transported by the cold water, hot water and steam and removed by the groundwater and vapor extraction wells (Davis 1998).

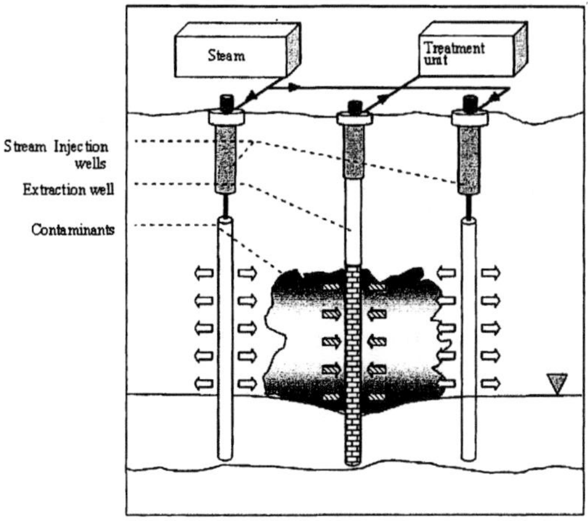

Figure 12.6 A diagram of an *in situ* thermal remediation system using steam-enhanced extraction.

12.5.4.2 Electrical Resistance Heating

Electrical resistance heating applies thermal energy generated from the natural resistance of soils or rocks in response to the flow of electric current. NAPL contaminants are volatilized and the steam generated serves as a carrier gas to sweep the contaminants to extraction wells. Electric current is applied to the geologic media through an array of electrodes placed in the subsurface throughout the remediation area. Electrical resistance heating provides relatively uniform heating throughout the remediation area regardless of lithologies encountered. Furthermore, *in situ* steam generation can occur in fractured or porous rock, and in all soil types irrespective of permeability (Beyke and Fleming 2005).

12.5.4.3 Thermal Conduction Heating

Thermal conduction heating is used for the remediation of contaminated soils. Heat energy is applied to the soils through arrays of vertical or horizontal wells containing heating elements. Heat flows through the soils from the heating elements primarily by thermal conduction. As the soils are heated, NAPL contaminants are vaporized and/or destroyed by a number of mechanisms including steam distillation, boiling, oxidation and pyrolysis. The vaporized water and contaminants are drawn from the subsurface via vacuum extraction wells (ITRC 2004).

12.5.4.4 Key Monitoring Parameters

Monitoring of *in situ* thermal remediation systems is implemented in key locations in the subsurface and throughout the remediation train, such as the steam injection/extraction wells, monitoring wells, electrodes, inlet and outlet piping of the surface treatment unit, and the effluent stack. Table 12.5 shows key parameters that need to be monitored during thermal remediation using steam-enhanced extraction, electrical resistance heating and thermal conduction heating.

Heating progress is monitored by measuring the temperature in subsurface using thermocouple strings placed between injection, extraction wells, various locations between electrodes or between heating elements. Temperature data obtained from thermocouple strings co-located with peripheral monitoring wells and monitoring wells situated inside and outside the remediation area can show if there is a loss of hydraulic control of contaminated groundwater and mobilized contaminants during the thermal remediation. For instance, temperature increases at a certain depth of the monitoring well outside the remediation area would indicate flow of heated water out of the remediation area. Understanding the direction where water is leaving the remediation zone allows adjustment to be made to improve the hydraulic control. The area beneath the contaminated zone is heated up first and establishes a hot barrier at the bottom of the contaminated zone. The hot barrier vaporizes the contaminants that move downwards. To ensure the establishment of a hot floor, thermocouples are installed below the contaminant zone. Subsurface pressure is also an important monitoring parameter of thermal remediation systems for the control of generated vapors. It can be measured using the piezometers located within and around the perimeter of the treatment area, and co-located with monitoring wells (ITRC 2004).

Monitoring of the vapor-phase contaminant concentration in soil gas or extracted vapors can provide valuable information related to progress towards attainment of remediation objectives. These measurements can also indicate if concentrations in the extracted vapor have reached an asymptotic limit. Contaminant concentrations in the soil should be monitored during installation of the remediation system and immediately after the system shutdown. Soil concentration data obtained during system installation can provide baseline information on the extent of contamination while concentrations obtained after completion of remediation can indicate whether remediation objectives have been met. Soil concentration data should also be obtained from locations where complete heating is difficult to achieve. Interim soil sampling can also be implemented to evaluate remediation progress. When groundwater is extracted during thermal remediation, monitoring of contaminant concentrations in the extracted fluids can provide data to calculate contaminant mass removal.

Table 12.5. Key monitoring parameters for *in situ* thermal remediation systems

Locations/Media		Parameters	Purposes
General			
Peripheral monitoring wells	Groundwater	Water levels and contaminant concentrations	Migration control
	Vapor	Subsurface pressure	Migration control
	Thermocouples	Temperature	Migration control
Remediation zone (before, during and after the remediation)	Thermocouples	Temperature	Heating progress and completeness
	Soil Groundwater Vapor	Contaminant concentration	Remedial performance/progress
Extraction and treatment systems	Extracted fluids (water, vapor and NAPL contaminants)	Temperature, pressure and flowrate	Mass and energy balance, cooling, and vacuum confirmation
		Contaminant concentrations	Remedial performance
Steam-Enhanced Extraction Only			
Injection wells	Steam, liquid and vapor lines	Temperature, pressure and flowrate	Confirmation of steam injection, injection safety and process control, energy balance and site balancing
Steam systems	Steam headers	Temperature, pressure and flowrate	Safety and process control, mass and energy balance
Electrical Resistance Heating Only			
Electrical heating systems	Electrodes	Power, current and voltage	Energy balance and site balancing
		Water flowrate	Power delivery maintenance
Thermal Conduction Heating Only			
Electrical heating systems	Thermal well circuits	Power, current and voltage	Confirmation of heat delivery and energy balance

12.5.5 *In Situ Chemical Oxidation Systems*

Unlike P&T, SVE/AS and surfactant/cosolvent flushing systems which extract contaminants from the subsurface for subsequent treatment above ground, *in situ* chemical oxidation (ISCO) systems directly destroy the contaminants in soil and groundwater. ISCO involves injection of oxidants and amendments directly into the contaminant source zone and the downgradient contaminant plume. When the oxidants come in contact with the contaminants, the latter are oxidized into benign products, such as carbon dioxide, water, and/or inorganic salts (ITRC 2005). The four most frequently used oxidants in ISCO include permanganate (MnO_4^-), sodium persulfate ($Na_2S_2O_8$), hydrogen peroxide (H_2O_2) and ozone gas (O_3). The oxidative power of these oxidants follows the order: $O_3 > Na_2S_2O_8 > H_2O_2 > MnO_4^-$ (ITRC 2004; ITRC 2005).

Permanganate is a stable oxidant and has a unique affinity for oxidizing organic compounds containing carbon-carbon double bonds, aldehyde groups or hydroxyl groups. Under normal groundwater pH and temperature, carbon-carbon double bonds of alkenes are broken spontaneously and converting the carbon to CO_2 through hydrolysis or further oxidation by the permanganate (Eq. 12.2)

$$4KMnO_4 + 3C_2Cl_4 + 4H_2O \rightarrow 6CO_2 + 4MnO_2(s) + 4K^+ + 12Cl^- + 8H^+ \quad \text{(Eq. 12.2)}$$

Hydrogen peroxide is an oxidant but it reacts slowly with organic contaminants at low concentrations. Ferrous salt is usually added to dramatically increase the oxidative strength of hydrogen peroxide. Such enhancement of the strength is ascribed to the production of hydroxyl radicals ($OH^\bullet$) (Eq. 12.3). The reaction of iron catalyzed peroxide oxidation at pH 2.5 to 3.5 is called Fenton's reaction and the iron/peroxide mixture is known as Fenton's reagent. At high concentrations, however, H_2O_2 itself produces sufficiently high concentrations of $OH^\bullet$ radicals without the need for amendments.

$$Fe^{2+} + H_2O_2 \rightarrow Fe^{3+} + OH^\bullet + OH^- \quad \text{(Eq. 12.3)}$$

Sodium persulfate or persulfate anion ($S_2O_8^{2-}$) is also kinetically slow in destroying many organic contaminants despite being a strong oxidant. Under dilute acid conditions, the hydrolysis of the $S_2O_8^{2-}$ yields bisulfate anion and H_2O_2. However, activation by heat or catalysis by chelated metals (e.g., iron, copper and silver) can substantially increase the oxidative strength of $S_2O_8^{2-}$ due to generation of sulfate free radicals ($SO_4^{-\bullet}$), as shown in Eq. 12.4.

$$Fe^{2+} + S_2O_8^{2-} \rightarrow Fe^{3+} + SO_4^{-\bullet} + SO_4^{2-} \quad \text{(Eq. 12.4)}$$

Ozone is the strongest oxidant available for ISCO. Ozone-based *in-situ* oxidation involves injection of ozone gas creating very different design and operational issues. In practice, O_3 is either injected into the vadose zone or sparged below the water table. Contaminants in the ozonated zone are oxidized directly by O_3 (Eq. 12.5) or indirectly by the generated $OH^{\bullet}$ radicals.

$$O_3 + RC = RC \rightarrow RCOCR + O_2 \qquad \text{(Eq. 12.5)}$$

12.5.5.1 Key Monitoring Parameters

Since ISCO is a destructive remediation technology, assessment of contaminant mass treatment is more difficult than for technologies involving extraction of contaminants. Since a precise estimation of the extent of contaminant degradation is often impractical, performance monitoring of ISCO systems customarily focuses on evaluating treatment progress of the source zone (ITRC 2004). Groundwater contaminant concentrations in monitoring wells located downgradient of injection wells, and soil contaminant concentrations are the main parameters measured before and after the injection of oxidant to assess remedial progress and/or effectiveness of an ISCO technology (Table 12.6). The monitoring data are also useful in evaluating the extent of displacement of contaminated groundwater away from the injection wells during oxidant injection. If contaminant concentrations in all the monitoring wells show declines after injection, the displacement is expected to be minimal. On the other hand, an increase in the concentration at one or more monitoring locations may indicate displacement of contaminated groundwater from the treatment zone. Measurement of oxidant concentrations in groundwater also provides information about persistence of the oxidant in the subsurface and the distribution of oxidant concentrations across the treatment zone. Improper application of Fenton's reagent or O_3 has the potential to generate heat or oxygen gas in the subsurface capable of volatilizing or stripping contaminants into the vapor phase and subsequently to the atmosphere. Transfer of contaminants to atmosphere can also occur when using ozone as an oxidant for the treatment of impaired vadose zones. Therefore, monitoring of the contaminant concentrations in the soil gas is often recommended to ensure that contaminants are destroyed in the subsurface rather than transferred to atmosphere.

During injection, the injection flow rate and oxidant concentration should be measured near injection wells to ensure that the oxidant concentration injected and the radius of influence created in the subsurface are consistent with design specifications. Monitoring other field parameters such as DO, pH, temperature, and redox potential can also help evaluate the system effectiveness. These field parameters can also help decide if site conditions after injection have reached post-treatment equilibrium, and whether they have improved from pre-injection conditions. Analysis of dissolved metal levels in groundwater during and after injection is also essential because certain redox-sensitive

Table 12.6. Key monitoring parameters of *in situ* chemical oxidation technologies.

Locations	Parameters	Purposes
Groundwater in monitoring wells downgradient of injection wells (pre- and post-injection)	Contaminant concentrations	Remedial progress and effectiveness
	Oxidant concentrations	Estimation of oxidant persistence and radial influence
Soil within contaminant zone (pre- and post-injection)	Contaminant concentrations	Remedial effectiveness
Soil vapour within the contaminant zone	Contaminant concentrations	Prevention of transfer of contaminants to vapour phase
Injection wells (during injection)	Injection flow rate and oxidant concentrations	Confirmation of the oxidant concentration and volumes injected, and radial influence
Extraction wells associated with recirculation system, if installed	Extraction flow rate and oxidant concentrations	Oxidant recovery
Groundwater in monitoring wells, downgradient of injection wells (pre-, during, and post-injection)	Field parameters (DO, pH, temperature, specific conductance and redox potential)	Evaluation of system efficiency
	Tracer movement	Observation of travel times and distribution of oxidants
	Metals (e.g., manganese, aluminium, arsenic, lead and chromium)	Mobilization of insoluble metals by injected oxidants and evaluation of potential for the formation clogging
	Water quality parameters (sulfate, chloride, calcium, natural organic matter, nitrate and alkalinity)	Progressing of the treatment, and influence of non-target oxidizable matters

metals, such as arsenic, barium, cadmium, chromium, copper, lead and selenium, can be oxidized to more soluble states during oxidant injection. Metal monitoring data can also be used to evaluate metal precipitation in subsurface that may clog the porous media and reduce the hydraulic conductivity of the aquifer. Monitoring of certain specific indicating parameters, such as chloride for chlorinated solvents, can also provide evidence of treatment by the injected oxidant. Groundwater analysis can also help evaluate the influence of non-target oxidizable materials in the subsurface on oxidant consumption (ITRC 2004; ITRC 2005).

12.5.6 In Situ Bioremediation

In situ groundwater bioremediation stimulates growth and reproduction of indigenous microorganisms to enhance biodegradation of organic contaminants inaquifers (Bedient et al. 1999). To stimulate and maintain the microbial activity in subsurface, a delivery system providing electron acceptors (e.g., oxygen, nitrate), nutrients (nitrogen, phosphorous) and/or energy sources (carbon) is required for bioremediation. During bioremediation, groundwater is typically removed from extraction wells and treated, if necessary, to remove dissolved contaminants. The treated groundwater is then mixed with electron acceptors, nutrients and/or other constituents and re-injected upgradient of or within the contaminant source, as illustrated in Figure 12.7. The major biological processes involved in degrading organic contaminants include aerobic, anaerobic, and co-metabolic degradation processes. Aerobic degradation is the most effective process in treating petroleum based aliphatic and aromatic hydrocarbons, while all three processes are important in the biotransformation of chlorinated contaminants (USEPA 2004b, ITRC 1998). Reductive dechlorination processes under anaerobic conditions can replace chlorine atoms in the chlorinated organics by hydrogen. In this process, an electron donor, either hydrogen gas or a precursor carbon compound, is required. Chlorinated hydrocarbons can also be degraded by cometabolic aerobic processes that transform the pollutants via biochemical pathways mediated by non-specific enzymes. Direct degradation or oxidation of less chlorinated organics can occur in both anaerobic and aerobic environments (ITRC 1998).

12.5.6.1 Key Monitoring Parameters

Monitoring parameters selected to evaluate the effectiveness and progress of bioremediation should examine whether subsurface conditions are suitable for bioremediation, measure contaminant biodegradation in the subsurface, and ensure appropriate deliver and distribution of nutrients and/or oxygen within the treatment zone (ITRC 2004; USEPA 2004b). Table 12.7 summarizes the key monitoring parameters of *in situ* bioremediation technologies. The volumes of water extracted and injected should be recorded periodically to ensure that stable and balanced extraction and injection flows are established for oxygen and nutrient delivery.

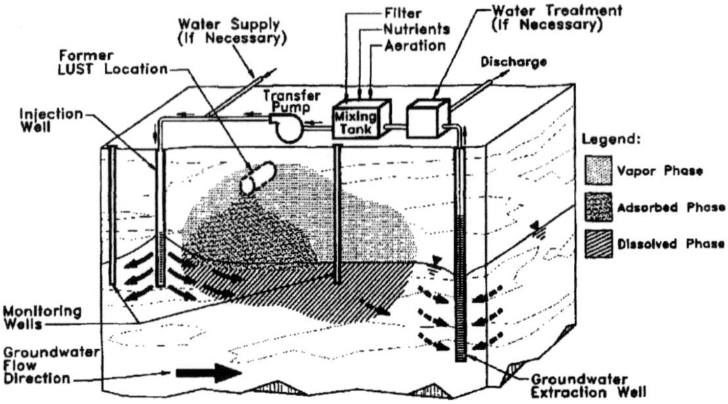

Figure 12.7. Schematic diagram of a typical *in situ* bioremediation system for groundwater remediation (USEPA 2004b)

Monitoring of the general water quality parameters, such as pH, temperature and specific conductance, is also required to ensure suitable subsurface conditions for bioremediation. Since oxygen is a key element for aerobic biodegradation of petroleum hydrocarbons, dissolved oxygen, H_2O_2 or ozone levels and distribution in aquifers should be measured. These monitoring data can also provide the information indicating the performance of the delivery system and the transmitted distance of H_2O_2 or ozone in aquifers. Furthermore, measurement of the CO_2 levels in the soil gas can help determine the progress of biodegradation. When *in situ* bioremediation is applied for the degradation of chlorinated hydrocarbons, monitoring of subsurface redox sensitive parameters, such as the redox potential, and concentrations of hydrogen, ferrous iron, sulfateand methane is required to ensure achievement of appropriate conditions for reductive dechlorination. The microbial community in the aquifer should also be studied to check for specific types of microbes and adequate microbial populations for reductive dechlorination. Monitoring of other indicating parameters such as chloride, carbon isotope ratios, and metabolites can also evince the progress of *in-situ* reductive dechlorination (ITRC 2004; USEPA 2004b).

Table 12.7. Key monitoring parameters for *in situ* bioremediation technologies applied for petroleum hydrocarbon and chlorinated hydrocarbon degradation.

Locations/media	Parameters	Purposes
General		
Extraction and injection wells	Extraction and injection volumes	Balanced and stabilized extraction and injection flows
Monitoring wells	Water levels	Determination of hydraulic conditions
Groundwater	pH, temperature, specific conductance and turbidity	Confirmation of suitable subsurface conditions for bioremediation
Groundwater	Bio-nutrients, such as ammonia and phosphate	Performance of delivery systems and remedial progress
Bioremediation of Petroleum Hydrocarbons		
Groundwater	Dissolved oxygen and redox potential	Delivery system performance on establishing aerobic environment
Groundwater	H_2O_2 or ozone	Their transmitted distance in subsurface by the delivery systems
Groundwater	Concentration of petroleum hydrocarbons and the byproducts	Remedial progress/effectiveness
Soil vapor	Carbon dioxide	Evidence showing occurrence of biodegradation
Soil vapor	Oxygen, H_2O_2 or ozone	Potential loss of injected oxygen, H_2O_2 or ozone
Soil vapor	Concentration of petroleum hydrocarbons and the byproducts	Possible escape of petroleum contaminants to vadose zone
Soil	Concentration of petroleum hydrocarbons and the byproducts	Remedial progress/effectiveness
Bioremediation of Chlorinated Hydrocarbons		
Groundwater	Concentration of chlorinated contaminants, chlorinated byproducts and benign byproducts	Remedial progress/effectiveness, and extent of dechlorination
Groundwater	Dechlorination indicating parameter (e.g., chloride)	Remedial progress and extent of dechlorination
Groundwater	Redox sensitive parameters, such as redox potential, dissolved H_2, DO, Fe(II), Mn(II), NO^{3-}, SO_4^{2-} and CH_4	Suitability of subsurface conditions for reductive dechlorination
Groundwater	Electron donor parameters, such as COD, TOC, volatile fatty acid and speciated electron donors	Measurement of availability and distribution of electron donors in subsurface
Groundwater	Respiration indicator, such as carbon dioxide and alkalinity	Evaluation of areas of increased biological activity
Groundwater	Microbial communities and molecular parameters	Presence of sufficient or suitable microbes for bioremediation
Groundwater	Stable carbon isotopes of chlorinated contaminants and the byproducts	Remedial progress and extent of dechlorination

12.5.7 Monitored Natural Attenuation

Monitored natural attenuation (MNA) relies on natural attenuation processes for soil and groundwater remediation. It can achieve site-specific remediation objectives within time frames that usually longer than those required by more active remediation technologies. Source control systems and long-term performance monitoring are the key elements of any MNA remedy (USEPA 1999). Natural attenuation processes involved in MNA include a variety of *in situ* physical, chemical or biological processes, such as dispersion, dilution, volatilization, decay, sorption, and chemical or biological stabilization, transformation or destruction of contaminants. These natural processes are capable of reducing the mass, toxicity, mobility, bioavailability, volume or concentration of contaminants in soil and groundwater without human intervention. *In situ* processes that result in contaminant reduction via dispersion, dilution, volatilization or sorption are called non-destructive processes and those resulting in contaminant transformation are considered destructive processes. Removal of petroleum-based contaminants, such as benzene, ethylbenzene, toluene and xylene (BTEX) has been observed to occur readily during MNA; degradation of chlorinated solvents happens under specific site conditions. Removal of inorganic contaminants during MNA is achieved by via sorption, precipitation and complexation processes. Oxidation-reduction reactions can transform some inorganics into less soluble or less toxic forms. Typically, MNA is used in areas with low contaminant concentrations while active remediation technologies may be applied at locations where contaminant concentrations are high. MNA is also used as a follow-up to the active remediation technologies.

12.5.7.1 Key Monitoring Parameters

Performance monitoring is a critical component of MNA because of the longer remediation timeframes, the potential for contaminant migration and the absence of human intervention. Parameters selected for monitoring MNA remedial performance should be able to:

- evince that natural attention is progressing as expected;
- detect changes in hydrologic, geochemical and microbiological conditions, or other changes that may reduce the effectiveness of MNA;
- identify potentially toxic transformation byproducts;
- testify that contaminant plumes are not expanding, either downgradient, or laterally and vertically;
- verify no unacceptable impact to downgradient receptors;
- detect the new release of contaminants to subsurface;
- demonstrate the effectiveness of controls placed by regulatory agencies to protect potential receptors; and
- verify the attainment of the remediation objectives.

Key monitoring parameters for MNA, and the suite of geochemical parameters that should be measured during application of MNA are listed in Tables 12.8 and 12.9, respectively. Contaminant transformation processes may produce more toxic, mobile or recalcitrant byproducts. Robust performance monitoring requires that the contaminants originally present in the contaminated site, as well as their transformation byproducts be measured at appropriate intervals of time. Monitoring of the groundwater and soil quality in upgradient and sidegradient monitoring wells can provide baseline data for the site and indicate if there is a plume expansion. Periodic analysis of soil and groundwater contaminant concentrations inside the contaminant source zone and the downgradient plume can illustrate changes in source characteristics and plume strength over time during the MNA. Contaminant concentration data from the monitoring wells located just and further downgradient from the plume can examine the possible migration of the plume to downgradient receptors. These data can be used to compute field-scale attenuation rates and contaminant mass reductions so that results can be compared with design specifications for the MNA operation. An increase in contaminant concentration or detection of new contaminants at monitoring points may indicate new releases to the subsurface. Furthermore, periodic monitoring of the contaminant distribution can ensure that appropriate progress is made towards the contaminant reduction objective (USEPA 2004a).

Geochemical parameters summarized in Table 12.9 should be measured throughout the plume for effective monitoring of MNA performance. Geochemical conditions may affect microbial populations and consequently the transformation processes resulting in contaminant destruction. Furthermore, the biological degradation of contaminants can cause specific geochemical changes in the subsurface. Monitoring geochemical parameters during the MNA can help establish a correlation between contaminant degradation and microbial activity. Geochemical monitoring should also be implemented in upgradient and sidegradient monitoring wells to determine the spatial variability of geochemical parameters and differentiate the changes in geochemistry caused specifically by natural attenuation processes (USEPA 2004a).

12.5.8 Permeable Reactive Barrier Technologies

Permeable reactive barriers (PRBs) is a passive *in situ* groundwater remediation technology. Unlike other physical barriers constraining plume migration, PRBs are designed as conduits for contaminated groundwater flow. PRB involves emplacement of reactive materials in subsurface designed to intercept a contaminant plume, provide a flow path through the reactive media, and transform the contaminants into environmentally acceptable forms to attain remediation goals downgradient of the barriers (Figure 12.8) (USEPA 1998). Zero-valent metals, such as zero-valent iron (Gillham and O'Hannesin 1992; Lai et al. 2006c), zero-valent tin (Su and Puls 1999), zero-valent zinc (Arnold and Roberts 1998), bimetallic reactive materials (Gavaskar et al. 2000), nanoscalezero-valent iron (Lien et al. 2001), and *Cercona*[TM] iron foam (Bostick et al. 1996) are some of the reactive materials available for PRBs. When contaminants come in contact with the reactive materials in a PRB, they are transformed into either benign products or immobile forms through the process of abiotic reduction,reductive precipitation or adsorption (USEPA 2002; Lai et al. 2006c; Lo et al. 2006b). Deposition of precipitates in the PRB eventually deteriorates PRB performance over time.

Table 12.8. Key monitoring parameters of monitored natural attenuation (MNA).

Locations	Parameters		Purposes
	Initial Samples	Subsequent Samples	
Upgradient wells Sidegradient wells	Contaminants, byproducts, and full suite of geochemical parameters	Pertinent geochemical parameters	Background water and soil quality
Source impact wells or monitoring wells within contaminant source zone	Contaminants, byproducts, and full suite of geochemical parameters	Contaminants, byproducts, and pertinent geochemical parameters	Changing source strength
Downgradient source impact wells or monitoring wells within the plume	Contaminants, byproducts, and full suite of geochemical parameters	Contaminants and byproducts	Contaminant plume behavior over time
Monitoring wells just and further downgradient of the plume	Contaminants, byproducts, and full suite of geochemical parameters	Contaminants and byproducts	Detection of plume migration
Farthest downgradient monitoring wells	Contaminants, byproducts, and full suite of geochemical parameters	Contaminants and byproducts	Compliance monitoring

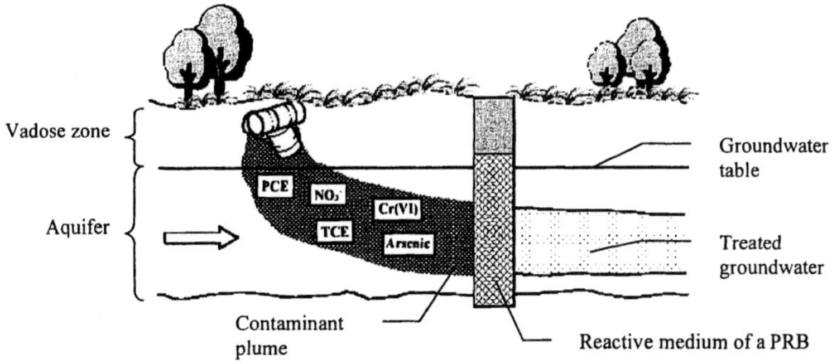

Figure 12.8. Schematic diagram showing groundwater remediation by permeable reactive barriers (PRBs) (Lo et al. 2006a).

12.5.8.1 Key Monitoring Parameters

Performance monitoring of PRBs includes evaluation of chemical, geochemical and mineralogical parameters in groundwater. Geochemical parameters to be measured generally include pH, specific conductance, redox potential, dissolved oxygen, hardness, alkalinity, total dissolved sulfide, ferrous iron and dissolved hydrogen. These parameters can indicate the progress of contaminant transformation mediated by the reactive media, and the extent of precipitate formation within the PRBs. For instance, application of zero-valent iron in the PRBs for the removal of chlorinated solvent lowers the redox potential, increases local pH and dissolved hydrogen gas concentrations, and produces a substantial drop in groundwater hardness and alkalinity (Lai et al. 2006c). Relatively small volumes of water are usually sampled within and around the PRBs during monitoring because of a comparatively small treatment zone within the aquifer (USEPA 1998; Gavaskar et al 2000).

Groundwater contaminant concentrations and geochemical parameters should be measured from the monitoring wells located within and immediately downgradient to the PRB reactive zone to confirm progress of contaminant degradation or removal and attainment of downgradient concentration goals (Table 12.10). Monitoring concentrations of transformation byproducts generated inside the PRB reactive zone is necessary to ensure that no toxic or harmful byproducts are mobilized towards downgradient receptors. The contaminant concentration profile can be used to determine field degradation rates, which can be utilized to ensure that the PRB is performing as designed and assess temporal changes in the PRB reactivity. Upgradient monitoring wells can warn of a sudden increase in upgradient contaminant concentrations that can cause contaminant breakthrough within the PRB. Monitoring hydraulic gradient, tracer movement and groundwater velocity across the PRB can help determine the hydraulic capture zone of the PRB and the residence time inside the reactive barriers for contaminant removal. Hydraulic conductivity within the PRB should be measured periodically to monitor any decrease in PRB permeability or clogging of the reactive media over time. Mineralogical and microbiological analyses of reactive medium cores taken from an emplaced PRB can provide useful information for assessing the effect of groundwater geochemistry and microbial activity on the reactivity and longevity of the PRB.

Table 12.9. A suite of geochemical parameters required for performance monitoring of monitored natural attenuation (MNA).

Geochemical Parameters	Purposes	
	Water	Soil
Dissolved oxygen	Determination of metabolic pathway and bioactivity	Determination of bioactivity in vadose zone
Iron (III)	/	Prediction iron reduction
Total organic compounds	Determination of extent of groundwater contamination	Determination of extent of soil contamination
Carbon dioxide	Determination of bioactivity in aquifers	Determination of bioactivity in vadose zones
Nitrate	Indication of respiration in the absence of oxygen	Electron acceptor for organic compounds oxidation under some conditions
Iron (II)	Indication of anaerobic degradation	Electron donor
Chloride	Byproduct from chlorinated solvent reduction	Byproduct from chlorinated solvent reduction
Sulfate	Indication of anaerobic microbial respiration	Electron acceptor for organic compounds oxidation under some conditions
Oxidation-reduction potential	Indication of nature of degradation	Indication of nature of degradation
Alkalinity	Indication of buffering capacity	Indication of buffering capacity
Methane	Methanogenesis	/
pH	Condition for some metabolic process to occur	Condition for some metabolic process to occur
Temperature	Indication of particular microbial species in subsurface and the approximate degradation rate	Indication of particular microbial species in subsurface and the approximate degradation rate
Conductivity	Water quality parameter	/

Table 12.10. Key monitoring parameters for permeable reactive barrier (PRB) technologies.

Locations	Parameters	Purposes
Monitoring wells located within or immediately downgradient to the PRB reactive zone	Contaminant and toxic byproduct concentrations	Remedial effectiveness, contaminant degradation rates and temporal changes in the PRB reactivity
Upgradient monitoring wells	Contaminant concentrations	Potential breakthrough of contaminants
Monitoring wells located within or immediately downgradient to the PRB reactive zone	Geochemical indicating parameters	Remedial progress and potential impact of precipitate formation on the PRB effectiveness
Monitoring wells inside the PRBs	Hydraulic conductivity	Potential clogging or decrease in the PRB permeability
Monitoring wells inside and across the PRBs	Hydraulic gradient	Hydraulic capture zone and residence time
Monitoring wells inside and across the PRBs	Tracer movement	Hydraulic capture zone and residence time
Monitoring wells inside and across the PRB	Groundwater velocity	Hydraulic capture zone and residence time
Reactive medium cores from PRB	Mineralogical information of the reactive medium	potential impact of precipitate formation on the PRB effectiveness
Reactive medium cores from PRB	Heterotrophic plate count and phospholipid fatty acid profiles	Microbial activity

12.5.9 Phytoremediation

Phytoremediation utilizes plants to degrade, extract, contain, or immobilize contaminants in soil and groundwater. Contaminant transformation is mediated by rhizodegradation or phytodegradation. Rhizodegradation is the breakdown of organic contaminants in soil by the enhanced microbial activity in the rhizosphere due to the presence of root exudates and aerobic conditions. Phytodegradation is the breakdown of contaminants that have been translocated within plants through metabolic processes, or degradation outside the plants via extracellular biomolecules produced by the plants. Plants utilize their roots to absorb contaminants and accumulate them in shoots and leaves via a process called phytoextraction. Plants also utilize rhizofiltration to extract contaminants from soil solution and deposit them on plant roots. Harvesting the crop and the roots removes the contaminants from the affected site. Phytovolatilization is the uptake and transpiration of contaminants by plants and their subsequent release into the atmosphere. When plants are used for containment of contaminants, they enable contaminant binding to the soil rendering them non-bioavailable. Plants can also arrest the migration of subsurface pollutants by removing the means of transport, i.e. groundwater, through uptake and consumption. Vegetative cover systems canbe deployed to reduce transport of rainwater into the subsurface, thereby minimizing infiltration-driven plume migration driven.

12.5.9.1 Key Monitoring Parameters

Climatic factors that affect plant growth are the major influencing parameters in phytoremediation technologies (Table 12.11). For example, phytovolatilization processes are affected by temperature, precipitation, humidity, solar radiation and wind velocity because these factors can affect the transpiration rate. Climatic data can help assess the need for maintenance actions, such as irrigation. Visual monitoring of plant characteristics, such as leave mass and damage from insects and animals can indicate if plant replacement, fertilizer and pesticide additions are required. The amount of contaminants and byproducts extracted can be determined by measuring plant tissue composition, transpiration gas composition, transpiration rate and root density.

Monitoring geochemical conditions is necessary to determine optimal conditions for the plant growth and microbial growth in the rhizosphere. Measuring the rhizosphere microbial population can ensure that there are sufficient microbes for rhizodegradation of contaminants. Monitoring data collected for contaminant and byproduct concentrations in soil and groundwater can be used to estimate the mass of contaminants extracted, destroyed or remaining in the treatment site and assess the progress achieved towards meeting the remediation objectives. The level of groundwater table and groundwater flow rate and direction require consistent monitoring since they are among the most significant factors affecting the success of the hydraulic control of the contaminant plume by the water consumption. In addition, when contaminants are translocated to edible parts of plants, such as leaves and seeds, monitoring of the food chain for bioaccumulation of the contaminants is needed (USEPA 2000).

12.6 Conclusion

Performance monitoring is an essential component of site remediation projects since it provides the data for assessing whether the remediation technologies are proceeding as expected towards the attainment of remediation objectives and whether contingency plans are required to implement further site treatment. Site managers utilize long-term monitoring data to decide the termination of the remediation projects.

Generally, performance monitoring involves assessment of the effectiveness and efficiency of the applied remediation technologies. Effectiveness refers to the capability of the remediation technologies to meet remediation objectives at contaminated sites. Efficiency refers to the optimization of time, energy and costs expended towards the attainment of remediation effectiveness and is typically assessed by comparing system operating parameters to the relevant design specifications. To assess remediation effectiveness and efficiency, each remediation technology has its own set of monitoring parameters. Generally, concentration of contaminants and/or the relevant transformation byproducts in groundwater, soil and soil gas are the overriding monitoring parameters since these monitoring data can be directly used to assess remediation progress, contaminant mass reduction, toxicity reduction, and the mobility or mass flux of contaminant sources.

Table 12.11. Summary of key monitoring parameters of phytoremediation technologies

Parameters	Purposes
Climatic Data	
Temperature Precipitation Relative humidity Solar radiation Wind speed and direction	Maintenance requirements (e.g., irrigation) Determination of water balance and evapotranspiration rate
Plants	
Visual characteristics (viability, damage from insects or animals, growth, leaf mass etc.)	Maintenance (plant replacement, fertilizer, pesticide application etc.)
Tissue compositions (roots, shoots, stems, leaves etc.) Transpiration gases Transpiration rates Root densities	Quantification of contaminants and byproducts extracted Quantification and/or prediction of system operation
Soil	
Geochemical parameters (pH, nutrient concentrations, water content, oxygen content etc.) Microbial population Contaminant and breakdown product levels	Optimization of vegetative, root or microbial growth Determination of water balance and evapotranspiration rates Remedial progress/effectiveness Quantification of contaminants and byproducts removed and remaining Quantification and/or prediction of system operation
Groundwater	
Aquifer information (direction and rate of groundwater flow and depth to groundwater etc.) Contaminant and breakdown product levels	Remedial progress/effectiveness Quantification of contaminants and byproducts removed and remaining Quantification and/or prediction of system operation

References

Arnold, W. A., and Roberts, A. L. (1998). "Pathways of chlorinated ethylene and chlorinated acetylene reaction with Zn(0)." *Environ. Sci. & Technol.*, 32(19), 3017-3025.

Bedient, P. B., Rifai, H. S., and Newell, C. J. (1999). *Ground water contamination transport and remediation*, 2nd Ed., Prentice Hall PTR, Upper Saddle River, NJ, United States.

Beyke, G., and Fleming, D. (2005). "*In situ* thermal remediation of DNAPL and LNAPL using electrical resistance heating." *Wiley, Periodicals, Inc.*, DO1: 10.1002/rem.20047, 5-22.

Bockelmann, A., Ptak, T., and Teutsch, G. (2001). "An analytical quantification of mass fluxes and natural attenuation rate constants at a former gasworks site." *J. Cont. Hydrol.*, 53, 429-453.

Bostick, W. D., Jarabek, R. J., Slover, W. A., Fiedor, J. N., Farrell, J., and Heferich, R. (1996). *Zero-valent iron and metal oxides for the removal of soluble regulated metals in contaminated groundwater at a DOE site, K/TSO-35-P,* Lockheed Martin Energy Systems, Inc., Oak Ridge, TN.

Davis, E. L. (1998). *Steam injection for soil and aquifer remediation*, EPA/540/S-97/505, Ground Water Issue, Office of Research and Development, United States Environmental Protection Agency.

Dayan, H., Abrajano, T., Sturchio, N. C., and Winsor, L. (1999). "Carbon isotopic fractionation during reductive dehalogenation of chlorinated ethenes by metallic iron." *Organic Geochemistry*, 30(8), 755-763.

Dwarakanath, V. (1997). *Characterization and remediation of aquifers contaminated by nonaqueous phase liquids using partitioning tracers and surfactants*, Ph. D. Thesis, University of Texas at Austin, Texas.

Eldho, T. I. (2003). "Remediation of contaminated groundwater – Scope of onsite pump and treat technology." *The Institution of Engineers (India) Journal: Civil Engineering*, 84, 11-17.

ESTCP (2003). *Demonstration and validation of a water and solute flux measuring device*, ESTCP CU-0114, Environmental Security Technology Certification Program.

Garabedian, S. P., LeBlanc, D. R., Gelhar, L. W., and Celia, M. A. (1991). "Large-scale natural gradient tracer test in sand and gravel, Cape Cod, Massachusetts 2. Analysis of spatial moments for a nonreactive tracer." *Water Resour. Res.*, 27(5), 911-924.

Gavaskar, A., Gupta, N., Sass, B., Janosy, R., and Hicks, J. (2000). *Design guidance for application of permeable reactive barriers for groundwater remediation*, Strategic Environmental Research and Development Program, Battelle Columbus, Ohio, <http://www.itrcweb.org/Documents/PRB-2a.pdf>, <http://www.itrcweb.org/Documents/PRB-2b.pdf>, <http://www.itrcweb.org/Documents/PRB-2c.pdf>.

Gillham, R. W., and O'Hannesin, S. F. (1992). "Metal-catalyzed abiotic degradation of halogenated organic compounds." In *Proceedings of the IAH Conference Modern Trends in Hydrogeology*, May 10-13, Hamilton, Ontario, Canada.

Heraty, L. J., Fuller, M. E., Huang, L., Abrajano Jr, T., and Sturchio, N. C. (1999). "Isotopic fractionation of carbon and chlorine by microbial degradation of dichloromethane." *Organic Geochemistry*, 30, 793-799.

ITRC (1998). *Technical and regulatory requirements for enhanced in situ bioremediation of chlorinated solvents in groundwater*, The Interstate Technology & Regulatory Cooperation Workgroup *In Situ* Bioremediation Subgroup.

ITRC (2003). *Technical/regulatory guidelines - Technical/regulatory guidance for surfactant/cosolvent flushing of DNAPL source zones*, The Interstate Technology & Regulatory Council Dense Nonaqueous-Phase Liquids Team.

ITRC (2004). *Technical/regulatory guidelines - Strategies for monitoring the performance of DNAPL source zone remedies*, The Interstate Technology & Regulatory Council Dense Nonaqueous-Phase Liquids Team.

ITRC (2005). *Technical/regulatory guidelines - Technical/regulatory guidance for in situ chemical oxidation of contaminated soil and groundwater*, 2nd Ed., The Interstate Technology & Regulatory Council *In Situ* Chemical Oxidation Team.

Jin, M. (1995). *Surfactant enhanced remediation and interwell partitioning tracer test for characterization of DNAPL contaminated aquifers*, Ph. D. Thesis, University of Texas at Austin, Texas.

Jin, M., Delshad, M., Dwarakanath, V., McKinney, D. C., Pope, G. A., Sepehrnoori, K., Tilburg, C. E., and Jackson R. E. (1995). "Partitioning tracer test for detection, estimation and remediation performance assessment of subsurface nonaqueous phase liquids." *Wat. Resour. Res.*, 31(5), 1201-1211.

Kao, C. M., and Wang, Y. S. (2001). "Field investigation of the natural attenuation and intrinsic biodegradation rates at an underground storage tank site." *Environmental Geology*, 40(4-5), 622-631.

Lai, K. C. K., Lo, I. M. C., and Kjeldsen, P. (2006a). "Natural gradient tracer test for the permeable reactive barrier in Denmark 2. Spatial moments analysis and dispersion of conservative tracer." *J. of Hazardous, Toxic and Radioactive Waste Management, ASCE*, in press.

Lai, K. C. K., Lo, I. M. C., and Kjeldsen, P. (2006b). "Natural gradient tracer test for the permeable reactive barrier in Denmark 1. Field study of tracer movement." *J. of Hazardous, Toxic and Radioactive Waste Management, ASCE*, in press.

Lai, K. C. K., Lo, I. M. C., Birkelund, V., and Kjeldsen, P. (2006c). "Field monitoring of a permeable reactive barrier for removal of chlorinated organics." *J. Environ. Eng. ASCE*, 132(2), 199-210.

Lien, H. L., and Zhang, W. X. (2001). "Nanoscale iron particles for complete reduction of chlorinated ethenes." *Colloids and Surfaces A: Physicochemical & Eng. Aspects*, 191, 97-106.

Lo, I. M. C., Surampalli, R., and Lai, K. C. K. (2006a). *Zero-valent iron reactive materials for hazardous waste and inorganic removal*, American Society of Civil Engineers, in press.

Lo, I. M. C., Lam, C. S. C., and Lai, K. C. K. (2006b). "Hardness and carbonate effects on the reactivity of zero-valent iron for Cr(VI) removal." *Wat. Res.*, 40(3), 595-605.

NRC (1994). *Alternatives for ground water cleanup*, National Research Council, National Academy Press, Washington, DC.

NRC (1997). *Innovatives in groundwater and soil cleanup: From concept to commercialization*, National Academy Press, Washington, DC.

Rao, P. C. S., Jawitz, J. W., Enfield, C. G., Falta, R. W., Annable, M. D., and Wood, A. L. (2001). "Technology integration for contaminated site remediation: Cleanup goals & performance criteria." In *Groundwater Quality: Natural and Enhanced Restoration of Groundwater Pollution*, IAHS, 275, 571-578.

Song, D. L., Conrad, M. E., Sorenson, K. S., and Alvarez-Cohen, L. (2002). "Stable carbon isotope fractionation during enahcned in situ bioremediation of trichloroethene." *Environ. Sci. & Technol.*, 36(10), 2262-2268.

Sturchio, N. C., Clausen, J. L., Heraty, L. J., Huang, L., Holt, B. D., and Abrajano Jr, T. A. (1998). "Chlorine isotope investigation of natural attenuation of trichloroethene in an aerobic aquifer." *Environ. Sci. & Technol.*, 32(20), 3037-3042.

Su, C., and Puls, R. W. (1999). "Kinetics of trichloroethene reduction by zero-valent iron and tin: Pretreatment effect, apparent activation energy, and intermediate products." *Environ. Sci. & Technol.*, 33(1), 163-168.

Teutsch, G., Rugner, H., Zamfirescu D., Finkel, M., and Bittens, M. (2001). "Source remediation vs plume management: Critical factors affecting cost-efficiency." *Land Contamination and Reclamation*, 9(1), 128-139.

USEPA (1992). *General methods for remedial operation performance evaluations*, EPA/600/R-92/002, Robert S. Kerr Environmental Research Laboratory, United States Environmental Protection Agency, Ada, OK 74820.

USEPA (1994). *Methods for monitoring pump-and-treat performance*, EPA/600/R-94/123, Office of Research and Development, United States Environmental Protection Agency, Washington DC 20460.

USEPA (1996). *Pump-and treat ground-water remediation: A guide for decision makers and practitioners*, EPA/625/R-95/005, Office of Research and Development, United States Environmental Protection Agency, Washington DC 20460.

USEPA (1997). *Rule of thumb for Superfuud remedy selection*, EPA/540/R-97/013, Solid Waste and Emergency Response, United States Environmental Protection Agency, Washington, DC 20460.

USEPA (1998). *Permeable reactive barrier technologies for contaminant remediation*, EPA/600/R-98/125, Office of Research and Development, United States Environmental Protection Agency, Washington, DC 20460.

USEPA (1999). *Use of monitored natural attenuation at superfund, RCRA corrective action, and underground storage tank sites*, OSWER Directive 9200.4-17P, April 21, United States Environmental Protection Agency.

USEPA (2000). *Introduction to phytoremediation*, EPA/600/R-99/107, Office of Research and Development, United States Environmental Protection Agency, Cincinnati, Ohio 45268.

USEPA (2001a) *Groundwater pump and treat systems: Summary of selected cost and performance information at superfund-financed sites*, EPA-542-R-01-021a, Solid Waste and Emergency Response (5102G), United States Environmental Protection Agency, Washington, DC 20460.

USEPA (2001b). *Development of recommendations and methods to support assessment of soil venting performance and closure*, EPA/600/R-01/070, Office of Research and Development, United States Environmental Protection Agency, Washington, DC 20460.

USEPA (2002). *Field applications of in situ remediation technologies: Permeable reactive barriers*, EPA/68/W-00/084, Technology Innovation Office, Office of Solid Waste and Emergency Response, United States Environmental Protection Agency, Washington, DC 20460.

USEPA (2004a). *Performance monitoring of MNA remedies for VOCs in ground water*, EPA/600/R-04/027, Ground water and Ecosystems Restoration Division, National Risk Management Research Laboratory, Office of Research and Development, United States Environmental Protection Agency, Ada, OK 74820.

USEPA (2004b). *How to evaluate alternative cleanup technologies for underground storage tank sites*, EPA/510/R-04/002, Solid Waste and Emergency Response 5401G, United States Environmental Protection Agency.

VanStone, N. A., Fotch, R. M., Mabury S. A., and Lollar, B. S. (2004). "Effect of iron type on kinetics and carbon isotopic enrichment of chlorinated ethylenes during abiotic reduction on Fe(0)." *Ground Water*, 42(2), 268-276.

Editor Biographies

Alok Bhandari is an Associate Professor of Civil Engineering at Kansas State University where he teaches undergraduate, graduate and distance education courses in environmental engineering, and conducts research on the fate and removal of organic contaminants associated with water and soil. In the past 12 years he has participated in 27 research/educational projects worth over $4.5 million, authored over 70 publications, and delivered more than 80 scientific/scholarly presentations. He is a recipient of the National Science Foundation CAREER Award, the KSTAR First Award and Virginia Tech Civil Engineering's Outstanding Young Alumni Achievement Award. He serves as Associate Editor of the *Journal of Environmental Engineering* and the *Practice Periodical of Hazardous, Toxic & Radioactive Waste Management*. Alok Bhandari obtained his Ph.D. in Civil Engineering from Virginia Tech and is a Registered Professional Engineer.

Rao Surampalli is an Engineer Director with the United States Environmental Protection Agency. He holds Adjunct Professor positions at the University of Nebraska-Lincoln, University of Missouri-Columbia, University of Missouri-Rolla, and Iowa State University. Rao Surampalli has published/presented over 300 papers in refereed journals and conferences. He is the author/editor of three books and twenty-two book chapters. Rao Surampalli is the Editor-in-chief of *Practice Periodical of Hazardous, Toxic & Radioactive Waste Management* and serves on the editorial boards of *Water Environment Research*. He is a recipient of the Founders Gold Medal awarded by the National Society of Professional Engineers and was named the society's 2001 National Federal Engineer of the Year. Rao Surampalli is a Fellow of the American Association for the Advancement of Science, Fellow of the American Society of Civil Engineers and a Diplomate of the American Academy of Environmental Engineering. He is a recipient of the Distinguished Engineering Alumnus Awards of Oklahoma State University (2001) and Iowa State University (2002). He conducts research on hazardous/solid waste management, soil and groundwater treatment, water/wastewater treatment, and sludge treatment/disposal. Rao Surampalli received his Ph.D. in Environmental Engineering from Iowa State University and is a Registered Professional Engineer.

Pascale Champagne is an Assistant Professor of civil engineering at Queen's University. She was previously a faculty member at Carleton University. Her areas of research and expertise include: passive and semi-passive wastewater treatment processes, alternative waste management approaches, extraction of value-added products from waste materials and sustainable environmental design strategies. Pascale Champagne holds a Biology degree with a Minor in Northern Studies from McGill University, a Water Resources Engineering degree from the University of Guelph, a Master's of Applied Science in Civil Engineering and Ph.D. in Environmental Engineering from Carleton University. She is a Registered Professional Engineer.

Say-Kee Ong is a Professor of Environmental Engineering and the Director of the Environmental Engineering program at Iowa State University. Prior to his academic career, he worked as an environmental engineer with Battelle Memorial Institute focusing on the development of innovative remediation technologies. Say-Kee Ong's publication record includes more than 40 archival papers, principally in the development of treatment technologies and contaminant fate of pollutants in natural systems (soil and groundwater) and engineered systems (wastewater treatment facilities). He is a recipient of the 2003 Distinguished Service Award by the Association of Environmental Engineering & Science Professors and the 2004 Outstanding Achievement in Teaching Award by the Iowa State University Foundation. Say-Kee Ong received his Ph.D. in Environmental Engineering from Cornell University.

R.D. Tyagi is the Canada Research Chair and a Professor at Institut National de la Recherche Scientifique of the University of Quebec. He is an Adjunct Professor at the University of Missouri-Columbia and has held research/academic appointments at the Indian Institute of Technology Delhi and the Universite de Technologie de Compiegne, France. R.D. Tyagi's research focuses on the applications of biochemical engineering and process control on municipal and industrial wastewater treatment, and soil and groundwater remediation. He is an owner of nine technology patents and has published over 200 papers. R.D. Tyagi is a multi-year recipient of the Best Performance Award by the Institut National de la Recherche Scientifique of the University of Quebec. He is an Associate Editor or Editorial Board Member of the *Journal of Microbial Biotechnology*, *Process Biochemistry*, and *Practice Periodical of Hazardous, Toxic and Radioactive Waste Management*. R.D. Tyagi received his Ph.D. from the Indian Institute of Technology, New Delhi.

Irene M. C. Lo is an Associate Professor of civil engineering and Associate Director of the Environmental Engineering Program at the Hong Kong University of Science and Technology. She has held Visiting Professor appointments at the Technical University of Denmark and the University of Wisconsin at Madison. She is a recipient of the 2002 Wong Kuan Cheng Scholar Award, the 2003 Peter H K Chan Best Paper Award by Hong Kong Institute of Engineers, the 2003 Best Paper award from the International Conference on Soil and Groundwater Contamination and Clean-up, the 2004 James Croes Medal by the American Society of Civil Engineers, and the 2004-05, Teaching Excellence Award by the Hong Kong University of Science and Technology. Irene Lo's research interests include remediation technologies for contaminated soils and groundwater, pollutant migration and transport behaviors in saturated and unsaturated soils, and innovative liner materials for waste containment facilities. She has published over 120 refereed journal and conference papers. Irene Lo received her Ph.D. in Civil Engineering from the University of Texas at Austin and is a chartered engineer of Hong Kong Institute of Engineers.

Index

Aberdeen Proving Grounds 336, 340–341
abiotic reduction 177–187
absorption 17, 97, 100, 101*f*
acid extraction 365
activated carbon 102
activation energy, chemical precipitation 142, 143*f*
adsorbents 102–103
adsorption 17, 97, 100–102, 101*f*, 183–184, 379
advection 8–9
aerobic biotransformation 241–242
agglomeration 143
air flow rates, air sparging 61
Air Force Center for Environmental Excellence (AFCEE) 233–234
air sparging 59–62, 60*f*, 229, 413–414; advantages and limitations 61*t*; duration of 61
algae 25–26, 300, 310
aliphatic hydrocarbons 200
anaerobic biotransformation 239–240
anaerobic transformations 165–166; bioventing 234
aquatic plants 300, 334–335
aquifers, redox processes 140–141, 141*t*
aquifers, restoration 410*t*, 411
aromatic hydrocarbons 202, 204*f*
arsenic 355
atrazine 13
autotrophic microbial metabolism 31, 32*t*

bacteria 25–26, 202, 204*f*, 207–208, 369–371; selection of 237–238; wetlands 311
binders 97–98
bioaccumulation 202, 335, 237–239
bioavailability 202
biodegradation 201, 203*t*; cyanide 215; rate of 211–213
bioemulsifiers 217
bioleaching 369–371
biological denitrification 153
biological fixation 242–244
biological processes 24–25; wetlands 305, 308–309
biological reactors 246–247, 377; limiting factors 246
biological volatilization 300–332
bioprecipitation 377
bioremediation 424*f*, 425*t*; monitoring 423–425; rate of 217
biosorption 33, 215, 379–380
biosparging 224, 225*f*; advantages and limitations 229–230, 231*t*; applications 230, system design 229, 230*t*
biostimulation 235–237
biosurfactants 368
biotests 360
biotransformation 201
bioventing 230, 231*f*, 232–235; advantages and limitations 232*t*; limiting factors 232, 233*t*
blast-enhanced fracturing 72
bromate, abiotic reduction 179
Brookhaven National Laboratory 329–330

cadmium 355–356
capture zones 52, 53*f*, 343–344
carbon dioxide 292–293
carbon fixation 243
case studies 163–165, 165–166, 213 217, 320–322; aerobic biotransformation 242; anaerobic biotransformation 239–240;

bioaugmentation 238–239; biological fixation 243; biological reactors 246–247; bioremediation, mycorrhizal fungi 274, 276; biostimulation 236–237; bioventing 233–235; enzyme catalyzed treatment 244–245; fungal metabolism 267–268, 277–278; granular activated carbon 105; metal bioremediation 273; natural attenuation 249–251; phytodegradation 333–337; phytostabilization 324–326; phytovolatilization 329–332, 340–341; rhizodegradation 337–340; rhizofiltration 326–329; white rot fungi 269–270
catabolic transformation 202, 204*f*
cement based stabilization 97–98
cementitious materials 98
chelating agents 366–368, 378–379; soil extractants 319
chemical adsorption 101
chemical leaching 110–115, 114*f*; applications 117–118*t*; cost of 115
chemical leaching agents 113*t*
chemical oxidation/reduction 86*t*, 422*t*; monitoring 420–423; objectives 83–84
chemical precipitation 79–83, 83*f*, 96, 141–143, 375–377; cost of 83; disadvantages 376
chemical processes, and electrochemical processes combined 382; separation 365–369, 365*f*; wetlands 305, 308
chlorate, abiotic reduction 179
chlorinated aliphatic hydrocarbons (CAHs) 155–158, 157*f*, 158*t*, 159*t*; abiotic reduction 178–179
chlorinated hydrocarbons, bioremediation 239–240; natural attenuation, monitoring 250
chlorinated solvents 401
chlorine 84–86
chromium 356, 368–369, 145–150; Pourbaix diagram 146; reductive precipitation 182–183. *See also* Cr(VI)
climatic changes 407
cometabolic air sparging (CAS) 62
common-ion effect 144
complexation 20–22
Comprehensive Environmental Response, Compensation and Liability Act (CERCLA) *See* Superfund
constructed treatment wetlands 303, 330–332, 334–335
contaminant evaporation 364
contaminant properties 56
contaminant source area 406
contaminant structure 212–213
contaminant transport 8–12, 116
contaminants, boiling point 228–229; characteristics 227–229; chemical properties 14*t*; chemical structures 227–228, 235–236; concentration 342, 399–400, 406, 421; containment, failure 411; degradation, plant-soil systems 298–299; mass destroyed, estimating 400–401; mass discharge rate 403; mass extracted, estimating 400; physicochemical processes 12; remaining after treatment 401–402; treatability 343
contaminant-soil interaction 17, 18*f*
contaminant confinement 363
contaminant isolation 363
copper, solubility products 14, 15*t*
cosolvents 122*t*, 414–416, 415*f*, 415*t*
Cr(VI), reduction of 145–149
Craney Islane Fuel Terminal 337–338
crude oil 268–270, 270; bioremediation 276
crystal growth 142–143
cyanide 85–86, 209–210, 210*f*, 211*f*; removal of 215

DCE 165–166
dechlorination 156, 158*t*, 160, 161*f*, 162*t*, 166, 167*f*
dehalogenzse enzyme 335–336
dense nonaqueous phase liquids. *See* DNAPLs
desorption 110, 335
deuteromycete fungus 270
diffusion 8
direct electrolysis 106, 107*f*
dispersive transport 8, 11
dissolution 13–16
dissolved oxygen 187–189
diurnal changes 305–306
DNAPLs 47, 68; degradation, monitoring 249-250; recovery complications 51
Donnan membrane 380
drawdown 69*f*
dual phase extraction 68
dyes 261–262, 274; removal of, case studies 214–215

ectomycorrhizal (ECM) basidiomycetes 278–279
Eh. *See* redox potential
electrical resistance heating 417
electrochemical cell design 107–108
electrochemical processes 105–110, 107*f*; advantages and limitations 105–106
electrochemical reactions 372–373, 381–382
electrochemical reactor, selection of 107
Electrochemical Remediation Technologies (ECRTs) process 108
electrocoagulation 381
electrodialysis 381
electrokinetics technology 108, 109*f*, 371–373, 372*f*
electrolytes 144
electron acceptors 29
electron activity 138–140
electro-osmosis 372
electrostatic adsorption 101
environmental remediation, cost of 1
enzyme catalyzed treatment 244–245; limiting factors 244
enzyme regulation 205–206
enzymes 268–269
equipotential lines 8, 9*f*
ETBE, anaerobic biotransformation 241
excavation and disposal 375
explosives, biodegradation 270–272; phytoremediation 215–216
extraction wells 52, 57–59, 403*f*, 417, 423; radius of influence 58

Fazenda Brasileiro mine, Bahia, Brazil 321–322
ferrous hydroxide 188
ferrous iron 157, 159*t*, 227
field-scale tests, rhizofiltration 327–329
filtration, wetlands 308
fixed bed systems 103
flushing agents 414–416, 415*f*, 415*t*; types of 121*t*
fracture initiation pressure 71
free product recovery 47–51; recoverable volume 49; system design phase 48–49
free water surface wetlands 303, 304*f*
fuel oil contaminated site, phytoremediation 339–340
fungi 25–26; bioremediation, case studies 269–270

geochemical behavior 427, 429, 430*t*; permeable reactive barriers 184–187
Gibbs free energy 136
gold phytomining 321–322
granular activated carbon 379
granular activated carbon reactors 103 104, 104*t*
groundwater, contaminant

concentration 399–400; monitoring 248*f*
groundwater circulation wells (GWC) 62–66; advantages and limitations 66; applicability 65; circulation patterns 63*f*, 64*f*
groundwater flow 6–8, 407
groundwater flow velocity 6–7; effect on chlorinated aliphatic hydrocarbon reduction 178–179
groundwater recharge rates 407
groundwater sampling 403–404, 404–408; location 406
groundwater temperatures 226, 418
groundwater treatment 375*t*

half reactions 137, 138, 139*t*, 150*f*
hazardous waste solidification 96–97
hazardous waste stabilization 96–97
hazardous waste treatment 94–96
heavy metals 20, 79–83, 111, 228, 325–326; chemical separation 365–369; geochemical distribution 360; groundwater 375; immobilization 323; natural attenuation, monitoring 250; phytoaccumulation 317–320; treatability 359 361
Henry's Law 13, 14*t*, 16, 229
heterotrophic microbial metabolism 31, 32*t*
hexavalent chromium 163–165, 164*f*
HMX, phytoremediation 215–216
horizontal flow wetlands 304, 305*f*
humification 243
hydraulic barriers, wetlands 315
hydraulic behavior, permeable reactive barriers 189–190
hydraulic conductivity 7–8, 7*t*
hydraulic containment 301–302, 341, 410*t*; failure 411; monitoring 409–410
hydraulic fracturing 72–73
hydrocarbon solvents 113
hydrocarbons, bioremediation 216–217
hydro-complexes 81
hydrogen peroxide 88–89, 420; and ultrasound, combined 91–92; and UV radiation, combined 89–91, 90*f*, 91*t*
hydrogenolysis 156, 157*f*, 178
hydrologic modelling 344
hydrolysis 12–13
hydroponic systems 316–317, 318*f*

immobilization, contaminants 301
indirect electrolysis 106
induced fracturing 70–73; types of 70–71
industrial sites 339–340, 353–354
infiltration control, vegetation 313–314, 314*t*
injection rates 409, 421
injection wells 115, 423
inorganic contaminants 329; phytoremediation 317
inorganic nutrients 212
integrated pump test 403
in-well air sparging. *See* groundwater circulation wells
in-well vapor stripping. *See* groundwater circulation wells
ion exchange 20, 100–102, 103, 104, 377–379, 378*f*
ionic strength effect 144
Iowa Army Ammunition Plant 333–335
iron rective barriers 189*f*
irrigation 344
irrigation wells 115
isotherms 102

Krebs' cycle 205

laboratory tests 213–214, 214
landfill leachate, natural attenuation 249

leachability testing 99
leachates 115, 366–367
lead 357; phytoremediation 320–321; solubility products 14, 15*t*
lead compounds 98–99; solubility products 80*t*
ligands 21–22, 22*f*
light nonaqueous phase liquids. *See* LNAPLs
ligninolytic fungi 259; oxidation 262, 266
LNAPLs 47, 67, 68

macroencapsulation 97
macrophytes 311
marine pollution 216–217
mass flux, decrease 402–404
mechanical separation 363–364; principal processes 364*t*
medium-term treatment 112
membrane separation 380–381, 380*f*
mercury 356
metabolic pathways 205–206, 205*f*
metal concentration, reductive precipitation 360*t*, regulations 359*t*
metal hydroxides, solubilities 81*f*
metal pollution, sources of 354–355
metal sulfides 370–371; solubilities 82*f*
metallic contaminants 21
metal-microbial interactions 370*f*
metals, biological fixation 243–244; bioremediation 272–273, 279–280; biotransformation 208; desorption of 110
microbial adaptability 27
microbial degradation, cyanide 209–210, 210*f*, 211*f*
microbial density 227
microbial metabolism 27–31, 28*f*
microbial processes 25–35, 309
microbial reduction 153
microbiological processes 33*t*
microencapsulation 96–97
microorganisms 25–27, 200, 295; taxonomic classification 27, 28*t*; wetlands 309, 311
mine drainage 321–322, 325–326; monitoring 249
mineralization 201
minerals, permeable reactive barriers 190*f*; precipitation 187–189
mixing 115, 125
mobility, contaminant source 402
monitoring network, groundwater 404, 405*f*
monitoring wells 406–407, 429
MTBE, anaerobic biotransformation 241
multi-phase extraction 66–69, 67*f*; advantages and limitations 69*t*
munitions waste, biodegradation 270–272; bioremediation 278–279
mycorrhizal fungi 259, 261*f*, 295; bioremediation, effect on pollutants 275–276*t*; processes 274

nanofiltration 380
naphthalene, aerobic bacterial degradation 207*f*
NAPLs 14–16, 15*f*, 67, 212
natural attenuation, monitoring 247–251, 248*f*, 426–427, 428*t*
natural reduction 163–165
Nernst equation 137*f*, 139, 140
nickel 356–357
nitrate 152–155, 154*t*, 208; abiotic reduction 179
nitrite 152–155, 154*t*
nitrogen compounds 208–210
nitrogen fixation 242
non-aqueous phase liquids. *See* NAPLs
nucleation 142
nutrients 227, 235

oil spill sites, bioremediation 216–217; biostimulation 236–237

organic compounds 111, 121; destruction by supercritical water oxidation 94; destruction by wet air oxidation 93*t*; microbial transformation 34–35
organic contaminants 332–333
organic pollutants, second order reaction rates 88*t*
organic wastes 92
oxidation 93
oxidation numbers, assigning to atoms 135*t*
oxidation-reduction reactions. *See* redox reactions
oxidative techniques 376, 420
oxidizing agents 84, 85*t*, 368–369
oxo-anions, abiotic reduction 179
oxygen 211–212; effect on biological processes 305, 306
oxygen concentration 226–227
oxygen transfer 294*f*
ozone 86–88, 87*f*, 421; and UV radiation, combined 89–91, 90*f*

palladized zero-valent iron 160–161, 162*t*
palladized zinc 161, 162*t*
Palmerton Zinc Superfund Site 324–325
particulate separation 113, 123–124
passive borehole flux meter 404
PCBs 109, 158, 160–162, 160*f*, 161*f*, 162*t*, 180*f*, 207–208, 274; abiotic reduction 179; balancing, case studies 213–214; degradative pathways 209*f*
permanganate 420
permeable reactive barriers 429*f*, 431*t*; examples of 177; geochemical behavior 184–187; hydraulic behavior 189–190; monitoring 427–431
pesticides 215, 261–262, 274
petroleum compounds 228
petroleum contaminants 59
petroleum hydrocarbons, biotransformation, case studies 214; bioventing 234
pH range 81, 185–187, 226; effect on biological processes 305, 307
phenenthrene, biodegradability 238
phosphorus, adsorption 183–184, 184*f*
phosphorus fixation 243
physical adsorption 101–102
phytoaccumulation 36, 38*f*, 300, 317–320
phytocontainment 43, 44*f*
phytodegradation 36–38, 39*f*, 314; organic contaminants 333–337
phytoremediation 35–43, 200–201, 332–333, 374*f*, 433*t*; advantages and limitations 344–348, 346*t*; applications 37*t*, 302; cost of 345, 347*t*; design 342; elements of 291–292; explosives by-products, case studies 215–216; fate of contaminants 291*t*; metal contaminants 373–374; monitoring 431–432
phytoremediation process 297–298, 298*t*
phytostabilization 38–39, 40*f*, 301, 322–326
phytovolatilization 40, 41*f*, 329–332, 340–341
plants 200–201, 337–338, 431–432; metal accumulators 308–319; selection of, 292 342–343, 373–374; wetlands 310–311
plant-soil interface. *See* root-soil interface
plume control 315
plume load 402
plume migration 120, 427
plumes 403*f*; hydraulic containment 409
pneumatic fracturing 72–73
pollutant concentration 228

polyaromatic hydrocarbons (PAHs), bioremediation 268–269
polychlorinated biphenyls (PCBs) 109, 158, 160–162, 160*f*, 161*f*, 162*t*, 180*f*, 207–208, 261–262, 274; abiotic reduction 179; biodegradation, case studies 213–214; degradative pathways 209*f*
polychlorinated methane, bioremediation 240
polycyclic aromatic hydrocarbons (PAHs) 199, 206, 207*f*; bioremediation 276; fungal metabolism of 266*f*
poplar trees 341
porosity 7*t*
porous media 5; properties 49*t*
pozzolan based stabilization 98
precipitation. *See* chemical precipitation
pressure response time, induced fracturing 71*f*
pretreatment 113
pump-and-treat 51–55, 316, 408–412, 409*f*; cost of 52*f*, 54–55; duration of 52*f*; monitoring 410*t*
pumping rates 409
pumping wells 50
pyrene, biodegradability 238
pyrometallurgical separation 364

qualitative evaluation 399
quantitative evaluation 398–399
quinoline compounds removal bioaugmentation 238

radio frequency heating (RFH) 74
radius of influence 68
rain interception 312–314, 313*t*
Raoult's Law 58*e*
RDX 333–335; phytoremediation 215–216
rebound 54, 55
re-circulation wells. *See* groundwater circulation wells
redox conditions, wetlands 306
redox equilibria 136–138
redox potential 137, 138, 139*t*
redox reactions 22–24, 24*t*, 134–136; balancing 136*t*
redox zones 28, 29*f*
reductants, comparison of 148, 149*t*
reductive precipitation 181–183, 376
regulations (laws) 358–359
remediation objectives, defining 396, 398
remediation technologies, assessment 61; demand for 2–3, 3*t*; monitoring 397*t*; types of 3, 4*t*
Resource Conservation and Recovery Act (RCRA) 1
reverse osmosis 380
rhizodegradation 41–42, 42*f*, 299, 314, 337–340
rhizofiltration 42–43, 43*f*, 301, 317, 326–329
rhizosphere 294–296
riparian buffers 315–316, 316*f*
root-soil interface 295, 296*f*, 343

salinity, effect on biological processes 305
sampling frequency, groundwater 407–408
saprotrophic fungal processes 260–261
Saprotrophs 259, 261*f*. *See also* white rot fungi
Savannah River Site 329–330
seasonal changes 305–306, 346–347, 407
sedimentation 375; wetlands 308
selenium 330–332, 357–358
short-term treatment 112
site assessment 60, 120
site characteristics 224–227
site geochemistry 120
site hydrogeology 120

six-phase soil heating (SPSH) 73–74
skimming systems 49, 50*t*
sodium persulfate 420
soil amendments 323, 324*t*
soil characteristics 112, 119, 224–227
soil flushing 115–123, 123*f*; advantages and limitations 120; applications 117–118*t*; cost of 123
soil gases 16, 418
soil heating 73–75, 73*f*, 416–208
soil organic carbon 147
soil permeability 225–226
soil properties 56
soil restoration 361
soil stabilization 322–326
soil stratification 226
soil structure 226, 235, 308
soil temperatures 418
soil treatment 362*t*
soil vapor, contaminants 400
soil vapor extraction 51, 55–59, 57*f*, 66–67, 73, 412*f*; advantages and limitations 56*t*; cost of 59; monitoring 412–413, 413*t*
soil venting 55. *See also* soil vapor extraction
soil washing 110, 123–126; cost of 125
soil washing units 125
soils, contaminant concentration 399
solidification, soils 361–362
solubility 212
solubility product 80, 119, 143–144
soluble metal ion complexes 21*f*
solute transport 5, 8–12
solvent extraction 110–115, 114*f*; applications 117–118*t*
sorption 17–19, 100, 212; wetlands 308
sorption isotherms 17–19, 18*f*
source zone 421; monitoring 426
St. Joseph, Michigan 165–166, 167*f*
stabilization, soils 361–362
stabilization/solidification 94–96; design 98–99, 99*f*; testing 99
steam injection 74–75, 416, 417*f*
streamlines 8, 9*f*
substrate systems, mixed 30–31
subsurface environment, characteristics 5, 6*t*
subsurface flow wetlands 303, 305*f*
sulfides 82, 376
sulfur compounds 146
sulfur fixation 243
sulfur-containing materials 157, 159*t*
supercritical water oxidation 94, 95*f*, 95*t*
Superfund 1
Superfund sites 163–165, 165–166, 320–321, 324–325, 340–341; number of 1
supersaturation 141, 144
surfactant modified zeolite (SMZ) 184; adsorption 185*f*
surfactants 116, 119, 122*t*, 240, 368, 414–416, 415*f*, 415*t*

tailing 54, 55*f*
TAME, anaerobic biotransformation 241
TCE. *See* trichloroethylene
technology assessment, groundwater circulation wells 64–65
temperatures 418; effect on biological processes 305
terrestrial systems 312
thermal conduction 417
thermal remediation 416–419, 419*t*
TNT 333–335; bioremediation 271, 273*f*
toxic pollutants, removal of, bioaugmentation 238–239; sources of 199
toxicity, contaminant source 402
TPH 337–338
transpiration 312, 313*t*, 343–344
treatability tests 112
treatment area 344

treatment methods 361*t*; selection of 360–361
tree stand systems 312
Trenton, NJ, Superfund site 320–321
trichloroethylene (TCE) 165–166, 167*f*, 180*f*, 335–336; effect on chromium removal 183*t*
Trinity Sand Aquifer 163–165, 163*f*, 164*f*, 166
tritium 329–330

ultrafiltration 380
ultrasound 91–92
underground storage tanks 59
United States Coast Guard Support Center, Elizabeth City, North Carolina 187*f*, 190*f*
uranium 150–151; redox processes 151*t*; reductive precipitation 183; rhizofiltration 327–329
UV radiation 89–91

vacuum extraction wells 417
vadose zone properties 56
Vapokon site, Denmark 185, 186*f*, 191*f*
vapor extraction 60*f*
vapor pressure 56, 228
vegetation, wetlands 309–310
vegetation-based treatment, characteristics 35–36
vegetative cover systems 312–315, 337–338
vertical flow systems 304, 306*f*
volatile organic compounds (VOCs) 62–63, 67, 340–341
volatilization 16, 55. *See also* air sparging; soil vapor extraction

waste water treatment, bioaugmentation 239; wetlands 303–304
water level fluctuation 306
water quality parameters, monitoring 424
water table depression 50
well screens 406–407
wells, location 121; radius of influence 63
wet air oxidation 92–94, 93*f*
wetland processes 304–307, 307*f*, 330–332
wetlands 303; types of 303–304
white rot fungi 260–261, 262*f*, 268–269; bioremediation, effect on pollutants 263–266*t*; explosives degradation 271; oxidation 262, 266
Whitewood Creek 325–326

zeolites 184
zero-valent iron 147, 151*t*, 152, 154*t*, 155–156, 157*f*, 158*t*, 159*t*, 177–178, 185–187, 188, 427
Zimpro Process 92
zinc 324–325, 358

